Strategic Issues in Air Transport

Ruwantissa Abeyratne

Strategic Issues in Air Transport

Legal, Economic and Technical Aspects

Dr. Ruwantissa Abeyratne
International Civil Aviation Organization (ICAO)
999 University Street
Montreal H3C 5H7, Quebec
Canada
tabeyratne@icao.int

ISBN 978-3-642-21959-7   e-ISBN 978-3-642-21960-3
DOI 10.1007/978-3-642-21960-3
Springer Heidelberg Dordrecht London New York

Library of Congress Control Number: 2011939308

© Springer-Verlag Berlin Heidelberg 2012

This work is subject to copyright. All rights are reserved, whether the whole or part of the material is concerned, specifically the rights of translation, reprinting, reuse of illustrations, recitation, broadcasting, reproduction on microfilm or in any other way, and storage in data banks. Duplication of this publication or parts thereof is permitted only under the provisions of the German Copyright Law of September 9, 1965, in its current version, and permission for use must always be obtained from Springer. Violations are liable to prosecution under the German Copyright Law.

The use of general descriptive names, registered names, trademarks, etc. in this publication does not imply, even in the absence of a specific statement, that such names are exempt from the relevant protective laws and regulations and therefore free for general use.

Printed on acid-free paper

Springer is part of Springer Science+Business Media (www.springer.com)

Preface

Currently, and arguably throughout the twenty-first century, along with new issues that might emerge in air transport, there are four strategic issues that will continue to apply to air transport: safety; security; environmental protection; and sustainable development.

Of these, safety, security and environmental protection were heavily discussed at the 37th Session of the Assembly convened by the International Civil Aviation Organization (ICAO) in Montreal from 28 September to 8 October 2011. The Assembly was attended by a record 1,588 participants from 176 Member States and 40 international organizations involved in civil aviation, achieved important advancements in aviation safety and aviation security assuring even greater safety performance in the already safest and most secure mode of mass transport in the world.

Reaffirming ICAO's its leadership role, the meeting adopted a comprehensive resolution to reduce the impact of aviation emissions on climate change. The agreement provides a roadmap for action through 2050 for the 190 Member States of the Organization. Solidifying its global influence, the Organization signed numerous international agreements, including cooperation agreements with regional civil aviation organizations and bodies from all regions of the world.

The Assembly endorsed a proactive safety strategy based on the sharing of critical safety information among governments and industry stakeholders. It also endorsed ICAO's plan to establish a multi-disciplinary approach to address the critical issue of runway safety. This will bring together representatives from airlines, airports, air navigation service providers and regulatory authorities. Following a successful diplomatic Conference in Beijing in August 2010, the Assembly built on this achievement by recognizing the need to strengthen aviation security worldwide. In a Declaration, unanimously adopted by participants, international commitment was reaffirmed to enhance aviation security collaboratively and proactively through screening technologies to detect prohibited articles, strengthening international standards, improving security information-sharing and providing capacity-building assistance to States in need. The Assembly also put its full support behind a comprehensive, new ICAO aviation security strategy.

The Resolution adopted by the Assembly on climate change makes ICAO the first United Nations Agency to lead a sector in the establishment of a globally harmonized agreement for addressing its CO_2 emissions. The resolution was adopted with some States expressing reservations and calling upon the ICAO Council to continue its work on specific aspects of the agreement. This remarkable accomplishment came only 2 months before negotiations are again undertaken by these very same States at the 16th Conference of the Parties of the United Nations Framework Convention on Climate Change (UNFCCC) meeting scheduled for December in Mexico.

Against this backdrop, this book addresses these four issues and the various complex areas that they present.

Montreal   Ruwantissa Abeyratne

Contents

1 **Strategic Issues and Regulation** 1
  1.1 Prelude to the Assembly 1
      1.1.1 Safety 2
      1.1.2 Security 3
      1.1.3 Environmental Protection 4
      1.1.4 Law 4
  1.2 The Assembly 5
      1.2.1 Safety 5
      1.2.2 Security 10
      1.2.3 Environmental Protection 13
      1.2.4 Law 15
  References 17

2 **Safety Issues** 19
  2.1 Safety Management Systems 19
      2.1.1 Responsibility of States 22
      2.1.2 Provision of Air Navigation Services 26
      2.1.3 Provision of Meteorological Information 29
      2.1.4 Provision of Air Traffic Services 31
      2.1.5 The Safety Roadmap 35
      2.1.6 Economic Liberalization and Safety 42
      2.1.7 Regulatory Aspects of Economic Liberalization
           and Safety 49
  2.2 The Use of Airspace 64
      2.2.1 Distinction Between Civil and Military Aviation 68
      2.2.2 The Use of Civil Aircraft for Military Purposes 70
      2.2.3 Some Recent Developments 72
      2.2.4 ICAO Initiatives 73
  2.3 Aviation Medicine 78
      2.3.1 The Aerotoxic Syndrome 78

		2.3.2 A Novel Approach to Liability	85
		2.3.3 Medical Issues of Technical Crew Members	88
		2.3.4 Liability Issues	93
	2.4	Meteorological Issues	98
		2.4.1 ICAO's Work on Mitigating the Effects of Volcanic Ash on Aviation	100
		2.4.2 Some Liability Issues	106
		2.4.3 State Liability	110
		2.4.4 Air Carrier Liability to Passengers	115
	2.5	Unmanned Aircraft Systems	118
		2.5.1 The UAS and Civil Aviation Issues	123
		2.5.2 Operations over the High Seas	129
		2.5.3 Air Traffic Services	132
		2.5.4 UAS as State Aircraft	132
	2.6	Regional Safety in Air Transport	139
		2.6.1 The Role of ICAO	141
		2.6.2 The Regional Safety Oversight Manual	142
		2.6.3 The ICAO AFI Comprehensive Implementation Programme (ACIP)	144
		2.6.4 Safety Oversight in Africa	145
		2.6.5 Safety Oversight in Central America (ACSA)	147
		2.6.6 Safety Oversight in the Caribbean	147
		2.6.7 Safety Oversight in South Asia	151
		2.6.8 Safety Oversight in Europe	151
	2.7	Safety Oversight Audits	154
		2.7.1 Blacklisting of Airlines	155
		2.7.2 Policy Aspects of Blacklisting	156
		2.7.3 Regulatory Oversight	160
	References		162
3	**Security Issues**		**165**
	3.1	Cyber Terrorism	165
		3.1.1 International Efforts	171
		3.1.2 National Efforts	175
	3.2	Attacks on the Integrity of Travel Documents	177
		3.2.1 Complicity	180
		3.2.2 Condonation	181
		3.2.3 Knowledge	184
		3.2.4 Security of the Passport	189
	3.3	Full Body Scanners and Emergent Issues	197
		3.3.1 The Right of Privacy of the Passenger	199
		3.3.2 Security of the State	205
		3.3.3 Flight NW 253	209
		3.3.4 The AVSEC Panel	212

	3.4 Suppressing the Financing of Terrorism	225
	3.4.1 Acts of International Terrorism	229
	3.4.2 Money Laundering	236
	3.5 Civil Unrest and Aviation	239
	3.5.1 Keeping Airports Open	241
	3.5.2 Airport and Aviation Security	243
	References	276
4	**Environmental Issues**	279
	4.1 The Reality of Climate Change	279
	4.2 Aspirations and Goals	283
	4.3 Trading and Market Based Measures	285
	4.4 Emissions Trading	294
	4.5 The Carbon Market	298
	4.6 Consequences of Carbon Trading	302
	4.7 The Emissions Trading Scheme of the European Union	306
	4.8 EU ETS and the Trading Mechanism	308
	4.9 Calculating Emissions	311
	References	318
5	**Sustainable Development of Air Transport**	319
	5.1 The Nature of Air Transport	319
	5.2 Market Access	320
	5.2.1 The Anomaly	324
	5.2.2 The Solution?	325
	5.2.3 Economic Considerations	329
	5.2.4 Some Basic Misconceptions	331
	5.2.5 Legal Considerations	333
	5.3 Open Skies	338
	5.3.1 The Issues Involved	342
	5.3.2 Effect of Open Skies Competition	346
	5.3.3 Commercial Considerations	350
	5.4 Slot Allocation	355
	5.4.1 Slot Allocation in Europe	360
	5.5 Corporate Foresight	368
	5.5.1 Airport Responsibility	370
	5.5.2 Elements of Corporate Foresight Planning	375
	5.6 The Use of Alternative Fuels as an Economic Measure	382
	5.6.1 The Use of Fossil Fuel and Its Effects	383
	5.6.2 The Availability of Sustainable Alternative Fuels and Costs Involved	385
	5.6.3 The Rio Conference of 2009	389
	References	393

6	**Conclusion**	395
	References	399

Appendix 401

Index 413

Table of Cases

A
Air India v. Wiggins [1980] 1WLR 815 t 819; 77 ILR 276 at 27, 304
Ashcroft v. Iqbal, 129 S.Ct. 1937, 85

B
Barrett v. Enfield LBC [2001] 2.A.C. 550, 113, 114
Barry v. Arnaud [1839] 10 Ad. & E. 646, 113
Bell Atlantic Corp. v. Twombly [2007] 550 U.S. 544, 85
Brantley v. Vaughn, 835F. Supp. 258., 376
British Caledonian Airways Ltd. v. Bond [1981] 665 F.3d.1153 (D.C. Cir. 1981), 159
Brooke Group Ltd. v. Brown and Williamson Tobacco Corp. [1993] 509 U.S. 209, 351
Brownlie v. Campbell [1880] 5 App. Cas. 925 (per Lord Blackburn), 95
Buergenthal [1969] at p. 121, 28

C
Campbell v. Tormey [1969] 1 All E.R. 961, cited in Smith & Hogan, Criminal Law, op.cit. at 432, 263, 264
Caparo Industries Plc v. Dickman [1990] 2 AC 605, 86, 97, 113
Centros Ltd. v. Erhvervs-og Selskabsstyrelsen [1999] Case no. C–212/97, ECR, 1–1459, 327
Charzow Factory [1928] Case PCIJ Series A No. 17, 33
Christy v. Leachinsky [1947] 1 All E.R. 567 supra, note 315, at 572–573, 264, 265
Commission v. French Republic [1974] Case 167/73, ECR 359 at 370, 349, 368
Conly v. Gibson [1957] 355 U.S. 41, 85
Corfu Channel Case [1949] I.C.J.R.1, 22, p. 4, 33, 185
Cox v. Dubois, 16F.Supp. 2d 861, 376
Curran v. Northern Ireland Co-ownership Housing Association Ltd. [1987] A.C.718, 114
Curtis v. Chemical Cleaning and Dyeing Company [1951] 1. K.B. 805 (C.A), 95
Cutler v. Wandsworth Stadium Ltd. [1949] AC 398, 113

D
Daves v. Hawaiian Dredging Co. [2009] (Justice Souter quoted) 114 F. Supp. 643 at 645, 85
Department of Water and Power of City of Los Angeles v. Anderson 95F.2d 577, 376
DOJ v. Reporters Comm. for Freedom of the Press [1988] 489 U.S. 749 AT 763, 203

E
El Al Isreal Airlines Limited v. Tseng [1999] Westlaw 7724, 84
Entick v. Carrington [1765] 19 St. Tr. 1029 at p. 1081, 112
Esso Petroleum Co. Ltd. v. Mardon [1976] Q.B. 801, 95

G
Goris v. Scott [1875] L.R. 9 ex. 125, 113
Gould Estate v. Stoddart Publishing Company [1996] O.J. No. 3288 (Gen. Div), 204
Gronau v. Schlamp Investments Ltd., [1974] 52 D.L.R. (3d) 631 (Man.Q.B.), 95
Groves v. Lord Winborne [1898] 2.Q.B.402, 113

H
Hamdan v. Rumsfeld [2006] 126S.Ct.2749, 263
Hartford Fiore Insurance Company v.California [1993] 113 S. Ct. 2891 per Souter J., 306
Hedley Byrne & Co. Ltd. v. Heller & Partners [1964] A.C. 465, 95
Herman v. Trans World Airlines [1972] 330 N.Y.S.D. 2nd 829 (Sup.Ct. 1972), 253
Hoffman-La Roche & Co. A.G. v. Commission [1979] Case 85/76, ECR. 461, 350, 369
Holmes v. Bangladesh Biman Corporation [1989] AC 1112 at 1126; 87 ILR 365 at 369, 304
Hood v. Dealers Transport Co., 459F.Supp. 684, 376

I
In Re. Chorzow Factory (Jurisdiction) [1927] PCIJ, Ser. A, no. 9 at 21, 138
International Transport Workers' Federation and Finnish Seamens' Union v. Viking Line ABP and OU Viking Line Easti [2007] Case C–438/05, ECR 1–10779, 327
Island of Palmas Case [1928] 11 U.N.R. I.A.A. at 829, 184

J
Jane's Case, Supra, note 11, at 92, 181
Johnson v. Cenac Towing Inc., [2006] 468 F. Supp, 2d. 815, 96
Johnson v. State of California [1968] 447 P. 2d. 352, 114

K
Konop v Hawaiian Airlines [1986] 302 F 3d 868, 176

L
Laker Airways v. Sabena [1984] 731 F.2d 909, 305
Laughlin v. Rose [1958] 200 Va. 127, 104S.E. 2d 782, 376

Table of Cases

Laura M.B. Janes (USA) v. United Mexican States [1925] 4 R Intl Arb Awards 82, 181, 206

Libyan Arab Jamahiriya v. United Kingdom [1988] regarding the PANAM 103 accident at Lockerbie, Scotland in 1988, 224

Libyan Arab Jamahiriya v. United States of America [1988] regarding the PANAM 103 accident at Lockerbie, Scotland in 1988, 201, 224

Link v. Schiable [1961] 27 D.L.R. (2d) 461 (B.C.C.A), 95

M

Mannington Mills v. Congoleum Corporation [1979] 595 F.2d 1287; 66 ILR, 487, 305

Matsushita Electric Industrial Co. Ltd., et al. v. Zenith Radio Corp. et al. [1986] 475 U.S. 574, 351

McCorpen v. Cental Gulf Steamship Corporation [1968] 396 F2d 547, 95

McGrath v. MacLean [1979] 95 D.L.R. (3d) 144, 95

Mercy Docks and Harbour Board Trustees v. Gibbs [1866] L.R.1. H.L, 114

Murray v. Ministry of Defence [1988] 2 All E.R. 521, 265

N

Nicaragua v. The United States [1986] ICJ Reports 1986, 14. Also, 76 ILR 349, 185

O

O'Rourke v. Campden LBC [1998] A.C. 158, 113

Olmsted v. United States [1928] 277 U.S. 438, 478, 203

Olsen v. Poirier [1978] 91 D.L.R. (3d) 123 (Ont. H.C.J), 95

P

Pan Am Case, Infra, note 130, 186

Peabody Donation Fund v. Sir Lindsay Parkinson and Co. Ltd. [1985] A.C. 210, 114

Pfug v. Egyptair [1992] 961F. 2d. 26 (2nd Cir.), 253

Phelps v. Hillingdon LBC [2001] 2 A.C. 619 at 652, 114

Phillips v. Britannia Hygenic Laundry Co. Ltd. [1923] 2 K. B. 832, 113

Pickering v. James [1873] L.R. 8 C.P. 489, 113

R

R. v. Mafart and Prieur [1985] New Zealand High Court, Auckland Registry, Per Davison CJ reported in 74 ILR 241, 193

R. v. McLaughlin [1980] 2 SCR 331 at 339, 168

Rahman v. Queen [1985] 81 Cr App Rep 349 at 353, 263

Rainbow Warrior Case [1985] 81 AJIL, 1987 at 325. Als in 74 ILR at 241, 185, 193

Roe v. Wade [1973] 410 U.S. 113, 202

Rookes v. Bernard [1964] (H.L. 1964) 1 All Eng. Rep. 367, 88

Rowley v. Isley [1951] 3 D.L.R. 766 (Ct.), 95

Rowning v. Goodchild [1772] 2 W. Black 906, 113

S
Schenk v. US [1919] 249 US 47, 168
Schinotti v. Bumstead [1796] 6. T.R. 646, 113
Spicer v. Holt [1976] 3 All E. R. 71 at 79, 265
Sterling v. Turner [1672] I. Ventris 200, 113
Stovin v. Wise [1996] A.C. 923, 113
Sutherland Shire Council v. Heyman [1985] C.L.R. 424 at 464, 86, 97, 114

T
Terry v. Ohio, 392 [1968] U.S.1, argued 12 Dec. 1967, decided 10 June 1968, 264
Tetra Pak Rausing SA. v. Commission [1990] Case T–51/89, II ECR 309, 4C.M.L. R. 334 para 31, 350, 369
The Island of Palmas Case [1928] 11 U.N.R. I.A.A. at 829, 184
The Mayagna (Sumo) Indigenous Community of Awas Tingini v. Nicaragua, Inter American Court of Human rights, Judgment of 31 August 2001 (Ser. C) No. 79, para 163, 33
The Spanish Zone of Morocco Case, 2 RIAA at p. 615, 33
Timberlane Lumber Company v. Bank of America [1976] 549 F.2d 597; 66 ILR, 270, 305
Turner v. Eastwest Airlines Limited [2009] Case No. CV–09–HGD–2193-s in the District Court of Northern District of Alabama, 46 N. W. 677, 95
Turner v. Eastwest Airlines Limited [2009] NSW DDT 10, 83
Turner v. Eastwest Airlines, Infra, note 164, 78

U
US v. Eustaquio [1999] 198 R.3d 1068 (8th Cir.1999), 199
US v. Favela [2001] 247 F.3d.838, 199
US. v. Aluminium Company of America [1945] 148 F.2d 416, 305
US. v. Smith [1998] 155 F 3d 1051, at 1055, 176

V
Victoria Vaughn Holsted and Valerie Vaughn, Petitioners, vs. Southwest Airlines Co., Case No. BS120400, 82

W
Wallentin-Hermann v. Alitalia [2006] Case C–344/04 ECR ZLW 58.Jg. 2/2009 224–231 at 224, 115–117
White v. Edwards Chevrolet Co. [1947] 186 Va. 669, 43S.E. 2d 870, 376
Whitehouse v. Jordan [1981] 1 All E.R. 267, 96

X
X (Minors) v. Bedfordshire CC [1995] 2 A.C. 633 at 730, 112

Chapter 1
Strategic Issues and Regulation

The genesis of regulation of strategic issues lies in the Assembly of the International Civil Aviation Organization (ICAO)[1] and the 37th Session of the Assembly, held in Montreal from 28 September to 8 October 2010, was a signal event in this regard. Many strategic issues which were addressed at the Assembly and were part of Resolutions adopted therein had earlier been the focus of ICAO events, meetings and conferences as the discussion below will elaborate.

1.1 Prelude to the Assembly

A hive of activity at ICAO in the first half of 2010 reflected the culmination of an energetic and active triennium (2008–2010) during which the Organization endeavoured to attain global recognition as a performance and results-driven and values-based Organization. This period also brought to bear ICAO's transition from being mostly a document spewing international body over the past several decades to becoming one which also implemented its policies and assisted its member States[2] as a priority in the areas of safety, security, environmental protection and sustainable development of air transport. In particular, the activities of the first half of 2010, as a countdown to the 37th session of the Assembly, were a fitting precursor to the Assembly which made useful inroads into these three areas.

---

[1] ICAO is the specialized agency of the United Nations handling issues of international civil aviation. ICAO was established by the Convention on International Civil Aviation, signed at Chicago on 7 December 1944 (Chicago Convention). One of the overarching objectives of ICAO, as contained in Article 43 of the Convention is to foster the planning and development of international air transport so as to meet the needs of the peoples for safe, regular, efficient and economical air transport. ICAO has 190 member States, who become members of ICAO by ratifying or otherwise issuing notice of adherence to the Chicago Convention. See ICAO Doc 7300/9 Ninth Edition 2008.
[2] See Abeyratne (2009).

1.1.1 Safety

From 29 March to 1 April, ICAO held a High-level Safety Conference which was attended by 551 participants, including Ministers and Directors General of Civil Aviation from 117 Member States as well as representatives from 32 international organizations. The Conference called on ICAO to facilitate the collection, analysis and dissemination of safety information provided by States and industry partners and resulted in a strong mandate for the Organization to create a strategy to further reduce the global accident rate through the sharing of safety-related information among Member States and the air transport industry.

A comprehensive systems approach (CSA) was continued with regard to the Universal Safety Oversight Programme (USOAP). In the first 6 months of 2010, 14 States received CSA audits, bringing the total number of completed safety oversight audits to 159. Under the direction of the Assembly and the Council of ICAO, development began of a continuous monitoring approach (CMA) for the continuation of the USOAP beyond 2010. Consistent with safety management principles, ICAO's Integrated Safety Trend Analysis and Reporting System (iSTARS) continued being developed throughout 2010. iSTARS offers analysis capability for monitoring the achievement of global safety objectives through the assessment of numerous criteria.

The ICAO Online Aircraft Safety Information System (OASIS) which contains pertinent information concerning all aircraft habitually involved in international civil aviation, including registration, ownership and control, in accordance with Article 21 of the Convention on International Civil Aviation[3] was developed with the capability to establish unique identifiers for aircraft using two fields of data that, when combined, uniquely identify all aircraft entered into the database, regardless of their current registration marks. This repository of information will contain a history of aircraft ownership and control will thus be available.

In 2010 ICAO commenced establishing an international register of Air Operator Certificates (AOCs) to facilitate the surveillance of foreign operators. The project will take place in two development phases. On another front, ICAO continued its close collaboration with IATA on work related to training and qualifications initiatives (ITQI). Work progressed specifically on the development of guidance

---

[3] *Convention on International Civil Aviation*, signed at Chicago on 7 December 1944 (hereafter referred to as the Chicago Convention), ICAO Doc 7300/9, 2006. Article 21 provides that each contracting State undertakes to supply to any other contracting State or to the International Civil Aviation Organization, on demand, information concerning the registration and ownership of any particular aircraft registered in that State. In addition, each contracting State is required to furnish reports to ICAO, under such regulations as the latter may prescribe, giving such pertinent data as can be made available concerning the ownership and control of aircraft registered in that State and habitually engaged in international air navigation. The data thus obtained by ICAO is made available by it on request to the other contracting States.

material for inclusion in the Procedures for Air Navigation Services – Training[4] on competency-based training and assessment of maintenance personnel, evidence-based training for flight crew, and instructor and examiner qualifications. A symposium on the next generation of aviation professionals (NGAP) was held in March 2010. The theme was *Looking beyond the economic crisis: mobilizing the aviation community to recruit, educate, train and retain the next generation of aviation professionals.*[5]

Also in 2010, A new dangerous goods training programme was launched. The programme consists of a training manual and a course which will assist States in complying with the broad principles governing the international transport of dangerous goods by air outlined in Annex 18 to the Chicago Convention which addresses issues related to the safe transport of dangerous goods by air and detailed in the *Technical Instructions for the Safe Transport of Dangerous Goods by Air.*[6] Also, with a view to promoting performance based navigation (PBN), and assisting States in their PBN implementation ICAO planned PBN airspace design workshops, PBN operational approval courses and continuous descent operations workshops.

1.1.2 Security

In response to the attempted sabotage of Northwest Airlines Flight 253 on 25 December 2009,[7] ICAO used the AVSEC Point of Contact (PoC) Network to communicate information and recommendations to participating States, numbering 99 as of 31 May 2010. States were encouraged to conduct risk assessments and implement appropriate screening measures in light of the incident, and were reminded of the need for cooperation in all matters related to aviation security. The twenty-first meeting of the AVSEC Panel was held at ICAO Headquarters from 22 to 26 March 2010. The Panel considered the threat and risk environment in light of the attempted sabotage of 25 December 2009 and issued a number of recommendations. Provisions in Annex 17 to the Chicago Convention on Security were updated and strengthened, and are expected to become applicable in 2011, following formal consultation with Member States and approval by the Council.

---

[4] *PANS-TRG*, Doc 9868.

[5] The event attracted 403 participants from 71 States and 14 international organizations. Especially noteworthy was the participation of over 80 students involved in aviation-related university and college programmes. Following the symposium, the NGAP Task Force focused its work on the development of competencies for flight crew, air traffic management professionals and maintenance personnel.

[6] Doc 9284.

[7] For a discussion of this incident, See Abeyratne(2010a).

The sixth meeting of the Facilitation Panel, held at ICAO Headquarters from 10 to 14 May 2010, recommended the introduction of a new Standard in Annex 9 on facilitation of air transport, obliging all States to adhere to internationally recognized requirements for the transmission of advance passenger information (API) data. The Facilitation Panel also agreed on a new set of guidelines for the passenger name record (PNR) data exchange that will serve to help States implement their national PNR programmes. It also agreed to commence work, on an urgent basis, on the development of new guidelines for advanced data exchange programmes in coordination with the World Customs Organization and IATA.

1.1.3 Environmental Protection

The eighth meeting of the Committee on Aviation Environmental Protection (CAEP/8) was held from 1 to 12 February. The meeting was attended by 184 participants nominated by 22 Member States and 13 international organizations. The meeting dealt with various alternatives for reducing and limiting the environmental impact of aviation. Standards, policies and guidance material on measures to address aircraft noise and engine emissions were developed, including technological improvements; operating procedures; proper organization of air traffic; appropriate airport and land-use planning; and the use of market based options.

ICAO held its third Environmental Colloquium from 11 to 14 May 2010 in Montréal. The objective of the Colloquium was to provide the most up-to-date information that will form the basis for discussions and high-level decisions at the 37th Session of the Assembly. A tutorial was arranged on the first day to familiarize the participants with vocabulary and concepts used in the description, measurement, regulation, and management of aviation greenhouse gas (GHG) emissions. The Colloquium addressed the latest developments on the assessment of aviation emissions and highlighted various solutions to address related environmental impacts. It also focused on related key developments emanating from the ICAO High-level Meeting on International Aviation and Climate Change held in October 2009, the ICAO Conference on Aviation and Alternative Fuels held in Rio de Janeiro in November 2009, the United Nations Framework Convention on Climate Change (UNFCCC) 15th Conference of Parties (COP/15) held in Copenhagen in December 2009 and CAEP/8. The Colloquium had wide regional participation from representatives of ICAO's Member States, international organizations, aviation industries and academic/research institutions.

1.1.4 Law

ICAO held the Diplomatic Conference on Aviation Security in Beijing, China, from 30 August to 10 September 2010. Seventy-six States and four international organizations participated in the Conference. The Conference adopted the *Convention on the Suppression of Unlawful Acts Relating to International Civil Aviation* (Beijing Convention) and the *Protocol Supplementary to the Convention for the Suppression of Unlawful Seizure of Aircraft* (Beijing Protocol). These treaties criminalize, *inter alia*, the act of using civil aircraft as a weapon, and of using dangerous materials to attack aircraft or other targets on the ground. The unlawful transport of biological, chemical and nuclear weapons and their related material has been made punishable. Moreover, the criminal liability of directors and organizers of an offence under the treaties is specifically covered. Making a threat to commit an offence under the treaties may also trigger criminal liability, when the circumstances indicate that the threat is credible. Each of the two treaties requires 22 ratifications to bring it into force. As of 1 October 2010, the Convention had been signed by 20 States and the Protocol by 21 States.

1.2 The Assembly

1.2.1 Safety

One of the first areas in safety that the Assembly considered was safety management systems and safety data. As a first step, the Assembly considered a report on the evolution of ICAO's proactive safety management approach. This report provided an overview of ICAO's safety analysis strategy, including the eventual integration of operational data generated through future implementation of the State safety programme (SSP) and safety management systems (SMS). Also discussed were ICAO's leadership role in SMS; the development of common safety metrics, analysis methods and interoperable database systems to support safety performance measurement and ensure effective sharing of safety information among States; common methods and processes related to SMS implementation, acceptance, performance measurement and oversight; the need to educate senior management regarding their respective roles in support of SMS implementation and to develop skills within States and aviation organizations to support safety risk management activities, in particular, the ability to investigate safety related events of low consequence; the need for a definition of global safety metrics necessary to support a harmonized approach to safety analysis and cited The Civil Air Navigation Services Organisation (CANSO)'s work in development of leading and lagging safety indicators; and the need for the development of an international standard for SMS terms and definitions, risk forecasting techniques and computer systems to support proactive safety analysis. The Assembly also considered the need for the development of a new Annex to the Chicago Convention which addressed safety management.

The Assembly adopted a Resolution on global planning for safety and sustainability which recognized *inter alia*: the importance of a global framework to support the Strategic Objectives of ICAO; the importance of regional and national plans and initiatives based on the global framework for effective implementation; and that further progress in improving global safety and efficiency of civil aviation is best achieved through a cooperative, collaborative and coordinated approach in partnership with all stakeholders under the leadership of ICAO. The Resolution calls upon ICAO to implement and keep current the Global Aviation Safety Plan (GASP) and the Global Air Navigation Plan (GANP) to support the relevant Strategic Objectives of the Organization; calls upon States and invites other stakeholders to cooperate in the development and implementation of regional, subregional and national plans based on the framework of the global plans; instructs the Council to provide a report on the implementation and evolution of the global plans to future regular sessions of the Assembly; and instructs the Secretary General to promote, make available and effectively communicate the GANP, GASP and its associated Global Aviation Safety Roadmap (GASR) global plans.

Another Resolution adopted by the Assembly pertained to the Global Aviation Safety Plan of ICAO which reaffirmed that the primary objective of the Organization continues to be the improvement of safety and an associated reduction in the number of accidents and related fatalities within the international civil aviation system. It also recognized that safety is a shared responsibility involving ICAO, Contracting States and all other stakeholders. The Resolution urges contracting States to support the GASP objectives by: implementing the State Safety Programme (SSP); expeditiously implementing safety management systems across the aviation industry to complement the existing regulatory framework; sharing operational safety intelligence among States and relevant aviation stakeholders; ensuring that the travelling public has access to easily understandable safety-related information to enable informed decisions; creating an environment in which the reporting and sharing of information is encouraged and facilitated and in which remedial action is undertaken in a timely fashion when deficiencies are reported; and reporting accident and incident data as required to ICAO.

Inter alia, the Resolution also urges contracting States, regional safety oversight organizations and international organizations concerned to work with all stakeholders to implement the GASP objectives and GASR methodology objectives and to implement these methodologies to reduce the number and rate of aircraft accidents. Contracting States are called upon to demonstrate the political will necessary for taking remedial actions to address deficiencies including those identified by Universal Safety Oversight Audit Programme (USOAP) audits and through the application of GASP objectives and the ICAO regional planning process and to fully exercise safety oversight of their operators in full compliance with applicable Standards and Recommended Practices (SARPs), and assure themselves that foreign operators flying in their territory receive adequate oversight from their own State and take appropriate action when necessary to preserve safety. For this purpose, States should develop sustainable safety solutions to fully exercise their safety oversight responsibilities. This can be achieved by sharing resources, utilizing internal and/or external resources, such as regional and sub-regional safety oversight organizations and the expertise of other States.

On the subject of runway safety, the Assembly discussed ICAO's Runway Safety Programme and adopted a Resolution which recognizes that runway accidents constitute a large portion of all accidents and have resulted in a great number of fatalities and that runway excursions are the highest single occurrence category of all accidents over the last 10 years for all commercial and general aviation operations of fixed-wing aircraft above 5,700 kg certified maximum take-off mass. It also recognizes that there are several areas of technological development underway in the aviation industry that shows great promise in the prevention and mitigation of runway accidents and serious incidents. It urges States to take measures to enhance runway safety, including the establishment of runway safety programmes using a multidisciplinary approach, that include at least regulators, aircraft operators, air navigation services providers, aerodrome operators and aircraft manufacturers to prevent and mitigate the effects of runway excursions, runway incursions and other occurrences related to runway safety; and resolves that ICAO shall actively pursue runway safety using a multidisciplinary approach. States are invited by this Resolution to monitor runway safety events and related precursors as part of the safety data collection and processing system established under their State Safety Programmes.

A Resolution was also adopted on the development of an up-to-date consolidated statement of continuing ICAO policies and practices related to a global Air Traffic Management (ATM) system and communications, navigation and surveillance/air traffic management (CNS/ATM) systems. This Resolution calls upon States and regional safety oversight organizations (RSOOs) to establish a framework for joint planning and cooperation at the sub-regional level for joint development of CNS/ATM systems. Another Resolution, on development of an up-to-date consolidated statement of continuing ICAO policies and associated practices related specifically to air navigation, contains Appendices on such areas as coordination and cooperation of civil and military air traffic[8]; the provision of adequate aerodromes; and cooperation among Contracting States in investigations of certain aircraft accidents. On cooperation in the use of civil and military airspace, the Resolution requires the common use by civil and military aviation of airspace and of certain facilities and services to be arranged so as to ensure the safety, regularity and efficiency of civil aviation as well as to ensure the requirements of military air traffic are met; and that the regulations and procedures established by Contracting States to govern the operation of their state aircraft over the high seas ensure that these operations do not compromise the safety, regularity and efficiency of international civil air traffic and that, to the extent practicable, these operations comply with the rules of the air in Annex 2 to the Chicago convention on Rules of the Air. The Secretary General is required to provide guidance on best practices for civil/military coordination and cooperation; and Contracting States may include, when appropriate, representatives of military authorities in their delegations to ICAO meetings. A significant

---

[8] For more information and a discussion on this subject, see Abeyratne (2010b).

pronouncement of the Resolution is that ICAO serves as an international forum that plays a role in facilitating improved civil/military cooperation, collaboration and the sharing of best practices, and in providing the necessary follow-up activities that build on the success of the Global Air Traffic Management Forum on Civil/Military Cooperation (2009) with the support of civil/military partners.

Another Resolution – on performance based navigation global goals – urges all States to implement Area Navigation (RNAV)[9] and Required Navigation Performance (RNP)[10] air traffic services (ATS) routes and approach procedures in accordance with the ICAO PBN concept laid down in the *Performance-based Navigation (PBN) Manual*,[11] while another – on ICAO global planning for safety and sustainability – instructs the Council to amend the GANP to include a framework that will allow ICAO to easily analyze the impact of States' air navigation modernization plans on the global system and then take appropriate action as needed to ensure global harmonization. It also calls upon States, planning and implementation regional groups (PIRGs) and the aviation industry to utilize the guidance provided in the GANP for planning and implementation activities and urges Contracting States, industry and financing institutions to provide the necessary support for coordinated implementation of the GANP, avoiding duplication of effort. The Resolution also calls upon States that are developing new generation plans for their own air navigation modernization to share their plans in a timely manner with ICAO to ensure global compatibility and harmonization, and instructs the Council to ensure that the GANP is continuously maintained up to date in light of further operational and technical developments, in close collaboration with States and other stakeholders.

A significant guideline issued to the Council by the Assembly through this Resolution is to organize a Twelfth Air Navigation Conference in 2012, with a

---

[9] Area Navigation (RNAV) can be defined as a method of navigation that permits aircraft operation on any desired course within the coverage of station-referenced navigation signals or within the limits of a self contained system capability, or a combination of these. RNAV was developed to provide more lateral freedom and thus more complete use of available airspace. This method of navigation does not require a track directly to or from any specific radio navigation aid, and has three principal applications: a route structure can be organized between any given departure and arrival point to reduce flight distance and traffic separation; aircraft can be flown into terminal areas on varied pre-programmed arrival and departure paths to expedite traffic flow; and instrument approaches can be developed and certified at certain airports, without local instrument landing aids at that airport.

[10] Required navigation performance (RNP) is a type of performance based navigation (PBN) that allows an aircraft to fly a specific path between two 3-dimensionally defined points in space. RNAV and RNP systems are fundamentally similar. The key difference between them is the requirement for on-board performance monitoring and alerting. A navigation specification that includes a requirement for on-board navigation performance monitoring and alerting is referred to as an RNP specification. One not having such a requirement is referred to as an RNAV specification.

[11] Doc 9613.

view to developing longer-term planning for ICAO based on an update of the GANP.

The prevention of communicable diseases through air travel was another topical issue discussed by the Assembly. It adopted a Resolution which, in the backdrop of Article 14[12] to the Chicago Convention, urges Contracting States and regional safety oversight organizations to ensure that the public health sector and the aviation sector collaborate to develop a national preparedness plan for aviation which addresses public health emergencies of international concern and which is integrated with the general national preparedness plan. It also urges Contracting States to develop a national preparedness plan for aviation that is in compliance with the World Health Organization International Health Regulations (2005) and which are based on scientific principles and on the guidelines from ICAO and the World Health Organization. Contracting States, and regional safety oversight organizations as appropriate are also urged to establish requirements for the involvement of stakeholders such as airport operators, aircraft operators and air navigation service providers in the development of a national preparedness plan for aviation; and Contracting States are requested to join and participate in the Cooperative Arrangement for the Prevention of Spread of Communicable Disease through Air Travel (CAPSCA) project, where available, to ensure that its goals are achieved, unless equivalent measures are already in place.

On a regional basis, the Assembly adopted two resolutions – one on *A Comprehensive Regional Implementation Plan for Aviation Safety in Africa* and the other on *Regional Safety Oversight Organizations (RSOOs)*. In the former, the Assembly *inter alia* urges Contracting States of the AFI Region to commit to and accelerate the establishment of regional safety oversight organizations and regional accident investigation agencies, where required, and strengthen cooperation across the region in order to make the optimum use of available resources; and instructs the Council to notify States, industry and donors of the priority projects arising from the gap analysis. Other requirements of this resolution are that States, industry and donors implement priority projects identified by the gap analysis, performed in accordance with the Global Aviation Safety Plan (GASP); States, industry and donors make contributions in cash and kind towards the implementation of the AFI Plan and the Council recognize all such contributions; African States, ICAO and AFCAC jointly address deficiencies identified through the safety oversight audits and implement the recommendations made by the ICAO/AFCAC joint meeting on aviation safety in Africa; the Council monitor the implementation of the

---

[12] Article 14 provides: "Each contracting State agrees to take effective measures to prevent the spread by means of air navigation of cholera, typhus (epidemic), smallpox, yellow fever, plague, and such other communicable diseases as the contracting States shall from time to time decide to designate, and to that end contracting States will keep in close consultation with the agencies concerned with international regulations relating to sanitary measures applicable to aircraft. Such consultation shall be without prejudice to the application of any existing inter-national convention on this subject to which the contracting States may be parties".

recommendations of the joint ICAO/AFCAC meeting on aviation safety in Africa; the Council ensure a stronger ICAO leadership role in coordinating activities, initiatives and implementation strategies aimed specifically at implementing priority projects to achieve sustainable improvement of flight safety in the AFI Region and to allocate resources to the relevant Regional Offices accordingly; and monitor and measure the status of implementation in the AFI Region throughout the triennium and to report to the next ordinary session of the Assembly on the progress made.

The Resolution on RSOOs directs the Council of ICAO to promote the concept of regional cooperation for the purpose of enhancing safety and safety oversight, including the establishment of regional safety oversight organizations and to continue to partner with Contracting States, industry and other stakeholders for coordinating and facilitating the provision of financial and technical assistance to States and sub-regional and regional safety and safety oversight bodies, including regional safety oversight organizations, in order to enhance safety and strengthen safety oversight capabilities. It also directs the Council to the Council to continue the analysis of relevant safety-critical information for determining effective means of providing assistance to States and sub-regional and regional safety and safety oversight bodies, including regional safety oversight organizations; and to continue implementing an Implementation Support and Development – Safety (ISD-Safety) Programme to provide assistance to States and sub-regional and regional safety and safety oversight bodies, including regional safety oversight organizations. Contracting States are urged to develop and further strengthen regional and sub-regional cooperation in order to promote the highest degree of aviation safety, and encouraged to foster the creation of regional or sub-regional partnerships to collaborate in the development of solutions to common problems to build State safety oversight capability, and to participate in, or provide tangible support for, the strengthening and furtherance of sub-regional and regional aviation safety and safety oversight bodies, including regional safety oversight organizations.

The Assembly also adopted a resolution on proficiency in the English language used for radiotelephony communications which *inter alia* urges Contracting States that have not complied with the language proficiency requirement by the applicability date to post their language proficiency implementation plans including their interim measures to mitigate risk, as required, for pilots, air traffic controllers and aeronautical station operators involved in international operations on the ICAO website as outlined in accordance with the associated practices below and ICAO guidance material.

1.2.2 Security

In considering ICAO's security policy, the Assembly adopted a consolidated statement on the continuing ICAO policies related to the safeguarding of international civil aviation against acts of unlawful interference which strongly condemns all acts of unlawful interference against civil aviation wherever and by whomsoever and for whatever reason they are perpetrated. The Resolution notes with abhorrence acts and attempted acts of unlawful interference aimed at the destruction in flight of civil aircraft in commercial service including any misuse of civil aircraft as a weapon of destruction and the death of persons on board and on the ground and reaffirms that aviation security must continue to be treated as a matter of highest priority and appropriate resources should be made available by ICAO and its Member States. It calls upon all Contracting States to confirm their resolute support for the established policy of ICAO by applying the most effective security measures, individually and in cooperation with one another, to prevent acts of unlawful interference and to punish the perpetrators, planners, sponsors, and financiers of conspirators in any such acts.

The Assembly makes reference in the Resolution to legal instruments pertaining to aviation security[13] and calls upon Contracting States to give special attention to the adoption of adequate measures against persons committing, planning, sponsoring, financing or facilitating acts of unlawful seizure of aircraft, acts of sabotage or attempted sabotage or other acts or attempted acts of unlawful interference against civil aviation, and in particular to include in their legislation rules for the severe punishment of such persons. It also calls upon Contracting States to take adequate measures relating to the extradition or prosecution of persons committing acts of unlawful seizure of aircraft, acts of sabotage or attempted sabotage or other acts or attempted acts of unlawful interference against civil aviation by adopting appropriate provisions in law or treaty for that purpose or by strengthening existing arrangements and by concluding appropriate agreements for the suppression of such acts which would provide for the extradition of persons committing criminal attacks on international civil aviation.

On the subject of technical security measures, the Assembly urges all States on an individual basis and in cooperation with other States to take all possible measures for the prevention of acts of unlawful interference, in particular, those required or recommended in Annex 17 to the *Chicago Convention* as well as those recommended by the Council. It also urges Contracting States to intensify their efforts for the implementation of existing Standards and Recommended Practices (SARPs), and procedures relating to aviation security, to monitor such implementation, to take all necessary steps to prevent acts of unlawful interference against international civil aviation and to give appropriate attention to the guidance

---

[13] *Convention on Offences and Certain Other Acts Committed on Board Aircraft* (Tokyo, 1963), the *Convention for the Suppression of Unlawful Seizure of Aircraft* (The Hague, 1970), the *Convention for the Suppression of Unlawful Acts Against the Safety of Civil Aviation* (Montréal, 1971), the *Protocol for the Suppression of Unlawful Acts of Violence at Airports Serving International Civil Aviation, Supplementary to the Convention for the Suppression of Unlawful Acts Against the Safety of Civil Aviation* (Montréal, 1988), the *Convention on the Marking of Plastic Explosives for the Purpose of Detection* (Montréal, 1991), the *Convention for the Suppression of Unlawful Acts Relating to International Civil Aviation* (Beijing, 2010), the *Protocol Supplementary to the Convention for the Suppression of Unlawful Seizure of Aircraft* (Beijing, 2010).

material contained in the ICAO *Security Manual for Safeguarding Civil Aviation Against Acts of Unlawful Interference*[14] and available on the ICAO restricted website. Finally, it encourages Contracting States to promote aviation security as a fundamental component of national, social and economic priorities, planning and operations.

On the subject of unlawful interference the Assembly, while recognizing *inter alia* that acts of unlawful interference continue seriously to compromise the safety, regularity and efficiency of international civil aviation, urges Contracting States to cooperate for the purpose of providing a joint response in connection with an act of unlawful interference, as well as utilizing, if necessary, the experience and capabilities of the State of the operator, the State of manufacture and the State of registration of an aircraft which has been subjected to an act of unlawful interference, while taking measures in their territory to free the passengers and crew members of that aircraft.

It also condemns any failure by a Contracting State to fulfil its obligations to return without delay an aircraft which is being illegally detained and to submit to competent authorities or extradite without delay the case of any person accused of an act of unlawful interference with civil aviation, along with the reporting of false threats to civil aviation and *calls* upon Contracting States to prosecute the perpetrators of such acts in order to prevent the disruption of civil aviation operations. Finally, it calls upon Contracting States to continue to assist in the investigation of such acts and in the apprehension and prosecution of those responsible.

On the ICAO Universal Security Audit Programme, the Assembly adopted a resolution which urges all Member States to give full support to ICAO by: accepting the audit missions as scheduled by the Organization, in coordination with relevant States; facilitating the work of the audit teams; preparing and submitting to ICAO the required pre-audit documentation; and preparing and submitting an appropriate corrective action plan to address deficiencies identified during the audit, as well as other post-audit documentation. The Resolution also urges all Member States, if requested by another State, to share the results of the audit carried out by ICAO and the corrective actions taken by the audited State, as appropriate and consistent with their sovereignty. It requests that the Council report to the next ordinary session of the Assembly on the overall implementation of the USAP, including its decision with regard to the study to assess the feasibility of extending the CMA to the USAP after the conclusion of the current audit cycle in 2013.

The Assembly also adopted a *Declaration on Aviation Security* which urges Contracting States to *inter alia*, strengthen and promote the effective application of ICAO Standards and Recommended Practices, with particular focus on Annex 17 – *Security,* and develop strategies to address current and emerging threats; strengthen security screening procedures, enhance human factors and utilize modern

---

[14] Doc 8973.

technologies to detect prohibited articles and support research and development of technology for the detection of explosives, weapons and prohibited articles in order to prevent acts of unlawful interference; develop enhanced security measures to protect airport facilities and improve in-flight security, with appropriate enhancements in technology and training; and develop and implement strengthened and harmonized measures and best practices for air cargo security, taking into account the need to protect the entire air cargo supply chain.

1.2.3 Environmental Protection

The most topical and arguably contentious issue at the Assembly was climate change and, it must be mentioned that it was an achievement for ICAO member States to adopt a Resolution at the Assembly, *albeit* in the midst of some reservations recorded by States. On this issue, the Assembly noted that, if the global community were to stabilize greenhouse gas emissions in the atmosphere and maintain it at a level that would prevent dangerous anthropogenic interference with the climate, the increase in global temperature would have to be maintained below 2°C. In order to achieve this target, deep cuts in global emissions would be needed, and all sectors of the economy were being looked to for their contribution – including international aviation, which is well known as representing a significant and growing source of emissions.

The Assembly considered a proposal[15] from the Secretary General of ICAO which suggested that the Assembly adopt a resolution that required ICAO to exercise continuous leadership on environmental issues relating to international civil aviation, including greenhouse gas emissions; to continue to study policy options to limit or reduce the environmental impact of aircraft engine emissions and to develop concrete proposals and provide advice as soon as possible to the Conference of the Parties of the UNFCCC, encompassing technical solutions and market-based measures, and taking into account potential implications of such measures for developing as well as developed countries; and to continue to cooperate with organizations involved in policy-making in this field, notably with the UNFCCC. This was an unusual step – for a proposal to come from the Chief Executive of the ICAO Secretariat rather than from the Council of ICAO which reports to the Assembly – and reflected unequivocally that the Council had unprecedentedly failed to reach consensus on a comprehensive approach to aviation and climate change. The underlying reason for this impasse was that developing States could not agree to ambitious emissions reductions suggested by developed States.

The proposed resolution, which was subsequently adopted by the Assembly, also suggested *inter alia* that States and relevant organizations work through ICAO to

---

[15] A37-WP/262 EX/53.

achieve a global annual average fuel efficiency improvement of 2% until 2020 and an aspirational global fuel efficiency improvement rate of 2% per annum from 2021 to 2050, calculated on the basis of volume of fuel used per revenue tonne kilometre performed. Also suggested was the fact that ICAO and its member States with relevant organizations work together to strive to achieve a collective medium term global aspirational goal of keeping the global net carbon emissions from international aviation from 2020 at the same level, taking into account the special circumstances and respective capabilities of developing countries, the maturity of aviation markets and the sustainable growth of the international aviation industry. In this regard the proposed resolution suggested that the Council of ICAO consider a *de-minimis* exception for States which do not have substantial international aviation activity levels, in the submission of action plans and regular reports on aviation CO_2 emissions to ICAO. It also invited the Assembly to recognize that in the short term, voluntary carbon offsetting schemes constitute a practical way to offset CO_2 emissions, and invited States to encourage their operators wishing to take early actions to use carbon offsetting, particularly through the use of credits generated from internationally recognized schemes such as the Clean Development Mechanism (CDM).[16]

In addition to the 2% annual improvement in fuel efficiency discussed above, the 37th Session of the Assembly also considered a proposal that the feasibility of more ambitious medium and long term goals, including carbon neutral growth and emissions reductions be further explored. There was also a proposal by three States that a more ambitious goal be set – of carbon neutral growth by 2020 compared to 2005 levels. In response, a developing State took the position that ICAO should be guided by the principle of common but differentiated responsibilities (CBDR) under the UNFCCC; the next task for ICAO is to assist States to achieve the goal of 2% annual fuel efficiency improvement; the goal of carbon neutral growth is not realistic and not fair for developing States; and no States should be allowed to take unilateral actions on market-based measures, which drew some support from other developing States.

The main argument of the developing States at the Assembly was that since the larger quantity of GHG emissions was caused by developed States and that developing States should not be called upon to pay for ambitious emissions reduction levels at the same level as developed States. Furthermore developing States claimed that stabilizing the climate should be based on the principles of equity and common but differentiated responsibilities and those obligations under the framework of UNFCC. They concluded that any measure taken should not unduly curb the development of aviation in developing States.

---

[16] The Clean Development Mechanism (CDM) allows a developed country with an emission-reduction or emission-limitation commitment under the Kyoto Protocol to implement an emission-reduction project in developing countries. Such projects can earn saleable certified emission reduction (CER) credits, each equivalent to one tonne of CO_2, which can be counted towards meeting Kyoto targets. See http://www.icao.int/icao/fr/env2010/ClimateChange/Finance_f.htm.

1.2 The Assembly

The challenge faced by the Assembly during its discussions was to achieve consensus on establishing guiding principles when designing new and implementing existing market based measures for international aviation, and to engage in constructive bilateral and/or multilateral consultations and negotiations with other States to reach an agreement on issues such as carbon neutral growth and market based measures; as well as on a *de minimis* threshold of international aviation activity, consistent with 1% of total revenue ton kilometres to market based measures as follow:

a) Commercial aircraft operators of States below the threshold should qualify for exemption for application of MBMs that are established on national, regional and global levels; and
b) States and regions implementing MBMs may wish to also consider an exemption for other small aircraft operators.

It is significant therefore that, notwithstanding the divergence of views between States on the abovementioned issues, and reservations of some developing States, the Assembly was successful in adopting a Resolution which paves the way forward to more understanding and progress in the years to come.

1.2.4 Law

The Assembly adopted a Resolution on the Beijing Diplomatic Conference and the Convention and Protocol that ensued which recalls its Resolution A36-26, Appendix C, relating to the ratification of instruments which have been developed and adopted under the auspices of the Organization; and recognizes the importance of broadening and strengthening the global aviation security regime to meet new and emerging threats. On this basis, the Resolution urges all States to support and encourage the universal adoption of the *Convention on the Suppression of Unlawful Acts Relating to International Civil Aviation* (Beijing Convention of 2010) and the *Protocol Supplementary to the Convention for the Suppression of Unlawful Seizure of Aircraft* (Beijing Protocol of 2010). It urges all States to sign and ratify the Beijing Convention and Beijing Protocol of 2010 as soon as possible; and directs the Secretary General to provide assistance, as appropriate, with the ratification process if so requested by a Member State.

In retrospect, the 37th Session of the Assembly will be remembered as a forum at which ICAO's leadership role in international civil aviation as the only global forum that can assist States was reiterated in the key areas of safety, security and environmental protection. The Assembly also implicitly recognized that ICAO stands at the threshold of a renewed vision under a new leadership inspired by innovative thinking, and that the Organization is at the defining crossroads of its continuing path towards achieving its aims and objectives as set out in the Chicago Convention. New leadership and new thinking have been catalysts in this process, and, through a fog of rhetoric which in the past tended to obfuscate the role of the

Organization, a flight path has cleared that enables the Organization to steer towards a more relevant role in the twenty-first century.

From a legal perspective, one fact that emerges – particularly with regard to the Resolution on climate change – is that Member States of ICAO seemingly ascribe to Assembly resolutions a force that they do not have. ICAO is a specialized agency of the United Nations and therefore the law applicable to the United Nations in general applies to ICAO, and to its resolutions.[17] The record of the United Nations over its six decades of history is that member States have on occasion, but in a consistent manner, refused to automatically comply with the corporate will of the Organization.[18] *Brownlie* has expressed the view that decisions by international conferences and organizations can in principle only bind those States accepting them.[19] Shaw, referring to the binding force of United Nations General; Assembly Resolutions states:

> ...one must be alive to the dangers in ascribing legal value to everything that emanates from the Assembly. Resolutions are often the results of political compromises and arrangements and, comprehended in that sense, never intended to constitute binding norms. Great care must be taken in moving from a plethora of practice to the identification of legal norms.[20]

With regard to the practice of other international organizations, a little more caution might be required, as a resolution might create a custom. Non binding instruments form a special category that is sometimes referred to as "soft law" which is definitely not law in the sense of enforceability.[21]

ICAO's conferred powers enable the Organization to adopt binding regulations by majority decision (which is usually unnecessary as most of ICAO policy is adopted through consensus). However, States could opt out of these policies or make reservations thereto, usually before such policy enters into force. This is because States have delegated power to ICAO to make decisions on the basis that they accept such decisions on the international plane. In such cases States could contract out and enter into binding agreements outside the purview of ICAO even on subjects on which ICAO has adopted policy. The only exception to this rule lies

---

[17] Article 57 of the United Nations Charter provides that the various specialized agencies, established by intergovernmental agreement and having wide international responsibilities, as defined in their basic instruments, in economic, social, cultural, educational, health, and related fields, shall be brought into relationship with the United Nations in accordance with the provisions of Article 63, which provides that the Economic and Social Council may enter into agreements with any of the agencies referred to in Article 57, defining the terms on which the agency concerned shall be brought into relationship with the United Nations. Such agreements shall be subject to approval by the General Assembly. It may co-ordinate the activities of the specialized agencies through consultation with and recommendations to such agencies and through recommendations to the General Assembly and to the Members of the United Nations.

[18] Zoller (1987).

[19] Brownlie (1990).

[20] Shaw (2003).

[21] *Id.* 111. See also Tammes (1958).

in the adoption of Standards in Annex 2 to the Chicago Convention on Rules of the Air, in particular navigation over the high seas and other overflight areas where freedom of flight prevails which all Contracting States are bound to follow in order to maintain global safety.

There is little room for doubt that issues pertaining to safety, security, environmental protection and sustainability of air transport will remain with us for a long time, if not for ever. It is hoped that the discussions in this book will form a repository of reference and resource in this context.

References

Abeyratne R (2009) The role of ICAO in the twenty first century. Ann Air Space Law XXXIV:529–544

Abeyratne R (2010a) The NW 253 flight and the global framework of aviation security. Air Space Law 35(2):167–181

Abeyratne R (2010b) Compromises in the use of airspace in civil and military aviation. Eur Transport Law XLV(2):129–144

Brownlie I (1990) Principles of public international law, 4th edn. Clarendon Press, Oxford, p 691

Shaw MN (2003) International law, 5th edn. Cambridge University Press, Cambridge, p 110

Tammes AJP (1958) Decisions of international organs as a source of international law. HR 94:265

Zoller E (1987) The corporate will of the United Nations and the rights of the minority contributors. Am J Int Law 81(3):32

Chapter 2
Safety Issues

2.1 Safety Management Systems

Safety management systems encapsulate a macrocosm of intrinsic safety issues that require strategic intervention and response. These issues are discussed in some detail later in this chapter. Many of ICAO's 190 member States are already facing problems with respect to safety oversight. A glaring fact emerging from safety audits conducted by ICAO on States is that the findings of the initial safety oversight audit conducted by ICAO relating to the three Annexes to the Chicago Convention[1] – Annex 1 – *Personnel Licensing*, Annex 6 – *Operation of Aircraft* and Annex 8 – *Airworthiness of Aircraft*, indicated that of the 181 Contracting States that were audited between March 1999 and July 2004, considerable numbers of States had deficiencies in respect of a number of requirements under these Annexes. Furthermore, audit follow-up missions have revealed that in many cases, significant deficiencies identified during the initial audits remain.

Several fatal accidents in August, October and December 2005 brought to bear the need for regulators and operators to re-focus on the safety of civil aviation. One of the reasons for ensuring the safety of air transport is the need to keep up with the pace at which liberalization of air transport is spreading. With liberalization of commercial air transport comes the opportunity for new carriers to compete for market access. It also gives existing carriers the opportunity to compete with each other as well as with the new entrants with a view to consolidating their positions in a liberalized market. There is an onerous burden on States to ensure that, hand in hand with unfettered competition among carriers, rigid enforcement of technical regulation ensuring the safety of flight is necessary. This calls for the need to put into operation coordinated global safety management systems. This challenge calls for renewed thinking by all concerned in the best interests of the entire aviation industry.

---

[1] Convention on International Civil Aviation, signed at Chicago on 7 December 1944. *Ibid.*

Safety Management Systems (SMS) are processes which proactively manage the projected increase in aircraft incidents and accidents brought about by the increase in air traffic movements. SMS require vigilance in the liberalization of air transport and the correspondent increase in capacity. At the Directors General of Civil Aviation Conference on a Global Strategy for Aviation Safety, convened by the International Civil Aviation Organization in Montreal from 20 to 22 March 2006, Canada defined a Safety Management System as a businesslike approach to safety. An SMS is a systematic, explicit and comprehensive process for the management of safety risks that integrates operations and technical systems with financial and human resource management, for all activities related to an air operator as an approved maintenance organization's certificate holder.[2]

Any management system, including that which involves aviation safety, would necessarily entail planning, goal setting, performance measurement and accountability. This process is systemic, in that there cannot be one factor without the other to complete the entire system of the safety management process. It also requires a symbiotic relationship between operator and regulator, who are jointly responsible for the determination of the parameters of regulations and their intent. While the regulator's role is to give a clear set of instructions to the operator reflecting the expectations of the regulator, the operator has to ensure their compliance as well as maintain close coordination with the regulator.

At the DGCA Conference, ICAO emphasized that, although the prescriptive aspect of safety management, complied with by adherence to Standards and Recommended Practices (SARPs) of the Annexes to the Chicago Convention were essential, safety management has to transcend the mere compliance with SARPs toward a performance based process.[3] In this context, ICAO suggests a three step management process: firstly, between the operator and the oversight authority where there must be compete agreement between operators and service providers on the one hand and the oversight authority on the other hand on the safety performance to be expected. Performance will be measured by both parties through established safety indicators and safety targets. Secondly, all parties concerned would agree on the safety requirements necessary to achieve the safety performance agreed upon in the first step. Finally, oversight authorities would verify achievement of the agreed safety performance or lack thereof and the operator/service provider would rectify any observed deviations.

At the DGCA Conference, ICAO reiterated the fact that the term safety management conveys the notion that the management of safety is a business process that should be considered at the same level and along the same lines as any other business process. The safety management systems approach, as proposed, would not only be reactive to some triggering event such as an incident or reportable event, but it would also involve a proactive process of ongoing and routine collection and

---

[2] Management of Aviation Safety, Presented by Canada, DGCA/06-WP/15, 4/2/06, at 2.

[3] Implementation of Safety Management Systems (SMS in States), Presented by the ICAO Secretariat, DGCA/06-WP/6, 9/01/06, at 1 and 2.

2.1 Safety Management Systems

analysis of safety data during the course of a safety oriented work programme that an organization must pursue on a daily basis while in the business of its core functions. Such a system must essentially be performance driven and proactive. Furthermore, the term 'system' involves an integrated and systemic set of processes calculated to actively manage safety transcending departmental boundaries, thus ensuring safety from a broad perspective.

A systemic safety management system would also require a well thought out performance measurement process and indicators of performance which would be determined through organizational structure and accountability established by policies and procedures. The accountability in safety management should be enforced throughout an organization including its senior management.

It is incontrovertible that an effective safety management system has to necessarily co-exist with an efficient air transport product. If the former were to affect the latter's development and its capacity to offer services to the consumer, the fundamental purpose of safety management and its supportive role would be lost, as there would be no longer an air transport system that could adequately cater to the exponential demands made on it. The key purpose of the SMS is to adapt the growth in air transport, however evolving the operations of air transport are. A typical approach was exemplified by the United States at the ICAO DGCA Conference when they advised the delegates of four key areas of the SMS which work synergistically to maintain harmony between safety and efficiency, two of which – safety assurance and safety promotion – are critical to the efficiency and continuity of air transport. According to the United States submission the four key elements to effective SMS are: *Safety Policy*, which covers the SMS requirements, responsibilities and accountabilities for system functions; *Safety Risk Management*, which involve procedures used to identify hazards, assess current and future safety risk of the system or operation, manage the safety risk, document the mitigations selected to manage the safety risk and verify and monitor the mitigations throughout its lifecycle; *Safety Assurance*, which calls for the assurance of the safety product or service provider's system, including audits, evaluations and inspections, as well as data tracking and analysis; and *Safety Promotion*, which entails training, communication and dissemination of safety information to strengthen the safety culture and support integration of the SMS into operations.[4]

ICAO has added a unique dimension to safety management which is called the Unified Strategy. At its 35th session, held in September-October 2004, the ICAO. Assembly adopted Resolution A 35-7 which recognizes that transparency and sharing of safety information is one of the fundamental tenets of a safe air transportation system and urges all Contracting States to share with other Contracting States critical safety information which may have an impact on the safety of international air navigation and to facilitate access to all relevant safety information. The Resolution also encourages Contracting States to make full use of

---

[4] Safety Management System Concept, DGCA/06-WP/12, 4/2/06, at 2.

available safety information when performing their safety oversight functions, including during inspections as provided for in Article 16 of the Chicago Convention, which empowers State authorities to search aircraft of other States on landing or departure, and to inspect the certificates and other documents prescribed by the Convention. The Resolution also directs the Council to further develop practical means to facilitate the sharing of such safety information among Contracting States and reminds Contracting States of the need for surveillance of all aircraft operations, including foreign aircraft within their territory and to take appropriate action when necessary to preserve safety. A salient feature of Resolution A35-7 is that it touches on Article 54 j) of the Chicago Convention which makes it a mandatory function of the Council to report to Contracting States any infraction of the Convention, as well as any failure to carry out recommendations and determinations of the Council, and directs the Council to develop a procedure to inform all Contracting States, within the scope of Article 54 j) in the case of a State having significant compliance shortcomings with respect to ICAO safety-related SARPs.

One of the critical elements of the Resolution is that it directs the Council to promote the concept of regional or sub-regional safety oversight organizations, while encouraging States to foster the creation of regional or sub-regional partnerships to collaborate in the development of solutions to common problems to build their individual safety oversight capability. It also requests the Secretary General to continue to foster coordination and cooperation between USOAP and audit programmes of other organizations related to aviation safety, and specifically with the International Air Transport Association (IATA) and the European Organization for the Safety of Air Navigation (Eurocontrol). Furthermore, Contracting States are urged to further develop regional and sub-regional cooperation and, whenever feasible, partnership initiatives with other States, industry, air navigation service providers, financial institutions and other stake holders to strengthen safety oversight capabilities in order to foster a safer international civil aviation system and to better discharge their individual responsibilities.

2.1.1 Responsibility of States

Under ICAO's Universal Safety Oversight Audit Programme, States are required to establish a safety programme where aircraft operators, maintenance organizations and services providers implement appropriate safety management systems. The Chicago Convention contains certain provisions demarcating the responsibility of Contracting States of the International Civil Aviation Organization.[5] Article 28 of

---

[5] The International Civil Aviation Organization (ICAO) is the specialized agency of the United Nations on the subject of international civil aviation and was established by Article 43 of the Chicago Convention. ICAO has 190 Contracting States who have become member States of ICAO by placing their instruments of ratification of the Chicago Convention. Supra. n. 1.

2.1 Safety Management Systems

the Convention obligates contracting States to provide in their territories airports, radio services, meteorological services and other air navigation facilities to facilitate international air navigation, in accordance with the standards and practices established from time to time pursuant to the Convention.[6] This fundamental concept of State responsibility has to be viewed from the perspective of modern exigencies of the supply and demand curve of air traffic services where such services are currently being provided both regionally and on a flight information region (FIR) basis. The need for a shift of focus of the modern air navigational system is determined by two factors: the growing air traffic demand and the need for enhanced and more efficient air traffic services; and the transition into a seamless air traffic management system calculated to obviate inconsistencies caused by boundaries.[7] The goals of a global seamless air traffic management system are: to provide greater flexibility and efficiency by accommodating user-preferred flight profiles; to improve existing levels of safety; to accommodate the full range of aircraft types and airborne capabilities; to improve the provision of information to users, including weather conditions, the traffic situation and the availability of facilities; to organize air space in accordance with air traffic management (ATM) provisions and procedures; to increase user involvement in ATM decision making, including air-ground computer dialogue for flight plan negotiation; to create, to the extent possible, a single continuum of airspace where boundaries are transparent to users; and to increase capacity to meet future traffic demand.[8]

States have to be mindful of the fact that their overall responsibility under the Chicago Convention in providing air navigation services extends to the air traffic controller, whose service is of a unique nature. The special feature in the provision of air traffic control is brought to bear by the nature of the service provided, be it in the relaying of information on meteorology or on traffic. Globally, air traffic control services offer information relayed by people by means of radio communication involving extremely short time periods and using a standard set of terminology in the English language, even in regions of the world where English is not the first language.[9] ICAO endorses this view when it says:

> The air traffic controller's job consists of complex tasks demanding a high degree of skill and active application of unique cognitive abilities such as spatial perception, information processing, reasoning and decision making. The controller must know where all the aircraft under his/her responsibility are, and determine how and when to take action to ensure that they remain separated from each other, while also seeing to their requests and needs for descent, climb, take off, departure etc.[10]

---

[6] Chicago Convention, *supra*, note 3, Article 28 a).
[7] Sudharshan (2003), at p. 2.
[8] *Global Air Navigation Plan for CNS/ATM Systems*, Second Edition: 2002, ICAO Doc 9750, AN/963, p. 1-4-3 at paragraph 4.12.
[9] Miyagi (2005), at p. 143.
[10] *Global Air navigation Plan for CNS/ATM Systems*, *supra*, note 6, at p. 1-4-7, paragraph 4.39.

In view of the arduous task performed by the air traffic controller, ICAO has stretched the responsibility of the State, from a fundamental statement of state responsibility reflected in Article 28 a) of the Chicago Convention, to a more detailed pronouncement in its guidelines, calling on States to make improvements to the air traffic management system through supporting software that could assist the controller with conflict prediction, detection, advisory and resolution.[11] ICAO's focus of concentration is on a unified strategy which establishes a mechanism integrating the efforts to increase transparency and disclosure of safety related information. Although the unified strategy extends to encompass all areas of safety of flight including airworthiness, it is incontrovertible that the overall philosophy of the strategy will apply to the provision of air navigation services as well. One of the most fundamental aims of ICAO, as enshrined in Article 44 a) of the Chicago Convention, is to ensure the safe and orderly development of international civil aviation. To this end, and as part of its unified strategy, ICAO is suggesting the establishment of regional safety oversight organizations[12] along the lines of European Aviation Safety Agency (EASA) of Europe.

At the 35th Session of the ICAO Assembly, contracting States adopted Resolution A35-7,[13] which urges all contracting States to share with other contracting States critical safety information which may have an impact on the safety of international air navigation and to facilitate access to all relevant safety information. This resolving clause is based on the premise, explicitly recognized in the Resolution, that the improvement of the safety of international civil aviation on a worldwide basis requires the active collaboration of all stakeholders. By adopting Resolution A35-7, the 190 ICAO contracting States have arrogated to themselves increased responsibility for the safety of international civil aviation, irrespective of the nature of the provider of services that ensure such safety. This strategic approach of global safety oversight was further endorsed at the ICAO Conference on a Global Strategy for Aviation Safety convened of the Directors General of Civil Aviation from 20 to 22 March 2006 in Montreal, where all States consensually agreed that there should be an effective implementation of Safety Management Systems (SMS) in States, established according to a performance based approach in three steps: Firstly, where oversight authorities and operators and service providers agree on the safety performance to be expected from them while conducting their core business functions; secondly, that oversight authorities and service providers agree on the safety requirements necessary to achieve the safety performance agreed upon in the first step; and thirdly and finally, that oversight authorities verify achievement of the agreed safety performance or its lack thereof, and operators and service providers correct observed deviations.[14]

---

[11] *Id.* paragraph 4.40.

[12] Dr. Assad Kotaite, President of the International Civil Aviation Organization *ICAO's Unified Strategy*, Airport 2005, 118 at 119.

[13] *Assembly Resolutions in Force* (as of 8 October 2004), ICAO Doc 9848, at 1–60.

[14] *Implementation of Safety Management Systems (SMS) in States*, Working paper submitted by the ICAO Secretariat, DGCA/06-WP/6, 9/01/06 at 1–2.

2.1 Safety Management Systems

All the above goes to demonstrate the heavy responsibility placed by States on themselves in undertaking that they and they alone will be accountable for the safety of aviation within their territories as well as cooperating on a global basis to implement a seamless air navigation system. Of course, any change in the provision of air traffic services, particularly with regard to the commercial nature of the service provider, will remain subservient to the fundamental concept of State sovereignty, which was endorsed by the Eleventh ICAO Air Navigation Conference held in Montreal from 22 September to 3 October 2003.[15]

However, although the overall responsibility of the State to ensure the provision of air navigation services is immutable, there is no legal impediment to a State handing over the physical task of provision of services to a private entity while retaining its oversight role. Accordingly, in the present context, it is common to see a State largely in a supervisory role retaining its ownership of air space, drafting national legislation; determining governance over air navigation service providers; continuing to hold responsibility for certification and designation of service providers a well as setting regulations, while the service provider provides a public function in managing airspace with the broad spectrum of safety and efficiency. Within the overarching umbrella of State responsibility, there are various models of air navigation service providers. The first category is the original one which has not changed, where a State instrumentality continues to provide air navigation services. Examples of these are the Federal Aviation Administration[16] of the United States and the national authority of France (DSNA). The second category is the privatized or profit service provider such as *NATS* and third is the privatised service provider for non profit purposes such as *NavCanada*. There is a fourth category of a corporatized service provider for profit and those that come into this group are New Zealand Airways and Air Services Australia. A fifth category also exists, corporatized for non profit such as the air navigation service providers of Continental Europe. All these types of service providers, with the exception of the first category are autonomous, but remain undisputedly under the administrative supervision of their respective States.

---

[15] See Eleventh Air Navigation Conference report, *ICAO Doc 9828, AN-Conf/*11, Montreal: 20 at para. 1.2.1.2, at p. 1–1, where it is recorded that the Conference agreed that the issue of sovereignty was paramount in the operation of global air traffic management, as interpreted through the Global ATM operational concept of sovereignty.

[16] The FAA has completed a Safety Management System Standard for aviation product and service providers. It outlines the key attributes of an SMS and establishes SMS requirements, and is calculated to be imposed by the FAAA on organization overseen by the FAA including aircraft operators, aircraft manufacturers and the FAA Air Traffic Organization (ATO) which is air navigation services provider of the United States. See *Safety Management System Concept*, Working Paper presented by the United States to the ICAO DGCA Conference of March 2006, DGCA/06-WP/12, 4/02/06 at 1.

2.1.2 Provision of Air Navigation Services

Article 12 of the Chicago Convention unambiguously states that over the high seas, the rules in force shall be those under the Convention and each Contracting State undertakes to insure the prosecution of all persons violating the regulations applicable. This peremptory principle,[17] of adherence by States and aircraft bearing their nationality to any Standards and Recommended Practices (SARPs) adopted in regard to the high seas, effectively precludes any possible reliance by States on Article 38 of the Convention which allows States to deviate from SARPs in general. In other words, Annex 2 on Rules of the Air, which contains provisions relating to the operation of aircraft over the high seas, is sacrosanct and inviolable. The first legal issue that would emerge from this clear principle is the question of applicability of Annexes (other than Annex 2) to the high seas and whether their provisions, if directly related to the principles of manoeuvre and navigation of aircraft over the high seas, would be binding with no flexibility offered by Article 38 of the Convention. Kaiser offers the opinion:

> Over the high seas, the rules of the air have binding effect under Article 12, Sentence 3 of the Chicago Convention. It should be clarified that rules of the air have a broader meaning than Annex 2 and encompass the Standards and Recommended Practices of all other Annexes as far as their application makes sense over the high seas.[18]

Kaiser is of course referring mainly to Annexes 10 and 11 to the Chicago Convention relating to air traffic services and air traffic management, while at the same time drawing the example of Annex 16 (on environmental) protection being applicable in a future date if extended beyond noise and engine emissions to the high seas under Article 12 of the Chicago Convention.[19] This argument, which would ascribe to the ICAO Council wider control over larger spans of the world's air space, would be acceptable only if provisions of other Annexes (other than those of Annex 2) would directly have a bearing on the manoeuvre and navigation of aircraft over the high seas, as exclusively provided for by Article 12 of the Chicago Convention.

The ICAO Council, in adopting Annex 2 in April 1948 and subsequently in November 1951 when Amendment 1 to the Annex was adopted, resolved that the Annex constitutes *rules* relating to the flight and manoeuvre of aircraft within the meaning of Article 12 of the Convention. Therefore, the Council explicitly recognized that the rules in the Annex applied to the manoeuvre and operation of aircraft without exception. Annex 2, in its Foreword, states that the Standards in the

---

[17] Bin Cheng confirms that over the high seas there is absolutely no option for States to deviate from rules established under the Chicago Convention for the manoeuvre and operations of aircraft. See Cheng (1962), at p. 148.

[18] Kaiser (1995) at p. 455. Bin Cheng states that contracting States are expected to be able to exercise control over all that takes place within their territories, but outside their respective territories only over aircraft bearing their nationality. Bin Cheng, *supra*, note 9, at 110.

[19] *Ibid.*

2.1 Safety Management Systems

Annex, together with the Standards and Recommended Practices of Annex 11, govern the application of the Procedures for Air Navigation Services Rules of the Air and Air Traffic Services, and the Regional Supplementary Procedures. The Regional Supplementary Procedures are subsidiary procedures of regional applicability. It is clear that by this introduction, there is established a distinct disparity between Annex 2 and Annex 11 where the provisions of the former remain unquestionably mandatory, and the provisions of the latter remain subject to Article 38 of the Chicago Convention and capable of being deviated from. However, it is clear that the purpose of Annex 11 is to ensure that flying on international routes is carried out under uniform conditions designed to improve the safety and efficiency of air operation and, therefore, provisions relating to air traffic control services, flight information services, and alerting services of Annex 11 when linked to the provisions of Annex 2, have a coercive effect that may in certain circumstances, transcend the parameters set in Article 38 of the Convention.

The second issue in determining the legal status of rules of the air over the high seas in relation to sovereignty is the element of control exercised by a State over aircraft operations over the high seas. Article 2.1.2 of Annex 2 provides that a Contracting State may deem to have accepted (unless ICAO is otherwise advised) that it provides air traffic services through an appropriate ATS Authority as designated to be responsible for providing air traffic services over parts of the high seas. An appropriate ATS Authority is defined in the foreword of the Annex as the relevant authority designated by the State responsible for providing air traffic services in the air space concerned. A contracting party accepts an Appropriate ATS Authority pursuant to a regional air navigation agreement, which is an agreement approved by the Council of ICAO usually based on the outcome of the findings of Regional Air Navigation Meetings.

It is somewhat disconcerting that neither the legal status of the regional air navigation plan or agreement, nor its definition is clear. In November 1996, at the 38th meeting of the European Air Navigation Planning Group, it was recorded that an Air Navigation Plan consisted of an authoritative internationally agreed reference document, which corresponded to a contract between States covered by the Plan regarding air navigation facilities to be provided, to be approved by the ICAO Council in accordance with the provisions of the Chicago Convention.[20] It was deemed that the Council, in any given instance, would be acting on behalf of all Contracting States, including those not covered by the Plan. There is a marked dichotomy in the terminology used, which refers to the Plan on the one hand as a contract between parties and on the other hand as a reference document. Buergenthal offers a more coherent view, by saying that ICAO Annexes, Plans, SUPPS[21] and Regional Air Navigation Plans constitute an integral body of aviation

---

[20] ICAO Doc. EANPG COG/2-WP/6, 12/03/1996 at 3.

[21] The ICAO Regional Supplementary Procedures (SUPPS) form the procedural part of the Air Navigation Plan developed by Regional Air Navigation (RAN) Meetings to meet those needs of specific areas which are not covered in the worldwide provisions. They complement the statement of requirements for facilities and services contained in the Air Navigation Plan publications.

legislation comparable both in structure and content to comprehensive domestic air navigation codes.[22] Yet another view is that the Regional Air Navigation Plan, not involving the process of ratification, signature or adoption, is a technical and operational document.[23] Confusion is further confounded by the fact that there is no direct consequence for any State which does not perform its obligations under a Regional Air Navigation Plan.

The provision of air navigation services are mainly regulated by three Annexes to the Chicago Convention, namely Annex 2 (Rules of the Air), Annex 3 (Meteorological Service for International Air Navigation) and Annex 11 (Air Traffic Services).[24] Of these, compliance with Annex 2 is mandatory[25] and does not give the States the flexibility provided in Article 38 of the Chicago Convention to register differences from any provisions of the Annex.

With regard to navigation over the high seas, the United Nations Convention on the Law of the Sea *UNCLOS*, Article 39, lays down the duties of ships and aircraft involved in transit navigation to the effect that ships and aircraft, while exercising the right of transit passage, should : proceed without delay through or over the strait; refrain from any threat or use of force against the sovereignty, territorial integrity or political independence of States bordering the strait, or in any other manner in violation of the principles of international law embodied in the Charter of

---

Procedures of worldwide applicability are included either in the Annexes to the Convention on International Civil Aviation as Standards or Recommended Practices, or in the Procedures for Air Navigation Services (PANS). See *ICAO Doc 7030*.

[22] Buergenthal (1969), at p. 121.

[23] Milde (2002).

[24] Article 54 1) of the Chicago Convention stipulates as a mandatory function of the Council the act of adopting, in accordance with Chapter VI of the Convention, international standards and recommended practices (SARPs) and for convenience designate them as Annexes to the Convention. Article 37 of the Convention reflects the areas in which SARPs should be developed and Annexes formed. Article 38 obliges contracting States to notify ICAO of any differences between their own regulations and practices and those established by international standards or procedures. The notification of differences however, does not absolve States from their continuing obligation under Article 37 to collaborate in securing the highest practicable degree of uniformity in international regulations, standards, and procedures.

[25] In October 1945, the Rules of the Air and Air Traffic Control (RAC) Division at its first session made recommendations for Standards, Practices and Procedures for the Rules of the Air. These were reviewed by the then Air Navigation Committee and approved by the Council on 25 February 1946. They were published as *Recommendations for Standards, Practices and Procedures – Rules of the Air* in the first part of Doc 2010, published in February 1946. The RAC Division, at its second session in December 1946–January 1947, reviewed Doc 2010 and proposed Standards and Recommended Practices for the Rules of the Air. These were adopted by the Council as Standards and Recommended Practices relating to Rules of the Air on 15 April 1948, pursuant to Article 37 of the Convention on International Civil Aviation (Chicago, 1944) and designated as Annex 2 to the Convention with the title *International Standards and Recommended Practices – Rules of the Air*. They became effective on 15 September 1948. On 27 November 1951, the Council adopted a complete new text of the Annex, which no longer contained Recommended Practices. The Standards of the amended Annex 2 (Amendment 1) became effective on 1 April 1952 and applicable on 1 September 1952.

the United Nations; refrain from any activities other than those incident to their normal modes of continuous and expeditious transit unless rendered necessary by force majeure or by distress; and comply with the relevant provisions of the Convention. Article 39 (3) explicitly states that aircraft in transit passage shall observe the Rules of the Air established by ICAO as they apply to civil aircraft and that state aircraft will normally comply with such safety measures and will at all times operate with due regard for the safety of navigation. The provision further states that at all times aircraft shall monitor the radio frequency assigned by the competent internationally designated air traffic control authority or the appropriate international distress radio frequency.

Standard 2.1.1 of Annex 2 to the Chicago Convention provides that the rules of the air shall apply to aircraft bearing the nationality and registration marks of a Contracting State, wherever they may be, to the extent that they do not conflict with the rules published by the State having jurisdiction over the territory over-flown.[26] The operation of an aircraft either in flight or on the movement area of an aerodrome shall be in compliance with the general rules and, in addition, when in flight, either with: visual flight rules (VFR); or the instrument flight rules (IFR).[27] Standard 2.3.1 further provides that the pilot-in-command of an aircraft shall, whether manipulating the controls or not, be responsible for the operation of the aircraft in accordance with the rules of the air, except that the pilot-in-command may depart from these rules in circumstances that render such departure absolutely necessary in the interests of safety.

2.1.3 Provision of Meteorological Information

Annex 3 to the Chicago Convention provides in Standard 2.1.1. that the objective of meteorological service for international air navigation shall be to contribute towards the safety, regularity and efficiency of international air navigation. This objective shall be achieved by supplying the following users: operators, flight crew members, air traffic services units, search and rescue services units, airport managements and others concerned with the conduct or development of international air navigation, with the meteorological information necessary for the performance of their respective functions.[28]

---

[26] The Council of the International Civil Aviation Organization resolved, in adopting Annex 2 in April 1948 and Amendment 1 to the said Annex in November 1951, that the Annex constitutes Rules relating to the flight and manoeuvre of aircraft within the meaning of Article 12 of the Convention. Over the high seas, therefore, these rules apply without exception.

[27] Information relevant to the services provided to aircraft operating in accordance with both visual flight rules and instrument flight rules in the seven ATS airspace classes is contained in 2.6.1 and 2.6.3 of Annex 11. A pilot may elect to fly in accordance with instrument flight rules in visual meteorological conditions or may be required to do so by the appropriate ATS authority.

[28] Standard 2.1.2.

The Annex, in Standard 2.1.3 calls for each Contracting State to determine the meteorological service which it will provide to meet the needs of international air navigation. This determination shall be made in accordance with the provisions of this Annex and with due regard to regional air navigation agreements; it shall include the determination of the meteorological service to be provided for international air navigation over international waters and other areas which lie outside the territory of the State concerned. Furthermore, each contracting State is required to designate the authority, hereinafter referred to as the meteorological authority, to provide or to arrange for the provision of meteorological service for international air navigation on its behalf. Details of the meteorological authority so designated shall be included in the State aeronautical information publication.

State responsibility for the provision of meteorological information is provided for in Standard 2.1.4 where each Contracting State is required to ensure that the designated meteorological authority complies with the requirements of the World Meteorological Organization in respect of qualifications and training of meteorological personnel providing service for international air navigation.[29]

It is also provided in the Annex that close liaison shall be maintained between those concerned with the supply and those concerned with the use of meteorological information on matters which affect the provision of meteorological service for international air navigation.[30] Furthermore, States have responsibility establish one or more aerodrome and/or other meteorological offices which shall be adequate for the provision of the meteorological service required to satisfy the needs of international air navigation.[31]

The Annex provides that an aerodrome meteorological office shall carry out all or some of the functions as necessary to meet the needs of flight operations at the aerodrome in the preparation of :forecasts and other relevant information for flights with which it is concerned; the extent of its responsibilities to prepare forecasts shall be related to the local availability and use of en-route and aerodrome forecast material received from other offices; preparation of and obtaining forecasts of local meteorological conditions; maintaining a continuous survey of meteorological conditions over the aerodromes for which it is designated to prepare forecasts; providing briefing, consultation and flight documentation to flight crew members and/or other flight operations personnel; supplying other meteorological information to aeronautical users; displaying the available meteorological information; exchanging meteorological information with other meteorological offices; and supplying information received on pre-eruption volcanic activity, a volcanic eruption or volcanic ash cloud, to its associated air traffic services unit, aeronautical

---

[29] Requirements concerning qualifications and training of meteorological personnel in aeronautical meteorology are given in WMO Publication No. 49, Technical Regulations, Volume I – General Meteorological Standards and Recommended Practices, Chapter B.4 – *Education and Training*.

[30] Standard 2.2.1.

[31] Standard 3.3.1.

information service unit and meteorological watch office as agreed between the meteorological, aeronautical information service and ATS authorities concerned.

Chapter 5 of the Annex includes further responsibilities of States. Standard 5.1 provides that each Contracting State shall arrange, according to the provisions of this chapter, for observations to be made by aircraft of its registry operating on international air routes and for the recording and reporting of these observations. Aircraft observations are required with regard to routine aircraft observations during en-route and climb-out phases of the flight; and special and other non-routine aircraft observations during any phase of the flight.

2.1.4 *Provision of Air Traffic Services*

The provision of air traffic services[32] is addressed in Annex 11 to the Chicago Convention which provides *in limine* that Contracting States shall determine, in accordance with the provisions of this Annex and for the territories over which they have jurisdiction, those portions of the airspace and those aerodromes where air traffic services will be provided. They shall thereafter arrange for such services to be established and provided in accordance with the provisions of this Annex, except that, by mutual agreement, a State may delegate to another State the responsibility for establishing and providing air traffic services in flight information regions, control areas or control zones extending over the territories of the former.[33]

The Standards and Recommended Practices in Annex 11, together with the Standards in Annex 2, govern the application of the *Procedures for Air Navigation Services – Air Traffic Management*[34] and the *Regional Supplementary Procedures – Rules of the Air and Air Traffic Services*, contained in Doc 7030, Annex 11 pertains

---

[32] According to Paragraph 2.2 of the Annex, The objectives of the air traffic services shall be to: a) prevent collisions between aircraft; b) prevent collisions between aircraft on the manoeuvring area and obstructions on that area; c) expedite and maintain an orderly flow of air traffic; d) provide advice and information useful for the safe and efficient conduct of flights; e) notify appropriate organizations regarding aircraft in need of search and rescue aid, and assist such organizations as required.

[33] 31 Standard 2.1.1. It is also provided in the Annex that if one State delegates to another State the responsibility for the provision of air traffic services over its territory, it does so without derogation of its national sovereignty. Similarly, the providing State's responsibility is limited to technical and operational considerations and does not extend beyond those pertaining to the safety and expedition of aircraft using the concerned airspace. Furthermore, the providing State in providing air traffic services within the territory of the delegating State will do so in accordance with the requirements of the latter which is expected to establish such facilities and services for the use of the providing State as are jointly agreed to be necessary. It is further expected that the delegating State would not withdraw or modify such facilities and services without prior consultation with the providing State. Both the delegating and providing States may terminate the agreement between them at any time.

[34] Doc 4444, PANS-ATM.

to the establishment of airspace, units and services necessary to promote a safe, orderly and expeditious flow of air traffic. A clear distinction is made between air traffic control service, flight information service and alerting service. Its purpose, together with Annex 2, is to ensure that flying on international air routes is carried out under uniform conditions designed to improve the safety and efficiency of air operation.

The Standards and Recommended Practices in Annex 11 apply in those parts of the airspace under the jurisdiction of a Contracting State wherein air traffic services are provided and also wherever a Contracting State accepts the responsibility of providing air traffic services over the high seas or in airspace of undetermined sovereignty. A Contracting State accepting such responsibility may apply the Standards and Recommended Practices in a manner consistent with that adopted for airspace under its jurisdiction.

Standard 2.1.2 of the Annex stipulates that those portions of the airspace over the high seas or in airspace of undetermined sovereignty where air traffic services will be provided shall be determined on the basis of regional air navigation agreements. A Contracting State having accepted the responsibility to provide air traffic services in such portions of airspace shall thereafter arrange for the services to be established and provided in accordance with the provisions of the Annex.[35] The Annex goes on to say that when it has been determined that air traffic services will be provided, the States concerned shall designate the authority[36] responsible for providing such services.[37] Situations which arise in respect of the establishment and provision of air traffic services to either part or whole of an international flight are as follows:

Situation 1: A route, or portion of a route, contained within airspace under the sovereignty of a State establishing and providing its own air traffic services.

Situation 2: A route, or portion of a route, contained within airspace under the sovereignty of a State which has, by mutual agreement, delegated to another State, responsibility for the establishment and provision of air traffic services.

Situation 3: A portion of a route contained within airspace over the high seas or in airspace of undetermined sovereignty for which a State has accepted the responsibility for the establishment and provision of air traffic services.

For the purpose of the Annex, the State which designates the authority responsible for establishing and providing the air traffic services is:

---

[35] 33 The phrase "regional air navigation agreements" refers to the agreements approved by the Council of ICAO normally on the advice of Regional Air Navigation Meetings. The Council, when approving the Foreword to this Annex, indicated that a Contracting State accepting the responsibility for providing air traffic services over the high seas or in airspace of undetermined sovereignty may apply the Standards and Recommended Practices in a manner consistent with that adopted for airspace under its jurisdiction.

[36] The authority responsible for establishing and providing the services may be a State or a suitable Agency.

[37] Standard 2.1.3.

2.1 Safety Management Systems

In *Situation 1*: the State having sovereignty over the relevant portion of the airspace;
In *Situation 2*: the State to whom responsibility for the establishment and provision of air traffic services has been delegated;
In *Situation 3*: the State which has accepted the responsibility for the establishment and provision of air traffic services.

As the overall liability of the State to provide air navigation services has been clearly identified by international treaty and, as already discussed, there are various kinds of air service providers ranging from State instrumentalities to private service providers, the liability regime could be varied and contentious. Clearly, liability of the State can be bifurcated into two areas under administrative law where liability of the State, its agency or a private body can be detained within the territory of a State, and international law, the latter involving principles of State responsibility and the liability of a State for causing injury to another State or its subjects.

Of these, the responsibility of providing airports and air navigation services to aircraft lies clearly with the State. Article 28 of the Chicago Convention is clear on this when it states that each Contracting State undertakes, as far as practicable to provide in its territory, airports, radio services, meteorological services and other air navigation facilities to facilitate air navigational facilities.

State responsibility is anchored on certain basic fundamental facts. Firstly, there should be international obligation between two States or more. Secondly, an act or omission by one State must violate that obligation which can be directly attributed or imputed to the State concerned. Thirdly, loss or damage must be incurred by the aggrieved State. These requirements fit well into the provision of air navigation services by one State to another which is recognized in Article 28 of the Chicago Convention. Therefore, the negligent act or omission of a service provider can be imputed to the State establishing or appointing such provider publicly or on a private basis. In the instance of an aircraft, whether national or foreign, carrying foreign nationals who are injured in an accident due to the negligent provision of air navigation would open a State to responsibility to make reparation to the claimants.[38] The objective responsibility theory suggests that a State's responsibility has to be one of strict liability, where good or bad faith is not a consideration, whereas the subjective responsibility test involves fault liability, involving negligence. In the 1926 *Neer* claim where an American superintendent of a Mexican mine was shot, the General Claims Commission hearing the claim of the United States applied the objective test and rejected the claim that Mexico was responsible for not pursuing the investigation diligently. Three years later, the French-Mexican

---

[38] Charzow Factory Case PCIJ Series A No. 17, 1928, where the Court held that it is a principle of international law, and even a greater concept of international law, that any breach of an enjoyment involves an obligation to make reparation. See also the *Corfu Channel Case*, ICJ Reports at pp. 4, 26, *The Spanish Zone of Morocco Case*, 2 RIAA at p. 615 and *The Mayagna (Sumo) Indigenous Community of Awas Tingini* v. *Nicaragua, Inter American Court of Human rights*, Judgment of 31 August 2001 (Ser. C) No. 79, para 163.

Claims Commission, when considering the shooting of a French citizen by Mexican soldiers, applied the objective test and held the Mexican Republic responsible to maker reparation to France.[39] The subjective approach involving an enquiry into negligence has been applied in the case of damage caused to one State by the rebels of another State as a result of a tax imposed by a third State where there was found to be no fault on the part of the third State.[40]

Apart from the direct attribution of responsibility to a State, particularly in instances where a State might be guilty of a breach of treaty provisions, or violate the territorial sovereignty of another State, there are instances where an act could be imputable to a State. Imputability depends upon the link that exists between the State and the legal person or persons actually responsible for the act in question. The legal possibility of imposing liability upon a State wherever an official could be linked to that State encourages a State to be more cautious of its responsibility in controlling those responsible for carrying out tasks for which the State could be ultimately held responsible. It is arguable that, in view of the responsibility imposed upon a State by the Chicago Convention on the provision of air navigation services, the principles of immutability in State responsibility could be applied to an instance of an act or omission of a public or private official providing air navigation services.

Economic liberalization as well as the evolution of business and operating practices may have serious implications for safety regulation, which need to be addressed properly. The most fundamental consideration is that safety of air transport must remain the ultimate responsibility of the State concerned and should not be compromised by economic considerations. While ICAO should continue its monitoring of liberalization within the parameter of safety, all players, including States, service providers and airlines should coordinate well in understanding their responsibilities and providing safety oversight. In this constantly evolving environment, due regard should be paid to the impact of a growing air transport industry's impact on safety so that coherent SMS are developed.

The operative criterion should be that safety of flight is a valid concern of the aviation industry whether or not air transport is liberalized and, as already discussed, liberalization and consequent competition between air carriers should alert States to the potential safety hazards that a highly competitive air transport industry brings to bear. Considering the elaborate work already accomplished by ICAO, in establishing guidance to States on the implementation of Article 83 *bis* to the Chicago Convention. In addition to Articles 12, 30, 31 and 32 a)[41] of the

---

[39] *Caire* Claim, 5 RIAA, p.516 (1929).

[40] Home Missionary Society Claim, 6 RIAA, p. 42 (1920).

[41] Article 12 stipulates that each contracting State is required to insure that every aircraft flying over or manoeuvring within its territory and that every aircraft carrying its nationality mark, wherever that aircraft might be, shall comply with the rules and regulations relating to flight and manoeuvre applicable there. It also provides that each Contracting States should keep its regulations consistent with the principles of the Chicago Convention. Over the high seas the rules will be as established under the Convention and States undertake, under Article 12, to

Convention and the up-to-date Standards and Recommended Practices adopted from time to time as necessary in Annexes 1, 6 and 8, there is no room for doubt that the regulatory provisions are in place.

Therefore what remains is for all players concerned to move toward a performance based SMS process where accountability will drive aviation safety to its desired goals.

Principles of modern treaty law demand that an ICAO contracting State, which has placed its instrument of ratification to the Chicago Convention has consented to be bound by the provisions of that treaty. The word "contracting State" refers legally to a State which is bound by the treaty concerned, irrespective of whether the treaty is in force or not.[42] This goes to the root of international responsibility as already discussed and it is improbable at common law that a court would consider otherwise and disregard a State's obligation in the provision of air navigation services, particularly in the context of governmental agencies and other instrumentalities of State. This however, does not completely exonerate the privatized service provider, who could be held liable at private law. Legally, as was discussed, neither the State nor the service provider can avoid liability on account of privatization. The State entails liability primarily at public international law and also at public law in general, and the provider incurs liability on a private basis in a private action that may be brought, under tort law principles or under contract law, as the case may be.

Paul Stephen Dempsey[43] sums it up well, when he says that the issue has two critical considerations, one relating to legal issues and the other impacting public policy. Dempsey states correctly that the skies belong to the public and the sovereign is but the trustee in this regard. Under any circumstances, whether on fiscal profit making or political motivation, States cannot abdicate or pass on their responsibility and accountability of their traditional function and fiduciary responsibility. Besides, holding governments responsible will ensure proper quality control in the provision of air navigation services.

2.1.5 The Safety Roadmap

In the field of aviation safety, there are three incontrovertible truths. The first is that the attainment of safety is the highest priority in aviation. The second is that safety cannot be de-regulated. The third is that safety is a global concern and therefore States cannot have their own individual plans for ensuring safety. The last brings to

---

prosecute any offender who does not adhere to these principles. Articles 30 and 31 refer to the need for aircraft to carry radio equipment as required and be issued with certificates of airworthiness by the State of Registry, respectively. Article 32 a) requires the pilot and other members of the crew to be provided with certificates of competency and licenses issued by the State of Registry.

[42] Aust (2000), at p. 75.

[43] See Dempsey (2003), at 118–119.

bear a compelling need for an action plan of global dimension that identifies the roles to be played by the regulatory and industry partners in ensuring safety.

Significant progress was made throughout 2007 in transforming ICAO into a performance-based and results oriented Organization, in keeping with the Organization's Business Plan. The most significant improvements are highlighted in this Annual Report which, in its more accessible format and with links to the ICAO website, is a graphic illustration of this new way of doing business. Responding to the outcome of the 36th Session of the ICAO Assembly and to facilitate the transition to results-based planning and results-based budgeting, the Council, in November, reviewed the Business Plan for the next triennium (2008–2010) to support the implementation of the Strategic Objectives. The task involved identifying and implementing additional ways and means of further increasing ICAO's efficiency as an ongoing process throughout the Organization.

One of ICAO's Strategic Objectives is to enhance aviation safety globally, and to this extent the ICAO Safety Roadmap serves as an effective tool to assist member States of the Organization in addressing their safety issues efficiently.

The Safety Roadmap (hereinafter referred to as the Roadmap), which is applicable to all these statements and responds to needs arising therefrom, was put together by a group of industry partners called the Industry Safety Strategy Group (ISSG), the establishment of which was inspired by the International Civil Aviation Organization (ICAO) at its Seventh Air Navigation Commission Industry Meeting in May 2005. Although the title of this article ascribes ownership of the Safety Roadmap to ICAO, it is a joint effort of the ISSG[44] which comprises Airports Council International, Airbus Industrie, Boeing, the Civil Air Navigation Services Organization, Flight Safety Foundation, International Air Transport Association and the International Federation of Airline Pilots Associations.

The Roadmap presents a phased approach that would ensure safety in aviation through a proactive modality using ICAO as the main protagonist. Part 1 of the Roadmap – *A Strategic Action Plan for Future Aviation Safety* – provides the framework for action by Contracting States of ICAO, regions and the industry to correct inconsistencies and weaknesses in 12 main focus areas, including implementation of international standards, regulatory oversight, incident and accident investigation, Safety Management Systems (SMS)[45] and sufficient qualified personnel. SMS are processes which proactively manage the projected increase in

---

[44] The ISSG worked with ICAO, the primary customer for their work, to produce the *Global Aviation Safety Roadmap Part 1 – A Strategic Action Plan for Future Aviation Safety,* which was handed over to ICAO in December 2005 and presented to the Directors General Civil Aviation (DGCA) Conference on a Global Strategy for Aviation Safety (DGCA/06) in March 2006. The conference welcomed the *Global Aviation Safety Roadmap* and recommended that:

> ICAO, in collaboration with all States & other stakeholders, continue development of an integrated approach to safety initiatives based on the Global Aviation Safety Roadmap - a global framework for the coordination of safety policies and initiatives.

[45] For a discussion on SMS see, Abeyratne (2007a).

2.1 Safety Management Systems

aircraft incidents and accidents brought about by the increase in air traffic movements. SMS require vigilance in the liberalization of air transport and the correspondent increase in capacity. As stated earlier, at the Directors General of Civil Aviation Conference on a Global Strategy for Aviation Safety, convened by ICAO in Montreal from 20 to 22 March 2006, Canada defined a Safety Management System as a businesslike approach to safety. An SMS is a systematic, explicit and comprehensive process for the management of safety risks that integrates operations and technical systems with financial and human resource management, for all activities related to an air operator as an approved maintenance organization's certificate holder.[46]

The Roadmap sets one or more short-term and medium-term objectives for each focus area over the next 10 years. Part 2 – *Implementing the Global Aviation Safety Roadmap* – describes and prioritizes specific coordinated actions by industry to reduce risk and improve safety worldwide. For each objective identified in Part 1, the Roadmap proposes best practices with related industry references and compliance metrics. Part 2 also includes Annexes containing recommendations on existing and proven technologies (and associated training programmes) to further enhance safety in flight operations, airport operations and air traffic control domains, as well as regional implementation through a knowledge-based regional assessment and deployment strategy.

The Roadmap provides a common frame of reference to all stakeholders[47] focussing on States' responsibilities in addressing inconsistent implementation of international standards; inconsistent regulatory oversight; impediments to reporting of errors and incidents and ineffective incident and accident investigation. The key focus areas for the industry are: impediments to reporting and analysing errors and incidents; inconsistent use of safety management systems; inconsistent compliance with regulatory requirements; inconsistent adoption of industry best practice; non-alignment of industry safety strategies; insufficient numbers of qualified personnel; and gaps in the use of technology to enhance safety.

ICAO has a grave responsibility and important role to play in implementing the plans contained in the Roadmap and making sure that the Roadmap proves to be an implementing tool. This responsibility can be attenuated from the fundamental aim of ICAO, enshrined in Article 44 of the Convention on International Civil Aviation[48] to develop the principles and techniques of international air navigation. The Air Navigation Commission of ICAO will keep the Roadmap in review with a view

---

[46] Management of Aviation Safety, Presented by Canada, DGCA/06-WP/15, 4/2/06, at 2.

[47] The stakeholders are identified as States, ICAO, airlines/operators, airports, air navigation sevice providers, aircraft and equipment manufacturers, maintenance and repair organizations and industry representatives.

[48] Convention on International Civil Aviation, signed at Chicago on 7 December 1944, Supra. n. 3.

to ensuring that it assists in implementing Strategic Objective A of ICAO's Business Plan, which is the enhancement of safety.[49]

2.1.5.1 The Role of ICAO

The Roadmap calls for leadership, particularly from ICAO as the global forum for international civil aviation. The hallmark of leadership is action. The action required of ICAO in this regard is to create a paradigm shift that would take key stakeholders away from the reactive role of implementing safety measures to a proactive role of actively reducing safety related accidents and incidents in aviation. ICAO as a true leader must possess the two recognized values of leadership – actual leadership, in giving guidance or direction, and potential leadership that carries the capacity or ability to lead. Therefore, ICAO's role should be that of a leader who influences people to strive willingly for the group objective of ensuring safety in aviation.

The simplest way to measure the effectiveness of leadership involves evaluating the size of the following that the leader can muster. To measure leadership more specifically, one may assess the extent of influence on the followers, that may involve testing the results of leadership activities against the goal, vision, or objective.

It is fair to say that ICAO is at the crossroads of re- defining its continuing role towards achieving its aims and objectives as set out in the Convention on International Civil Aviation. With a view to setting its course in line with rapidly evolving trends of globalization and regionalization, the Organization has embarked on implementing an aggressive business plan that calls for a cultural transition and change of mind-set that rids ICAO from the shackles of the Convention. New leadership and new thinking have been catalysts in this process, and, through a fog of rhetoric which sometimes accused the Organization of being a bureaucracy that was rapidly headed towards obsolescence, a flight path seems to have cleared that enables the Organization to steer towards a more relevant role in the new century. The Roadmap is a crucial test for ICAO in playing that role.

Today, ICAO is a results-based, performance and values driven Organization guided by its own business plan. The Business Plan translates its Strategic Objectives into action plans and establishes a link between planned activities, organizational cost and performance assessment. A vital dimension of this approach is the integration of programmes and activities of Bureaus and Regional Offices for optimum allocation of resources based on agreed priorities.

---

[49] It must be noted that ICAO's Business Plan is driven by six Strategic Objectives: Safety – Enhance global civil aviation safety; Security – Enhance global civil aviation security; Environmental Protection – Minimize the adverse effect of global civil aviation on the environment; Efficiency – Enhance the efficiency of aviation operations; Continuity – Maintain the continuity of aviation operations; and Rule of Law – Strengthen law governing international civil aviation.

2.1 Safety Management Systems

Together, the Business Plan and the related budget provide the basis for a reporting framework that unites strategies, activities, funds, human resources and time frames into a coherent and effective means of monitoring and evaluating outcomes. By engaging staff at all levels in the performance improvement process, highlighting responsibilities, and by holding managers accountable for their performance and regularly measuring, monitoring and evaluating results, the Organization will strengthen accountability, demonstrate value for money and improve overall performance at the operational and strategic levels. Moving from concept to action and results also involves a set of Supporting Implementation Strategies and the successful Technical Co-operation Programme which has a long tradition of supporting Contracting States in the implementation of ICAO regulations, procedures and policies.

Many of ICAO's 190 member States are already facing problems with respect to safety oversight. A glaring fact emerging from safety audits conducted by ICAO on States is that the findings of the initial safety oversight audit conducted by ICAO relating to the three Annexes to the Chicago Convention[50] – Annex 1– *Personnel Licensing*, Annex 6 – *Operation of Aircraft* and Annex 8 – *Airworthiness of Aircraft*, indicated that of the 181 Contracting States that were audited between March 1999 and July 2004, considerable numbers of States had deficiencies in respect of a number of requirements under these Annexes. Furthermore, audit follow-up missions have revealed that in many cases, significant deficiencies identified during the initial audits remain.

Under ICAO's Universal Safety Oversight Audit Programme (USOAP), States are required to establish a safety programme where aircraft operators, maintenance organizations and services providers implement appropriate safety management systems.

The first ICAO safety audit cycle was conducted between 1999 and 2004 where almost all Contracting States (except a few who could not be accessed due to adverse conditions in their territories) were audited. These audits were conducted on the basis of Annexes 1, 6 and 8 of the Chicago Convention. From January 2005, ICAO started its 6 year cycle of audits based on 16 of the 18 ICAO Annexes (all except Annex 9 and Annex 17) and by the end of April 2008 had audited 90 States.

The responsibilities of ICAO in the area of safety are grave and compelling. For the past 60 years or so, ICAO has been active in its standardization role, which has been blended in recent years with a burgeoning implementation role that is gradually blurring the former. In a world that is becoming largely globalized and regionalized, ICAO has vastly to focus on not so much what it does but how it does its work. In this context, ICAO has a dual role to play. The first is to act as a global forum for aviation, which is primarily the role expected of ICAO by the developed nations which are largely self reliant in regulatory matters. However, they need ICAO to set global standards that could apply to all ICAO's 190 member

---

[50] Convention on International Civil Aviation, *supra*, note 3.

States. On the other hand, ICAO has to be both a global forum and a mentor to the developing world which expects ICAO to assist and guide them.

In order to serve its 190 member States, irrespective of whether they are in the developed or developing category, ICAO has to justify its performance and values based stature. In other words, ICAO needs to undergo a whole system change. For the Business Plan to be implemented and results to be produced, firstly ICAO's leaders (its Council and the senior managers of the Secretariat) have to drive the process of transition from service to performance. They need to be the ambassadors of the Organization's mission and vision statements and set values and behaviours. They must "talk the talk" and "walk the walk".

The next step is to ensure that the mission and vision statement influences all decision making. This should permeate right to the bottom of the ICAO Secretariat. Thirdly, the new culture and its results must be measured by causal performance indicators. In other words, ICAO's new culture should be constantly monitored. The final measure would be to ensure that the values of the Organization's culture pervade and drive every aspect of decision making and be seen in every system and process.

In a way, ICAO is already undergoing a cultural transformation. It has come a long way in developing a mission and vision statement and a business plan driven by strategic objectives. There is a leadership that is committed to its work. There is also every indication that the leadership is ready and willing to involve the entire Organization in defining the mission, vision and values of the Organization. However, in order to achieve this successfully, strong tools and aggressive goals have to be in place through a robust and energized operational plan that is not disaggregated among the Organization's constituent bureaux and other offices.

Such an operational plan must have objectives and key performance indicators, as in the end it is measurement that matters. For this there must be targets set, not just improvement of performance. The Roadmap has to be part of target setting. In this regard it must be noted that ICAO's Business plan is on the right track, as it has all three types of indicators : *causal indicators* – which relate to values and behaviours (which are known in other words as core competencies); *output indicators* – which measure performance in terms if efficiency and productivity; and *outcome indicators*, which relate to the result or effect on clients and stakeholders.

Of these indicators, ICAO's concentration should be mainly on output indicators that measure productivity, efficiency, quality, innovation, creativity of the Organization as a whole and ensuing customer satisfaction. Innovation and creativity are key factors that serve to promote ICAO's contribution to its member States. Just as an example, since many States do not have the volume and scope of aviation activities which generate the resources and the base-line activity necessary to support a workable safety oversight system, ICAO's role must be to take the leadership in providing States with templates of different models of safety oversight and recommend what is best suited for them. ICAO could also further the involvement of regional safety oversight organizations that are successful; and provide guidance to States as to the modalities of the transfer of responsibilities or tasks, depending on the model used, from participating State to regional safety oversight organizations.

Leadership is the key to ICAO's role in the Twenty-first Century and the Roadmap is a tangible example of how this leadership can be demonstrated. The first step in driving the Roadmap is for ICAO to make a philosophical adjustment and ensure that it keep abreast of the new world order where States are increasingly being disaggregated into components which act in global networks, linking the world together in a manner that enables global trends to permeate the local environment.[51] In other words, ICAO should facilitate interaction between States and their components that interact in matters of civil aviation. For example, in many member States, aviation has numerous players in different areas such as customs and immigration, medical and quarantine, tourism, police, airports and air navigation service providers. In most instances these players do not act in accord, thus resulting in disharmony in the ultimate delivery of an efficient air transport product. ICAO's Mission and Vision Statement[52] exhorts ICAO to do just what is needed – to act as the global forum in the key areas of concern to international civil aviation through cooperation between its member States.

While promoting fluid dialogue and cooperation among its member States, ICAO should take the initiative to assist States both in technical issues. This assistance is not confined to providing technical assistance through projects administered by the Technical Cooperation Bureau but should also extend to providing guidance, mainly to States which still look up to ICAO as the global forum of aviation experts.

A critical area that has not been addressed in the Roadmap equation is the funding of safety. There is seemingly no conscious awareness on the part of States that a roadmap would be useless if there are no resources to correct the various deficiencies addressed by the safety audits and the overall application of the Roadmap. Neither ICAO nor its member States have aggressively addressed this issue, except for the instance of the International Financial Facility for Aviation Safety (IFFAS)[53] – a funding mechanism (now defunct) established under the ICAO legal umbrella but operating completely outside the budget of the Organization. It is operated on the basis of voluntary contributions from States and other parties, a fact which speaks for itself in terms of the magnitude of funding available under IFFAS.

---

[51] In the new world order, judges of a particular jurisdiction keep in close contact with their counterparts in other jurisdictions across the seas, which enables them to infuse global trends into their judgements; ministers keep in touch with their counterparts overseas, making them aware of the single thread of statecraft that comes with globalization; and police investigators collaborate closely with other police officers across the world to make inroads into the solving of crime. See Slaughter (2004), Introduction at 1.

[52] ICAO's Mission and Vision Statement calls for ICAO to work to achieve its vision of safe, secure and sustainable development of civil aviation through cooperation amongst its Contracting States.

[53] For some discussion on IFFAS see Abeyratne (2002). Also, Saba (2003), Fall 2003.

IFFAS was not tied to any of ICAO's programmes including the Roadmap, and one wonders whether, through the collective will of States, the possibility of redefining the scope of IFFAS should not be considered. In any event, It does not seem to be the responsibility of ICAO to identify safety deficiencies in its member States' territories and also to correct them, the latter responsibility clearly falling on the States, either individually or collectively.

In implementing the Roadmap, ICAO's leadership role hinges on two key factors: an aggressive operational plan with key performance indicators and targets; and the realization that organizational culture, which is an intangible asset, is the new frontier of competitive advantage. The latter is particularly important under the current circumstances of ICAO where human resources and expertise are in short supply. Cultural transformation starts with the leadership and individual and leadership values. When one looks at ICAO's current leadership structure, there is no room for doubt that this is not in short supply. However the trick is to motivate the staff sufficiently so that they would be impelled to follow their leaders in the transformation and forge ICAO's leadership forward in its various areas of work.

All this leads one to the bottom line, which is the need for change in the mindset of the Organization, from its service role to a role of implementation and assistance. The human factor is an essential consideration in this metamorphosis. The key and the starting point, however, is to recognize the need for the transition, which ICAO has already done. The next step is to recognize that ICAO needs its peoples' best efforts, both individually and collectively. ICAO's image and the perception of the outside world of ICAO as an effective Organization is anchored on the extent to which its workers represent themselves as good stewards of ICAO's business. They should therefore work together in the overarching interest of the Organization. When all these factors are considered together, there is nothing to suggest that ICAO is headed in the wrong direction. However, what seems to be badly needed is funding.

2.1.6 Economic Liberalization and Safety

In order to devise successful safety management systems, it is necessary to identify the exigencies of air transport that might pose threats to safety. Operations carried out by foreign registered aircraft reflects the first area of concern regarding safety. Within the basic principle enunciated in Article 17 of the Chicago Convention, which provides that aircraft have the nationality of the State in which they are registered, air operators have, over the past two decades, increasingly employed foreign registered aircraft for various reasons. Aircraft being leased or otherwise interchanged and operated outside the State of Registry, sometimes for long periods of time, is common. There are also instances where a foreign registered aircraft might be leased or sub-leased or chartered from one country to another. While such arrangements are legitimate from an economic and regulatory perspective, they can present problems from a safety viewpoint because of the bifurcation of the State of

Registry and State of the Operator. For example, this could result in a situation where operators can be subject to different interpretations of SARPs of ICAO as implemented by different States.

A major safety concern is the problem of "flags of convenience" associated with foreign registered aircraft. "flags of convenience" is a term derived from the maritime industry which denotes a situation in which commercial vessels owned by nationals of a State, but registered in another State, are allowed to operate freely between and among other States. When an aircraft rarely, if ever, returns to the State of Registry, its airworthiness oversight becomes an issue in the absence of safety oversight arrangements between the State of Registry and the State of the Operator. There are broadly two groups of foreign registered aircraft that can be deemed to operate under a flag of convenience: those done for fiscal purposes and those done to take advantage of a system with minimal economic or technical oversight or no oversight at all. The first group may not pose a serious problem if arrangements are made between concerned States to ensure proper oversight, for example through bilateral agreements under Article 83 *bis*, which permits States to transfer all or a part of certain safety oversight responsibilities under the Convention. Even for this group, the reality remains far from satisfactory in that relatively few bilateral agreements implementing Article 83 *bis* have been notified to ICAO (by 1 May 2006, 148 States had ratified the provision), and numerous aircraft of all types all over the world are still subject to split oversight responsibility. It is the second group that creates a major safety problem which needs to be addressed. (ICAO: AN-WP/8015).

Another area that might impinge on the safety of flight concerns operations involving foreign flight crew. Split oversight problems could also occur in respect of foreign-licensed flight crew. For example, dry leases (i.e. the lease of an aircraft without crew) raise the problem of validation of foreign crew licences by the State of Registry. The issue becomes complicated when the rules and requirements for crew licences in the State of Registry are at variance with the corresponding rules in the State that initially issued the licences. Differences between the laws and regulations of the State of Registry and those of the State of the Operator may also exist in the case of wet leases (i.e. a lease of aircraft with crew). While the lessor usually remains the official operator in such cases, the lessee may already operate aircraft of a similar type under its Air Operators Certificate. It may happen then that the wet-leased aircraft are operated under the lessee's Air Operator's Certificate and, consequently, the State of the lessee becomes the State of the Operator. In such circumstances, proper surveillance of the operating crew may become difficult. The situation could become more complicated if the operation involves a mixed crew (e.g. the cabin crew from the lessee carrier and the cockpit crew from a foreign lessor carrier).

"Off-shore" operations involving flight operations away from the designating State, (State of Registry or State of the Operator) may also have safety implications. In a situation where the designated airlines of a bilateral agreement are granted the so-called 7th freedom rights (i.e. *to carry traffic from the second State to/from third State(s) without the need for the service to connect the home State*), such airlines may set up an operational base in a second country for services to/from third countries. Where cabotage or right of establishment is permitted, air carriers may operate in the territory of the granting State. Such a situation could raise the question as to how the required safety oversight should be handled between the State of the Operator and the State in which the operation is based.

Growing commercial trends in air transport suggest that operations involving multiple parties and the use of another's brand, such as codesharing and franchising, could adversely affect safety unless checked by the States concerned. Codesharing has been the most prevalent element in transnational airline alliance arrangements and can take a variety of forms.[54] Although it is usually treated as a commercial arrangement, because of the complexity of some codesharing arrangements (e.g. a flight using the codes of several carriers from different countries), the safety/security authorities may find it difficult to determine their level of involvement *vis-à-vis* other authorities. In such circumstances, the questions of responsibility and accountability for safety/security can lead to uncertainty[55] (ICAO:2004). Also, since such arrangements allow an operator to use the name or assume the public face of another carrier (e.g. in the case of franchising), the need to safeguard reputation in terms of service/safety quality have led to some regulatory action on safety/security. For example, some States require foreign airlines with which their national airlines have codesharing arrangements to meet a similar level of safety. This could also raise a question as to whether all States whose airlines are involved in a codesharing operation should be involved in such safety oversight, and to what extent each should be involved. Another concern arising from codesharing relates to the security implications caused by the potential transfer of a security threat, which may exist against one airline and be spread to its partner or partners in a codesharing arrangement, and any subsequent additional security measures imposed by the appropriate authorities. Since technical and operational regulations may vary considerably from one partner airline/State to the other, this raises the question as to how the accountability and responsibility for safety/security should be handled amongst the partner airlines and States.

Cross-border airline merger/acquisition, where allowed, could lead to such companies having operations or places of business in different States, or operating mainly outside the State in which their registered offices and/or owners are located. This situation could raise questions regarding the attribution of regulatory oversight responsibility amongst the States concerned (e.g. in the case of the merged airline having two principal places of business), or on the application of whose standards, where they differ between the countries concerned.

Outsourcing of activity affecting aircraft operation is now a regular practice in the air transport industry and can take various forms. For example, airlines may out source their ground handling; send their aircraft to be repaired and/maintained in

---

[54] R.I.R. Abeyratne, Legal and Regulatory Issues of Computer Reservation Systems and Code Sharing Agreements in Air Transport, Editions Frontieres, 195, at 119–176.

[55] ICAO Circular 269, *Implications of Airline Code-sharing*.

foreign countries, and contract out certain flight operations and/or crew administration to another airline or company. In each of these cases, multinational industries have emerged to provide such services. Some States also encounter situations where an Air Operator's Certificate applicant had only a corporate skeleton with most of the proposed operational activities to be performed/provided by foreign companies (including the aircraft and flight crews). This situation could present challenges for the licensing and safety oversight authorities from both the State issuing the Air Operator's Certificate and the State of the outsourced activity on how to ensure that such practice or entity properly meet the safety and security requirements.

While some of the above situations already make it difficult individually for identifying or attributing the responsibility for safety/security compliance and oversight, it could become even more problematic when dealing with a complex situation that combines many or all of the above features. As reflected in the above discussion, there is an increasing number of situations in which one is dealing with a cascade of States, each having a share of responsibility in an air transport operation. The challenge for States is how to ensure that, regardless of the form of regulatory or commercial arrangement, there should always be a clear point of contact for the safety and security oversight responsibility in a clearly identified State or its delegated authority for any given aircraft operation.

Another situation that could have some implications for safety regulation relates to regionalism. Along with liberalization, many States have taken a regional approach in the regulation of air transport services. Substantial steps have also been taken on a regional basis to strengthen safety regulation. For example, the programme of the European Civil Aviation Conference (ECAC) includes safety assessments not only of aircraft of its member States but also of other air carriers operating into Europe. The European Union has established the European Aviation Safety Agency (EASA), whose functions already include certification of aeronautical products and may extend to approval of air operations and personnel licensing. Similar approaches to safety coordination are also being pursued in other regions (e.g. Autorité Africaine et Malgache de l'Aviation Civile (AAMAC); the Regional System for Cooperation on Operational Safety Oversight (SRVSOP) of the Latin American Civil Aviation Commission (LACAC); the Regional Aviation Safety Oversight System for the Caribbean (RASOS); the Central American Agency for Aviation Safety (ACSA); and the Pacific Aviation Safety Organization (PASO)). While these regional arrangements have many advantages and can bring benefits, chiefly including economies of scale and the promotion of uniformity within the region, they vary a great deal in the extent to which they have been delegated the execution of national responsibilities. This situation could present a challenge in terms of harmonization on a broader scale (e.g. the assessment of compliance by one body may differ from that of another). There is also a need for transparency so that all parties affected, especially third parties, may know exactly what functions have been delegated to the regional body and what remains with the State. In addition, there could be a situation where a regional body may be asked to assist a State or States of another region in safety/security regulation, which adds another element of complexity for identifying States' responsibility.

2.1.6.1 Performance Management and Accountability

The key words to management of safety are coordination, performance and accountability. Of these, performance is the most critical. While coordination ensures global interoperability of safety measures and performance measurement, accountability ensures the implementation of the safety management system worldwide. Performance measurement is inextricably linked to the most critical factors – the performance indicators. In SMS, performance should be measured within the basic premise that only what is important for safety management should be measured. One must not measure too much; and factors that impact customer satisfaction should also be measured.

A critical first step towards implementing performance management in any organization is to define a set of conceptual, high-level, key performance indicators (KPIs). In view of the level at which they are identified, KPIs do not identify targets to be achieved, deadlines to be met or any other details regarding execution of tasks. They would flow from an organization's strategic objective of ensuring safety and are described at three levels:

- *Outcomes*: through which the decision making bodies of the organization can objectively validate the strategies of the Organization, in other words: "Is the Organization doing the right things and do the overall effects of performance bring on expected results?";
- *Outputs*: through which the senior management of the organization can objectively assess the effectiveness of the Organization, in other words: "How well is the Organization doing what it set out to do?"; and
- *Measurement:* through which those accountable for key activities can independently assess the proper alignment of their resources and timely achievement of outputs.

Once the KPIs have been approved, the organization needs to develop and implement the other key components of its performance management system to enable the indicators to drive better utilization of the available resources. These components include, but are not limited to:

- The infusion of the performance management philosophy into the management culture and practices of the organization;
- The determination of realistic and measurable targets for each of the indicators;
- The alignment of data collection and interpretation mechanisms with the defined indicators and targets; and
- The traceability of operational plans and milestones to the targets.

Once implemented, the performance measurement system will be subject in its entirety to regular monitoring and use by linking it to planned budgetary allocations. The outcomes will also serve as a barometer to redefine measures, targets, outputs and resources if necessary. In order to derive full benefits from the performance measurement system, SMS may have to be realigned periodically with the changing external environment and newly identified critical objectives. This in turn will involve some realignment on the designed performance measurement system and the operational plans that are in place.

The above-mentioned approach will result in a management system where the allocation of resources and measurement of performance will bring about the timely achievement of outcomes that are most valuable to stakeholders.

The key performance indicators in a SMS should be also developed on the basis that the most critically important factor is employee involvement. The first step toward employee involvement is identification of the process flow. The fundamental issue is that if employees cannot agree on their process(es), they could not effectively measure them or utilize the output of what they have measured.

Another important consideration is to identify the critical activity to be measured. The critical activity is that culminating conduct where it makes the most sense to locate a sensor and define an individual performance measure within a process. This should immediately be followed by the recognition that the next step would be to establish performance goal(s) or standards. All performance measures should be tied to a predefined goal or standard, even if the goal is at first somewhat subjective. Having goals and standards is the only way to meaningfully interpret the results of measurements and gauge the success of management systems.

Consequent to establishing goals and standards, it is necessary to establish performance measurement(s). This helps build the performance measurement system by identifying individual measures.

Once measures are identified, the next step would be to identify responsible parties. A specific entity (as in a team or an individual) needs to be assigned the responsibilities for each of the steps in the performance measurement process. Thereafter, collection of data becomes the next consideration. In addition to writing down the numbers, the data need to be pre-analyzed in a timely fashion to observe any early trends and confirm the adequacy of your data collection system.

A corollary to the identification of goals, measurements and the collection of data is analysis of reports of actual performance. The raw data are formally converted into performance measures, displayed in an understandable form, and disseminated in the form of a report. Comparison actual performance to goal(s) is a key step, where the Organization will compare performance, as presented in the report, to predetermined goals or standards and determine the variation (if any).

Finally, to complete the development of key performance indicators, consideration of new goals is important. Even in highly successful safety management systems, changes may need to be revised in order to establish ones that challenge an organization's resources but do not overtax them. Goals and standards need periodic evaluation to keep up with the latest organizational processes. Management of the SMS and in particular the performance measurement side of things is a key factor, bringing to bear the importance of the question as to who will manage the PM system. The person responsible for measurement of performance would have to prepare performance reports and decide on the periodicity of measurement.

Good performance in SMS requires the maintenance of public confidence in the safety of international civil aviation. This is primarily achieved by the decrease in incidents and accidents involving passenger fatalities and fatalities on the ground. The measurements would evaluate the capability of the organization concerned in collection and analysis of safety information, leading to the output of implementation of a global safety data collection and analysis capability. This would in turn lead to the outcome that ensures global resources are aligned to address the most critical safety risks.

Economic liberalization as well as the evolution of business and operating practices may have serious implications for safety regulation, which need to be addressed properly. The most fundamental consideration is that safety of air transport must remain the ultimate responsibility of the State concerned and should not be compromised by economic considerations. While ICAO should continue its monitoring of liberalization within the parameter of safety, all players, including States, service providers and airlines should coordinate well in understanding their responsibilities and providing safety oversight. In this constantly evolving environment, due regard should be paid to the impact of a growing air transport industry's impact on safety so that coherent SMS are developed.

The operative criterion should be that safety of flight is a valid concern of the aviation industry whether or not air transport is liberalized and, as already discussed, liberalization and consequent competition between air carriers should alert States to the potential safety hazards that a highly competitive air transport industry brings to bear. Considering the elaborate work already accomplished by ICAO, in establishing guidance to States on the implementation of Article 83 *bis* to the Chicago Convention. In addition to Articles 12, 30, 31 and 32 a)[56] of the Convention and the up-to-date Standards and Recommended Practices adopted from time to time as necessary in Annexes 1, 6 and 8, there is no room for doubt that the regulatory provisions are in place.

Therefore what remains is for all players concerned to move toward a performance based SMS process where accountability will drive aviation safety to its desired goals.

---

[56] Article 12 stipulates that each contracting State is required to insure that every aircraft flying over or manoeuvring within its territory and that every aircraft carrying its nationality mark, wherever that aircraft might be, shall comply with the rules and regulations relating to flight and manoeuvre applicable there. It also provides that each Contracting States should keep its regulations consistent with the principles of the Chicago Convention. Over the high seas the rules will be as established under the Convention and States undertake, under Article 12, to prosecute any offender who does not adhere to these principles. Articles 30 and 31 refer to the need for aircraft to carry radio equipment as required and be issued with certificates of airworthiness by the State of Registry, respectively. Article 32 a) requires the pilot and other members of the crew to be provided with certificates of competency and licenses issued by the State of Registry.

2.1.7 Regulatory Aspects of Economic Liberalization and Safety

At the Fifth Worldwide Air Transport Conference (ATConf/5) of the ICAO, which was held in Montreal from 24 to 28 March 2003[57], the Conference observed the need to ensure that liberalization did not compromise safety and security. The Conference unanimously recognized that safety and security should be be paramount, regardless of the type of economic regulation employed, and that both safety and security were indicators gauging public confidence in international air transport. The Conference was also of the view that liberalization efforts could be destabilized where there were insufficient resources for States to meet their safety and security responsibilities. One of the key issues addressed by the Conference concerned financing of safety and security to ensure their global application, together with the realization that a safety and security system based on the single State concept was becoming increasingly complex given regional liberalization and the global organization of the world economy. It was recognized that the ultimate responsibility for safety and security lies with the State, although it was thought that regional agreements for safety oversight might also prove effective.

The Conference noted that when the Chicago Convention first came into force, the typical carrier in international aviation was either state-owned or owned by a company that was clearly a citizen of the State of Registry (the flag State) in every respect. All aircraft used by the carrier was registered at home and crews and management comprised national citizens. At the inception of regulation in the commercial air transport field, typical international carriers were totally independent of economic ties with other operators, domestic or foreign, and many did not even have much of a domestic route structure. With the increase in demand for capacity and expansion of route structures, particularly in the seventies and eighties, operators increasingly employed aircraft registered in countries other than their own bearing foreign flags for various reasons. This trend continued with unceasing regularity, particularly in the area of leasing where aircraft were commonly leased or otherwise interchanged and operated outside the State of Registry, sometimes for long periods of time. This deviation by many national carriers, from being a State owned aircraft fleet to having a composition of aircraft bearing foreign nationality, prompted ICAO to respond by developing the concept of the State of the Operator, a term which was not envisioned at the time of the Chicago Conference of 1944 as being different from the regular concept of nationality. Therefore it was not included in the Convention. However, ICAO included a definition of the term in Annex 6 to the Convention as "the State in which the operator's principal place of business is located or, if there is no such place of business, the operator's permanent

---

[57] The Conference had the theme "Challenges and Opportunities of Liberalization" and examined the key issues of liberalization such as air carrier ownership and control; market access; faior competition and safeguards; consumer interests; product distribution; dispute resolution; and transparency. See *Report of the Worldwide Air Transport Conference*, Doc 9819, ATConf/5 2003 at p. 1 and 2.

residence". The State of the Operator is responsible under the Annexes for issuance of the Air Operator Certificate (Air Operator's Certificate) and largely for operational safety and security oversight. The State of Registry, if it is different from the State of the Operator, is responsible for aspects of the aircraft's airworthiness. The result is that, operators can be subject to ICAO Standards and Recommended Practices (SARPs) as implemented by different States. In addition, States of Registry may be responsible for airworthiness oversight of aircraft that rarely, if ever, enter their territory.

A recurring concern of the Conference was that some types of liberalization could lead to flags of convenience and efforts to enhance safety and security through, for example, the use of agreements under Article 83 *bis*[58] of the Convention on International Civil Aviation (Chicago Convention)[59] on the lease, charter and interchange of aircraft must take this into account. Article 83 *bis* essentially allows the State of Registry of an aircraft which is operating under a lease, charter or interchange or any similar arrangement to transfer responsibility of the State of Registry to a State in which an aircraft operator has either his principle place of business or residence, provided the latter State consents to the transfer. As will be seen later in this article, this may have far reaching consequences for safety of operations. Article 83 *bis* stands as ICAO's principal response to the bifurcation of nationality and national responsibility for airworthiness and safety of aircraft which effectively amends the Convention to provide for voluntary transfers of safety oversight authority from the State of Registry to the State of the Operator, thus bringing to bear a clear devolution of responsibility and reunifying the responsibilities that had split over the years through changes in business practices. Article 83 *bis* allows States voluntarily entering into bilateral agreements to transfer all or a part of certain identified oversight responsibilities under the Convention. It is worthy of note that although Article 83 *bis* came into force in 1997, it has not been widely implemented. Aircraft of all types-all over the world-are still subject to split oversight responsibility.[60]

---

[58] The Protocol relating to amending the Chicago Convention to introduce Article 83 *bis*, signed at Montreal on 6 October 1980, entered into force on 20 June 1997. As at 5 May 2005, when this article was written, 144 States parties had deposited their instruments of ratification.

[59] Convention on International Civil Aviation, Supra n. 3. The Chicago Convention is an international treaty which lays down principles of State conduct on matters of international civil aviation; establishes the International Civil Aviation Organization and identifies ICAO's functions. The Chicago Convention has been ratified by 190 States which, *ipso facto*, become members of ICAO. Hererafter, any reference to an Annex would be to one of the 18 Annexes to the Chicago Convention.

[60] The Conference noted that operators continued to increasingly built economic ties to non-homeland carriers in the creation of code-sharing, alliances and other business relationships. They have increasingly engaged in a variety of domestic and international leasing arrangements-wet (with crew and operational management), dry (without crew), and so-called "damp" leases (various intermediate forms). See Safeguarding Safety and Security Oversight Through the Course of Economic Liberalization, *ATConf/5-WP/96*, 11/3/03, paper presented to the Conference by the United States.

2.1 Safety Management Systems

In Europe, the system under Community rules is structured toward minimizing carriers who might use "flags of convenience" for purposes of registration thus endangering safety. Inside the European Community, where the need for bilateral agreements between Member States has been obviated by collective Community law, aviation law still adheres to international norms. Council Regulation 2407/92 requires that the granting and validity of an operating licence shall be dependent on the possession of a valid Air Operator's Certificate which complies with the relevant Council Regulation or, until such time as one is applicable, national regulations. All EU States, being signatories to the Chicago Convention, require their national laws and regulations to meet ICAO standards.

Furthermore, Regulation 2407/92 requires that a Member State shall not grant an operating licence to an undertaking unless its principal place of business and, if any, its registered office, are located in that Member State. This allows a Member State to query the granting of an operating licence and Air Operator's Certificate by another Member State where the operator concerned seems to want to operate predominantly from inside the first Member State regarding which the first Member State might have serious safety concerns.

European Community laws have ensured that the traditional ownership and control 'link' between airlines and individual States cements and reinforces the international safety regulation system – a system which relies on mutual interdependence, whereby all States sign the Chicago Convention and undertake to comply with the Annexes to that Convention establishing minimum standards for the safe operation of aircraft. Signatory States are required to accept certificates issued by other Signatories and allow their aircraft and operators access to their airspace and airports on the basis that those minimum standards have been equalled or exceeded. If standards are not met, States can refuse access. Above all, the European Community has ensured that "flags of convenience" have not developed to any marked extent in the aviation sector, in contrast to maritime experience.[61] However, full liberalization of commercial arrangements would not sit so comfortably with nationally based safety regulation, bringing to bear the question as to whether safety oversight may be compromised or weakened by economic liberalization.

The perceived threats to safety in aviation resulting from a liberalized air transport system stem from various factors. The first, the emergence of "flags of convenience" resulting from liberalized ownership and control of aircraft, has already been mentioned. Secondly, the commonly known airline alliance called "code sharing"where two or more airlines enter into a commercial partnership to provide "seamless"through carriage simulating carriage by a single carrier, has caused apprehension in governments, particularly wen they are unsure as to the code sharing partner of its national carrier conforms to prudent safety requirements. Thirdly, liberal provisions for ground handling in bilateral air transport agreements

---

[61] Safety Aspects of Liberalization, *ATConf/5-WP/68*, 28/2/03, paper presented by the European Community, at 2.

often give rise to outsourcing by airlines to specialized companies of whom States authorities might not have adequate supervision, creating another area of concern.

The fourth area which might give rise to a comprise in safety through liberalization is leasing, where a leased aircraft, registered in a State other than that of the operator using the aircraft in international operations. The various scenarios of leasing, often involving various States with regard to equipment and crew, will be discussed in some detail later on in this work. Privatization of airports and air navigation is another area requiring State supervision, in order to ensure adherence by operators to international regulations relating to safety.

The Conference concluded that economic liberalization has implications for safety and security regulation, which needed to be properly addressed at the national, bilateral, regional and global levels, as appropriate, in order to ensure continued safe, secure and orderly development of civil aviation. Integral to this conclusion is the fact that the Chicago Convention imposes responsibility on Contracting States for compliance with standards and practices related to safety and security. Therefore, the Conference was of the view that safety and security regulation must not only be maintained but should also be strengthened. An essential consideration in this regard was that safeguards for liberalizationmust be in the form of measures to ensure compliance with applicable safety and security standards and enhance regulatory oversight. The Conference advised States that when introducing economic liberalization, they should ensure that safety and security not be compromised by commercial considerations, and that clear lines of responsibility and accountability for safety and security must be established for the parties involved in any liberalized arrangements. It was also the view of the Conference that, regardless of the form of economic regulatory arrangements, there should be a clear point of contact for the safety and security oversight responsibility in a clearly identified Contracting State or other regulatory authority designated by that State for any given aircraft operation.

The Conference called upon ICAO to continue to play a leading role in developing global strategies for the regulation and oversight of aviation safety and security, both definitively and in the context of facilitating economic regulatory reform. The consideration that the changing regulatory and operating environment in international air transport calls for the development of new regulatory devices capable of adapting to the changes and addressing related concerns was paramount to the Conference.

2.1.7.1 Some Specific Instances of Commercial Aviation Which Might Affect Safety

With Liberalization of commercial air transport comes the opportunity for new carriers to compete for market access. It also gives existing carriers the opportunity to compete with each other as well as with the new entrants with a view to consolidating their positions in a liberalized market. There is an onerous burden on States to ensure that, hand in hand with unfettered competition among carriers, rigid enforcement of technical regulation ensuring the safety of flight is necessary. The issue of the potential impact of liberalization and consequent competition between carriers on safety and security and their interrelationship dates back some years. It was the subject of discussion at the two most recent ICAO air transport conferences held in 1994 and 2003. The discussions at both conferences and conclusions flowing therefrom reflected a consensus that liberalization is a desirable goal that should be pursued by each State at its own choice and own pace. At the same time, the conferences were unequivocal that safety and security must remain of paramount importance, irrespective of any change in regulatory or commercial arrangements arising from demand for air services and competition. The widespread view was that, as liberalization spreads, and competition among carriers pervade the international market, there continues to be a need to address existing as well as potential concerns over its implications on safety and security. Therefore, States are faced with the challenge of maintaining uncompromising standards of safety whilst maximizing the benefits of economic liberalization.

In the air transport field, liberalized policies (e.g. on market access, airline designation, capacity, pricing, and commercial opportunities) drive the development of the industry, bringing about many economic benefits for States, the industry and consumers, such as growth in traffic (both in terms of passenger/cargo traffic and aircraft movements). Competition as a by-product of liberalization has spawned multiple air carriers (including low-cost carriers) who have entered the market offering competitive pricing which the legacy carriers find difficult to match. In addition, increased service options and development of travel and tourism have created jobs in the aviation and allied industries, making liberalization indispensable to the air transport industry. Although there is no indication that safety standards in many liberalized markets have not been maintained, nonetheless, it is also clear that the resulting growth in air transport activity and complex commercial arrangements stemming from increased business and operating practices could put additional pressure on the State in terms of its capacity in safety regulation. A State is required to provide safety oversight both to its own aircraft operators and those foreign operators that operate in its airspace. This puts an enormous burden on a State just to cope with the consequences of liberalization requiring that State to have the necessary legal, regulatory and organizational infrastructure and human and financial resources in place in order to perform the required safety/security regulatory functions.

Many of ICAO's 190 member States are already facing problems with respect to safety oversight. A glaring fact emerging from safety audits conducted by ICAO on States is that the findings of the initial safety oversight audit conducted by ICAO relating to Annex 1– *Personnel Licensing*, Annex 6 – *Operation of Aircraft* and Annex 8 – *Airworthiness of Aircraft*, indicated that of the 181 Contracting States that were audited between March 1999 and July 2004, considerable numbers of States had deficiencies in respect of a number of requirements under these Annexes. Furthermore, audit follow-up missions have revealed that in many cases, significant deficiencies identified during the initial audits remain. Therefore, the ineluctable conclusion is that, when considering liberalization, States should be concerned not merely of the economic benefits that would result but also its potential impact on safety and security regulation,. States must ensure their continued capacity to meet those requirements so that, as prescribed in the Chicago Convention, civil aviation develops in a safe and orderly development manner.

2.1.7.2 Specific Commercial Arrangements

Concerns for safety arise mainly from those commercial arrangements or practices which impinge on the operation of aircraft or the operating personnel. Such concerns can be bifurcated into two areas: one concerning commercial arrangements that could have an impact on safety regulation (such as increasing the pressure on licensing and oversight) but do not pose a problem in terms of identifying the State's responsibility; and the other regarding situations involving multiple States which could raise questions regarding the delineation of accountability or responsibility for safety oversight under the existing regulatory system based on ICAO provisions. The former concerns mostly activity taking place within a single State, for example, operations involving non-traditional, new entrant operators or services providers; airlines having financial difficulties; transfer of government operations as a result of commercialization or privatization of airports or air navigation services providers Following are some examples illustrating such situations.

Operations carried out by foreign registered aircraft reflects the first area of concern regarding safety. Within the basic principle enunciated in Article 17 of the Chicago Convention, which provides that aircraft have the nationality of the State in which they are registered,. air operators have, over the past two decades, increasingly employed foreign registered aircraft for various reasons. Aircraft being leased or otherwise interchanged and operated outside the State of Registry, sometimes for long periods of time, is common. There are also instances where a foreign registered aircraft might be leased or sub-leased or chartered from one country to another. While such arrangements are legitimate from an economic and regulatory perspective, they can present problems from a safety viewpoint because of the bifurcation of the State of Registry and State of the Operator. For example, this could result in a situation where operators can be subject to different interpretations of the Standards and Recommended Practices (SARPs) of ICAO as implemented by different States.

A major safety concern is the problem of "flags of convenience"[62] associated with foreign registered aircraft. When an aircraft rarely, if ever, returns to the State of Registry, its airworthiness oversight becomes an issue in the absence of safety oversight arrangements between the State of Registry and the State of the Operator.

---

[62] "Flags of convenience" is a term derived from the maritime industry which denotes a situation in which commercial vessels owned by nationals of a State, but registered in another State, are allowed to operate freely between and among other States.

2.1 Safety Management Systems

There are broadly two groups of foreign registered aircraft that can be deemed to operate under a flag of convenience: those done for fiscal purposes and those done to take advantage of a system with no or minimal economic or technical oversight. The first group may not pose a serious problem if arrangements are made between concerned States to ensure proper oversight, for example through bilateral agreements under Article 83 *bis*, which permits States to transfer all or a part of certain safety oversight responsibilities under the Convention. Even for this group, the reality remains far from satisfactory in that relatively few bilateral agreements implementing Article 83 *bis* have been notified to ICAO (by March 2005, 114 agreements are in force involving only 34 States), and numerous aircraft of all types all over the world are still subject to split oversight responsibility. It is the second group that creates a major safety problem which needs to be addressed.[63]

Another area that might impinge on the safety of flight concerns operations involving foreign flight crew. Split oversight problems could also occur in respect of foreign-licensed flight crew. For example, dry leases (i.e. the lease of an aircraft without crew) raise the problem of validation of foreign crew licences by the State of Registry. The issue becomes complicated when the rules and requirements for crew licences in the State of Registry are at variance with the corresponding rules in the State that initially issued the licences. Differences between the laws and regulations of the State of Registry and those of the State of the Operator may also exist in the case of wet leases (i.e. a lease of aircraft with crew). While the lessor usually remains the official operator in such cases, the lessee may already operate aircraft of a similar type under its Air Operators Certificate. It may happen then that the wet-leased aircraft are operated under the lessee's Air Operator's Certificate and, consequently, the State of the lessee becomes the State of the Operator. In such circumstances, proper surveillance of the operating crew may become difficult. The situation could become more complicated if the operation involves a mixed crew (e.g. the cabin crew from the lessee carrier and the cockpit crew from a foreign lessor carrier).

"Off-shore" operations involving flight operations away from the designating State, (State of Registry or State of the Operator) may also have safety implications. In a situation where the designated airlines of a bilateral agreement are granted the so-called 7th freedom rights (i.e. *to carry traffic from the second State to/from third State(s) without the need for the service to connect the home State*), such airlines may set up an operational base in a second country for services to/from third countries. Where cabotage or right of establishment is permitted, air carriers may operate in the territory of the granting State. Such a situation could raise the

---

[63] This problem is currently being addressed separately by the ICAO Air Navigation Commission and the Council in association with the Unified Strategy to resolve safety-related deficiencies within the scope of Article 54 j) of the Chicago Convention, which requires the Council to report to States any infraction of the Convention, as well as any failure to carry out recommendations or determinations of the Council. See ICAO Air Navigation Commission working paper AN-WP/8015.

question as to how the required safety oversight should be handled between the State of the Operator and the State in which the operation is based.

Growing commercial trends in air transport suggest that operations involving multiple parties and the use of another's brand, such as codesharing and franchising, could affect safety adversely unless checked by the States concerned. Codesharing has been the most prevalent element in transnational airline alliance arrangements and can take a variety of forms. Although it is usually treated as a commercial arrangement, because of the complexity of some codesharing arrangements (e.g. a flight using the codes of several carriers from different countries), the safety/security authorities may find it difficult to determine their level of involvement *vis-à-vis* other authorities. In such circumstances, the questions of responsibility and accountability for safety/security can lead to uncertainty.[64] Also, since such arrangements allow an operator to use the name or assume the public face of another carrier (e.g. in the case of franchising), the need to safeguard reputation in terms of service/safety quality have led to some regulatory action on safety/security. For example, some States require foreign airlines with which their national airlines have codesharing arrangements to meet a similar level of safety. This could also raise a question as to whether all States whose airlines are involved in a codesharing operation should be involved in such safety oversight, and to what extent each should be involved. Another concern arising from codesharing relates to the security implications caused by the potential transfer of a security threat, which may exist against one airline and be spread to its partner or partners in a codesharing arrangement, and any subsequent additional security measures imposed by the appropriate authorities. Since technical and operational regulations may vary considerably from one partner airline/State to the other, this raises the question as to how the accountability and responsibility for safety/security should be handled amongst the partner airlines and States.

Cross-border airline merger/acquisition, where allowed, could lead to such companies having operations or places of business in different States, or operating mainly outside the State in which their registered offices and/or owners are located. This situation could raise questions regarding the attribution of regulatory oversight responsibility amongst the States concerned (e.g. in the case of the merged airline having two principal places of business), or on the application of whose standards, where they differ between the countries concerned.

Outsourcing of activity affecting aircraft operation is now a regular practice in the air transport industry and can take various forms. For example, airlines may out source their ground handling; send their aircraft to be repaired and/maintained in foreign countries, and contract out certain flight operations and/or crew administration to another airline or company. In each of these cases, multinational industries have emerged to provide such services. Some States also encounter situations where an Air Operator's Certificate applicant had only a corporate skeleton with

---

[64] See ICAO Circular 269, *Implications of Airline Codesharing*.

most of the proposed operational activities to be performed/provided by foreign companies (including the aircraft and flight crews). This situation could present challenges for the licensing and safety oversight authorities from both the State issuing the Air Operator's Certificate and the State of the outsourced activity on how to ensure that such practice or entity properly meet the safety and security requirements.

While some of the above situations already make it difficult individually for identifying or attributing the responsibility for safety/security compliance and oversight, it could become even more problematic when dealing with a complex situation that combines many or all of the above features. As reflected in the above discussion, there is an increasing number of situations in which one is dealing with a cascade of States, each having a share of responsibility in an air transport operation. The challenge for States is how to ensure that, regardless of the form of regulatory or commercial arrangement, there should always be a clear point of contact for the safety and security oversight responsibility in a clearly identified State or its delegated authority for any given aircraft operation.[65]

Another situation that could have some implications for safety regulation relates to regionalism. Along with liberalization, many States have taken a regional approach in the regulation of air transport services. Substantial steps have also been taken on a regional basis to strengthen safety regulation. For example, the programme of the European Civil Aviation Conference (ECAC) includes safety assessments not only of aircraft of its member States but also of other air carriers operating into Europe. The European Union has established a European Aviation Safety Agency (EASA), whose functions already include certification of aeronautical products and may extend to approval of air operations and personnel licensing. Similar approaches to safety coordination are also being pursued in other regions (e. g. Autorité Africaine et Malgache de l'Aviation Civile (AAMAC); the Regional System for Cooperation on Operational Safety Oversight (SRVSOP) of the Latin American Civil Aviation Commission (LACAC); the Regional Aviation Safety Oversight System for the Caribbean (RASOS); the Central American Agency for Aviation Safety (ACSA); and the Pacific Aviation Safety Organization (PASO)). While these regional arrangements have many advantages and can bring benefits, chiefly including economies of scale and the promotion of uniformity within the region, they vary a great deal in the extent to which they have been delegated the execution of national responsibilities. This situation could present a challenge in terms of harmonization on a broader scale (e.g. the assessment of compliance by one body may differ from that of another). There is also a need for transparency so that all parties affected, especially third parties, may know exactly what functions have been delegated to the regional body and what remains with the State. In addition, there could be a situation where a regional body may be asked to assist

---

[65] Doc 9819 note.78 at paragraph 1.2.3.1 c).

a State or States of another region in safety/security regulation, which adds another element of complexity for identifying States' responsibility.

2.1.7.3 Regulation of Safety

The Chicago Convention and its Annexes provide the legal and operational framework for Contracting States to build and maintain a civil aviation safety/security system based on mutual trust and recognition. From a strictly legal viewpoint, the system is designed to ensure that international civil aviation operates in a safe and secure manner independently of the air transport policy and economic regulations that Contracting States may follow. Accordingly, economic liberalization should not pose any legal problem *vis à vis* the functioning of the system if the relevant provisions, SARPs and ICAO guidance material are fully implemented. Regardless of any change in economic arrangements, the responsibility for safety/security compliance and oversight remains vested in the Contracting States. States implement their safety and security oversight obligations imposed by the Convention and its Annexes through relevant national laws and regulations, as well as provisions in bilateral air services agreements.

Against the backdrop of globalization and liberalization, it is important for ICAO to make sure that the SARPs and guidance material it has developed for safety and security remain effective and capable of handling the changes. ICAO has carried out a review of the existing provisions contained in the Convention and relevant Annexes against the situations identified. It was found that, as far as establishing the respective responsibilities of involved States are concerned, existing SARPs and guidance material are deemed to be generally adequate. However, more work could be done to improve the existing SARPs and/or guidance material to address the new challenges brought about by the evolution of business practices in international air transport.

More specifically, for situations involving service providers with a permanent base (such as ground handling companies, airport operators and air navigation service providers), it is clear that the State in which such companies are based shall be responsible for safety and security oversight in accordance with the requirements set out in the applicable Annexes (e.g. regarding certification and surveillance of aerodrome operators and ground handling companies). For those situations involving the operation of aircraft, the safety aspects are addressed by Annexes 6 and 8. Annex 6 Part I is particularly relevant, as it pertains to operations of aircraft engaged in international air transport. Annex 8 contains Standards and Recommended Practices on airworthiness of aircraft. There are three levels of responsibility referred to in Annex 6 regarding the operation of aircraft which are assigned respectively to the State of Registry, the State of the Operator, and the Operator of the aircraft. The logical trail of responsibility is easy to follow in a situation where all three are part of the same State. In this case the operator is responsible to the State of the Operator, which is also the State of Registry. There are, however, situations that are more complex, which are described below.

Under Annex 6 provisions, an air operator is responsible for conducting the commercial operations in accordance with the Air Operator's Certificate issued by the State of the Operator. Therefore, codesharing or franchising flights are conducted under the responsibility of the operator that is actually operating the flight no matter what the aircraft livery or flight number might be. The oversight of such operation is normally conducted by the State of the Operator. However, if the operator uses aircraft registered in a State other than that of the operator, oversight may be required by the State of Registry if an agreement such as Article 83 *bis* or a bilateral agreement is not in place between the States concerned. It should be noted that any operator, codesharing partner or not, is expected to meet the applicable requirements of the ICAO SARPs when engaged in international operations. In leasing situations, the aircraft can only be operated under an Air Operator's Certificate issued by the State of the Operator. In the case of a dry lease, the lessee State will always be the State of the Operator, and will always be responsible for issuing the Air Operator's Certificate. In the case of a wet lease, the aircraft will generally be operated under the lessor's Air Operator's Certificate and the State of the Operator responsible for the Air Operator's Certificate is the lessor State. However, depending on the provisions and circumstances of the lease, the lessee State may become the State of the Operator, and therefore will be responsible for the Air Operator's Certificate. In addition, the lessee State must ensure that the flight crew, licensed in the lessor State, are trained and demonstrate competency in accordance with applicable regulatory requirements and conditions of the Air Operator's Certificate issued by the lessee State. Note that certain regulatory authorities will not enter into this type of agreement, as the training of flight crews to satisfy the requirements incumbent on the lessee can present difficulties and potential safety problems. Guidance material concerning lease, charter and interchange agreements is contained in secveral ICAO publications.[66]

Where the State of the Operator is different from the State of Registry, Annexes 6 and 8 establish the respective responsibilities for the safety of operations and airworthiness of the aircraft. In terms of Annex 6 requirements, the operator has the responsibility of maintaining adequate organization, control and supervision of flight operation. It has also the responsibility to establish and maintain appropriate maintenance arrangements to ensure that the aircraft, under its control, meets all the applicable airworthiness requirements that are under the responsibility of the State of Registry. The State of the Operator has therefore the ultimate oversight responsibility for the safety of flight operations conducted by the operator, and the State of Registry has the responsibility for the airworthiness of each individual aircraft on its registry. While the respective responsibilities of the State of the Operator and the State of Registry are clearly spelled out in the Annexes, the actual situation may be complex and lead to some fragmentation of responsibilities. For example, several

---

[66] *Manual of Procedures for Operations Inspection, Certification and Continued Surveillance* (Doc 8335), *the Airworthiness Manual* (Doc 9760) and *Guidance on the Implementation of Article 83 bis of the Convention on International Civil Aviation*, Cir 295, LE/2, February 2003.

States of Registry may be involved if an operator's fleet includes aircraft registered in different States. An additional potential level of complexity is that the State of Registry may validate a certificate issued by another State rather than issuing its own Certificate of Airworthiness. In most instances, the sharing of responsibility between the State of the Operator and the State(s) of Registry can be handled through well-established rules and procedures, even in complex cases. However, it does complicate the accountability for safety oversight and, in the absence of proper implementation of the rules, may be a potential area of weakness of the existing system.

The allocation of responsibility between the State of the Operator and the State of Registry derives to a large extent from the Convention that assigns the responsibility for aircraft airworthiness and flight crew licences to the State of Registry and only recognizes the role of the State of Operator in Article 83 *bis*. As a result, there are only limited ways in which the potential fragmentation of responsibility described in the previous paragraph can be avoided without changing the Convention. In this context, the transfer of certain functions from the State of Registry to the State of the Operator by way of implementing Article 83 *bis*, in respect of lease, charter and interchange of aircraft, provides an effective solution but one that is nevertheless limited by the voluntary nature of such agreement. Another course of action that can be considered is an amendment to Annex 6 that would require that a certified true copy of the Air Operator's Certificate under which the aircraft is operated, and in which it should be listed, be carried on board on international flights. This would help in identifying the States responsible for safety oversight on the occasion of any verification process such as ramp inspections. This provision could also be complemented by a Standard specifying that a given aircraft can only be operated under one Air Operator's Certificate at any given time. Additional clarification in the form of guidance material on the relationship between the State of Registry, the State of the Operator, and the Operator could be developed. This guidance should address the responsibilities of each party involved in relation to the Convention and its Annexes, and in relation to each other.

With regard to surveillance and inspection by States other than the State of Registry or the State of the Operator. Article 16 of the Convention gives the right to States to search, without unreasonable delay, aircraft of the other Contracting States on landing or departure, and to inspect the certificates and other documents prescribed by this Convention which include the licence of the flight crew and the certificate of airworthiness. There are, however, some practical limits to what can be achieved through the application of Article 16, which are mainly due to the fact that a valid certificate of airworthiness does not necessarily mean that the aircraft is airworthy and to the absence of a requirement in the Convention or in the Annexes on the carriage of a copy of the Air Operator's Certificate. With regard to the former, the reason is that a temporary loss of airworthiness, caused by a malfunction or other event, is normally dealt with by the operating regulations requiring an aircraft to be airworthy before it is operated (e.g. Annex 8, Part I, Paragraph 3.5 or Annex 6, Part I, Paragraph 8.1.1 a)) rather than by a suspension or revocation of a certificate of airworthiness. However, Annex 8, Part I, Paragraph 3.6.2, enables the authorities of a Contracting State to detain a damaged airplane registered in another Contracting State, provided that the State of Registry is advised immediately and given all of the necessary information to enable the State of Registry to determine the airworthiness of the aircraft. Amendment 100 to Annex 8, which will become applicable on 13 December 2007, clarifies the responsibilities of the respective States in this situation by introducing a requirement for the State of Registry to consider limitations proposed by the Contracting State that detained the aircraft, when authorizing a ferry flight to an aerodrome where the necessary maintenance can be carried out.

On the subject of the interchange of personnel through wet leasing of aircraft, the provisions in Article 32 of the Chicago Convention on the grant of licences of competence to crew members by the State of Registry and Annex 1 and Annex 6, Part I, are considered generally adequate for addressing the various situations involving flight crew members. The responsibility for validation or conversion of the licences and for maintaining the licence validity lies with the State of Registry while the responsibility for maintaining the competence of the crew lies with the State of the Operator. Although the maintenance of validity of the licence under Annex 1 on personnel licensing. and the maintenance of competency under Annex 6 are technically independent, the proficiency check of Annex 6 is accepted in practice for maintaining a valid pilot licence and there is a note to that effect in Annex 1. One particular safety aspect of economic liberalization is the increasing use of validation for flight crew licences. The ICAO safety oversight audits have indicated a certain number of problems with validations that relate to the traceability of the original licence (in particular to the limitation or restriction that may have been attached to it) and to extension of privilege of the original licence (type ratings in particular). These issues were reviewed by the ICAO Flight Crew Licensing and Training Panel as part of its global revision of flight crew licensing SARPs. The Panel has proposed some changes to Annex 1 and to the guidance material that will be presented to the Air Navigation Commission of ICAO during the second quarter of 2005.

The International Air Transport Association (IATA) has recorded that 2004 was the safest year since World War II for air transport,[67] both in terms of fatalities and aircraft irreparably damaged or destroyed. ICAO statistics reveal that in 2004 there was a large decrease in the number of fatalities in scheduled operations from 466 in 2003 to 266. This notwithstanding, the operative criterion should be that safety of flight is a valid concern of the aviation industry whether or not air transport is liberalized and, as already discussed, liberalization and consequent competition between air carriers should alert States to the potential safety hazards that a highly competitive air transport industry brings to bear. Considering the elaborate work

---

[67] Safest Year for Air Transport: IATA, *Air Letter*, Thursday, 10 March 2005, No. 15,695, at 1. IATA records that data shows the chances of dying in an airline accident in 2004 were one in every 10 million people flying against three in 2002 and over seven in 1996. For a description of IATA see *infra* note 107.

already accomplished by ICAO, in establishing guidance to States on the implementation of Article 83 *bis*,[68]in addition to Articles 12, 30, 31 and 32 a)[69] of the Chicago Convention and the up-to-date Standards and Recommended Practices adopted fro time to time as necessary in Annexes 1, 6 and 8, there is no room for doubt that the regulatory provisions are in place.

As for basic principles, States, in interpreting rules governing an incipient air transport industry in the 1940s followed the practice that flight crew licencing and certificate of airworthiness of the aircraft was the responsibility of the carrier, handed down by the State of Registry of the aircraft which had overall responsibilities for safety. Recognition of this practice was sustained through the 1970s until deregulation and liberalization became a regulatory as well as commercial trend. The change in practice, which is applicable today, ensured that more responsibilities were added on to the carrier or operator on whom devolved the task of obtaining flight crew licences, the air operator's certificate and certificate of airworthiness from the regulators who were either the State of the Operator and or the State of Registry depending upon who was designated. Training and maintenance also became the responsibility of the operator. Under Article 83 *bis*, the State of Registry transfers to the State of the Operator the responsibility of issuing the air operator's certificate and overall control and supervision of flight operations, together with ground handling personnel training and aircraft maintenance. This transfers to the State of the Operator supervisory responsibility from a regulatory perspective for arrangements made regarding commercial aspects of ground handling, engineering and maintenance while operational responsibility for such arrangements vests with the operator. The State of the Operator also assumes responsibility for the issuance of flight crew licences and certificates of airworthiness. However, such matters as validity of licences issued and extension of privileges (e.g. type rating) remain with the State of Registry. Annex 6 provides that an operator shall establish and maintain a ground and flight training programme approved by the State of the Operator, which is required to ensure that all flight crew members are adequately trained to perform their assigned duties.[70] The maintenance of competency of flight crew members engaged in commercial air

---

[68] *Supra.* note 58.

[69] Article 12 stipulates that each contracting State is required to insure that every aircraft flying over or maneuvering within its territory and that every aircraft carrying its nationality mark, wherever that aircraft might be, shall comply with the rules and regulations relating to flight and maneuver applicable there. It also provides that each Contracting States should keep its regulations consistent with the principles of the Chicago Convention. Over the high seas the rules will be as established under the Convention and States undertake, under Article 12, to prosecute any offender who does not adhere to these principles. Articles 30 and 31 refer to the need for aircraft to carry radio equipment as required and be issued with certificates of airworthiness by the State of Registry, respectively. Article 32 a) requires the pilot and other members of the crew to be provided with certificates of competency and licenses issued by the State of Registry.

[70] Annex 6, Part I paragraph 9.3.1.

2.1 Safety Management Systems

transport operations may be satisfactorily established by demonstration of skill during proficiency flight checks completed in accordance with Annex 6.[71]

In terms of maintenance of aircraft, maintenance arrangement entered into by the operator should be acceptable to the State of the Operator[72] while the maintenance programme followed would be approved by the State of Registry.[73] The maintenance control manual used by the operator should be acceptable to the State of Registry[74] and should include material required by both the State of Registry as well as the State of the Operator[75] Other shared responsibilities between the State of Registry and the State of the Operator are seen in the area of flight operations which are generally under the oversight of the State of the Operator but the State of Registry retains responsibility for the code of performance,[76] and performance data of the flight manual.[77]

In code sharing and franchising situations, the aircraft is required to be operated under the Air Operator's Certificateof the operator of the flight (*i.e.* the operator physically operating the flight in question).[78] In the case of dry lease and wet lease of aircraft the aircraft is operated under the Air Operator's Certificate of the operator (in the event of dry lease). With regard to a wet lease from an air operator's certificate holder, the operator may be the lessor or the lessee depending on the conditions contained in the lease. One shortcoming in Annex 6 is that there is no mandatory provision requiring the operator to carry a true copy of the Air Operator's Certification on board. This is indeed a necessity.

With such global regulations in place, the only drawback is the absence of guidance material in the form of a legal manual that would give both States and operators guidelines on responsibilities of individual parties, as the inadequacy of regulations in this regard reflects a certain perceived fragmentation of responsibility. For example, in a situation where an aircraft is registered in State A, is dry-leased (without crew) to an operator in State B, who in turn sub leases it to an operator in State C, who in turn wet leases (with crew from the operator of State C) to an operator in State D, the respective responsibilities of all four States and their operators involved in this complex transaction must be clearly identified by cogent expositions and interpretations of the regulatory material involved. In another instance involving a complex commercial arrangement such as code sharing, franchising or leasing, where an operator of State A grants a franchise to an operator of State B to fly its aircraft which are registered in State C, and where the repair and

---

[71] Annex 6, Note 1 to paragraph 1.2.5.1.1.

[72] Annex 6, Part 1 paragraph 4.2.1.3.

[73] Annex 6, Part1, paragraph. 8.3.1.

[74] Annex 6, Part 1, paragraph 8.2.1.

[75] Annex 6, Part 1, paragraph 8.2.4.

[76] Annex 6, Part 1, paragraph 5.1.1.

[77] Annex 6, Part 1, paragraph 6.1.2.

[78] Annex 6, Part 1, Chapter 2 and paragraph 4.2.1.

maintenance of the aircraft are conducted in State D, and the crew is obtained from State E, it might be difficult at first hand to determine individual responsibilities of the operators and States concerned. Again, in a cross-border merger situation where two airlines from State A and State B respectively merge, the question arises as to which would be the State of the Operator and which State should issue the air operator's certificate. Also, should there be a regional policy for mergers? Although as discussed above, answers to these questions – as to who would be responsible for what in the above instances – are contained in the various Annexes to the Chicago Convention and other guidance material already referred to, they are not readily ascertainable and might well be open to interpretation if a dispute were to arise. The international community owes it to itself to put together a compilation of all possible commercial arrangements being entered into as a result of liberalization and identify those responsible in each instance. Such material could also serve as guidelines to the courts in instances of adjudication of disputes and claims.

Finally, since ICAO safety audit reports and accident reports reflect that in many instances safety is compromised in States which do not adhere to ICAO Standards and Recommended practices, making it a critical issue, States should collectively recognize the worth of adhering to ICAO policy and also recognize Article 83 *bis* as a bilateral tool to be used in the instances described therein.

2.2 The Use of Airspace

Military aviation and civil aviation are intrinsically different from each other in their nature and functions. However, both operate in the same air traffic management environment and therefore use common airspace which needs to be stringently managed, not only for safety reasons but also for reasons of efficiency. While military aviation is essential for national security and defence and therefore is a legitimate and indispensable activity, civil air transport is not only necessary for global interaction between nations but it also makes a significant contribution to the global economy.[79] These two equally important activities call for uncompromising cooperation between one another in the shared use of airspace and an enduring understanding of each other's needs. Military aviation not only includes the operation of conventional aircraft for military purposes but also involves the use of unmanned aerial systems (UAS)[80] and missile testing, all of which call for a close look at the use of airspace in the modern context.

---

[79] Abeyratne (2007b), at pp. 25–47.

[80] The potential explosion of unmanned aircraft Systems (commonly called UASs) in airspace also brings to bear the need to have a closer look at the civil-military aviation airspace demarcation. UASs are commonly associated with military operations in many parts of the world. The question that would arise in this context is how would a State feel about sharing airspace over contiguous

2.2 The Use of Airspace

A grave concern confronting the civil aviation community is that, with the proliferation of military activity will inevitably come the issue of endangerment of air routes.

The consequences of the nuclear missile firings of 5 July 2006 by the Democratic Peoples' Republic of Korea (DPRK)[81] brought to bear the hazards and grave dangers such activities pose to civil aviation. In this instance, missiles launched by DPRK crossed several international air routes over the high seas. It was revealed that, when extrapolating the projected paths of some of the missiles, it appeared that they could have interfered with many more air routes, both over Japan and the air space of the North Pacific Ocean.

This is not the first instance of its kind. A similar incident took place on 31 August 1998 in the same vicinity in which the North Korean missiles were fired in July 2006. An object propelled by rockets was launched by North Korea and a part of the object hit the sea in the Pacific Ocean off the coast of Sanriku in north-eastern Japan. The impact area of the object was in the vicinity of the international airway A590 which is known as composing NOPAC Composite Route System, a trunk route connecting Asia and North America where some 180 flights of various countries fly every day.

From an aeronautical perspective, Annex 11 to the Chicago Convention,[82] which deals with the subject of air traffic services, lays down requirements for coordination of activities that are potentially hazardous to civil aircraft. The International Standards and Recommended Practices in the Annex, Chapter 2, (2.17 and 2.18 in particular) contain provisions for co-ordination between military authorities and air traffic services and co-ordination of activities potentially hazardous to civil aircraft. These provisions specify that air traffic services authorities shall establish and maintain close co-operation with military authorities responsible for activities that may affect flights of civil aircraft. The provisions also prescribe that the arrangements for activities potentially hazardous to civil aircraft shall be co-ordinated with the appropriate air traffic services authorities and that the objective of this co-ordination shall be to achieve the best arrangements which will avoid hazards to civil aircraft and minimize interference with the normal operations of such aircraft. Standard 2.17.1 stipulates that arrangements for activities potentially hazardous to civil aircraft, whether over the territory of a State or over the high seas, shall be coordinated with the appropriate air traffic services authorities, such coordination to be effected early enough to permit timely promulgation of information regarding the activities in accordance with the provisions of Annex 15 to the Chicago Convention. Standard 2.17.2 of Annex 11 explains that the objective of the coordination referred to in the earlier provision shall be to achieve the best

---

States with a swarm of UASs operated by a mix of military/law enforcement and commercial enterprises? For more information see Abeyratne (2009a).

[81] http://au.china-embassy.org/eng/xw/t261698.htm See *infra* text pertaining to notes 49 and 50.

[82] Convention on International Civil Aviation, signed at Chicago on 7 December 1944 Supra. n. 3.

arrangements that are calculated to avoid hazards to civil aircraft and minimize interference with the normal operations of aircraft. One must of course hasten to add that Article 89 of the Convention stipulates that in case of war, the provisions of the Convention (and, by implication its Annexes) shall not affect the freedom of action of any of ICAO's member States affected, whether as belligerents or as neutrals.[83] The same principle would apply in the case of any member State which declares a state of national emergency and notifies the fact to the ICAO Council.

The above considerations of safety notwithstanding, it is incontrovertible that cooperation in the activities of military and civil aviation is not only about sharing airspace. It is also about the efficient allocation of airspace to both categories of activity in separating such flights, particularly in the context of military flights which operate in special use airspace and those proceeding to special use airspace across civilian air routes. This brings to bear the inevitable conclusion that there must essentially be coordination between military authorities and air navigation service authorities.

At the Global Air Traffic Management Forum on Civil and Military Cooperation,[84] convened by ICAO on 19 October 2009, the International Air Transport

---

[83] In October 1949, on the occasion of the adherence of Israel to the Chicago Convention, the Government of Egypt advised ICAO that, in view of considerations of fact and law which at that time affected Egypt's special position with regard to Israel and in pursuance of Article 89, Israeli aircraft may not claim the privilege of flying over Egyptian territory. See letter dated 16 October 1949 reproduced in Annex A to Doc 6922-C/803 at p. 125. It was Egypt's claim, as was later clarified by Egypt upon a query of the Secretary General of ICAO, that a state of war existed between the two countries. The Government of Iraq also advised ICAO along similar lines, that a state of emergency had been declared on 14 May 1848 and therefore Article 89 was applicable and all Israeli aircraft were denied the privilege of flying over the territory of Iraq. On 28 November 1962 the Government of India informed ICAO that as a result of external aggression into Indian Territory by the People's Republic of China a state of grave emergency existed and the Government of India may not find it possible to comply with any or all of the provisions of the Chicago Convention. On 6 September 1965 the Government of Pakistan notified ICAO of the state of emergency under Article 89. In all instances, ICAO relayed the communications received to all its member States.

[84] The theme of the Forum was *"Time to take it global: Meeting each other's needs without compromising the Mission"*. The event was held as a follow up to recommendations of the Eleventh Air Navigation Conference (Doc 9828, Rec. 1/2) concerning coordination with military authorities with a view to achieving enhanced airspace organization and management and as an integral supporting mechanism of the successful series of civil/military air traffic management summits instituted by the Air Traffic Control Association (ATCA). It was also a follow up to ICAO Assembly Resolution A36-13, Appendix O, *Coordination of civil and military air traffic* wherein States are asked to take appropriate action to coordinate with military authorities to implement a flexible and cooperative approach to airspace organization and management. The Forum was intended to create awareness among civil and military policy makers and regulators, civil and military air navigation service providers (ANSPs) and civil and military airspace users, on the need to improve civil/military cooperation and coordination in support of an optimum use of airspace by all users.

2.2 The Use of Airspace

Association (IATA)[85] noted that, given the equal importance of civil and military aviation, it was imperative that airspace, which is an international and national resource, be managed as a whole, as a continuum and one common source and not a collection of segregated areas. This called for minimal restrictions on the use of airspace by both users, which in turn called for a structured and systematic management of the scope and duration of the use of airspace.

At the ICAO Forum, the Civil Air Navigation Services Organization (CANSO)[86] underscored the fact that increasing growth in civil air transport and traffic was putting pressure on limited airspace resources and that civil-military cooperation was becoming imperative. CANSO, while calling for a global platform of cooperation, emphasized that the key to successful cooperation is the establishment of trust, respect, transparency and flexibility on all key players and that States could play a lead role in developing a framework of cooperation. It also stated that a regional approach (as against a national approach) was essential, citing EUROCONTROL[87] as a true civil military agency which involved both civil and military offices at policy making level. In summing up, CANSO called for a fully integrated Civil-Military ATM, leading to the complete union of Civil-Military partners at national, regional and global level.[88]

A good example of the management system called for by IATA, and balanced cooperation as referred to by CANSO is the establishment of a Single European Sky (SES) legislation that is aimed at ensuring a harmonized regulatory framework for air traffic management and which uniformly and harmoniously applies in all 27 member States of the EU and 28 other associated States surrounding the Union.

---

[85] The International Air Transport Association, an association of air carriers, was formed in 1919 as the International Air Traffic Association. Encapsulated in IATA's overall mission are seven core objectives: to promote safe, reliable and secure air services; to achieve recognition of the importance of a healthy air transport industry to worldwide social and economic development; to assist the air transport industry in achieving adequate levels of profitability; to provide high quality, value for money, industry-required products and services that meet the needs of the customer; to develop cost effective, environmentally-friendly standards and procedures to facilitate the operation of international air transport; to identify and articulate common industry positions and support the resolution of key industry issues; and to provide a working environment which attracts, retains and develops committed employees.

[86] CANSO is the global voice of the air traffic management profession. Its members comprise over 50 air navigation service providers who control more than 85% of global air traffic movements. CANSO seeks to promote best practices within the industry.

[87] EUROCONTROL, the European Organisation for the Safety of Air Navigation, is an intergovernmental organisation made up of 38 Member States and the European Community. Its primary objective is the development of a seamless, pan-European air traffic management (ATM) system. EUROCONTROL contributes to making European aviation safer, performance-driven **and** environmentally sustainable. It was originally founded in 1960 as **a** civil-military organisation to deal with air traffic control for civil and military users in the upper airspace of its six founding European Member States. EUROCONTROL has developed into a vital European repository of ATM excellence, both leading and supporting ATM improvements across Europe.

[88] See Civil-Military Cooperation – The CANSO Perspective, October 2009.

This legislation is accompanied by a technology programme called Single European Sky Air Traffic Management Research (SESAR) which modernizes and helps run the European air traffic control infrastructure modernization programme making SES and SESAR the essential components of the full air transport policy of Europe.

The outcome of this merger between policy and infrastructure technology has resulted in a robust civil-military aviation cooperation enabling all EU member States to be represented by a civilian and a military officer in the EU Single Sky Committee (which, *inter alia*, develops legislation) and military officers to be included in other bodies working on SES and SESAR.

The counterpart of SESAR in the United States is the Next Generation Air Transport System (NextGen). Next Gen, which is scheduled to be effective from 2012 to 2025, calls for a shift in airspace management to a trajectory-based system. It will have the following five attributes: Automatic dependent surveillance broadcast (ADS-B) which will use the Global Positioning System (GPS) satellite signals to provide air traffic controllers and pilots with much more accurate information that will help to keep aircraft safely separated both in the air and op runways; System-wide Information Management System (SWIM) which will provide a single infrastructure and information management system to deliver high quality, timely data to many users and applications; Next Generation Data Communications which will provide an additional means of two-way communication for air traffic control clearances, instructions, advisories, flight crew requests and reports; Next Generation Network Enabled Weather (NNEW) which will cut weather-related delays at least in half; and NAS Voice Switch which will replace existing voice systems with a single air/ground and ground/ground voice communications system.

Both SESAR and NextGen, which are targeted for post 2020, would improve the performance of the air traffic management system by combining increased automation with new procedures that improve and achieve benefits related to safety, economic efficiency, capacity and environmental protection.

2.2.1 Distinction Between Civil and Military Aviation

A simplistic but apt definition of civil aviation is "aviation activities carried out by civil aircraft". A civil aircraft has been defined as any aircraft, excluding government and military aircraft, used for the carriage of passengers, baggage, cargo and mail.[89] However, civil aviation comprises in general all aviation activities other than government and military air services which can be divided into three main categories: commercial air transport provided to the public by scheduled or non

---

[89] Groenewege (1999), at p. 437. It must also be noted that an aircraft has been defined in Annexes 6, 7 and 8 to the Chicago Convention as any machine which can derive support in the atmosphere from the reactions of air other than the reactions of air on the Earth's surface.

2.2 The Use of Airspace

scheduled carriers; private flying for business or pleasure; and a wide range of specialized services commonly called aerial work, such as agriculture, construction, photography, surveying, observation and patrol, search and rescue, aerial advertisement et al.[90] By the same token, military aviation must be aviation activities carried out by military aircraft. Military aircraft have been defined as aircraft that are designed or modified for highly specialized use by the armed services of a nation.[91]

Military aviation therefore can be identified as the use of aircraft and other flying machines for the purposes of conducting or enabling warfare, which could include the carriage of military personnel and cargo used in military activities such as the logistical supply to forces stationed along a front. Usually these aircraft include bombers, fighters, fighter bombers and reconnaissance and unmanned attack aircraft such as drones.[92] These varied types of aircraft allow for the completion of a wide variety of objectives.

Arguably, the most fundamental difference between the operation of civil and military aircraft lies in the fact that, although they are expected to share the same skies, the procedures by which they do this vary greatly. Civil aircraft depend entirely on predetermined flight paths and code of commercial conduct which vary depending on aircraft type and types of traffic carried, whereas military aircraft operate in line with the exigency of a situation and are not necessarily always guided by predetermined flight paths. This dichotomy led to the adoption, at the 10th Session of the ICAO Assembly (Caracas, 19 June to 16 July 1956) of Resolution A10-19 which, while recognizing that the skies (airspace) as well as many other facilities and services were commonly shared between civil and military aviation, focused on ICAO's mandate to promote the safety of flight.[93]

The preponderance of weight in prioritizing civil and military aviation seems therefore to be in favour of civil aviation, particularly when taking into consideration this Resolution and the earlier discussion on Annex 11 to the Chicago Convention, thus attenuating the principle that military aviation should, of necessity, consider the compelling need to protect civil aviation from the spontaneous risks that the former may carry with it.

The above notwithstanding, a glaring example of conflict in the civil and military aviation environment can be seen in the ongoing conflict between Greece and Turkey (Aegean crisis). According to reports,[94] the core of the conflict is the

---

[90] *Ibid.*

[91] http://www.answers.com/topic/military-aircraft.

[92] In a report released on 21 December 2009, Venezuelan President Hugo Chavez is reported to have announced that, on Sunday 20 December, military drones had penetrated Venezuelan airspace along the North-western border with Colombia He had warned that Venezuela was prepared to defend itself if any State were to violate its sovereignty. See http://www.venezuelanalysis.com/news/5022. On 4 January 2010, it was reported that a US drone had fired two missiles in Pakistan, flattening an extremist hideout in Pakistan's lawless tribal belt on Sunday 3 January 2010, killing five militants in a recent spike in drone attacks. See http://www.channelnewsasia.com/stories/afp_asiapacific/view/1028351/1/.html.

[93] As per Article 44 of the Chicago Convention.

[94] http://www.aegeancrisis.org/category/air-space/.

persistent abuse of "Flight Information Region" (FIR)[95] responsibility by Greece. FIR responsibility over the Aegean international airspace was assumed in 1952 by Greece. The report goes on to say that Greece considers the FIR as a national boundary line and a defence perimeter (i.e. Western boundary of Turkey and Eastern boundary of Greece, embracing all international airspace in the Aegean beyond Turkish territorial sea within Greek sovereignty area) and that consequently, Greece maintains the view that military aircraft entering into Athens FIR and flying in international airspace should submit flight plans and come under control of Greek air traffic control authorities.

Against Greece's alleged claim that non-submission of flight plans by Turkish military aircraft constitutes a violation of the Greek FIR, some have argued that there is no need for Turkish military aircraft to file flight plans under the Chicago Convention as the Convention explicitly states in Article 3 that it would not apply to State aircraft (which includes military aircraft) and that there is no possibility of one violating an FIR which has nothing to do with the territorial sovereignty of a State, thus leading to the conclusion that Greek abuse of FIR responsibility is yet another manifestation of her claim of "de facto sovereignty" over the whole Aegean airspace.

2.2.2 The Use of Civil Aircraft for Military Purposes

The distinction between civil and military aviation cannot be made without addressing the purpose for which an aircraft is employed. This is particularly significant in instances where civil aircraft are used for military purposes. The fact that military strategists have come to expect support services from civil aviation is becoming more evident with the increasing need for military operations both in war situations and in instances of human tragedy brought about by civil conflict or natural disasters. There have been many such instances, ranging from the use by British military of chartered commercial cargo aircraft in the Falklands in 1982 to earlier practices of India and Pakistan in 1971 when both countries used civilian passenger aircraft for the transportation of their troops during the Indo-Pakistan war.

The use of civil aircraft for military purposes brings to bear issues of identification of aircraft and the status of aircraft under article 3 of the Chicago Convention.[96]

---

[95] FIRs were devised by ICAO in the 1950s to provide facilities and services to the civilian aircraft in the international airspace. FIR arrangements solely entail technical responsibility. It does not change the free status of the airspace over the high seas under international law.

[96] Article 3 of the Chicago Convention states that the Convention applies only to civil aircraft and not to State aircraft, and goes on to explain that aircraft used in military, customs and police services shall be deemed to be State aircraft. Article 3 c) prohibits State aircraft of one State from

2.2 The Use of Airspace

The question as to whether civil aviation and military aviation have demarcated operational regimes or whether they can still function in symbiosis has become an argumentative one in view of developments in the air transport industry which have occurred over the years. There are some determinants in this regard. Firstly, the nature of the cargo carried. Are they supplies or equipment for the military, customs or police services of a State? Article 35 of the Chicago Convention recognizes that the mere carriage "of munitions or implements of war" does not by itself make an aircraft a state aircraft. Then there is the question of ownership of the aircraft. Is it owned privately or by the State? The degree of control and supervision of the operation of the aircraft by the specified services are also factors to be considered in this equation. The nature of the passengers or personnel carried is also a consideration. Are they military, customs or police officials, or members of the public at large? Is the particular flight open for use by members of the public? Do aircraft registration and nationality markings become relevant? Will a usual civil (ICAO) flight plan be submitted and the usual air traffic clearances obtained? What is the nature of crew? Are the crew civilian, or are they military, customs or police personnel, or employed by these services? Who is the operator? Is the operator a military, customs or police agent? What sort of documentation is carried in the aircraft? Are the documents required by the Chicago Convention and its Annexes to be carried on civil aircraft in fact being carried (e.g. certificate of registration, certificate of airworthiness, licences for the crew, journey log book, etc.)? What would the area of operations be? Will the aircraft fly to, or over, areas in a situation of on-going or imminent armed conflict? What about customs clearances? Will the normal clearances be obtained?

The broad answer to all these questions would lie in the fact that, in the ultimate analysis, the responsibility of using civil aircraft and crew for military purposes rests with the State concerned. The fundamental legal premise which applies in such situations is that, in international relations, the erosion of one's legal interests by another brings to bear the latter's responsibility. State responsibility is a recognized principle of international law in the current context. The law of international responsibility involves the incidence and consequence of acts which are irregular at international law, leading to the payment of compensation for the loss caused. It might therefore just be worthwhile to inquire as to whether Article 89[97] of the Chicago Convention should be reviewed so that the international community and ICAO could be given more flexibility in the determination of propriety in the use of civil aircraft for military purposes.

---

flying over the territory of another State or landing thereon without special agreement or otherwise.

[97] *Supra.*

2.2.3 Some Recent Developments

At the Global Air Traffic Management Forum on Civil and Military Cooperation[98] ICAO subsumed its position by stating that airspace is a natural resource with finite capacity for which demand from all users is constantly expanding and that there has been an increased requirement on airspace use to meet a fast-growing aviation demand. States elected to be parties to the Chicago Convention in order that international civil aviation may be developed in a safe and orderly manner and that international air transport services may be established on the basis of equality of opportunity and operated soundly and economically. To achieve these objectives in recent years and to take due account of current and future needs in aviation, ICAO developed its vision of a seamless air traffic management (ATM) system.[99] ICAO further advised that, although the Chicago Convention governs international civil aviation and is not applicable to State aircraft (aircraft used in military, customs and police services)[100] State aircraft as well as military CNS/ATM systems and services are an integral part of the aviation community. A much closer cooperation between civil and military organizations will contribute to the vision encapsulated in the preamble to the Chicago Convention, leading to the optimum use of the airspace and balancing State requirements for both civil and military aviation.

ICAO drew the attention of the Forum to Assembly Resolution A 36-13[101] adopted at the 36th ICAO Assembly (Montreal, 18–28 September 2007), Appendix O of which recognizes that the airspace as well as many facilities and services should be used in common by civil aviation and military aviation and that the ICAO Global ATB Operational Concept[102] states that all airspace should be a usable resource and that therefore any restriction on the use of any particular volume of airspace should be considered transitory, and all airspace should be managed flexibly. It was noted by the Forum that, through A36-13, the Assembly resolved that the common use by civil and military aviation of airspace and of certain facilities and services shall be arranged so as to ensure safety, regularity and efficiency of international civil air traffic and that the regulations established by ICAO member States to govern the operation of their State aircraft over the high seas shall ensure that these operations did not compromise the safety, regularity and efficiency of international civil air traffic and to the extent possible such operations conformed to the Rules of the Air contained in Annex 2 to the Chicago Convention.

---

[98] *Supra.*

[99] Global Air Traffic Management Operational Concept, *ICAO Doc 9854*, AN/458, First Edition-2005, Chapter 1, Para 1.1.

[100] *Supra.*

[101] Consolidated statement of continuing ICAO policies and associated practices related specifically to air navigation, Assembly Resolutions in Force (as of 28 September 2007), *Doc 9902*, II-2.

[102] *Supra.*

2.2 The Use of Airspace

The resolution also requested the Council of ICAO to provide guidance and advice to States that wished to establish civil/military agreements.

Against this backdrop, ICAO advised the Forum of the need for a strengthened civil/military cooperation and coordination which called upon ICAO Member States to initiate as necessary or improve the coordination between their civil and military air traffic services. It was important that States, in view of the increasing need to cooperate with multiple airspace users, developed an integrated and cohesive civil-military coordination strategy with a roadmap indicating short, mid and long term objectives. ICAO further advised that the benefits of enhancing civil-military cooperation should be considered at the global level with a view to identifying best practices through dialogue and exchange of information. Effective civil/military cooperation and coordination is required not only to meet future civil and military air traffic requirements for increased safety, security, capacity, efficiency, environmental sustainability but also to achieve interoperability, seamlessness and harmonisation through sound policy, a structured framework, effective liaisons and management at working level.[103]

2.2.4 ICAO Initiatives

One of the recent initiatives is the ICAO global air traffic management operational concept[104] which visualizes an integrated, harmonized and global interoperable ATM[105] system. The broad vision of this concept is to achieve an interoperable global air traffic management system for all users during all phases of flight that meets agreed levels of safety, provides for optimum economic operations, is environmentally sustainable and meets national security requirements.[106] The ATM system is based on the provision of services, through a framework which involves airspace, aerodromes, aircraft and persons which are part of the ATM system. The benefits accruing to all members of the ATM community are greater equity in airspace activity; greater access to timely and meaningful information for decision support and more autonomy in decision making including conflict

---

[103] In its briefing, ICAO emphasized that cooperation between civil and military authorities should be aimed at achieving optimal use of the airspace resulting in increased airspace capacity, operational flexibility, and savings in flying time, fuel and CO2 emissions. The Forum noted that safety, economical impact, efficiency and interoperability are objectives shared by both civil and military aviation communities.

[104] An operational concept is a statement of what is envisioned.

[105] Air traffic management is the dynamic, integrated management of air traffic and airspace – safely, economically and efficiently – through the provision of facilities in collaboration with all parties. See *ICAO Doc 9854, supra* at 1. 1.

[106] *Ibid.*

management, and the opportunity to better deliver business and individual outcomes within an appropriate safety framework.

ICAO has also issued guidelines on the coordination between military authorities and air traffic services (ATS) authorities which recognize *in limine* that coordination between the responsible military authorities and appropriate ATS authorities is essential to the safety of civil aircraft operations whenever activities potentially hazardous to such operations are planned and conducted by any military units.[107] These guidelines go on to state that in the event that a sudden outbreak of armed hostilities or any other factors preclude this normal coordination process, appropriate State and ATS authorities, civil aircraft operators and pilots-in-command of aircraft must assess the situation based on the information available and plan their actions so as not to jeopardize safety.[108]

The Guidelines recommend that, in the event that a military unit observes that a civil aircraft is entering, or is about to enter, a designated prohibited, restricted or danger area or any other area of activity which constitutes potential hazards, a warning to the aircraft should be issued through the responsible ATS unit. The warning should include advice on the change of heading required to leave, or circumvent, the area.[109] If the military unit is unable to contact the responsible ATS unit immediately and the situation is deemed to be a genuine emergency, an appropriate warning to the aircraft may be transmitted on the VHF emergency channel 121.5 MHz. If the identity of the aircraft is not known, it is important that the warning include the SSR code, if observed, and describe the position of the aircraft in a form meaningful to the pilot, e.g. by reference to an ATS route and/or the direction and distance from an airport or an aeronautical radio navigation aid, an established waypoint or reporting point.[110] In the case where an unauthorized aircraft is observed visually to be flying in, or about to enter a prohibited, restricted or danger area, the following visual signal is prescribed by the International Standards in Annex 2 to the Chicago Convention – *Rules of the Air*, Appendix 1 to indicate that the aircraft is to take such remedial action as is necessary.[111] The Guidelines caution that the importance of co-ordinating with the responsible ATS unit(s), whenever possible, the issuance of any warnings and advice to civil aircraft regarding changes of flight path should be emphasized in any briefings or

---

[107] Manual Concerning Safety Measures Relating to Military Activities Potentially Hazardous to Civil Aircraft Operations, *ICAO Doc 9554-AN/932* First Edition, 1990, paragraph 3.1.

[108] *Id*. Paragraph 3.1.1. Examples of military activities which may pose a threat to civil aircraft and which should be coordinated with ATS authorities include practice firings or testing of any weapons air-to air, air to surface, surface to air in an area or in a manner that could affect civil air traffic; certain military aircraft operations such as air displays, training exercises and the intentional dropping of objects and paratroopers; launch and recovery of space vehicles; and operations in areas of conflict, when such operations include a potential threat to civil air traffic. See Paragraph 3.2. of Doc 9554.

[109] *Id*. Paragraph 8.1.

[110] *Id*. Paragraph 8.2.

[111] *Id*. Paragraph 8.3.

2.2 The Use of Airspace

instructions given by military authorities to their units, since uncoordinated warnings and associated navigational advice, when followed, may result in a potential risk of collision with other aircraft in the area.[112]

The objective of the co-ordination between the military authorities planning activities potentially hazardous to civil aircraft and the responsible ATS authorities is to reach agreement on the best arrangements which will avoid hazards to civil aircraft and minimize interference with the normal operations of civil aircraft. Ideally, this means the selection of locations outside promulgated ATS routes and controlled airspace for the conduct of the potentially hazardous activities. If the selection of such locations is not possible due to the nature and scope of the planned activities, temporary restrictions imposed on civil air traffic should be kept to a minimum through close co-ordination between the military units and the ATS unit.[113]

The Guidelines are clear on the fact that although Article 89 of the Chicago Convention[114] provides that in the event of armed conflict or the potential for armed conflict, the Convention does not affect the freedom of action of any Contracting State affected, whether as belligerents or as neutrals). Nonetheless, the need for close co-ordination between civil and military authorities and units is even more critical. The responsibility for initiating the co-ordination process rests with the States whose military forces are engaged in the conflict. The responsibility for instituting special measures to ensure the safety of international civil aircraft operations remains with the States responsible for providing air traffic services in the airspace affected by the conflict, even in cases where co-ordination is not initiated or completed.[115] Based on the information which is available, the State responsible for providing air traffic services should identify the geographical area of the conflict, assess the hazards or potential hazards to international civil aircraft operations, and determine whether such operations in or through the area of conflict should be avoided or may be continued under specified conditions. An international NOTAM containing the necessary information, advice and safety measures to be taken should then be issued and subsequently updated in the light of developments. All those concerned with initiating and issuing of NOTAM should be aware of the provisions governing the duration of the published NOTAM. Annex 15, Standard 5.3.1.2 states that a NOTAM given Class I distribution shall be superseded by a NOTAM given Class II distribution when the duration of the circumstances notified is likely to exceed 3 months or the NOTAM has remained in force for 3 months. A copy of the NOTAM should be forwarded to the appropriate regional office of ICAO.[116] If the necessary information is not forthcoming from the States whose

---

[112] *Id.* Paragraph 8.4.
[113] *Id.* Paragraphs 9.1 and 9.2.
[114] *Supra.*
[115] *Doc 9554*, Paragraph 10.2.
[116] *Id.* Paragraph 10.3.

military authorities are engaged in the armed conflict, the State responsible for providing air traffic services should ascertain the nature and scope of the hazards or potential hazards from other sources, such as aircraft operators, the International Air Transport Association (IATA) and the International Federation of Air Line Pilots' Associations (IFALPA), adjacent States or in some cases the relevant ICAO regional office.[117]

Separate guidelines[118] issued by ICAO provide that aircraft shall not be flown in a prohibited area, or in a restricted area, the particulars of which have been duly published, except in accordance with the conditions of the restrictions or by permission of the State over whose territory the areas are established.[119] The same guidelines also provide that special procedures shall be established with a view to ensuring that air traffic units are notified if a military unit observes that an aircraft which is, or might be a civil aircraft is approaching, or has entered any area in which interception might be necessary. In such an event all possible efforts should be made to confirm the identity of the aircraft and to provide it with the navigational guidance necessary to avoid the need for interception.[120] There is also a requirement to the effect that air traffic services authorities establish and maintain close cooperation with military authorities responsible for activities that may affect flights of civil aircraft.[121] As soon as an air traffic services unit becomes aware of an unidentified aircraft in its area, it is required to establish the identity of the aircraft whenever this is necessary for the provision of air traffic services or required by the appropriate military authorities in accordance with locally agreed procedures.[122]

As the foregoing discussion indicates, there is ample regulatory guidance from a civil aviation perspective to ensure a seamless and interoperable sharing of airspace between civil and military aviation activities. However, there remain some weak spots, the first being the perceived inadequacy and lack of clarity of Article 89 of the Chicago Convention which renders the legal structure in this context destitute of certainty and effect.[123] Another contentious area is missile testing involving airspace and air routes used by civil aircraft as was demonstrated by the DPRK issue of 2006.[124] Many concerned parties voiced their perturbation over the incident, including ICAO which, through the President of the Council sent a letter to the DPRK authorities voicing the grave concern of the international aviation community that Standards 2.17 and 2.18 of Annex 11 to the Chicago Convention were not followed by the military authorities of DPRK. Concurrently, Chinese Foreign

---

[117] *Id.* Paragraph 10.4

[118] See Manual Concerning Interception of Civil Aircraft, *ICAO Doc 9433-AN/926* Second Edition – 1990.

[119] *Id.* Paragraph 3.2.4.1.

[120] *Id.* Paragraph 3.2.6.1.

[121] *Id.* Paragraph 3.1.7.1.

[122] *Id.* Paragraph 3.1.9.1.

[123] *Supra.*

[124] See *supra.*

Minister Li Zhaoxing held phone talks with U.S. Secretary of State Condoleezza Rice, Japanese Foreign Minister Aso Taro, Minister of Foreign Affairs and Trade of the Republic of Korea Ban Ki-Moon and Australian Foreign Minister Alexander Downer respectively. Countries across the world joined in the protest, and the United Nations Security Council met for an emergency meeting to discuss the missile tests.

The United Nations Security Council condemned the test firing by DPRK of missiles and adopted Resolution 1695 which requested all Member States to prevent the transfer of missile and missile-related items, materials, goods and technology to the Democratic People's Republic of Korea's missile or weapons of mass destruction programmes, as well as procurement of such items and technology from that country. It also addressed the transfer of financial resources in relation to those programmes.

The resolution affirmed that such launches jeopardize peace, stability and security in the region and beyond, particularly in light of the country's claim that it has developed nuclear weapons. The Council underlined that DPRK needed to show restraint and refrain from any action that might aggravate tension, and continue to work on the resolution of non-proliferation concerns, through political and diplomatic efforts. In that connection, the Security Council strongly urged the country to return immediately to the six-party talks without precondition, to work towards expeditious implementation of the September 2005 joint statement and return to the Treaty on the Non-Proliferation of Nuclear Weapons (NPT) and International Atomic Energy Agency (IAEA) safeguards.[125]

In May 2009, DPRK test fired another short-range missile, apparently in clear violation of Resolution 1695 and, it is reported,[126] stated that it would take self defence action if the United Nations Security Council were to impose tougher sanctions. This missile, which was fired from the Masudan-ni site on DPRK's east coast, was the latest in the series of missiles the DPRK test fired since conducting a major nuclear test a few days before the firing.

If the response of State authorities who fire missiles into the air without paying heed to applicable regulations and guidelines were to be that, since the State concerned has sovereignty over its airspace (as recognized by Article 1 of the Chicago Convention) and that it does over its airspace is its concern, it must be pointed out that air routes are used by airlines of various nationalities carrying persons of various nationalities and that there must be recognition that the concept of sovereignty, in its pristine purity and simplistic interpretation cannot be sustained in this instance. One commentator states very aptly:

> The role of the State in the modern world is a complex one. According to legal theory, each State is sovereign and equal. In reality, with the phenomenal growth in communications and consciousness, and with the constant reminder of global rivalries, not even the most powerful of States can be entirely sovereign. Interdependence and the close knit character

---

[125] http://www.un.org/News/Press/docs/2006/sc8778.doc.htm.

[126] http://www.chinadaily.com.cn/cndy/2009-05/30/content_7953420.htm.

of contemporary international commercial and political society ensures that virtually any action of a State could well have profound repercussions upon the system as a whole and the decisions under consideration by other States.[127]

Therefore, in the ultimate analysis, cooperation between civil and military authorities, in accordance with the existing regulations and guidelines is essential, with the underlying consideration that civil aviation, with 15,000 aircraft airborne at any given time carrying 2.2 billion passengers annually, should not under any circumstances be compromised.

2.3 Aviation Medicine

2.3.1 The Aerotoxic Syndrome

The seminal case of *Turner* v. *Eastwest Airlines*[128] portrays well the significance of the aerotoxic syndrome as a safety issue in air transport. There have been some studies which have been conducted on the subject of the aerotoxic syndrome, that enanles one to make the distinction between the liability of an air carrier for passenger death or injury and death or injury to caused to crew members. Airlines will very likely in the future face increased actions from plaintiffs for toxic air injury and suggests that it would be prudent for airlines to ensure that maintenance procedures concerning engine seals are affected stringently and rigorously.

It is a platitude to say that air travel is the safest means of transport. Snug in this belief, a passenger takes comfort in the fact that he is just one of 2.2 billion passengers that are transported by air every year and that no danger would be lurking to threaten his life during the flight. The air traveller, and in particular the passenger, also believes that the modern aircraft is extremely sophisticated in design and structure and that it has passed rigorous and stringent tests before it has been certified to fly across oceans and continents. The glamorous captain, and the humble chap in overalls who performs the daily maintenance check on the aircraft also assure him that no ill can befall the aircraft in which he travels. It is hard to think otherwise, when billions of people have travelled by air comfortably over the years at 7 miles a minute, while the temperature outside is twice as cold as the freezer in one's fridge at home.

The only snag seems to be in what one breathes while travelling.[129] Under normal circumstances, the aircraft cabin environment is bad enough. An aircraft

---

[127] Shaw (2003), at p. 120.

[128] *Infra*, note 164.

[129] The ventilation system plays a critical part in this regard and therefore, it is crucial to an airline's conduct to determine the manner in which that carrier decides on ventilation systems in its aircraft. For instance, early jet aircraft until the last decade offered 100% fresh air in the cabin.

2.3 Aviation Medicine

in flight is a pressurised, airborne, air-conditioned, densely populated tourist and business facility at a high altitude with a relative humidity similar to that of Antarctica.[130] Inside the aircraft, humans release "on occasion, hostile viruses and bacteria, shed dead skin particles, fungal spores and emit body odours".[131]

Additionally, materials used in the operation of aircraft may contain hazardous ingredients, some with significant toxicities. Aircraft material such as jet fuel, de-icing fluids, engine oil and hydraulic fluids contain a range of ingredients, some of which are toxic.[132] Engine oil, hydraulic fluids and other materials which have in common the chemical presence of toxic ingredients such as organo-phosphates are used as of necessity by the aviation industry. Although these chemicals are usually contained within the engines and equipment into which they have been added, they can sometimes find their way into cabin air where crew and passengers are located. Most common causes of leakages of this kind are engine oil leaks, fluid ingestion by the Auxiliary Power Unit (APU) and engines, and by failure of seals designed to preclude seepage and leakage through ventilation systems. It is a fact that the air inhaled by those in the aircraft in flight is a mixture of "bleed air" (fresh air taken in from the atmosphere through the jet engines, part of which is used for pressurization of the cabin and the other part of which is used for purposes of inhalation) and air that is re-circulated after use.[133] If the engine seals (that seal off the lubricating oil) are not working properly, there could be an oil leak contaminating the air.[134] The leaks do not usually occur spontaneously as the seals do not leak suddenly but take a while to give way, spilling oil into the engine over a period of time.[135] This can happen under certain phases of flight where there is a great load on the engine than at other phases or when the aircraft is descending where the load on the engine is less than when there is less load.

Researchers have carried out tests over several years and concluded that passengers and crew have, immediately after flights, suffered from a range of common and similar symptoms which leave them debilitated, weak and suffering from coughing and difficulty to breathe. This basket of symptoms, which is yet to be identified as a single disease, is called the aerotoxic syndrome.[136] The toxicity and

---

However, in the 90s, ironically with more evolved technology, ventilation systems in aircraft were built in such a way as to recycle stale air, thus increasing the chances of survival of bacteria and deleterious particles in the aircraft cabin. Even if such a practice were ineluctable, in that recycling is a universal practice which is calculated to conserve fuel, a prudent airline would take other measures, such as change of air filters through which ventilation is provided.

[130] Crawford (1989).

[131] Holcomb (1988).

[132] See Rayman and McNaughten (1983); Smith et al. (1997).

[133] *Aviation Contaminated Air References Manual*, Captain Susan Michaelis ed. Published by Captain Susan Michaelis and printed by CPI Antony Rowe, Eastbourne, England, 2007 at iii.

[134] Tristan Lorraine, *Toxic Airlines*, DFT Enterprises: London, 2007 at 19–20.

[135] Captain John Hoyte, Aerotoxic Syndrome – Aviation's Best Kept Secret, http://www.aerotoxic.org/download/docs/news_and_articles/NEXUS-Aerotoxic-Syndrome.pdf.

[136] The term "aerotoxic syndrome" was first suggested in 1999. See Winder et al. (2002).

irritation causing this syndrome is due to neurotoxic organophosphates which contaminate the air that circulates in the cabins of aircraft propelled by jet engines.[137] The cause for this contamination has been identified as the use of lubricating oils and hydraulic fluids in jet engines and flawed designing of the air intake mechanisms.[138]

Pilots have also been affected by the ill-effects of this syndrome, causing widespread illness[139] and compelling some of them to retire prematurely.[140] Since the first known instance of a study on this subject in 1977[141] where a 34 year old pilot was examined for inhaling oil fumes and subsequently developing mental disorientation and neuromuscular discomfort, there have been several studies that have borne witness to illness in pilots due to the inhalation of contaminated air in the cockpit. For example, an inquiry into this issue in 2000 by the Australian Senate revealed that pilots had been disoriented and unable to concentrate on flying the aircraft, due to a feeling akin to being drunk.[142] In the same year there were four bulletins issued by the Civil Aviation Authority (CAA) of the United Kingdom, warning of the danger of pilot incapacitation by contaminated cabin air and suggesting procedures to counter the problem.[143] A Report published by the UK CAA in 2001 resulting from research into pilot incapacitation by contaminated air concluded that engine oil fumes could have probably caused the incapacitation.[144] In the same year, a similar conclusion was reached by the Swedish air safety

---

[137] Hale and Al-Seffar (2008), at 107.

[138] S. Myhill, Aerotoxic Syndrome, www.aerotoxic.org/articles/20071118.

[139] The symptoms are said to vary from fatigue, sleep deprivation, blackouts, seizures, neuromuscular pain and weakness. See Winder and Balouet (2001).

[140] J. Hoyte, Captain Hoyte's Account, www.aerotoxic.org/articles 20071114. Also, Toxic Free Airlines,(TFA) Poisoned Pilots Launch Campaign at Parliamentary Meeting: The Aerotoxic Association and Toxic Free Airlines to Expose Massive Public Health Scandal and Support Victims, www.toxicfreeairlines.com.

[141] Montgomery et al (1977).

[142] Technical Report on Air Safety and Cabin Air Quality in the BAe146 Aircraft, Parliament of the Commonwealth of Australia, Senate Rural and Regional Affairs and Transport Legislation Committee, Senate Printing Unit: Canberra, Australia, 2000 at pp. 115–128.

[143] CAA (2008) 'Flight Operations Department Communications (FODCOM) 17/2008' UK Civil Aviation Authority, Safety Regulation Group, Aviation House, Gatwick, West Sussex, England; CAA (2002) 'Flight Operations Department Communications (FODCOM) 21/2002' UK Civil Aviation Authority, Safety Regulation Group, Aviation House, Gatwick, West Sussex, England; CAA (2001) 'Flight Operations Department Communication (FODCOM) 14/2001' UK Civil Aviation Authority, Safety Regulation Group, Aviation House, Gatwick, West Sussex, England. CAA (2000) 'Flight Operations Department Communication (FODCOM) 17/2000' UK Civil Aviation Authority, Safety Regulation Group, Aviation House, Gatwick, West Sussex, England.

[144] Cabin air quality' CAA Paper 2004/04, Research Management Department, Safety Regulation Group, UK Civil Aviation Authority, Aviation House, Gatwick Airport South, West Sussex, UK.

6. Cabin air quality' CAA Paper 2004/04, Research Management Department, Safety Regulation Group, UK Civil Aviation Authority, Aviation House, Gatwick Airport South, West Sussex, UK.

authorities.[145] Three years later, in the United States, the Federal Aviation Administration issued a directive that required the operators of BAe 146 aircraft to preclude oil residue from accumulating in the air system ductwork of the aircraft.[146]

Although it is generally accepted that this problem is not confined to any particular type of aircraft and that all jet aircraft remain vulnerable to the seepage of oil from the engines and possible contamination of bleed air, a view has been expressed that the Boeing 787, which will come into operation in 2010, will not have this problem since supply air in this aircraft is processed in electrically generated compressors independent of the jet engines.[147] This design, called the "no bleed architecture", relies on electrically driven compressors to provide cabin pressure where fresh air is brought on board dedicated cabin air inlets.[148]

2.3.1.1 Legal and Regulatory Issues

The *Chicago Convention* of 1944[149] in its Preamble refers to the development of international civil aviation in a safe and orderly manner. The Convention also secures the undertaking from contracting States to adopt international Standards and Recommended Practices *inter alia* on matters concerned with the safety of air navigation.[150] ICAO, which was established by Article 43 of the Chicago Convention, has, as one of its aims and objectives, the ensuring of the safe and orderly growth of international civil aviation throughout the world,[151] to meet the needs of the peoples of the world for safe, regular, efficient and economical air transport,[152] and promote safety of flight in international air navigation.[153]

The Convention in its Annex 8 (Airworthiness of Aircraft)[154] stipulates that a certificate of airworthiness shall be issued by an ICAO member State concerning an aircraft, conditional upon and based on satisfactory evidence being received that the

---

[145] 'Report RL 2001:41e 'Accident investigation into incident onboard aircraft SE-DRE during flight between Stockholm and Malmo M County, Sweden,' Statens Haverikommission Board of Accident Investigation, Stockholm, Sweden.

[146] Airworthiness Directive 2004-12-05: BAE Systems (Operations) Limited Model BAe 146 Series Airplanes' Docket No. 2003-NR-94-AD, Federal Aviation Administration, Washington, DC.

[147] Submission by Susan Michaelis (Capt): To accompany all sections of A-NPA comments made by EASA CRT, 8/1/10, RE: A-NPA No. 2009-10 'Cabin air quality onboard large aeroplanes' at p. 9.

[148] See Sinnett (2007).

[149] Convention on International Civil Aviation, *supra*, note 3.

[150] *Id.* Article 37.

[151] *Id* Article 44(a).

[152] *Id* Article 44(d).

[153] *Id* Article 44(h).

[154] Annex 8 to the Convention on International Civil Aviation – *Airworthiness of Aircraft* – Tenth Edition, April 2005.

aircraft complies with the design aspects of the appropriate requirements.[155] The Annex goes on to state that a Contracting State shall not issue or render valid a Certificate of Airworthiness for which it intends to claim recognition pursuant to Article 33[156] of the Convention on International Civil Aviation unless it has satisfactory evidence that the aircraft complies with the applicable Standards of the Annex through compliance with appropriate airworthiness requirements.[157] The Annex also requires States to establish a safety programme with a view to achieving an acceptable level of safety in civil aviation,[158] while going on to say that the design of the airplane shall take into consideration The design of the aeroplane shall take into consideration the flight crew operating environment including: a) effect of aeromedical factors such as level of oxygen, temperature, humidity, noise and vibration; b) effect of physical forces during normal flight; c) effect of prolonged operation at high altitude; and d) physical comfort.[159]

ICAO Assembly Resolution A35-12[160] declares that the protection of the health of passengers and crews on international flights is an integral element of safe air travel and that conditions should be in place to ensure its preservation in a timely and cost effective manner. It also requests the Council of ICAO to support further research on the consequences of air transport on the health of passengers and crews. Resolution A 27-13[161] reaffirms the public-service character of the service provided by air transport operators, recognizing that the essential purpose of such a service is to satisfy the common good of peoples in whose development States, carriers and users are all equally interested.

The last mentioned regulatory requirement is especially relevant to some recent cases, the first being *Victoria Vaughn Holsted and Valerie Vaughn*, Petitioners, *vs. Southwest Airlines Co.*[162] The facts of this case were as follow: on January 27, 2009, petitioners Victoria and Valerie boarded Southwest Airlines Flight 1705, which departed at approximately 10:00 AM from Los Angeles International Airport with stops scheduled for Nashville International Airport in Nashville, TN, then

---

[155] *Id.* Standard 3.2.1.

[156] Article 33 provides that certificates of airworthiness and certificates of competency and licenses issued or rendered valid by the contracting State in which the aircraft is registered, shall be recognized as valid by the other contracting States, provided that the requirements under which such certificates or licences were issued or rendered valid are equal to or above the minimum standards which may be established from time to time pursuant to the Convention.

[157] *Id.* Standard 3.2.2.

[158] *Id.* Standard 5.1.

[159] Id. Part III B, Sub Part J, J-4 (Operating Environmental Factors).

[160] Resolution A35-12 *Protection of the Health of Passengers and Crews and Prevention of the Spread of Communicable Disease through International Travel*, Assembly Resolutions in Force (as of 28 September 2007), ICAO Doc 9902, at 1–77.

[161] *Id.* 1–79.

[162] Case No. BS120400, http://www.finanznachrichten.de/nachrichten-2009-04/13758467-southwest-airlines-flight-1705-passengers-file-petition-against-the-airline-to-preserve-evidence-of-onboard-exposure-to-contaminated-air-causing-them-004.htm.

Birmingham-Shuttlesworth International Airport in Birmingham, AL and finally Baltimore/Washington International Airport in Baltimore, MD. About 1-h into the flight they, along with the other passengers, began to experience hypoxia (oxygen deprivation), among other problems. Once alerted to the air quality problem in the cabin, the pilot then engaged the engines at full thrust and entered a steep ascent. As this occurred, super heated air began to surge out of the ventilation system and onto the passengers. Also present was the appearance of a mist.

In their petition to the Los Angeles Superior Court for the preservation of evidence against the defendant airline, the petitioners claimed that exposure to contaminated air had caused them to suffer serious and debilitating health problems, among them motor skill deficiencies, loss of balance, vision impairment and uncontrollable tremors. The plane in question was a Boeing 737–300 jet aircraft. They also claimed that, despite repeated requests, the defendant airline was ambivalent and vague in its responses to questions posed by the petitioners that were calculated to assist the physicians of the petitioners so that the doctors would know how to best treat them.

Accordingly, the petitioners prayed for damages on the ground that the defendant was negligent or wanton in that it failed to follow relevant safety, operation, maintenance, repair, service and inspection procedures with regard to the subject aircraft and that the defendant failed to provide its passengers with an aircraft that was in good mechanical condition and free of defects. The petitioners also claimed damages in breach of contract on the ground that the defendant agreed to transport them for compensation and the contract of carriage included the agreement express and/or implied, to transport them safely, in a non-negligent manner.[163]

The second case which is of relevance is *Turner* v. *Eastwest Airlines Limited*[164] which involved an action instituted by an employee of an airline who claimed that, on descent, there was smoke in the cabin of the aircraft in the nature of a thick cloud of smoke which she inhaled, causing her to cough and break out in sore eyes, a burning throat accompanied by a headache. The cough had persisted thereafter, causing periods of paroxysms of coughing. The tribunal hearing the case sought answers to such questions as "what was the plaintiff's condition and cause of her illness?" "Was the injury foreseeable" "did the airline have a reasonable response to the problem when it arose?" "Was there economic loss" and "is the plaintiff entitled to damages?" On various counts, the New South Wales Dust Diseases Tribunal, which heard the case, awarded the plaintiff $ 137,757, the counts being *inter alia*, non economic loss, loss of earnings, future loss of earnings, past out of pocket expenses, future out of pocket expenses.

---

[163] Case No. CV-09-HGD-2193-s in the District Court of Northern District of Alabama, 28 October 2009. 46 N. W. 677.

[164] [2009] NSW DDT 10, 5 May 2009. Also,New South Wales Dust Diseases Tribunal 10 (5 May 2009) Matter Number 428 of 2001, discussed in *ZLW*, 58 Jg 4/2009, at pp. 705–717.

The defendant appealed against the award. In September 2010, the High Court upheld the decision of the Dust and Diseases Tribunal.[165]

When there is incontrovertible evidence of a person contracting a disease as a result of being contaminated in an aircraft whilst on board, liability issues pertaining to the airline arising from the incident may involve principles of private air carrier liability. The *Montreal Convention* of 1999[166] which emerged consequent to the Diplomatic Conference on Private Air Law of the International Civil Aviation Organization held from 10 to 28 May 1999, provides that the carrier is liable for damage sustained in the event of death or bodily injury of a passenger upon condition only that the accident which caused the damage so sustained took place on board the aircraft or in the course of any of the operations of embarking or disembarking. The *Warsaw Convention* of 1929[167] provides that the carrier is liable for damage sustained in the event of death or wounding of a passenger or any other bodily injury suffered by a passenger, if the accident which caused the damage so sustained took place on board the aircraft or in the course of any of the operations of embarking or disembarking. Both these Conventions have similar wording, admitting only of death or bodily injury or wounding. Of course, on the face of the provision, the words "wounding" and "bodily injury" do not necessarily lend themselves to be associated with infection. *A fortiori*, according to the Montreal Convention, the bodily injury must be caused as a result of an accident, and, according to the Warsaw Convention, the wounding or injury must be caused by accident which is not typically a synonym for "infection" in both cases. However, the recent decision in *El Al Isreal Airlines Limited* v. *Tseng*[168] introduced a new dimension to the word "accident" under the Warsaw Convention by giving it pervasive scope to include such acts as security body searches performed by the airlines. In this context, the word "accident" loses its fortuity and it becomes applicable to an expected or calculated act. Thus, if an airline knows or ought to have known that there could be oil leakage from its engines that would mix with bleed air and make the passengers sick, it may well mean that the act of the airline would be construed by the courts as an accident within the purview of the Warsaw Convention.

---

[165] See, Flight attendant wins toxic cabin air damages, *Air Letter*, No. 17,074 Thursday 16 September 2010 at p.3. This article states that A University of New South Wales survey has found that about 25% of pilots who flew the BAe 146 aircraft suffered long term health degradation that deprived them of their pilot licences and that an Australian Senate inquiry had found East-West and Ansett Airlines had been paid more than $2 million by BAe Systems (British Aerospace's successor) to drop complaints about the aircraft.

[166] Convention for the Unification of Certain Rules for International Carriage by Air, signed at Montreal on 28 May 1999, ICAO Doc 9740.

[167] Convention for the Unification of Certain Rules Relating to International Carriage by Air, signed at Warsaw on 12 October 1929.

[168] 1999 Westlaw 7724 (January 12 1999).

2.3.2 A Novel Approach to Liability

The particular instance of the aerotoxic syndrome brings to bear the need to look at the parties responsible and the degree to which they are responsible. Arguably, the designers and manufacturers of the components involved as well as the air carriers who use such components know, or ought to know the potential danger to those who travel in aircraft whose components are prone to oil leaks into the cabin. The next issue is the rights of the plaintiff, particularly against the backdrop of a perceived absence of proven medical evidence linking contamination of air in the aircraft cabin to the aerotoxic syndrome. Can the plaintiff have access to a court and petition for discovery of facts?

There has been a recent trend in pleadings for dismissal of motions, particularly in the United States, which changed the existing judicial practice that went on the basis that the plaintiff only needed to provide a short and plain statement of the claim showing that he was entitled to relief and that the failure to do so did not give justification to dismiss the claim *in limine*.[169] The new trend goes on the basis that no longer would plaintiffs have it easy to gain access to the discovery phase of litigation. This trend started in 2007 with the case of *Atlantic* v. *Twombly*[170] in which Justice Souter, writing the majority judgment held that:

> When the allegations in a complaint, however true, could not raise a claim for entitlement of relief, this basic deficiency should...be exposed at the point of minimum expenditure of time and money by the parties and the court.[171]

In other words, the *Twombly* decision recognizes that the plaintiff has to make his pleading at least plausible to expect entitlement of relief. This makes it easier for a court to dismiss claims which are seemingly without merit. This judicial breakthrough was followed by the US Supreme Court in May 2009 in the decision in *Ashcroft* v. *Iqbal*[172] which considered the arrest of the plaintiff, a Pakistani Muslim, by federal officials following the attacks of 11 September 2001 on criminal charges. Iqbal claimed that he was discriminated upon by the officials with the label "high interest" categorizing him in relation to the investigation of 9/11. He also claimed that he was shackled, kicked, punched, strip-searched and denied medical attention during his imprisonment. Despite the dissimilarity between the *Twombly* and *Iqbal* circumstances, the majority decision of the latter case followed the principles of *Twombly* and dismissed Iqbal's petition, implicitly stating that the *Twombly* rationale applied to all civil actions.

Although the *Iqbal* decision has since been questioned and even criticized,[173] it still remains the law, which brings to bear a salient fact in litigation pertaining to the

---

[169] *Conly* v. *Gibson*, 355 U.S. 41 (1957).

[170] *Bell Atlantic Corp.* v. *Twombly*, 550 U.S. 544 (2007).

[171] *Id* at 11. Justice Souter quoted *Daves* v. *Hawaiian Dredging Co.* 114 F. Supp. 643 at 645.

[172] 129 S.Ct. 1937.

[173] Justice Ginsburg, a dissenting voice in the case, later criticized the majority decision as "messing up the federal rules" and Justice Souter himself (who delivered the majority *Twombly*

aerotoxic syndrome, which is that most allegations made with regard to the syndrome are anecdotal and hypothetical. This imposes upon the plaintiff a much higher burden of proof than that which existed prior to the *Twombly* decision.

Another issue is whether, if a link can be drawn between contamination of cabin air and the aerotoxic syndrome, it grounds an action in tort, and if so, against whom can the action be brought. Another question is, are those involved in providing services directly related to the transport by air of a person, such as manufacturers of aircraft components, the airlines and their pilots the defendants? If such were to be the case, can they be included, for purposes of tort liability, in a special category of skilled personnel, such as surgeons and medical specialists, who warrant their special expertise to the public? It is submitted that the answer to the last question is yes.[174]

In applying the analogy of medical experts to the context of aviation professionals whose negligence causes the aerotoxic syndrome in crew and passengers, it is noteworthy that, in the United States, patients' rights are enforced against medical malpractice or negligence of physicians or medical institutions such as hospitals and clinics. Medical malpractice is a failure to meet generally accepted professional standards of medical practice that causes injury to a plaintiff. Medical malpractices are based on diverse grounds including non-diagnosis, misdiagnosis delay of treatment and incompetent medical procedures.[175] English law has accepted the concept of incrementalism,[176] which rejects generalization in relation to the duty of care, in favor of a cautious development of law founded on analogies to similar fact situations, but espousing and applying *fairness and justice* to each case. This approach has its genesis in the judgment of Justice Brennan in the 1985 case of *Sutherland Shire Council* v. *Heyman*[177] handed down by the High Court of Australia where His Honour said:

> It is preferable...that the law should develop novel categories of negligence incrementally and by analogy with existing categories, rather than by massive extension of a prima facie duty of care restrained only by the indefinable...considerations which ought to negative, or to reduce or limit the scope of the duty or the class of person to whom it is owed.[178]

The House of Lords found it fit to import this approach to the United Kingdom in the leading 1990 case of *Caparo Industries Plc* v. *Dickman*[179] where Lord Bridge stated:

---

judgment) was observed stating that the *Iqbal* decision had misapplied the pleading standard set in the *Twombly* decision. For a good discussion on this subject, see the Summer 2009 Aviation Round up of Condon & Forsyth LLP, at p. 3.

[174] See Abeyratne (1998).

[175] Koenig and Rustad (2005).

[176] See Stanton. Incremental Approaches to the Duty of Care, Chapter 2 of Mullany, *Torts in the Nineties*, Law Book Company:1997.

[177] (1985) 157 *CLR* 424.

[178] *Id.* 481.

[179] [1990] 2 *AC* 605.

> Whilst recognizing, of course, the importance of the underlying general principles common to the whole field of negligence, I think the law has now moved in the direction of attaching greater significance to the more traditional categorization of distinct and recognizable situation, as guides to the existence, the scope and the limits of the varied duties of care which the law imposes.[180]

Therefore, there is no simple formula or touchstone to which recourse can be had in order to provide in every case a ready answer to questions as to whether the law will or will not grant recourse based on traditional rules of negligence. In this context, one wonders whether the use of such catch phrases as "reasonably skilled professional", without any attendant criteria to define the phrase is practical anymore.

Given the aforesaid trend in tort liability favoring recognition that the law should develop novel categories of negligence incrementally, and the curbing of the plaintiff's right to access unless a plausible explanation can be given as to why he is seeking redress, the defendant's position seems to be slightly more advantageous than that of the plaintiff considering the limitation of the plaintiff's rights and the possibility of the aerotoxic syndrome being recognized as the effect of a wrongful act on the part of the plaintiff. What would eventually tip the scale in the defendant's favour would be the fact that there has been no structured, supervised and documented study which concludes that there is a link between the aerotoxic syndrome and contaminated cabin air. It has been stated that:

> [D]iscomfort does not automatically pose a health risk as it is sometimes suggested. In fact, it is mechanically impossible to meet the subjective comfort level expected by cabin crew and passengers at all times based on the different activity alone. This does not mean that a situation creating discomfort does not deserve attention, but simply that it can be approached differently.[181]

This being said, potential defendants could ensure that the question of negligence would not arise by evaluating oils and fluids for toxicity; and by establishing standards appropriate for use in the environment of the cabin in case of possible leakage. They could also ensure that oil seals in the engines that prevent the leakage of oil used for lubrication are secure and there is no danger of leakage while the aircraft is in flight. Proper filters could be established to filter bleed air. For their part, regulators could strictly ensure that the ICAO standards of Annex 8 and Assembly Resolutions are met by the aircraft operators. They should also address the issue of oil leakage through appropriate research and assessment.

As Prosser has stated: "perhaps more than any other branch of the law, the law of torts is a battleground of social theory".[182] An example in more recent times is the true story of Erin Brockovich, a young woman who successfully launched a toxic torts law suit against a California utility after having discovered Chromium 6 in the well water of a California town. The judgment awarded was $ 333 million in this

---

[180] *Id.* at 618.
[181] Thibeault (1997).
[182] Prosser (1964).

class action.[183] Modern tort law has the resilience of applying old actions to new threats. This ensures that no one, including multimillion dollar enterprises, can operate beyond the reach of the law. The punitive aspect of tort law, which punishes the wrongdoer, not only acts as a deterrent but also ensures compensation for the injured giving one the message that one cannot escape justice after having committed a civil wrong on another.[184] This could be the ultimate message to a negligent defendant.

2.3.3 Medical Issues of Technical Crew Members

International regulations propounded under the auspices of the International Civil Aviation Organization (ICAO) require that a pilot has to have a certificate of competence issued by the State in which the aircraft he flies is registered if he were to undertake flying an aircraft. Medical certification is an essential component in the licensing process and conditions and guidelines for the issuance of such certificates are provided in detail in ICAO documents. The overall responsibility of the pilot for his flight and persons therein which is legally recognized by international treaty, has necessitated the grounding of pilots for many reasons where their health did not reach the standards required, which in turn has resulted often in the concealment during their medical examinations by pilots of pre-existing medical conditions.

The pilot[185] operates in a highly complex environment, particularly in single pilot operations. Contemporary commercial airline practice and the tenets of air law attribute to the pilot of an aircraft absolute responsibility for the safe operation of his aircraft. This responsibility can be carried out only if the pilot is not negligent or if he enjoys basic health as required by applicable regulations. The Airline Pilots Association International (ALPA)[186] has a code of ethics for airline pilots which

---

[183] Lawrence (2000).

[184] *Rookes* v. *Bernard* (H.L. 1964) 1 All Eng. Rep. 367.

[185] For the purposes of this paper, a pilot is a person who engages in such flying as makes it necessary that he or she holds a valid airline transport pilot's license (ATPL). This paper will therefore not address the professional conduct of persons who are holders of private pilots' licenses and use such licences for non remunerative flights, or flights which do not involve payment of monies for services rendered in flying aircraft. Also excluded are holders of ATPLs who happen to fly on particular occasions under circumstances where a private pilot's license would be adequate, for instance, when a person who flies a recreational flight or on personal business although he may hold an ATPL.

[186] The Air Line Pilots Association, International (ALPA) is the largest airline pilot union in the world and represents nearly 53,000 pilots at 38 airlines of the United States and Canada. ALPA was founded in 1931, and the Association is chartered by the AFL-CIO and the Canadian Labour Congress. Known internationally as US-ALPA, it is a member of the International Federation of Air Line Pilot Associations (IFALPA).

2.3 Aviation Medicine

stipulates *inter alia* that a pilot will not knowingly falsify any log or record nor will he condone such action by other crew members. Furthermore, the code requires the pilot to keep uppermost in his mind that the safety, comfort, and well-being of the passengers who entrust their lives to him are his first and greatest responsibility. The pilot also undertakes that he will never permit external pressures or personal desires to influence his judgment, nor will he knowingly do anything that could jeopardize flight safety and that he will remember that an act of omission can be as hazardous as a deliberate act of commission, and he will not neglect any detail that contributes to the safety of his flight, or perform any operation in a negligent or careless manner.[187]

These ethics impliedly require a pilot to divulge to his employer and insurer any medical condition and medications taken to treat that condition. The typical pilot's disability insurance coverage is given upon the pilots assurance *inter alia* that he is not aware of any deterioration in general health, hearing, eyesight or blood pressure, and that in the event of any fraud, misstatement, concealment or failure to disclose information in response to any question, whether intentional or inadvertent, the coverage given will become void and no benefits will be payable.[188] There have been several instances where pilots have either falsified or concealed their medical history. One commentator records instances of 46 pilots in Northern California in the United States who did not disclose their debilitating health to the Federal Aviation Administration (FAA) that would have disqualified them from obtaining their pilots licences.[189] The pilots in question had claimed to be medically fit to fly, yet at the same time were receiving social security payments for medical disabilities. This discovery was a result of an investigation started by the FAA In July 2003, when, the Department of Transportation Office of Inspector General (DOT-OIG) and Social Security Administration Office of Inspector General (SSA-OIG), citing safety and security concerns, initiated a joint investigation to identify pilots misusing Social Security numbers. During the course of the investigation, Social Security records identified individuals who also held FAA medical certificates and who were receiving Social Security (SSA) disability benefits. The DOT-OIG and SSA-OIG launched an 18-month probe termed "Operation Safe Pilot" in coordination with the U.S. Attorney's Office (USAO) to look into possible fraudulent activity. 40,000 pilots were suspected of lying or falsifying their medical history[190]

---

[187] http://www.alpa.org/Home/WhoWeAre/CodeofEthics/tabid/2262/Default.aspx.

[188] See pdf brochure at http://www.insubuy.com/piu/pilotsdisabilityinsurance/brochure.jsp.

[189] See John Alan Cohan, Aero Legal Analysis: Pilots Accused of Medical Certification Fraud, http://espanol.groups.yahoo.com/group/Seguridad-Aerea/message/1600.

[190] It has been recorded that a surprisingly significant number of pilots face denial of medical certification because they are taking antidepressant or serotonin blocker drugs (SSRI's) such as Prozac, which could imperil a pilot's functions. Medical certification requires that airmen be able to exercise the duties privileges of a pilot in the class applied for. In addition, numerous medical conditions will disqualify a pilot from obtaining medical certification, including, *inter alia*: diabetes mellitus, myocardial infarction, cardiac valve replacement, permanent cardiac pacemaker, personality disorders that are severe enough to have repeatedly manifested itself by overt

and it was discovered that 3,220 pilots with current medical certificates were collecting SSA benefits, including disability benefits.[191]

In March 2007, The United States House Transportation and Infrastructure Committee Chairman James L. Oberstar released a committee oversight report which documented "widespread fraud" among pilots who do not disclose and deliberately hide from examining physicians medical conditions that would critically impact their ability to fly an aircraft so that they could retain medical certification for their FAA pilot certificates.[192] To counter this dangerous trend, The Aircraft Owners and Pilots Association (AOPA)[193] offered the US Congress a plan that would encourage pilots to disclose their infirmities at the medical examination,[194] adding that AOPA does not condone false statements on a pilot's medical application.[195]

---

acts, substance dependence or abuse, and epilepsy. Another cause for denial of certification of a pilot is organ transplant, on the basis that there could be a risk of a pilot suffering side effects of rejection during flight from immunosuppressant drugs, or that the organ might be rejected. Medical certification of a pilot and the consequent award of a licence hinge both on the health of the pilot as well as the welfare of the persons carried by him in the aircraft. *Id.*

[191] http://www.aopa.org/whatsnew/regulatory/operation-safe-pilot.html In February of 2009, a commuter plane crashed near the Buffalo Niagara International Airport. Forty-nine passengers and crew were killed as well as one person on the ground. Initially, ice build-up was suspected in the crash, but a report by The New York Times indicates that the National Traffic Safety Board's (NTSB) analysis shows ice was not a prominent factor in the crash. Instead, it appears pilot error is to blame. The NTSB report concluded that the pilot of Continental Flight 3407 did not react properly to a warning that his plane was slowing down too quickly and entering a stall. The report noted that the pilot's reaction was "consistent with startle and confusion" as the pilot pulled on the plane's control mechanism when pushing on it would have kept the plane in flight. Seehttp://www.hg.org/article.asp?id=20681.

[192] http://www.avweb.com/avwebflash/news/House_Committee_Probes_Aviation_Medical_Fraud_194776-1.html. also, http://transportation.house.gov/Media/File/Aviation/Safe Pilot Committee Report.pdf.

[193] The Aircraft Owners and Pilots Association, a not-for-profit organization dedicated to general aviation, with a membership of more than 412,000 pilots, was incorporated on May 15, 1939. From the start, AOPA has fought to keep general aviation fun, safe, and affordable. AOPA records that growth in the early years was slow, but by mid-1995, membership in AOPA had reached about 335,000. http://www.aopa.org/info/history.html.

[194] AOPA suggested that the FAA should add a statement to the medical certification application warning pilots that some medical data would be shared with other agencies. It also recommended a one-year amnesty program to encourage pilots to report all medical visits and conditions to the FAA. Also suggested was that the FAA should establish a data-sharing program with other public agencies such as the Social Security Administration within the limits permitted by the Privacy Act. With these three suggestions, AOPA offered to work with the Congress subcommittee and the FAA to educate pilots about properly reporting all pertinent medical information, and about the severe penalties and safety consequences of failing to do so. See http://www.aopa.org/whatsnew/newsitems/2007/070717medical.html.

[195] Statement of Phil Boyer, President, Aircraft Owners and Pilots Association Before the Transportation & Infrastructure Subcommittee on Aviation, U/S. House of Representatives concerning Falsified Pilot Medical Certificates, http://www.aopa.org/whatsnew/regulatory/operation-safe-pilot.html.

2.3 Aviation Medicine

In April 2010, the FAA announced the possibility of a special medical certificate being issued to pilots who are under medication for mild to moderate depression, so that they could be exempt from conditions that prohibit them from all flying duties.[196] This measure is consistent with the findings of a 12 year study conducted by a team of aviation medicine specialists in Australia which was released in 2007 which said that taking the drugs does not increase the risk of accidents, while banning them could increase risks by encouraging depressed pilots not to seek treatment.[197]

Also in April, 2010, both Houses of Congress passed the *Airline Safety and Pilot Training Improvement Act*, section 206 a. 1. A of which provides *inter alia* that flight crewmember mentoring programs will be established under which an air carrier will pair highly experienced flight crewmembers who will serve as mentor pilots and be paired with newly employed flight crewmembers. The provision states further that mentor pilots should be provided, at a minimum, specific instruction on techniques for instilling and reinforcing the highest standards of technical performance, airmanship, and professionalism in newly employed flight crewmembers. Section 212 a (1). Requires the Administrator of the Federal Aviation Administration to issue regulations, based on the best available scientific information, to specify limitations on the hours of flight and duty time allowed for pilots to address problems relating to pilot fatigue. The FAA relies on pilots to tell the truth about their physical and mental condition during the medical examination process.

In December 2010, The European Union (EU) Commission debated a revision to its regulations on medical certification of pilots which provided *inter alia* that licence holders will not exercise the privileges of their licence and related ratings or certificates at any time when they: are aware of any decrease in their medical fitness which might render them unable to safely exercise those privileges; take or use any prescribed or non-prescribed medication which is likely to interfere with the safe exercise of the privileges of the applicable licence; receive any medical, surgical or other treatment that is likely to interfere with flight safety.[198] The

---

[196] http://www.faa.gov/news/press_releases/news_story.cfm?newsId=11293. On April 2nd CNN covered a story titled "FAA to allow pilots to fly on antidepressants" The FAA announced that it will be lifting an outdated policy and will now allow pilots with mild depression to fly commercial airplanes. The pilots will have to be able to prove that they have been successfully treated for the past 12 months. The FAA says that this new policy will improve the safety in the air because pilots will be properly medicated and or not hiding the fact that they some are self medicating to avoid being grounded from the sky. The FAA considerably went public with this policy change because it is the public that these medicated pilots are trusting to fly them safely. See http://cspc10.wordpress.com/2010/04/04/faa-goes-public-on-pilot-depression/.

[197] See Pilots on Antidepressants Pose No Safety Risk http://www.newscientist.com/article/dn12981-pilots-on-antidepressants-pose-no-safety-risk.html. Over the 12-year study period, 481 pilots who were prescribed antidepressants had 11 accidents due to pilot error and 22 near misses. The researchers say this was not significantly different to the five accidents and 26 near misses of the similar number of pilots who did not take antidepressants, but who were matched by age, sex, and flying experience.

[198] http://easa.europa.eu/agency-measures/opinions.php.

European Aviation Safety Agency (EASA) in its opinion[199] on the draft revised regulation, drew the attention of the Commission to the diversity of medical practices and regulations in the various member Sates of the EU and exhorted the Commission to consider Regulations of ICAO as the common basis for such a revision.

The Convention on International Civil Aviation[200](hereafter referred to as the Chicago Convention) in Article 32 requires that the pilot of every aircraft and the other members of the operating crew of every aircraft engaged in international navigation be provided with certificates of competency and licences issued or rendered valid by the State in which the aircraft is registered. ICAO's global medical standards for the issuance of a pilot's license are contained in Chapter 6 of Annex 1[201] to the Chicago Convention. The Annex provides that if the medical standards prescribed in Chapter 6 for a particular license are not met, the appropriate medical assessment shall not be issued or renewed unless there were special circumstances that led to the applicant's failure to meet such requirements and that the special abilities, skill and experience of the applicant are given due consideration and that the license is appropriately endorsed with any special limitation when the safe performance of the license holder's duties is dependent upon that limitation.[202] The Annex further goes on to provide that license holders shall not exercise the privilege of their licenses and related ratings at any time when they are aware of any decrease in their medical fitness which might render them unable to safely and properly exercise their privileges.[203] A recommendation follows, that license holders should inform the licensing authority of confirmed pregnancy or any decrease in medical fitness of a duration of more than 20 days or which requires continued treatment with prescribed medication or which has required hospital treatment.[204] Another relevant provision prescribes that license holders shall not exercise the privilege of their licenses and related ratings at any time when they are under the influence of any psychoactive substance which might render them unable to safely and properly exercise their privileges.[205] Detailed guidance for the implementation of Annex 1 is contained in the *ICAO Manual of Civil Aviation Medicine*.[206]

---

[199] Opinion No 07/2010 of the European Safety Agency 13 December 2010.
[200] Supra. n. 3.
[201] Personnel Licensing, 10th Edition, July 2006.
[202] *Id*, Standard 1.2.4.8.
[203] *Id*. Standard 1.2.6.1.
[204] *Id*. Recommendation 1.2.6.1.1.
[205] *Id*. Standard 1.2.7.1.
[206] *Doc 8984-AN/895*, Second Edition, 1985.

2.3 Aviation Medicine

2.3.4 Liability Issues

The most pertinent fact of concealment of vital medical facts by the pilot is grounded on the principle *suggestio falsi*[207] or *suppressio veri*[208] (making false statements or suppressing the truth). The seriousness of a concealment by a pilot of a pre existing medical condition can be distinguished from such a condition which concerns any other, in that a pilot occupies a special position of responsibility. According to accepted principles of law as laid down by international convention, it is incontrovertible that the final responsibility for the safe operation of an aircraft lies with the pilot. Annex 6 to the Chicago Convention provides that:

> The pilot-in-command shall be responsible for the operation and safety of the aeroplane and for the safety of all persons on board, during flight time.[209]

This presumption of responsibility has influenced some States which have signed or ratified the Convention and is reflected clearly in their air navigation laws.[210] These laws have been have been observed to list requirements which any pilot with a sense of good airmanship would naturally comply with. Failure to comply with such regulations has been clearly interpreted to be bad airmanship which renders the pilot liable for prosecution on a criminal charge.[211] In any event, the fundamental postulate which imposes *prima facie* responsibility on the pilot has been accepted as a general principle of liability of the pilot which sets the base for determining his legal status and responsibility.[212]

The legal responsibility placed on the commander of the aircraft is therefore inextricably linked with the expectation of good airmanship. Airmanship has been

---

[207] A statement of a falsehood. This amounts to a fraud whenever the party making it was bound to disclose the truth or whenever the party making it was bound to disclose the truth. See http://www.law-dictionary.org/SUGGESTIO+FALSI.asp?q=SUGGESTIO+FALSI.

[208] In law, an undue concealment or non-disclosure of facts and circumstances which one party is under a legal or equitable obligation to communicate, and which the other party has a right – not merely in conscience, but juris et de jure – to know. See http://www.wordnik.com/words/suppressio%20veri.

[209] See Annex 6 to the Convention on International Civil Aviation signed in Chicago on 7 December 1944, Para 4.5.1.

[210] *See The British Air Navigation* Order (1985) Article 32, which states, *inter alia:* The Commander of an aircraft registered in the United Kingdom shall satisfy himself before the aircraft takes off – that the flight can safely be made, taking into account the latest information available as to the route and aerodromes to be used, the weather reports and forecasts available and any alternative course of action which can be adopted in case the flight cannot be completed as planned; See also generally, *U.S. Federal Aviation Regulations* FAR 91.3 (a), Australian Air Navigation Regulations, Regulation 219 and New Zealand Civil Aviation Regulations (1953), Regulation 59.

[211] See N. Price, *Pilot Error* (1976) at pp. 238–239. See also generally the findings of the *New Zealand Royal Commission of Inquiry into the 1979 Aft. Erebus DCIO Disaster.*

[212] Abeyratne (1998), See also N.M. Matte, *The International Legal Status of Aircraft Commander* (1975) at 34 and Videla Escalada, *Aeronautical Law* (1979) at 210–211, Speiser and Krause(1978).

regarded as an indefinable quality and has been used to describe the intuitive faculty of the pilot where he concerns himself with what is right or wrong in the operation of an aircraft which is acquired by sustained experience in flying.[213] Needless to say, a pre-existing medical condition such as depression could adversely affect the judgment of a pilot and preclude him from exercising good airmanship.

Aircraft accidents involving human error can happen due to two reasons: pilot error and pilot incapacitation. It was believed by the British Civil Aviation Administration (CAA) in the 1990s that pilot incapacitation comprise only 10% of the overall risk of pilot failure.[214] Although physiological health issues such as hypertension, acute diabetes and cardiovascular disease are common causes for a pilot to lose his license, these can mostly be detected at a medical examination in the certification process. The more difficult to detect and which are only known to the pilot applying for his license are psychiatric and neurological conditions.[215] The loss of a Royal Air Maroc commuter aircraft in 1994, the Egypt Air B767 crash in 1999 and the December 1997 Silk Air B 737 crash[216] which brought to bear suspicions in the aviation community that they were caused by emotionally disturbed pilots prompted some experts to claim that it was necessary for pilots to be psychiatrically evaluated for emotional maladjustment. They went on to say that pilots who have suicidal tendencies could then be identified and their licenses taken away. This theory has been found flawed on the ground that there could be no scientific validity for determining a possible suicidal maniac in the flight deck and that such a theory is purely speculative and conjectural.[217] One commentator has gone to the extent of claiming that introducing specific testing for pilots for mental disorders[218] would be like "a solution looking for a problem".[219]

---

[213] Burridge (1977).

[214] Evans (1995).

[215] Steenblick (1995), at 31.

[216] In December 1997, a *Silk Air* B737-300 operating a flight from Jakarta to Singapore crashed in Indonesia, killing 104 passengers and crew on board. Two theories regarding the accident were propounded: that some catastrophic malfunction of the aircraft led to the crash; or that the crew committed suicide. The aircraft plunged vertically from its cruising altitude directly into the Musi River in Southern Sumatra without any distress signal being emitted or any indication from the cockpit to air traffic control that the aircraft was experiencing problems. Unfortunately, much of the debris had been washed away by the fast flowing waters of the river when accident investigation commenced, leaving little, if any, information for deduction as to the cause of the accident. A significant factor, however, was that both the cockpit voice recorder and flight data recorder had stopped functioning immediately prior to the commencement of descent. It is reported that the Captain had been demoted from his position of flight instructor shortly before the crash and had allegedly been heavily in debt. The Singapore police investigating the crash had reportedly classified the case as involving suicide-cum murder by the pilot. See Abeyratne (2000).

[217] Besco (2000), at 3. The author goes on to say that pilots who suffer from depression or such tendencies would eventually find a way of trick the evaluation process. *Ibid.*

[218] Many types of mental disorders have been cited. Broad categories include anxiety disorders, mood disorders, personality disorders, schizophrenia, substance abuse and dependence and several others. See Snyder (1988), at 17.

[219] Phillips (1999), at 43.

2.3 Aviation Medicine

The disturbing factor in aviation accidents involving injury and death to passengers is the distinction between an accident which causes death or injury to passengers through a mechanical defect in the aircraft or other extraneous reason such as adverse weather conditions and an accident of an aircraft which is found to have been mechanically defect-free and travelling in turbulent free weather. This is significant if in the latter instance, the aircraft meets with an accident caused by ill health of the technical crew, where the law could construe the accident as having been caused by the wilful misconduct of the carrier. In this instance, the dependents of passengers killed in the crash could argue that the airline concerned was guilty of wilful misconduct in allowing a crew member whose mental state did not admit of his ability to fly an aircraft to do so.

The concealment of a pre-existing medical condition by a pilot could annul his medical insurance contract and even make his contract with his employer void or voidable. Although a misrepresentation, negligently made or withheld would ground an action in tort for damages when a close relationship exists between the parties,[220] a 1976 The Court of Appeals in a British case has held that a negligent misrepresentation inducing any contract gives rise to actions either in tort or contract.[221] In this context conduct may be construed or amount to a misrepresentation as was held in the 1974 Canadian case of *Gronau* v. *Schlamp Investments Ltd*[222] where a vendor, who concealed a crack in the building he sold the buyer, was found to have misrepresented facts. Similarly, an incomplete statement could mislead a party to a contract, in as much as a half truth or a false statement can be considered misrepresentations.[223] Silence could amount to a claim that there is nothing of significant to reveal.[224]

An analogy can be drawn with admiralty law[225] in which claims for injury are brought under the *Jones Act*.[226] In the 1968 decision of *McCorpen* v. *Cental Gulf*

---

[220] *Hedley Byrne & Co. Ltd.* v. *Heller & Partners* [1964] A.C. 465.

[221] *Esso Petroleum Co. Ltd.* v. *Mardon* [1976] Q.B. 801.

[222] (1974) 52 D.L.R. (3d) 631 (Man.Q.B.). See also, See *Rowley* v. *Isley* [1951] 3 D.L.R. 766 (Ct.) and *McGrath* v. *MacLean* (1979) 95 D.L.R. (3d) 144.

[223] *Curtis* v. *Chemical Cleaning and Dyeing Company* [1951] 1. K.B. 805 (C.A). Also, *Link* v. *Schiable* (1961) 27 D.L.R. (2d) 461 (B.C.C.A) and *Olsen* v. *Poirier* (1978) 91 D.L.R. (3d) 123 (Ont. H.C.J).

[224] *Brownlie* v. *Campbell* (1880), 5 App. Cas. 925 (per Lord Blackburn).

[225] Admiralty law (also referred to as maritime law) is a distinct body of law which governs maritime questions and offenses. It is a body of both domestic law governing maritime activities, and private international law governing the relationships between private entities which operate vessels on the oceans. It deals with matters including marine commerce, marine, navigation, shipping, sailors, and the transportation of passengers and goods by sea. Admiralty law also covers many commercial activities, although land based or occurring wholly on land, that are maritime in character. Admiralty law is distinguished from the Law of the Sea, which is a body of public international law dealing with navigational rights, mining rights, jurisdiction over coastal waters and international law governing relationships between nations.

[226] The Jones Act was an important piece of United States legislation passed in 1920. It supported the American merchant Marine, while also providing additional protections for sailors and ship's

Steamship Corporation[227] it was held that although wilful concealment by a seaman of a pre-existing medical condition would preclude recovery for medical expenses, it would not be a bar to recovery for damages caused to his health as a result of working in the employer's premises. In admiralty law under the *Jones Act*, it is a fundamental duty of an employer to provide the employee with a safe place to work irrespective of whether the employee disclosed a medical condition or not. In the 2006 case of *Johnson* v. *Cenac Towing Inc.*[228] the United States District Court for the Eastern District of Louisiana denied a seaman-plaintiff maintenance and cure benefit for wilful concealment of a pre-existing medical condition but awarded damages for injuries and illness suffered as a result of his employer's negligence.[229]

One recommendation for responding to instances of concealment by pilots of pre existing medical conditions is preventive in nature in that it advocates a pilot performance monitoring programme for airlines which is designed to ensure the fitness of crew members to perform their duties.[230] Degrading performance levels could be monitored instead of psychiatric screening. In the context of liability of the pilot for concealment, such must be viewed with grave concern as, unlike a seaman or sailor, the pilot has the lives of his passengers in his hands. Any hint of doubt on the part of a pilot in his incapability of ensuring the safety of his flight and that of his passengers should result in his abstaining from taking control of a flight. A mere claim of misjudgement that led to non-disclosure of the medical condition by the pilot should not be considered acceptable. The House of Lords in the 1988 case of *Whitehouse* v.*Jordan*[231] rejected the idea that mere errors of judgments cannot amount to negligence. Lord Fraser observed:

> Merely to describe something as an error of judgment tells us nothing about whether it is negligent or not. The true position is that an error of judgment may, or may not, be negligent; it depends on the nature of the error. If it is one that would not have been made by a reasonable competent professional man professing to have the standard and type of skill that the defendant held himself out as having, and acting with ordinary care, then it is negligent.[232]

The profession of aeronautics, particularly relating to the piloting of aircraft, remains one of the most responsible, particularly in the context of the many lives that are entrusted to the airline pilot at any given time. The realization that pilots should be fit and well to perform their professional duties has gradually evolved, from the initial requirement of a medical certificate issued according to the

---

crew. Several clauses in the Jones Act set a precedent, since they went above and beyond similar protective clauses under admiralty law.

[227] 396 F2d 547.
[228] 468 F. Supp, 2d. 815.
[229] *Id*, at 826.
[230] Besco, *supra*, note 217. ibid.
[231] [1981] 1 All E.R. 267.
[232] *Id.* at 276.

standards laid down by ICAO[233] to a ground breaking concept introduced in 1982 by Cardiologist, Professor Hugh Turnstall-Pedoe, who, while working at Ninewalls Hospital in Dundee, suggested that the health of pilots should be assessed the same way as engines are assessed by engineers.[234] One of the measures suggested was the permanent requirement of having a co-pilot in the cockpit. The hypothesis submitted in support of this requirement was that if an average flight lasted 60 min and the critical take-off and landing phases are taken as the first and last 3 min of a flight, having a second pilot reduces the risk of incapacitation of the first pilot causing a fatal accident by one-thousand fold.[235]

The durability of tort law lies in its enduring ability to adapt old remedies to new civil wrongs. Against this backdrop, the concealment by a pilot of a preexisting medical condition should be considered as being beyond the scope of the commonly accepted norms of negligence and consequently considered incrementally proportionate to the risk and danger his act of concealment portends. English law has accepted the concept of *incrementalism*,[236] which rejects generalization in relation to the duty of care, in favor of a cautious development of law founded on analogies to similar fact situations, but espousing and applying *fairness and justice* to each case. This approach has its genesis in the judgment of Justice Brennan in the 1985 case of *Sutherland Shire Council* v. *Heyman*[237] handed down by the High Court of Australia where His Honour said:

> It is preferable...that the law should develop novel categories of negligence incrementally and by analogy with existing categories, rather than by massive extension of a prima facie duty of care restrained only by the indefinable...considerations which ought to negative, or to reduce or limit the scope of the duty or the class of person to whom it is owed.[238]

The House of Lords found it fit to import this approach to the United Kingdom in the leading 1990 case of *Caparo Industries Plc* v. *Dickman*[239] where Lord Bridge stated:

> Whilst recognizing, of course, the importance of the underlying general principles common to the whole field of negligence, I think the law has now moved in the direction of attaching greater significance to the more traditional categorization of distinct and recognizable situation, as guides to the existence, the scope and the limits of the varied duties of care which the law imposes.[240]

---

[233] *Supra.* n. 1.

[234] Evans, How Fit is the Pilot, Doctor? *Supra* note 235 at p. 9.

[235] *Ibid.*

[236] See Stanton, Incremental Approaches to the Duty of Care, Chapter 2 of Mullany, *Torts in the Nineties*, Law Book Company:1997.

[237] (1985) 157 *CLR* 424.

[238] *Id.* 481.

[239] [1990] 2 *AC* 605.

[240] *Id.* at 618.

Therefore, there is no simple formula or touchstone to which recourse can be had in order to provide in every case a ready answer to questions as to whether the law will or will not grant recourse based on traditional rules of negligence. In this context, one wonders whether the use of such catch phrases as "reasonably skilled professional", without any attendant criteria to define the phrase is practical anymore. It would certainly be interesting if the "incrementalism" approach were to be applied along with established rules of law to instances of adjudication on negligence of a pilot in this context.

2.4 Meteorological Issues

On 12 May 2010, ICAO European and North Atlantic Volcanic Ash Task Force (EUR/NAT VATF),[241] established by ICAO in response to the volcanic eruption in Eyjafjallajökull, Iceland on 20 March 2010, agreed on a common working agenda to improve contingency plans in the European and North Atlantic Regions. The Task Force identified key areas for improving the efficiency of responses to volcanic ash emergencies, including harmonizing procedures used by North Atlantic States with those adopted in the European airspace, new requirements for services from the Volcanic Ash Advisory Centres, and the harmonization of format and content of aeronautical information on volcanic ash concentration. Karsten Theil, ICAO Regional Director for Europe and the North Atlantic region observed at the meeting:

> The fact that all concerned parties are involved in the discussions shows a high level of resolve to propose options that will significantly improve responses to future volcanic ash emergencies, with emphasis on both safety and efficiency.[242]

This agreement is part of ICAO's ongoing work on mitigating the effects of volcanic eruptions on aviation and is a particular response to seismic activity in Eyjafjallajökull, Iceland which commenced at the end of 2009, and resulted in its eruption.[243] The resultant volcanic ash which spewed over European airspace

---

[241] The group was convened to establish a coordinated region-wide operational approach by aviation to volcanic ash emergencies.

[242] http://icaopressroom.wordpress.com/.

[243] The eruption was followed by a large "explosivity" on 14 April 2010, which spewed ash 30,000 feet into the air. The unique damaging effect of this eruption was that the volcano was covered with a large mass of ice which melted spontaneously, transforming the lava and ice into aerosol form, which stayed airborne for days and travelled long distances. Although volcanic eruptions are classified in numerous ways according to the individual context in which it is referred to, the word commonly used in aviation parlance is "explosivity" which provides an evaluation of the volume of volcanic ash released and the likely height of the column. See *Manual on Volcanic Ash, Radioactive Material and Toxic Chemical Clouds*, ICAO Doc 9691, AN/954 Second Edition, 2007 at I-1-1.

2.4 Meteorological Issues

caused much of airspace to be closed from 15 April for a week.[244] Airspace was finally opened on 21 April after a week of aircraft on the ground and hundreds of thousands of passengers being stranded both in Europe and across the world. It was estimated that the disruption cost airlines at least $200 million a day and $184 million in airport revenue was lost due to the closure of airspace. More than 6.8 million passengers were affected.[245]

The International Airways Volcano Watch (IAVW) of ICAO provides international arrangements for the monitoring of volcanic ash in the atmosphere and for providing warnings to the aviation community, in accordance with Annex 3[246] to the Convention on International Civil Aviation[247] to the Convention in International Civil Aviation. The IAVW[248] is based on the cooperation of aviation and non-aviation operational units using information derived from observing sources and networks that are provided by States. The IAVW is coordinated by ICAO with the cooperation of other concerned international organizations. There are nine Volcanic Ash Advisory Centres (VAAC) that constitute IAVW, and each are strategically located around the world. These centres are responsible for coordinating and disseminating information on volcanic ash that may endanger aircraft which would fly through the ash cloud formed from a volcanic eruption. Each individual VAAC is operated as part of national weather forecasting organizations of the country in which it is located.

IAVW was formed by ICAO largely as a response to a serious incident of ingestion of volcanic ash which occurred when British Airways Flight 9, sometimes referred to as the Speedbird 9 or Jakarta incident, flew into a cloud of volcanic ash thrown up by the eruption of Mount Galunggung (circa 180 km (110 mi) south-east

---

[244] All civilian air traffic was grounded across the United Kingdom on 15 April 2010. See Volcanic ash cloud grounds flights over UK, *The Air Letter*, No. 16,967, Thursday, 15 April 2010 at 2. The problem was unprecedented as, unlike in earlier volcanic eruptions elsewhere, this eruption affected airspace over some European States which is usually congested. In Britain alone, there are approximately 5,000 daily inbound and outbound flights, with 1,300 flights and 180,000 passengers going through London Heathrow. See Airlines set to lose more than 100 million Pounds in ash chaos, *The Air Letter*, No. 16, 968, Friday, 16 April 2010 at 1.

[245] Some have attributed the opening of airspace to intense lobbying from the airlines which complained that they had lost $1 billion over the week of closure and cancelled 80,000 flights. See Airlines argue case, Airspace opens, *The Air Letter*, No. 16,970, Tuesday 20 April 2010 at 1.

[246] *Meteorological Service for International Air Navigation.*

[247] Convention on International Civil Aviation, signed at Chicago on 7 December 1944. See Doc 7300/9 Ninth Edition, 2006.

[248] The IAVW Operations Group (IAVWOPSG) was established in 2002 by the ICAO MET Divisional Meeting. The IAVWOPSG assists ICAO in the coordination and development of the IAVW. The IAVWOPSG is composed of experts from nine regional Volcanic Ash Advisory Centres (VAACs), other user States, the International Atomic Energy Agency (IAEA), the International Air Transport Association (IATA), the International Federation of Air Line Pilots' Associations (IFALPA), the International Union of Geodesy and Geophysics (IUGG) (covering the World Organization of Volcano Observatories) and the World Meteorological Organization (WMO).

of Jakarta, Indonesia), resulting in the failure of all four engines. BA 9 was a scheduled British Airways flight from London Heathrow to Auckland, with stops in Bombay, Madras, Kuala Lumpur, Perth, and Melbourne. On 24 June 1982, the route was flown by the City of Edinburgh, a 747-236B. The reason for the failure was not immediately apparent to the crew or ground control. The aircraft was diverted to Jakarta in the hope that enough engines could be restarted to allow it to land there. The aircraft was able to glide far enough to exit the ash cloud, and all engines were restarted (although one failed again soon after), allowing the aircraft to land safely.[249]

2.4.1 *ICAO's Work on Mitigating the Effects of Volcanic Ash on Aviation*

There is no room for doubt that the IAVW system, which was established by ICAO in coordination with the World Meteorological Organization (WMO), proved effective in ensuring the safety of air transport following the eruption of the volcano Eyjafjalla in Iceland on 14 April 2010. No incidents or accidents were reported due to volcanic ash, although initially disruptions in air traffic were being experienced in the United Kingdom and Scandinavia. It is quite obvious that the economic cost of volcanic ash to international civil aviation is staggering, involving complete engine changes, engine overhauls, airframe refurbishing, window re-polishing and/or replacement and pitot-static system repair, etc. Combined with maintenance downtimes, delays and rerouting issues, as well as volcanic ash effects to airport equipment and buildings, estimates generally put the cost of ash to aviation well in excess of $250 million since 1982. Given the safety and economic implications of volcanic ash to aircraft operations, it is necessary to maintain the IAVW facilities much in the same way that airport fire services are maintained.

The IAVW system is designed to detect and track the movement of volcanic ash in the atmosphere and to warn aircraft in flight about this hazard. The damage that can be caused by volcanic ash to the engines of an aircraft are significant, since volcanic ash is composed of very abrasive silica materials, and can therefore damage the airframe and flight surfaces, clog different systems, abrade cockpit windows and flame-out jet engines constituting a serious safety hazard. Volcanic ash can also have a serious effect on aerodromes located downwind of a volcanic ash plume since it contaminates runways, ground equipment and aircraft parked or taxiing around the aerodrome.

---

[249] A nearly identical incident occurred on 15 December 1989 when KLM Flight 867, a B747-400 plying from Amsterdam to Anchorage, Alaska, flew into the plume of the erupting Mount Redoubt, causing all four engines to fail due to compressor stall. Once the flight cleared the ash cloud, the crew was able to restart each engine and then make a safe landing at Anchorage.

2.4 Meteorological Issues

The main components of the IAVW system are comprised of nine volcanic ash advisory centres responsible for the provision of information related to areas and flight levels affected by volcanic ash and its future movement. These centres are strategically distributed around the globe. They are Anchorage, Buenos Aires, Darwin, London, Montreal, Tokyo, Toulouse, Washington and Wellington. Another important component of the system is the role of area control centres of each State to warn aircraft in flight about the location of the volcanic ash. They work together with selected State Volcano Observatories and the corresponding meteorological watch offices that are tasked with disseminating the corresponding warnings which are called SIGMETs.[250]

In response to the volcanic eruption in Eyjafjallajökull, the Council of ICAO met on 19 April 2010 to review the situation and welcomed the decision by the Ministers of Transport of the European Union (EU) to gradually reopen the European airspace, in a safe and coordinated manner. The following day, the Organization's Air Navigation Commission considered near-term initiatives to advance the science of aviation safety and airspace contaminated by volcanic ash. These meetings were held against the backdrop of existing ICAO Standards as prescribed in Annex 3[251] to the Chicago Convention which provides that a member State of ICAO, having accepted consequent upon a regional air navigation agreement the responsibility for providing a Volcanic Ash Advisory Centre (VAAC) within the framework of the international airways volcano watch, is required to arrange for that centre to respond to a notification that a volcano has erupted or is expected to erupt or volcanic ash is reported in the area of responsibility, by arranging for that centre to monitor relevant geo-stationary and polar orbiting satellite data. The purpose of this provision is to ensure the detection of the existence and extent of volcanic ash in the atmosphere in the area concerned and activate the volcanic ash numerical trajectory/dispersion model in order to forecast the movement of any ash "cloud "which has been detected or reported. The provision in the Annex also requires the State concerned to issue advisory information regarding the extent and forecast movement of the volcanic ash cloud and issue

---

[250] SIGMET, or Significant Meteorological Information, is a weather advisory that contains meteorological information concerning the safety of all aircraft. There are two types of SIGMETs, convective and non-convective. The criteria for a non-convective SIGMET to be issued are severe or greater turbulence over a 3,000-square-mile (7,800 km^2) area, or severe or greater icing over a 3,000-square-mile (7,800 km^2) area or IMC conditions over a 3,000-square-mile (7,800 km^2) area due to dust, sand or volcanic ash.

[251] Annex 3, *Meteorological Service for International Air Navigation* Sixteenth Edition, July 2007. The object of the meteorological service outlined in Annex 3 is to contribute to the safety, efficiency and regularity of air navigation. This is achieved by providing necessary meteorological information to operators, flight crew members, air traffic services units, search and rescue units, airport management and others concerned with aviation. Close liaison is essential between those supplying meteorological information and those using it.

updated advisory information to the meteorological watch offices, area control centres flight information centres and VAACs.[252]

There is also an obligation for member States of ICAO that maintain volcano observatories monitoring active volcanoes to arrange that selected State volcano observatories, as designated by regional air navigation agreement, observing: significant pre-eruption volcanic activity, or a cessation thereof; a volcanic eruption, or a cessation thereof; and/or volcanic ash in the atmosphere to send that information as quickly as practicable to its associated area coordination centre (ACC), meteorological watch office (MWO) and VAAC[253].Cloud amount, cloud type and height of cloud base are observed and reported as necessary to describe the clouds of operational significance. When the sky is obscured, vertical visibility is to be observed and reported, where measured, in lieu of cloud amount, cloud type and height of cloud base. The height of cloud base and vertical visibility is required to be reported in metres (or feet).[254] Once the Volcanic Ash Advisory Centres detect that an eruption hastaken place and can discern the geographic and meteorological features of the event in question, they generate a forecast of how it is going to disperse into the atmosphere. Standard advisories include eruption location, the specific volcano producing the event, wind speed and direction and, most importantly, flight levels affected by ash so that aircraft en route or scheduled to depart can be advised to fly well-above or around the affected areas.

The Annex recommends that the occurrence of pre-eruption volcanic activity, volcanic eruptions and volcanic ash cloud should be reported without delay to the associated air traffic services unit, aeronautical information services unit and meteorological watch office.[255] It also calls for special observations by aircraft whenever volcanic ash *inter alia* is encountered or observed[256] and requires that each meteorological authority, on request and to the extent practicable, shall make available to any other meteorological authority, to operators and to others

---

[252] Annex 3, *id,* at Standard 3.5. During the past few years a number of incidents have occurred due to aircraft encounters with volcanic ash clouds following volcanic eruptions. In order to provide for the observation and reporting of volcanic ash clouds and the issuance of warnings to pilots and airlines, ICAO, with the assistance of other international organizations, has established an international airways volcano watch (IAVW). The cornerstones of the IAVW are nine volcanic ash advisory centres which issue advisory information on volcanic ash globally, both to aviation users and meteorological offices concerned.

[253] *Id.* Standard 3.6.

[254] *Id* Standard 4.6.5.

[255] *Id.* Recommended Practice 4.8. The report should be made in the form of a volcanic activity report comprising the following information in the order indicated: (a) message type, Volcanic Activity Report; (b) station identifier, location indicator or name of station; (c) date/time of message; (d) location of volcano and name if known; and (e) concise description of event including, as appropriate, level of intensity of volcanic activity, occurrence of an eruption and its date and time, and the existence of a volcanic ash cloud in the area together with direction of ash cloud movement and height.

[256] *Id.* Standard 5.5.

2.4 Meteorological Issues

concerned with the application of meteorology to international air navigation, meteorological observational data required for research, investigation or operational analysis.[257] States are required to make available suitable telecommunications facilities to permit aerodrome meteorological offices and, as necessary, aeronautical meteorological stations to supply the required meteorological information to air traffic services units on the aerodromes for which those offices and stations are responsible, and in particular to aerodrome control towers, approach control units and the aeronautical telecommunications stations serving these aerodromes.[258]

The Chicago Convention, in Article 9 provides that each contracting State may, for reasons of military necessity or public safety, restrict or prohibit uniformly the aircraft of other States from flying over certain areas of its territory, provided that no distinction in this respect is made between the aircraft of the State whose territory is involved, engaged in international scheduled airline services, and the aircraft of the other contracting States likewise engaged. Such prohibited areas would be of reasonable extent and location so as not to interfere unnecessarily with air navigation. These prohibited areas and their descriptions have to be notified to ICAO and other member States of ICAO.[259] Furthermore each member State of ICAO is entitled to reserve the right, in exceptional circumstances or during a period of emergency, or in the interest of public safety, and with immediate effect, temporarily to restrict or prohibit flying over the whole or any parts of its territory, provided that such a restriction does not discriminate as to nationality of aircraft of other member States.[260]

At the Task Force meeting, which was mentioned at the outset of this paper, the view was put forward that the initial eruption and ash contamination severely disrupted air operations throughout the UK and north-western Europe. The "zero tolerance" of ash from the ICAO guidance led directly to a sequence of decisions that reduced air traffic flow through UK, Irish and Continental European airspace to a "zero rate". This "zero rate" was applied in those sectors identified as falling within the London Volcanic Ash Advisory Centre (VAAC) contaminated area (the level at this stage, the red line of the VAAC charts, equates to concentrations of ash greater than 2×10^{-4} g/m^3). The imposition of a "zero tolerance" was questioned as to whether it was necessary to maintain flight safety.

The meeting was advised that, on 16 April, the UK CAA hosted the first of a series of teleconferences with major aircraft and engine manufacturers, Air Traffic Service experts, airspace managers along with leading scientific experts in fields of

---

[257] *Id.* Standard 8.4. Furthermore, the meteorological authority is required to designate a meteorological office to be associated with each air traffic services unit. The associated meteorological office shall, after coordination with the air traffic services unit, supply, or arrange for the supply of, up-to-date meteorological information to the unit as necessary for the conduct of its functions.
[258] See Standard 11.1.1.
[259] Chicago Convention, *supra* note 3, Article 9a).
[260] *Id.* Article 9b).

meteorology, geology and volcanology. The outcome of these teleconferences resulted in agreement from the aircraft engine manufacturers to redefine a tolerable ash density level of 2×10^{-3} g/m^3. The manufacturers group also considered how the tolerable level could be used to aid flight planning. This was augmented by the ad-hoc teleconference, 19 April 2010, which included Members of the provisional Council and CEOs of the ANSPs, and effectively lead to the adoption of this new tolerance level and resultant "no-fly zone" with a buffer of 60 nautical miles, throughout European airspace.[261]

This in effect had raised the tolerance level of ash concentration one order of magnitude greater than the defined ICAO limit. The result being a three-area approach, which consisted of:

a) An area of no contamination and no flying restrictions;
b) An area defined being from the ICAO threshold to the new defined threshold of 2×10^{-3} g/m^3, which requires operators to implement enhanced procedures; and,
c) An area above the tolerance of 2×10^{-3} g/m^3 plus a buffer, which was deemed not safe for air traffic operations.

Thus in UK and European airspace, operations were occurring, where previously a "zero rate" had been applied.[262]

This raises the fundamental issue as to whether ICAOs role in assisting States in a crisis situation concerning the presence of volcanic ash in the airways used by aircraft is a prescriptive one. In other words, does ICAO have a leadership role to bail out States, or for those matter airlines from a natural disaster? Does ICAO have a mandate to prescribe rules of conduct for States? The answer has to be a resounding "no" for the following reasons.

a) The decision to close sovereign airspace is not a prerogative of ICAO. It is a matter for the States concerned. The Chicago Convention, in Article 9 provides that each contracting State may, for reasons of military necessity or public safety, restrict or prohibit uniformly the aircraft of other States from flying over certain areas of its territory, provided that no distinction in this respect is made between the aircraft of the State whose territory is involved, engaged in international scheduled airline services, and the aircraft of the other contracting States likewise engaged. Such prohibited areas would be of reasonable extent and location so as not to interfere unnecessarily with air navigation. These prohibited areas and their descriptions have to be notified to ICAO and other member States of ICAO.[263] Furthermore each member State of ICAO is entitled to reserve the right, in exceptional circumstances or during a period of emergency, or in the interest of public safety, and with immediate effect, temporarily

---

[261] NATSPG Derogation to Volcanic Ash Contingency Plans, EUR/NAT VATF/1 – WP/03, 11/05/2010, at 2.
[262] Ibid.
[263] Chicago Convention, supra note 3, Article 9a).

2.4 Meteorological Issues

to restrict or prohibit flying over the whole or any parts of its territory, provided that such a restriction does not discriminate as to nationality of aircraft of other member States.[264]

b) ICAO's Mission and Vision Statements, as adopted and endorsed by the Council of ICAO, which is a "club" of member States of the Organization, clearly states that ICAO acts as a global forum for civil aviation and that ICAO works to achieve its vision of safe, secure and sustainable development of civil aviation through the cooperation of its member States. The fact that ICAO's main objective is to ensure cooperation of its member States to achieve its objectives is evidence that without such cooperation ICAO would be destitute of any authority.

c) ICAO's contribution is largely manifested in its guidelines, contained in the *Manual on Volcanic Ash, Radioactive Material and Toxic Chemical Clouds*[265] which is a set of guidelines developed under the auspices of the Air Navigation Commission of the ICAO Council with a view to assisting States in the dissemination of information on volcanic ash to pilots and the development of contingency arrangements for the diversion of aircraft around affected areas.

d) In Annex 3 to the Chicago Convention clear responsibility devolves upon the member States of ICAO. Standard 2.1.1. provides that the objective of meteorological service for international air navigation, (which has to be provided by the State under Article 28 a) of the Chicago Convention[266]) shall be to contribute towards the safety, regularity and efficiency of international air navigation. This objective has to be achieved by supplying the following users: operators, flight crew members, air traffic services units, search and rescue services units, airport managements and others concerned with the conduct or development of international air navigation, with the meteorological information necessary for the performance of their respective functions.[267]

The above notwithstanding, it is quite evident that ICAO has played its expected role by convening meetings and providing guidance material, including the establishment of the IAVW system in collaboration with the WMO. As far back as 7 November 1963, a Flight Safety Foundation Award for Distinguished Service was awarded to ICAO personnel for weather ships which manned ICAO's Ocean Station Network in the North Atlantic, for assistance in aviation throughout the world.[268] Fast forwarding to more recent activity, the new Volcanic Ash Task

---

[264] *Id.* Article 9b).

[265] Doc 9691 AN/954, Second Edition 2007.

[266] Article 28 a) provides that Each Contracting State undertakes, as far as it may find practicable, to provide in its territory *inter alia*, airports, radio services, meteorological services and other air navigation facilities.

[267] Standard 2.1.2.

[268] http://www.icao.int/icao/en/nr/1963/pio196317_e.pdf. The network consisted of nne floating ocean stations manned through an agreement coordinated by ICAO and participated in by 19 stations whose aircraft flew across the North Atlantic Ocean.

Force[269] created by ICAO, will by 1 August 2010, deliver a report on the collapse of European airspace in the days following the first eruption. This Report will identify contingency plans and a framework for action if similar events unfold, and will be followed by a roadmap for establishing globally harmonized ash concentration thresholds, options for improved detection systems of volcanic ash, as well as recommendations to improve notification and warning systems that should be completed by May 2011.[270]

Finally, it must be noted that international organizations such as ICAO can generally only work on the basis of legal powers that are attributed to them. Presumably, these powers emanate from the sovereign States that form the membership of such organizations.[271] Therefore, the logical conclusion is that if ICAO were to act beyond the powers accorded to it, ICAO would be presumed to act *ultra vires*.[272] With regard to the conferral of powers by States to ICAO, States have followed the classic approach of doing so through an international treaty. However, neither is there explicit mention of such a conferral on ICAO in the Chicago Convention nor is there any description of ICAO's powers, except for an exposition of ICAO's aims and objectives. One can only draw some link between the status of ICAO and the opinion of the International Court of Justice in the 1996 *WHO Advisory Opinion* case[273] which was that the powers conferred on international organizations are normally the subject of express statement in their constituent instruments.[274] Considering that ICAO has, as according to one of its aims and objects, has to ensure the safe and orderly growth of international civil aviation throughout the world,[275] it cannot be disputed that in response to the Eyjafjallajökull volcanic eruption and its fallout, ICAO has amply played its constituent role.

2.4.2 Some Liability Issues

The fallout from the Icelandic volcano in April 2010 brought to bear yet another burden on the airline industry and issues and questions on responsibility for

---

[269] The task force will comprise representatives from Argentina, Australia, Canada, France, Japan, New Zealand, the U.K. and the U.S., where the world's nine Volcanic Ash Advisory Centers are located, as well as Brazil, for its role in commercial aircraft production, and Spain, which currently holds the European Union presidency.

[270] http://www.aviationweek.com/aw/generic/story_channel.jsp?channel=comm&id=news/avd/2010/04/30/06.xml.

[271] See de Witte (1998), at pp. 277–304.

[272] Klabbers (2002), at p. 60.

[273] *Legality of the Threat or Use of Nuclear Weapons, Advisory Opinion*, ICJ Reports, 1996, p. 64.

[274] *Id.* p. 79.

[275] Chicago Convention, *supra*, note 3, Article 44 a).

stranded passengers worldwide. The most curious outcome of the closure of airspace over some European States and the subsequent grounding of flights worldwide, resulting in hundreds of thousands of passengers being left to their own devices in instances where the airlines did not, or could not provide assistance, was the noticeable silence on the part of States concerned with regard to their responsibilities for travelling citizens. Although under ordinary circumstances airlines are liable to look after their passengers, particularly in circumstances within their control, the volcano was completely unexpected and air carriers were faced with State action which closed the air space of several European States.

Seismic activity in Eyjafjallajökull, Iceland commenced at the end of 2009, resulting in an eruption on 20 March 2010. This was followed by a large "explosivity"[276] on 14 April, which spewed ash 30,000 feet into the air. The unique damaging effect of this eruption was that the volcano was covered with a large mass of ice which melted spontaneously, transforming the lava and ice into aerosol form, which stayed airborne for days and travelled long distances. This in turn caused much of airspace over European States to be closed from 15 April for a week.[277] Airspace was finally opened on 21 April after a week of aircraft on the ground and hundreds of thousands of passengers being stranded both in Europe and across the world. The disruption cost airlines at least $ 200 million a day and $184 million in airport revenue was lost due to the closure of airspace. More than 6.8 million passengers were affected. Some have attributed the opening of airspace to intense lobbying from the airlines which complained that they had lost $1 billion over the week of closure and cancelled 80,000 flights.[278] The International Air Transport Association (IATA) was critical of European governments for what it called a decision to shut down airspace without proper risk assessment, consultation, coordination and leadership,[279] implicitly bringing to bear some justification for possible claims for compensation from stranded passengers.

The plumes of a volcano are filled with abrasive silica based material which could cause engines to be clogged and windscreens to be sandblasted. In addition, pulverized rock in the volcanic emission is powerful enough to strip away paint,

---

[276] Although volcanic eruptions are classified in numerous ways according to the individual context in which it is referred to, the word commonly used in aviation parlance is "explosivity" which provides an evaluation of the volume of volcanic ash released and the likely height of the column. See *Manual on Volcanic Ash, Radioactive Material and Toxic Chemical Clouds*, ICAO Doc 9691, AN/954 Second Edition, 2007 at I-1-1.

[277] All civilian air traffic was grounded across the United Kingdom on 15 April 2010. See Volcanic ash cloud grounds flights over UK, *The Air Letter*, No. 16,967, Thursday, 15 April 2010 at 2. The problem was unprecedented as, unlike in earlier volcanic eruptions elsewhere, this eruption affected airspace over some European States which is usually congested. In Britain alone, there are approximately 5,000 daily inbound and outbound flights, with 1,300 flights and 180,000 passengers going through London Heathrow. See Airlines set to lose more than 100 million Pounds in ash chaos, *The Air Letter*, No. 16, 968, Friday, 16 April 2010 at 1.

[278] Airlines argue case, Airspace opens, *The Air Letter*, No. 16,970, Tuesday 20 April 2010 at 1.

[279] http://www.airtransportnews.aero/print_article.pl?id=23057.

disrupt the airstream and seriously hinder or block speed sensors. An Airbus Industrie flight operations briefing note states that volcanic particles have a lower melting point than the temperature in a turbine engine of an aircraft in flight. Therefore the entry of such particles into an aircraft engine could cause them to melt, clogging turbine vanes and disturbing high flow turbine gases, eventually stalling the engines.[280] At high altitudes volcanic ash could cause what is known as "St. Elmo's fire", a bright blue or violet glow, appearing like fire in some circumstances, from tall, sharply pointed structures such as aircraft wings.[281]

One of the landmark events of an aircraft being affected by volcanic ash occurred on 24 June 1982 when a British Airways B747 aircraft lost power on all its engines while flying at an altitude of 37,000 feet on its way from Kuala Lumpur to Perth.[282] For 16 min, the aircraft descended with no power from 37,000 feet to 12,000 feet at which point the pilot was able to rejuvenate three of the aircraft's engines and land safely in Jakarta. The intense inspection and investigation which followed, involving civil aviation authorities, engine manufacturers and British Airways, revealed that the airframe and engines were seemingly "sandblasted" in the wing and engine inlet surfaces, the radome[283] and cockpit windows. A nearly identical incident occurred on 15 December 1989 when KLM Flight 867, a B747-400 plying from Amsterdam to Anchorage, Alaska, flew into the plume of the erupting Mount Redoubt, causing all four engines to fail due to compressor stall. Once the flight cleared the ash cloud, the crew was able to restart each engine and then make a safe landing at Anchorage.

The Chicago Convention, in Article 9 provides that each contracting State may, for reasons of military necessity or public safety, restrict or prohibit uniformly the aircraft of other States from flying over certain areas of its territory, provided that no distinction in this respect is made between the aircraft of the State whose territory is involved, engaged in international scheduled airline services, and the aircraft of the other contracting States likewise engaged. Such prohibited areas would be of reasonable extent and location so as not to interfere unnecessarily with

---

[280] http://www.bloomberg.com/apps/news?pid=20601110&sid=ar0RLkiY9Dkc.

[281] The phenomenon is scientifically known as a *corona* or *point discharge*. It occurs on objects, especially pointed ones, when the electrical field potential strength reaches about one thousand volts per centimetre.

[282] British Airways Flight 9, sometimes referred to as the Speedbird 9 or Jakarta incident, was a scheduled British Airways flight from London Heathrow to Auckland, with stops in Bombay, Madras, Kuala Lumpur, Perth, and Melbourne. On 24 June 1982, the route was flown by the City of Edinburgh, a 747-236B. The aircraft flew into a cloud of volcanic ash thrown up by the eruption of Mount Galunggung (circa 180 kilometres (110 mi) south-east of Jakarta, Indonesia), resulting in the failure of all four engines. The reason for the failure was not immediately apparent to the crew or ground control. The aircraft was diverted to Jakarta in the hope that enough engines could be restarted to allow it to land there. The aircraft was able to glide far enough to exit the ash cloud, and all engines were restarted (although one failed again soon after), allowing the aircraft to land safely.

[283] A radome is a structural, weatherproof enclosure that protects a microwave or radar antenna.

2.4 Meteorological Issues

air navigation. These prohibited areas and their descriptions have to be notified to ICAO[284] and other member States of ICAO. Furthermore each member State of ICAO is entitled to reserve the right, in exceptional circumstances or during a period of emergency, or in the interest of public safety, and with immediate effect, temporarily to restrict or prohibit flying over the whole or any parts of its territory, provided that such a restriction does not discriminate as to nationality of aircraft of other member States.[285]

Annex 3[286] to the Chicago Convention provides that a member State of ICAO, having accepted consequent upon a regional air navigation agreement the responsibility for providing a Volcanic Ash Advisory Centre (VAAC) within the framework of the international airways volcano watch, is required to arrange for that centre to respond to a notification that a volcano has erupted or is expected to erupt or volcanic ash is reported in the area of responsibility, by arranging for that centre to monitor relevant geo-stationary and polar orbiting satellite data. The purpose of this provision is to ensure the detection of the existence and extent of volcanic ash in the atmosphere in the area concerned and activate the volcanic ash numerical trajectory/dispersion model in order to forecast the movement of any ash "cloud" which has been detected or reported. The provision in the Annex also requires the State concerned to issue advisory information regarding the extent and forecast movement of the volcanic ash cloud and issue updated advisory information to the meteorological watch offices, area control centres flight information centres and VAACs.[287]

There is also an obligation for member States of ICAO that maintain volcano observatories monitoring active volcanoes to arrange that selected State volcano observatories, as designated by regional air navigation agreement, observing: significant pre-eruption volcanic activity, or a cessation thereof; a volcanic eruption, or a cessation thereof; and/or volcanic ash in the atmosphere to send that information as quickly as practicable to its associated area coordination centre (ACC), meteorological watch office (MWO) and VAAC.[288] Cloud amount, cloud

---

[284] Supra, note 1.

[285] Id. Article 9b).

[286] Annex 3, *Meteorological Service for International Air Navigation* Sixteenth Edition, July 2007. The object of the meteorological service outlined in Annex 3 is to contribute to the safety, efficiency and regularity of air navigation. This is achieved by providing necessary meteorological information to operators, flight crew members, air traffic services units, search and rescue units, airport management and others concerned with aviation. Close liaison is essential between those supplying meteorological information and those using it.

[287] Annex 3, id. at Standard 3.5. During the past few years a number of incidents have occurred due to aircraft encounters with volcanic ash clouds following volcanic eruptions. In order to provide for the observation and reporting of volcanic ash clouds and the issuance of warnings to pilots and airlines, ICAO, with the assistance of other international organizations, has established an international airways volcano watch (IAVW). The cornerstones of the IAVW are nine volcanic ash advisory centres which issue advisory information on volcanic ash globally, both to aviation users and meteorological offices concerned.

[288] Id. Standard 3.6.

type and height of cloud base shall be observed and reported as necessary to describe the clouds of operational significance. When the sky is obscured, vertical visibility is to be observed and reported, where measured, in lieu of cloud amount, cloud type and height of cloud base. The height of cloud base and vertical visibility is required to be reported in metres (or feet).[289]

The Annex recommends that the occurrence of pre-eruption volcanic activity, volcanic eruptions and volcanic ash cloud should be reported without delay to the associated air traffic services unit, aeronautical information services unit and meteorological watch office.[290] It also calls for special observations by aircraft whenever volcanic ash *inter alia* is encountered or observed[291] and requires that each meteorological authority, on request and to the extent practicable, shall make available to any other meteorological authority, to operators and to others concerned with the application of meteorology to international air navigation, meteorological observational data required for research, investigation or operational analysis.[292] States are required to make available suitable telecommunications facilities to permit aerodrome meteorological offices and, as necessary, aeronautical meteorological stations to supply the required meteorological information to air traffic services units on the aerodromes for which those offices and stations are responsible, and in particular to aerodrome control towers, approach control units and the aeronautical telecommunications stations serving these aerodromes.[293]

2.4.3 State Liability

The first issue is whether a State is liable to compensate an airline for loss of business caused as a result of the closure of airspace. IATA's claim that the European Union closed airspace over European States without logical reason and proper evaluation of the threat posed by volcanic ash in airspace over European

---

[289] *Id* Standard 4.6.5.

[290] *Id.* Recommended Practice 4.8. The report should be made in the form of a volcanic activity report comprising the following information in the order indicated: (a) message type, Volcanic Activity Report; (b) station identifier, location indicator or name of station; (c) date/time of message; (d) location of volcano and name if known; and (e) concise description of event including, as appropriate, level of intensity of volcanic activity, occurrence of an eruption and its date and time, and the existence of a volcanic ash cloud in the area together with direction of ash cloud movement and height.

[291] *Id.* Standard 5.5.

[292] *Id.* Standard 8.4. Furthermore, the meteorological authority is required to designate a meteorological office to be associated with each air traffic services unit. The associated meteorological office shall, after coordination with the air traffic services unit, supply, or arrange for the supply of, up-to-date meteorological information to the unit as necessary for the conduct of its functions.

[293] See Standard 10.1.1.

2.4 Meteorological Issues

States hints at such liability. Giovanni Bisignani, Director General and CEO of IATA[294] said:

> We are far enough into this crisis to express our dissatisfaction on how governments have managed it - with no risk assessment, no consultation, no coordination, and no leadership. This crisis is costing airlines at least $200 million a day in lost revenues and the European economy is suffering billions of dollars in lost business. In the face of such dire economic consequences, it is incredible that Europe's transport ministers have taken five days to organize a teleconference.[295]

British Airways requested financial compensation from the European Union and the British Government for the closure of airspace, basing its claim on the fact that its test flight through the designated no-fly zone revealed that there were no variations in normal operational performance.[296] British Airways Chief Executive Willie Walsh is alleged to have said:

> We believe airlines are best positioned to assess all information and determine what, if any, risk exists to aircraft, crew and passengers.[297]

This statement has to be tested both for its justification and verity in the light of the paramount duty of a State to ensure the safety of flight and, *a fortiori*, the safety of crew and passengers. In addition to the earlier discussion in this article on the responsibility and right of a State to close airspace to all aircraft in the case of an emergency (under the Chicago Convention), there is incontrovertible recognition that it is the State concerned that is primarily responsible for aviation safety. Regulation (EC) 261/2004 of the European Parliament and Council recognizes that action by the European Community in the field of air transport should aim, among other things, at ensuring a high level of protection for passengers, and moreover, full account should be taken of the requirements of consumer protection.[298] From a public international law perspective, Article 28 of the Chicago Convention obligates contracting States to provide in their territories airports, radio services, meteorological services and other air navigation facilities to facilitate international air navigation, in accordance with the standards and practices established from time to time pursuant to the Convention.[299]

---

[294] For a description of IATA see *supra* note 85. The mission of IATA is to promote safe, reliable and secure air services for the benefit of the peoples of the world; provide means of collaboration among airlines engaged directly or indirectly in air transport; and cooperate with the International Civil Aviation Organization and other relevant organizations. See *Manual on the Regulation of International Air Transport*, ICAO Doc 9626, Second Edition −2004, at 3.8-1.

[295] http://airtransportnews.aero/print_article.pl?id=23057.

[296] BA seeks compensation for ash chaos, *The Air Letter*, Monday 19 April 2010, No. 16,969 at 2. British Airways claimed that it cost $23-30 million a day as a result of the closure of airspace.

[297] *Ibid.*

[298] Regulation (EC) No 261/2004 of the European Parliament and of the Council of 11 February 2004 establishing common rules on compensation or long delay of flights repealed Regulation (EEC) No 295/91 *Whereas* clause No. 1.

[299] Chicago Convention, *supra*, note 3, Article 28 a).

Annex 3 to the Chicago Convention provides in Standard 2.1.1. that the objective of meteorological service for international air navigation shall be to contribute towards the safety, regularity and efficiency of international air navigation. This objective has to be achieved by supplying the following users: operators, flight crew members, air traffic services units, search and rescue services units, airport managements and others concerned with the conduct or development of international air navigation, with the meteorological information necessary for the performance of their respective functions.[300]

The Annex, in Standard 2.1.3 calls for each Contracting State to determine the meteorological service which it will provide to meet the needs of international air navigation. This determination has to be made in accordance with the provisions of this Annex and with due regard to regional air navigation agreements; The State is required to include the determination of the meteorological service to be provided for international air navigation over international waters and other areas which lie outside the territory of the State concerned. Furthermore, each contracting State is required to designate the authority, hereinafter referred to as the meteorological authority, to provide or to arrange for the provision of meteorological service for international air navigation on its behalf. Details of the meteorological authority so designated have to be included in the State aeronautical information publication.

Leaving aside the vexatious issue of who has the right to decide to operate aircraft over the skies of a State, there is the question as to whether the United Kingdom government was liable in tort for the loss of legitimate business of the airline. This has to be viewed against the backdrop of IATA's statement that implied negligence. With regard to tortious liability of a State, the basic premise is that a public body acting *ultra vires* is as liable for an act committed beyond tis powers as is a private individual.[301] In the 1995 Case of *X (Minors)* v. *Bedfordshire CC*,[302] Lord Browne Wilkinson envisioned that there would be three possible causes of action where a plaintiff might bring a cause of action against a public entity: a breach of statutory duty without the necessity to prove carelessness; a common law breach of duty of care resulting from a breach of a statutory duty or a performance of a duty; and misfeasance in a public office. His Lordship ruled out any liability for carelessness in the performance of a duty imposed by statute. This leaves the question open as to whether there exists liability for an arbitrary or baseless exercise (as claimed) of a treaty provision – such as Article 9 of the Chicago Convention – on the part of a State or public authority.

Where a breach of statutory duty imposed on a State authority was an issue, the remedy in early times was predicated upon the premise that if a person responsible for the act in question had a statutory duty of care which was breached, an action lay

---

[300] Standard 2.1.2.
[301] *Entick* v. Carrington (1765) 19 St. Tr. 1029 at p. 1081.
[302] [1995] 2 A.C. 633 at 730.

2.4 Meteorological Issues
113

at the suit of the aggrieved person.[303] In the early twentieth Century, courts slightly adapted their fundamental approach to look at the intent of the statute on legislation particularly whether the law intended to create a cause of action for its breach. The overall judicial philosophy was that if there was room for awarding compensation under the principles of tort law, there would be no room for action under the statute itself.[304] Additionally, for there to be compensation for a breach of statutory duty, the Courts looked for a prescribed penalty within the Statute, together with a link between harm suffered and the risk which the statute intended to prevent.[305] A social interest statute enacted for the benefit of society at large would usually not give rise to a breach of duty action, unless negligence can be proved.[306] This principle was based on the fact that social legislation affects a class of persons and accommodation of an individual within that legislation was grounded on the discretion of the authority vested with power under the legislature concerned.[307]

As the *Bedfordshire*[308] case illustrated, no action would lie for mere carelessness in the performance of a statutory duty unless there is a specific common law right of action and that action is predicated upon a common law duty of care to be established by the plaintiff. Courts cannot impose their judgment on a discretion vested in an authority because such discretion is purely for the authority to exercise and decide upon. The discretion can only be impugned if the plaintiff shows manifest unreasonableness on the part of the authority vested with discretion, making the action fall outside the purview of the discretion.[309] In the 1990 case of *Caparo Industries plc* v. *Dickman*[310] the court set specific standards for determining whether a challenged decision fell outside a common law duty of care. They were: whether injury caused by the defendant was foreseeable; was there sufficient proximity between the parties; and whether the imposition of a duty of care on the authority was just, equitable and fair.[311]

In the 1996 case of *Stovin* v. *Wise*[312] the Court established that, on the issue of breach of statutory authority, the consideration of a public duty to act under statute was not the only criterion. There was also the consideration as to whether the action in question breaches a private law duty to act which would ground a claim in

---

[303] *Sterling* v. *Turner* (1672) I. Ventris 200. Rowning v. Goodchild (1772) 2 W. Black 906, Schinotti v. Bumstead (1796) 6. T.R. 646, Barry v. Arnaud, (1839) 10 Ad. & E. 646, *Pickering* v. *James* (1873) L.R. 8 C.P. 489.

[304] *Phillips* v. *Britannia Hygenic Laundry Co. Ltd.* [1923] 2 K. B. 832.

[305] *Cutler* v. *Wandsworth Stadium Ltd.* [1949] AC 398. See Also *Groves* v. *Lord Winborne* [1898] 2.Q.B.402 and *Goris* v. *Scott* [1875] L.R. 9 ex. 125.

[306] *Barrett* v. *Enfield* LBC [2001] 2.A.C. 550.

[307] *O'Rourke* v. *Campden LBC* [1998] A.C. 158.

[308] *Supra*, note 302.

[309] *Id* 736 A – 737 A.

[310] [1990] 2.A.C.605.

[311] *Id.* 611.

[312] [1996] A.C. 923.

damages.[313] An omission to exercise statutory power was actionable if the authority was proved to have been irrational in not exercising the power and that there was provision in the statute for the award of compensation if an action prescribed under statute was not taken. The reliance on policy explicit or implicit in a statute under question, that compensation would accrue to a plaintiff in the instance of a breach of statutory duty, was to be based on the fact that the purpose of the statute was to devolve responsibility on an authority whose actions within the statute would accomplish what members of a society could not accomplish by themselves.[314]

With regard to negligence, in the early case of *Mercy Docks and Harbour Board Trustees* v. *Gibbs*, decided in 1866, the court of first instance held that a public body could be held liable in negligence when exercising a statutory power. Blackburn J. rejected the argument of the defendant that a remedy lay only within statutory bounds, a decision which was upheld later by the House of Lords.[315] The duty of care lay pursuant to a statutory power but was not prescribed both in terms of content and compensability within it. However, the scope of the statute and the persons it was meant to protect is important.[316]

Discretion and negligence are not mutually exclusive.[317] The operational criterion is whether, in the exercise of discretion, there was a breach of a duty of care. This liability and consequent consideration that grounds an action based on a breach of a duty of care depends entirely on policy considerations whether it would be fair, just and reasonable to impose such a duty. Clearly on the basis of *Barrett*[318] and *Phelps*[319] one could conclude that a duty of care is owed by a State or instrumentality of that State (whether he is an agent of the government or a private body)[320] to both the operators of the aircraft as well as those who use the operator's services in the given instance.

---

[313] *Id.* 949–950.

[314] This policy was enunciated by Lord Hoffman in *Stovin*, supra, note 41 where His Lordship qualified the Australian Case *Sutherland Shire Council* v. *Heyman* (1985) C.L.R. 424 at 464 in which Mason J. established a doctrine of general reliance.

[315] (1866) L.R.1. H.L.

[316] Governors of the *Peabody Donation Fund* v. Sir *Lindsay Parkinson and Co. Ltd.* [1985] A.C. 210. Also, *Curran* v. *Northern Ireland Co-ownership Housing Association Ltd.* [1987] A.C.718.

[317] This is a principle applicable both in the United Kingdom and the United States. See *Johnson* v. *State of California*, 447 P. 2d. 352 (1968).

[318] In *Barrett* v. *Enfield LBC*, [2001] 2. A.C. 550, the claim for breach of statutory duty per se was not pursued before the Court of Appeal or the House of Lords, in the case of a local authority sued for negligence in caring for a child.

[319] . In *Phelps* v. *Hillingdon LBC* [2001] 2 A.C. 619 at 652, the House of Lords held that duties cast on local authorities in the context of specific education needs were for the benefit of all children in a particular geographic location or area and therefore did not come under an action for breach of statutory duty in the case of a specific person.

[320] Street (1953).

2.4.4 Air Carrier Liability to Passengers

Regulation 261/2004,[321] which does not apply to non-European carriers' flights into the European Union, but applies to passengers departing from an airport located in the territory of a member State,[322] recognizes that denied boarding and cancellation or long delay of flights cause serious trouble and inconvenience to passengers. It also acknowledges that passengers whose flights are cancelled should be able either to obtain reimbursement of their tickets or to obtain re-routing under satisfactory conditions and should be adequately cared for while awaiting a later flight.[323] However, the Regulation qualifies this provision with the recognition that obligations on operating carriers should be limited or excluded in cases where an event has been caused by extraordinary circumstances which could not have been avoided even if all reasonable measures had been taken. Meteorological conditions incompatible with the operation of a flight concerned is one of the extraordinary circumstances identified in the Regulation.[324] Extraordinary circumstances are deemed to exist where the impact of an air traffic management decision in relation to a particular aircraft on a particular day gives rise to a long delay, an overnight delay, or the cancellation of one or more flights by that aircraft, even though all reasonable measures had been taken by the air carrier concerned to avoid the delays or cancellations.[325] The Regulation further strengthens its thrust against non-liability in the case of adverse meteorological conditions when it provides that "denied boarding" means a refusal to carry passengers on a flight, although they have presented themselves for boarding under the conditions laid down in Article 3(2), except where there are reasonable grounds to deny them boarding, such as reasons of health, safety or security, or inadequate travel documentation.[326]

It must be noted that the "extraordinary circumstances" defense is not an absolute one. It must be accompanied by the fact that the extraordinary circumstances could not have been avoided even if all reasonable measures had been taken. In the instance of the closure of airspace due to the presence therein of volcanic ash, the maxim in tort *Res ipsa Loquitur* (facts speak for themselves) would apply and the airlines who cancelled their flights as a result of the closure need not go to any lengths defending the fact that the circumstances could not be

---

[321] *Supra*, note 27.

[322] *Id.* Article 3.1.a). The Regulation also applies to passengers departing from an airport located in a third country to an airport situated in the territory of a Member State to which the Treaty applies, unless they have received benefits or compensation and were given assistance in that third country, if the operating air carrier of the flight concerned is a Community carrier. See Article 3 1. b).

[323] *Id*, 14th *Whereas* clause.

[324] *Ibid.*

[325] Regulation 261/2004 *supra note 298*, see 15th Whereas clause.

[326] *Id.* Article 2 (j).

avoided. A similar provision exists in the Montreal Convention of 1999,[327] where Article 19 on compensation for delay provides that the carrier is liable for damage occasioned by delay in the carriage by air of passengers, baggage and cargo. Nevertheless, the carrier is not liable for damage occasioned by delay if it proves that it and its servants and agents took all measures that could reasonably be required to avoid the damage or that it was impossible for it or them to take such measures. As carriers found it impossible to resume their cancelled flights due to the closure of European airports following the volcanic eruption, there is no room for doubt that this provision would also exonerate them from any liability. This notwithstanding, it would be useful to examine the judicial interpretation of the term "extraordinary circumstances" in the light of the inability of the carrier to avoid delay or cancellation of flights.

In the case of *Wallentin-Hermann* v. *Alitalia*[328] where the plaintiff invoked Regulation 261/2004, the Court, which did not attempt to clarify the notion of "extraordinary circumstances", was of the view that the term should be considered as applying only to circumstances which are not inherent in the normal exercise of the activity of the air carrier concerned and are beyond the actual control of that carrier on account of its nature or origin. This clearly brings the circumstances caused by the Eyjafjallajökull volcano within the defence under Regulation 261/2004 and exempts the air carrier from liability. The Court noted that air carriers are often called upon as a matter of course in the exercise of their activity to cope with various technical problems to which the operation of those aircraft inevitably gives rise. The resolution of a technical problem which comes to light during aircraft maintenance or is caused by failure to maintain an aircraft must therefore be regarded as inherent in the normal exercise of an air carrier's activity and cannot therefore constitute as such an "extraordinary circumstance" within the meaning of Article 5(3) of the Regulation.

---

[327] Convention for the Unification of Certain Rules for International Carriage by Air, signed at Montreal on 28 May 1999. See ICAO Doc 9740. The European Community acceded to this Convention on 29 April 2004 and the Convention entered into force on 28 June 2004.

[328] Case C-344/04 [2006]ECR. This was a case where Mrs. Wallentin-Hermann booked three seats on a flight with Alitalia from Vienna to Brindisi (Italy) via Rome for herself, her husband and her daughter. The flight was scheduled to depart from Vienna on 28 June 2005 at 6.45 a.m. and to arrive at Brindisi on the same day at 10.35 a.m. After checking in, the three passengers were informed, 5 min before the scheduled departure time, that their flight had been cancelled. They were subsequently transferred to an Austrian Airlines flight to Rome, where they arrived at 9.40 a. m., that is 20 min after the time of departure of their connecting flight to Brindisi, which they therefore missed. Mrs. Wallentin-Hermann and her family arrived at Brindisi at 14.15 p.m. The cancellation of the Alitalia flight from Vienna resulted from a complex engine defect in the turbine which had been discovered the day before during a check. Alitalia had been informed of the defect during the night preceding that flight. The repair of the aircraft, which necessitated the dispatch of spare parts and engineers, was completed on 8 July 2005. Following Alitalia's refusal to pay her compensation of EUR 250 and EUR 10 for telephone charges, Mrs. Wallentin-Hermann brought legal proceedings.

2.4 Meteorological Issues

Another significantly relevant fact is that Regulation 261/2004 does not totally exonerate carriers in the context of extraordinary circumstances. It limits or excludes obligations of carriers, which means that there may be certain obligations which the carrier might have to fulfil with regard to stranded passengers. On the basic argument that the contract of carriage imposes an obligation on the carrier to carry the passenger (irrespective of date and time) safely to his destination, the carrier could be held responsible under the Regulation to reimburse the fare,[329] offer meals and refreshments in a reasonable relation to the waiting time and also hotel accommodation[330] and transport between the airport and place of accommodation.[331]

Although these provisions are not directly applicable to air carrier liability which resulted from the cancellation of flights owing to the closure of airspace over European States, it nonetheless is interesting to note that they, and in particular Article 6 of Regulation 261/2004 have been brought into question in the European courts by such Organizations as IATA and the European Low Fares Airline Association (ELFAA). Article 6 prescribes that air carriers should render assistance for delay in various categories of chronology and distance involved. IATA and ELFAA based their arguments on the grounds that there was inconsistency between the provisions of Regulation 261/2004 and the Montreal Convention and that the former failed to satisfy and respect principles of legal certainty, proportionality and non-discrimination. In an *au fait* article[332] written by Arnold and Mendes de Leon, which provides information on this issue and on the increasing *cursus curiae* questioning the legal validity of Regulation 261/2004, the authors arrive at the conclusion that the Regulation is a failure as it is not sufficiently effective in protecting the rights of the passenger on the one hand while hindering the operations of the airline by imposing unconscionable costs on them. They also argue that the Regulation adversely affects the interests of safety and lacks clarity and consistency that would make for easy judicial interpretation and that its principles may not mesh with the principles of international law.[333]

The above notwithstanding, it is worthy of note that in general, the "extraordinary circumstances" defence in its entirety would work to the advantage of European carriers such as British Airways in terms of flights to and from Europe in the context of the Eyjafjallajökull fallout, and save them 600 Euro per passenger for long haul flights, which makes a saving of up to 200,000 per fairly full long haul flight.[334]

Air transport is a beleaguered and complex business, which barely survives at the best of times. Natural disasters, economic recession, adverse governmental policy, restrictive economic policy pertaining to the operation of commercial flights, all present obstacles in various proportions and forms. Therefore, the last

---

[329] Regulation 261/2004 *supra* note 298 at Article 8.

[330] *Id.* Article 9.

[331] *Ibid.*

[332] Arnold and de Leon (2010).

[333] *Id.* 110.

[334] John Balfour, The "Extraordinary Circumstances" Defence in EC Regulation 261/2004 after Wallentin-Hermann v. Alitalia, *ZLW* 58.Jg. 2/2009 224-231 at 224.

118   2 Safety Issues

thing airlines need is liability for circumstances beyond their control and the responsibility to bear the burden of consequences wrought by governmental policy. Giovanni Bisignani, the down-to-earth, hard-hitting Director General and CEO of IATA,[335] in his address to the Royal Aeronautical Society, Montreal Branch on 1 December 2009 observed that the aviation industry is facing a crisis and that psast experience showed that airlines had lost US$53 billion in 8 years.[336]

Against this backdrop, Eyjafjallajökull, and perhaps other natural phenomena to come offer the aviation community some food for thought. A natural disaster is no one's fault, and least of all the airlines' fault. Some consideration must be given towards protecting the interests of airlines in the face of such interruptions to business. This would involve a delicate balance between economic theory, contract law principles, international law and regulation and social justice. In the ultimate analysis a natural phenomenon such as Eyjafjallajökull calls upon States to look after their citizens around the world if they are stranded for inordinate periods of time which require expenses for sustenance beyond their budget. Diplomatic theory calls for embassies to have consular services providing assistance in the case of an accident, serious illness or death, and arrangements for family to be informed; helping citizens that are arrested or stranded in a host country and arranging for family to be informed if wished; and providing emergency assistance in the case of civil disturbances and natural disasters. If circumstances are beyond their control, airlines should not be expected to look after their passengers abroad, which is really the duty of the respective States of which they are nationals.

2.5 Unmanned Aircraft Systems

An *Unmanned Aircraft System* (UAS) is an aircraft[337] and its associated elements which are operated with no pilot on board. UAS is an overarching term for the entire system comprising an *Unmanned Aerial Vehicle* (UAV)[338] which is applied to describe a self piloted or remotely piloted aircraft that can carry cameras, sensors, communications equipment or other payloads,[339] as well those which support

---

[335] *Supra*, note 6.

[336] http://www.iata.org/pressroom/speeches/2009-12-01-01.htm.

[337] An aircraft is defined as "any machine that can derive support in the atmosphere from the reactions of the air other than the reactions of the air against the earth's surface." This definition appears in Annexes 1,2,3,7,8,11,13,16 and 17 to the Convention on International Civil Aviation, signed at Chicago on 7 December 1944. See ICAO Doc 7300/9 Ninth Edition, 2006.

[338] For more details of UAS operations and their nature visit www.uvs-info.com.

[339] In January 2007, the Air Navigation Commission (ANC) of the International Civil Aviation Organization (ICAO) consulted with States and appropriate international organizations on convening an Accident Investigation and Prevention Group (AIG) meeting in 2008 to discuss subjects in the field of accident investigation. One of the proposed subjects for discussion is the amendment of the definition of accident in Chapter 1 of Annex 13 to include events involving unmanned aerial

2.5 Unmanned Aircraft Systems

unmanned flights such as air traffic management and remote controllers of such aircraft. The United States Department of Defence defines a UAV as "a powered aerial vehicle that does not carry a human operator, uses aerodynamic forces to provide vehicle lift, can fly autonomously or be piloted remotely, can be expendable or recoverable, and carry a lethal or non-lethal payload".[340] Ballistic or semi-ballistic vehicles, cruise missiles and artillery projectiles were not considered UAVs by this definition.[341] The Federal Aviation Administration of the United States defines a UAS as "a device that is used or intended to be used for flight in the air that has no onboard pilot. This includes all classes of airplanes, helicopters, airships, and translational lift aircraft that have no onboard pilot".[342] All references to UAS that follow in this article therefore necessarily include UAVs.

UAS are used to serve different purposes and therefore come in a variety of models, shapes and sizes. Their sizes may differ from having as wide a wing span as a Boeing 737 aircraft to that of a radio-controlled model airplane. A UAS has of necessity to be guided and operated by a pilot on the ground. A strategic use of a UAS is military reconnaissance and attack where they are commonly called drones. However, they now also serve to increase efficiency, and be cost effective, enhance safety and even save lives. They could also be used in aerial photography, surveying land and crops, monitoring forest fires and environmental conditions, and protecting borders and ports against intruders.[343]

UAS have been used to conduct reconnaissance and intelligence-gathering for nearly 60 years (since the 1950s). The future role of the UAS is a more challenging one which, in addition to its current uses will include involvement in combat missions.[344] The issues and challenges that UAS bring to civil aviation can be bifurcated into two main areas. The first concerns airworthiness regulations which are required to ensure that a UAS is built, maintained and operated at high standards that ensure the safety of all involved including crew and passengers of manned civilian and military aircraft with which UAS will share de-segregated airspace as

---

vehicles (UAV). See Addressing Unmanned Aircraft System Accident Investigation and Prevention (Paper presented by the United States at the 36th ICAO Assembly) A36-WP/217 TE/70 18/09/07.

[340] *Unmanned Aerial Vehicles: Background and Issues for Congress*, Report for Congress written by Elizabeth Bone and Christopher Bolkcom, Congressional Research Service: The Library of Congress, April 25 2003 CRS 1.

[341] *Ibid.*

[342] Unmanned Aircraft System Regulation Review, September 2009, Final Report, DOT/FAA/AR-09/7, 14.

[343] See FAA Fact Sheet – Unmanned Aircraft Systems (UAS), December 1, 2010 at http://www.faa.gov/news/fact_sheets/news_story.cfm?newsId=6287.

[344] Since 1964 the US Defense Department has developed 11 different UAS, though due to acquisition and development problems only 3 entered production. The US Navy has studied the feasibility of operating Vertical Take off and Landing (VTOL) UAS since the early 1960s, the QH-50 Gyrodyne torpedo-delivery drone being an early example. However, high cost and technological immaturity have precluded acquiring and fielding operational VTOL UAS systems.

well as persons and property on the ground.[345] The International Civil Aviation Organization[346] began addressing issues concerned with the operation of UAS and principles applicable thereto[347] since UAS were increasingly requiring access to all categories of airspace including non segregated airspace.

The second challenge is more far reaching and concerns the possibility of the UAS encroaching on air traffic control (ATC) functions in non segregated airspace. In doing so, UAS should not place an added burden and demands on airspace management and the flow of general air traffic within the en-route air space structure which must not be impeded by the presence of UAS. In this context, the priority would lie in collision avoidance, primarily through effective separation of aircraft by which aircraft could be kept apart by the application of appropriate separation minima. The two key players in this exercise would be the pilot of the manned aircraft involved and the air navigation service provider who would be jointly or severally liable if separation minima were compromised.

Although there are international regulations in place that address the operation of UAS in non segregated airspace, there is provision under ICAO regulations for the appropriate procedure to be followed. Annex 11 to the Chicago Convention,[348] which deals with the subject of air traffic services, lays down requirements for coordination of activities that are potentially hazardous to civil aircraft. Standard 2.17.1 stipulates that arrangements for activities potentially hazardous to civil aircraft, whether over the territory of a State or over the high seas, shall be coordinated with the appropriate air traffic services authorities, such coordination to be effected early enough to permit timely promulgation of information regarding the activities in accordance with the provisions of Annex 15 to the Chicago Convention.[349] Standard 2.17.2 of Annex 11 explains that the objective of the coordination referred to in the earlier provision shall be to achieve the best

---

[345] The main concern of the International Civil Aviation Organization in its role as regulator in this context is with international civil UAS operations and those standards that affect such operations.

[346] An ICAO Exploratory Meeting on Unmanned Aerial Vehicles (UAS) was held at ICAO Headquarters in Montreal from 23 to 24 May 2006. The primary objective of the meeting was to explore the current state of affairs with respect to development of regulatory material related to UAS and to discuss the possible role of ICAO in the regulatory process. The meeting was informed that the ICAO Secretariat would use the results of the meeting as the basis for developing a report to the ICAO Air Navigation Commission (ANC) along with recommendations on an ICAO work programme.

[347] At least four States: Australia; France; South Africa; and the United States are known to have commenced a programme developing standards for UAS operations. See Alexander ter Kuille, UAV and the ATM Community: The CANSO Policy of Engagement, *UAS Systems, The Global Perspective, 2006/2007* Blyenburgh & Co.: France, p. 24 at 25.

[348] *Convention on International Civil Aviation, supra,* note 3. Air traffic Services: Annex 11 to the Convention on International Civil Aviation, Thirteenth Edition, July 2001.

[349] Annex 15 contains Standards and Recommended Practices relating to Aeronautical Information Services.

arrangements that are calculated to avoid hazards to civil aircraft and minimize interference with the normal operations of aircraft.

The Chicago Convention[350] is focused on civil aviation, and applies to civil aircraft. The Convention does not apply to State aircraft, which are identified as aircraft engaged in police, military an customs services[351]. Therefore, principles of the Convention will apply only to UAS not engaged in such activities as are excluded. One of the provisions which may have a bearing on UAS in the Convention is Article 8 which stipulates that no aircraft capable of being flown without a pilot shall be flown without a pilot over the territory of a contracting State without special authorization by that State. Furthermore states allowing the operation of aircraft that do not have a pilot in air space open to civil aircraft are required to ensure that they are so controlled as to obviate danger to civil aircraft. One of the common usages of UAS – aerial photography- is affected by Article 36 of the Chicago Convention which empowers contracting states to prohibit or regulate the use of photographic apparatus in aircraft over its territory. Presumably this provision can be tagged on to Article 1 of the Convention whereby every State has complete and exclusive sovereignty over the airspace above its territory. Another important consideration could lie in Finally Annex 17[352] to the Chicago Convention, on the subject of aviation security where. Article 2.1.2 of the Annex states that each Contracting State shall establish an organization and develop and implement regulations, practices and procedures to safeguard civil aviation against acts of unlawful interference taking into account the safety, regularity and efficiency of flights. This could impel States to develop regulations and practices addressing the interference of control signals or even the hostile takeover of the command of an UAS which is a very common hazard to the operation of UAS.

Another challenge in the operation of UAS is licensing of personnel in charge of the operation of the vehicle and certification of the UAS. Article 31 of the Convention provides that every aircraft engaged in international navigation shall be provided with a certificate of airworthiness issued or rendered valid by the State in which it is registered. The Standards and Recommended Practices (SARP) for the issuance of an airworthiness certificate are laid down in Annex 8[353] to the Chicago Convention. Annex 8 (in its 9th Edition) only addresses aeroplanes[354] over 5,700 kg certificated take-off mass and helicopters[355] without a limitation on the mass of an

---

[350] Supra n 3.

[351] Id Article 3.

[352] Annex 2 to the Convention on International Civil Aviation (note 15), "Safety – Safeguarding International Civil Aviation Against Acts of Unlawful Interference", 8th edition, April 2006.

[353] Annex 8 to the Convention on International Civil Aviation (note 15), "Airworthiness of Aircraft", 10th edition, April 2005.

[354] "A power-driven heavier-than-air aircraft, deriving its lift in flight chiefly from aerodynamic reactions on surfaces which remain fixed under given conditions of flight.", see definitions in note 18.

[355] "A heavier-than-air aircraft supported in flight chiefly by the reactions of the air on one or more power driven rotors on substantially vertical axes.", see definitions in note 18.

aircraft which is intended for the carriage of passengers or cargo or mail in international air navigation[356] This might provoke the argument that Annex 8 would not usually apply to UAS since only large UAS exceed the weight of 5,700 kg. The lack of internationally recommended and accepted standards and practices for smaller aeroplanes is a challenge for the operation of UAS as well as for aeroplanes with a pilot on board. This point is covered in the 10th edition of Annex 8 which, in addition to the provisions in part VI on helicopters has been amended to be applicable for helicopters with a certificated take-off mass over 750 kg only. In terms of licensing it has to be noted that Annex 1[357] to the Chicago convention, defines SARPs for personnel licensing, in that a person shall not act as an air crew member unless a valid license is held[358] by that person. Pilots are considered not only flight crew but as well flight navigators, flight engineers and radiotelephone operators.[359] Implicitly, this means that not only is the remote pilot of UAS subject to licensing, but also personnel who are involved in the navigation and technical operation of UAS should be licensed as well. Furthermore mechanics of UAS be should also be licensed according the provisions in chapter 4.1 and 4.2 of Annex 1 to the Chicago Convention. Article 29 of the Chicago Convention requires the carriage of documents in aircraft such certificates of registration and airworthiness but also the appropriate licenses for each member of the crew. Although certificates of airworthiness can be carries in an aircraft in the manner required, the carriage of other documents may pose difficulties as some UAS are designed to operate over extended periods of time, up to several months, and the specific operators who would operate for such long periods may not be known at the initial stage of the flight. One potential solution could be to electronically store the data and electronic licenses (be it in the form of scanned documents or other forms) of the current crew on board of the vehicle, but this would need in depth assessment in regards to the legal validity of such a form.

Annex 2 to the Chicago Convention, detailing the rules of the air referred to in Article 12 of the Convention, states *inter alia* that the rules of the air shall apply to aircraft bearing the nationality and registration marks of a Contracting State. These rules applicable to UAS as well. Two main categories of rules of the air exist: visual flight rules and instrument flight rules. The note to article 2.2. of Annex 2 states inter alia that a pilot may elect to fly in accordance with instrument flight rules in visual meteorological conditions. The rules of the air adhered to are thus distinct and separate from the metrological conditions prevailing in the area of operation, except for instrument metrological conditions, requiring instrument flight rules to be applied. Chapter 3.1 of Annex 2, contains an article on unmanned free balloons, stating that they shall be operated in such a manner as to minimize hazards to

---

[356] Supra note 15, part IV, article 1.1.2 (wording identical to 9th edition).

[357] Annex 1 to the Convention on International Civil Aviation (note 15), "Personnel Licensing", 10th edition, July 2006.

[358] *Supra* note 206, Article 1.2.1.

[359] *Supra* note 394, Chapter 3.

2.5 Unmanned Aircraft Systems

persons, property or other aircraft and in accordance with the conditions specified in Appendix 4. Appendix 4 states inter alia that heavy balloons[360] need to comply with similar provisions like normal aeroplanes, inter alia minimum height over "congested areas of cities, towns or settlements or an open-air assembly of persons not associated with the operation",[361] SSR equipment,[362] and lightening.[363] Article 3.3 of Appendix 4 to the Annex 2 to the Chicago convention contains a remarkable requirement to unmanned balloons. Such vehicles shall be equipped with at least two payload flight termination devices or systems. It may well be argued that such devises or systems are required for UAS as well. An analogy to the operation of UAS exists in Annex 2 which requires obliges pilots-in-command to take action as will best avert collision. The Annex also requires that vigilance for the purpose of detecting potential collisions be exercised on board an aircraft, regardless of the type of flight or the class of airspace in which the aircraft is operating. Therefore, it can be concluded that pilots flying according instrument flight rules are required to scan the environment visually in order to detect potentially conflicting traffic. This task may prove difficult in the case of UAS in that although many UAS are equipped with video cameras, it would be difficult for UAS operators to detect vehicles nearby, to assess the potential for conflict and to initiate appropriate actions. This inability might result in infringement of article 3.2.1 of Annex 2, which provides that an aircraft shall not be operated in such proximity to other aircraft as to create a collision hazard. A potential solution to this problem could be that movement sensors, based on radar or ultrasound devices, similar to parking assistants for cars, are built into UAS. The drawback of such a measure would be the cost involved and the additional weight that has to be carried by the UAS.

2.5.1 The UAS and Civil Aviation Issues

2.5.1.1 Work of ICAO

In early 2011, as a result of sustained work carried out on UAS[364] by the International Civil Aviation Organization (ICAO),[365] the Organization released a circular

---

[360] *Supra*, note 14, Appendix 4, article 1 c).

[361] Id, article 3.2.

[362] Id, article 3.4.

[363] Id, article 3.6.

[364] In November 2007 the Air Navigation Commission of ICAO established the unmanned Aircraft Systems Study Group comprising Australia, Austria, Brazil, China, Czech Republic, France, Germany, Italy, Netherlands, New Zealand, Russian Federation, Singapore, South Africa, Sweden, U.K., U.S., CANSO, EASA, EUROCAE, EUROCONTROL, IAOPA, ICCAIA, IFALPA, IFATCA, UVS Intl.

[365] Supra n. 1. The main objectives of ICAO are to develop the principles and techniques of international air navigation and to foster the planning and development of air transport. ICAO has

entitled Unmanned Aircraft Systems (UAS)[366] the purposes of which was to: apprise States of the emerging ICAO perspective on the integration of UAS into non-segregated airspace and at aerodromes; consider the fundamental differences from manned aviation that such integration will involve; and encourage States to help with the development of ICAO policy on UAS by providing information on their own experiences associated with UAS.[367] The fundamental premise that ICAO follows in this regard is that, since unmanned aircraft fall within the definition of "aircraft" all SARPs of the Annexes to the Chicago Convention applicable to aircraft would apply to UAS as well.[368]

It must be underscored that the preliminary aim of the guidance material in the ICAO Circular is to ensure aviation safety[369] based on the fact that the risk of midair collisions between UAS and aircraft manned by pilots on board is a critical safety concern for UAS operations worldwide. Accident investigation therefore becomes crucial both in cases where accidents cause death or injury to persons and damage to property and in instances where no collision occurs between UAS and manned aircraft. In order to determine what aspect of the operation failed, whether additional, previously unanticipated hazards were contributory, and what deficiencies need to be corrected to prevent such an event from progressing to a more serious outcome in the future.

To begin with, the Circular makes explicit mention of Article 8 of the Chicago Convention, which, as earlier discussed, requires special authorization by the State

---

190 Contracting States. ICAO's Mission and Vision Statement is "to achieve its mission of safe, secure and sustainable development of civil aviation through cooperation amongst its member States". In December 2004, following a decision by the 35th Session of the ICAO Assembly, the Council of ICAO approved six Strategic Objectives for 2005-2010: They were: safety; security; environmental protection; efficiency; continuity; and rule of law. From 2011, ICAO's Strategic Objectives will be based on safety; security; environmental protection and the sustainable development of air transport. However, during the 37th Session of the Assembly, the Strategic Objectives were reduced to three: safety; security; environmental protection and sustainable development of air transport.

[366] *Unmanned Aircraft Systems* (UAS) Cir 328- AN/190.

[367] *Id*. Paragraph 1.6.

[368] *Id*. Paragraph 1.7. Model aircraft, which are outside the purview of the Chicago Convention are not included in within this principle. *Id*. 2.4. It must be noted that a number of Civil Aviation Administrations (CAAs) have adopted the policy that UAS must meet the equivalent levels of safety as manned aircraft. UAS operations must be as safe as manned aircraft insofar as they must not present a hazard to persons or property on the ground or in the air that is any greater than that attributable to the operation of manned aircraft of equivalent class or category. In general, UAS should be operated in accordance with the rules governing the flight of manned aircraft and meet equipment requirements applicable to the class of airspace within which they intend to operate. UAS must be able to comply with ATC instructions.

[369] Safety is defined as: "The state in which the possibility of harm to persons or of property is reduced to, and maintained at or below, an acceptable level through a continuing process of hazard identification and safety risk management". See Cir 328- AN/190 *supra*, note 33 at p. 5.

flown over by an aircraft capable of being flown without a pilot which is in fact flown without a pilot. In this context the Circular clarifies any obfuscation that may arise as to what a "pilotless aircraft" is by quoting another ICAO document which states that Article 8 refers to an aircraft which is flown without a pilot in command on board the aircraft but which is either remotely or fully controlled from another place (ground, another aircraft or space).[370]

One of the main issues addressed by the Circular is that aircraft operating without a pilot on board present a wide array of hazards to the civil aviation system. These hazards must be identified and the safety risks mitigated,[371] just as with introduction of an airspace redesign, new equipment or procedures. In this regard, States are required to establish a State Security Programme (SSP) to include safety rulemaking, policy development and oversight. The operation of UAS in desegregated airspace would not only affect operations carried by commercial air carriers but would also affect general aviation. The International Council of Aircraft Owner and Pilot Association (IAOPA) has commented on the issue of UAS that operating rules for UAS must take into account their potential impact on general aviation aircraft operating in un-segregated airspace. IAOPA added that while segregated airspace contains operations subject to air traffic control, un-segregated airspace depends almost entirely on certain Annex 2 (to the Chicago Convention) cruising altitude conventions and mutual self-separation methods. Because self-separation methods for UAS are still in the conceptual stage and will likely require some time to perfect, there will be a temptation to impose un-segregated airspace restrictions on manned aircraft to accommodate RPA. Since un-segregated airspace is almost entirely the domain of general aviation, we do not want this to occur.

The second point raised by IAOPA is that State or military UAS must abide by whatever UAS operating rules are devised to ensure safe, hazard-free operations. Because non-civil UAS operations may wish to use lower altitude un-segregated airspace, there could be a tendency for States and the military to pre-empt conventional flight rules in these areas, either on a temporary or permanent basis.[372]

The answer to these two points seemingly rests with the fact that UAS will operate in accordance with ICAO Standards that exist for manned aircraft as well as any special and specific standards that address the operational, legal and safety

---

[370] *The Global Air Traffic Management Concept*, Doc 9854 referred to in Cir 328-AN/190 in 2.2. It must be noted that ICAO recognizes many categories of aircraft, including but not limited to balloons, gliders, aeroplanes and rotorcraft whether they operate from land or water.

[371] The term "safety management" includes two key concepts. First is the concept of a State safety programme (SSP), which is an integrated set of regulations and activities aimed at improving safety. Second is the concept of a safety management system (SMS) which is a systematic approach to managing safety, including the necessary organizational structures, accountabilities, policies and procedures.

[372] See http://www.iaopa.org/news/uas.html.

differences between manned and unmanned aircraft operations.[373] This includes applicable environmental rules and guidelines as well.[374] In order for UAS to integrate into non-segregated airspace and at non-segregated aerodromes, there will be a pilot responsible for the UAS operation. Pilots may utilize equipment such as an autopilot to assist in the performance of their duties; however, under no circumstances will the pilot responsibility be replaced by technologies in the foreseeable future. For greater fluidity in identifying who the pilot is, the Circular has introduced the concept of "remotely-piloted aircraft" (RPA) – which is a subset of unmanned aircraft – into the lexicon. An RPA[375] is an aircraft piloted by a licensed "remote pilot" situated at a "remote pilot station" located external to the aircraft (i.e. ground, ship, another aircraft, space) who monitors the aircraft at all times and can respond to instructions issued by ATC, communicates via voice or data link as appropriate to the airspace or operation, and has direct responsibility for the safe conduct of the aircraft throughout its flight. An RPA may possess various types of auto-pilot technology but at any time the remote pilot can intervene in the management of the flight. This equates to the ability of the pilot of a manned aircraft being flown by its auto flight system to take prompt control of the aircraft.

From a legal perspective, and in accordance with Article 3 *bis* of the Chicago Convention which stipulates *inter alia* that contracting States recognize that every State, in the exercise of its sovereignty, is entitled to require the landing at some designated airport of a civil aircraft flying above its territory without authority, contracting States are entitled, in certain circumstances, to require civil aircraft flying above their territory to land at designated aerodromes, Therefore the pilot of the RPA will have to be able to comply with instructions provided by the State, including through electronic or visual means, and have the ability to divert to the

---

[373] The principles so applicable in the context of the Chicago Convention are: Article 12 pertaining to rules of the air; Article 15 in the context of airport and air navigation services charges; Article 29 on documents carried on board aircraft; Article 31 which stipulates that every aircraft engaged in international navigation shall be provided with a certificate of airworthiness issued or rendered valid by the State in which it is registered; Article 32 which provides that the pilot of every aircraft and the other members of the operating crew of every aircraft engaged in international navigation shall be provided with certificates of competency and licenses issued or rendered valid by the State in which the aircraft is registered; and Article 33 which requires that Certificates of airworthiness and certificates of competency and licenses issued or rendered valid by the contracting State in which the aircraft is registered, shall be recognized as valid by the other contracting States, provided that the requirements under which such certificates or licences were issued or rendered valid are equal to or above the minimum standards which may be established from time to time pursuant to the Convention.

[374] *Supra*, note 391 paragraphs 6.48–6.51.

[375] RPA may have the same phases of flight – taxi, departure, en-route and arrival – as manned aircraft or they may be launched/recovered and/or conduct aerial work. The aircraft performance characteristics may be significantly different from traditional manned aircraft. Regardless, the remote pilot will operate the aircraft in accordance with the rules of the air for the State and airspace in which the RPA is operating. This will include complying with directions and instructions provided by the air traffic services (ATS) unit.

specified airport at the State's request. The requirement to respond to instructions based on such visual means may place significant requirements on certification of RPAS detection systems for international flight operations.

In terms of collision avoidance, the Circular makes the pilot in command of a UAS as responsible as a pilot of a manned aircraft for detecting and avoiding potential collisions and other hazards. Furthermore it provides that technology to provide the remote pilot with sufficient knowledge of the aircraft's environment to fulfil the responsibility must be incorporated into the aircraft with counterpart components located at the remote pilot station. Also, remote pilots, despite not being on board the aircraft, will be subject to the same requirements as aircraft pilots who are required to observe, interpret and heed a diverse range of visual signals intended to attract their attention and/or convey information. Such signals can range from lights and pyrotechnic signals for aerodrome traffic to signals used by intercepting aircraft. This would necessitate development and approval of alternate means of compliance with this requirement.

With regard to air traffic management (ATM) the Circular prescribes that whether the aircraft is piloted from on board or remotely, the provision of air traffic services (ATS) should, to the greatest practicable extent, be one and the same.[376] It further states that the introduction of RPA must not increase the risk to other aircraft or third parties and should not prevent or restrict access to airspace. ATM procedures for handling RPA should mirror those for manned aircraft whenever possible. There will be some instances where the remote pilot cannot respond in the same manner as could an on-board pilot and the Circular calls for ATM procedures to be able to take account of these differences. For this purpose, ATS/remote pilot communication requirements must be assessed in the context of an ATM function, taking into account human interactions, procedures and environmental characteristics. A safety management system (SMS) approach should be employed to determine the adequacy of any communications solutions.[377] The information exchange between ATC and the remote pilot will likely require the same levels of reliability, continuity and integrity, referred to as QOS, that are required to support operations with manned aircraft in the airspace in which a UA is intended to operate.[378]

The exchange of control information between the aircraft and its remote pilot station will require an extremely high level of availability, reliability, continuity and integrity. The determination of required communication performance and associated QOS levels will be based on functionality considering the level of ATS being provided.[379]

---

[376] ICAO Circular 328-AN/190, at paragraph 5.10.
[377] Id. Paragraph 5.14.
[378] Id. Paragraph 6.33.
[379] Id. Paragraph 6.34.

In terms of aerodrome operations and UAS, the Circular recognizes that integration of RPA into aerodrome operations will prove to be among the greatest challenges. At issue are provisions for the remote pilot to identify, in real-time, the physical layout of the aerodrome and associated equipment such as aerodrome lighting and markings so as to manoeuvre the aircraft safely and correctly. The Circular provides that RPA must be able to work within existing aerodrome parameters. Aerodrome standards should not be significantly changed, and the equipment developed for RPA must be able to comply with existing provisions to the greatest extent practicable. Moreover, where RPA are operated alongside manned aircraft, there needs to be harmonization in the provision of ATS.[380]

Meteorology is another important element that needs to be properly coordinated in the operation of UAS. The Circular provides that meteorological information plays a role in the safety, regularity and efficiency of international air navigation and is provided to users as required for the performance of their respective functions. Meteorological information supplied to operators and flight/remote crew members covers the flight in respect of time, altitude, and geographical area. Accordingly, the information relates to appropriate fixed times, or periods of time, and extends to the aerodrome of intended landing. It also covers meteorological conditions expected between the aerodrome of intended landing and alternate aerodromes designated by the operator.[381]

Meteorological services are critical for the planning, execution and safe operation of international aviation. Since the remote pilot is not on board the aircraft and may not be able to determine meteorological conditions and their real-time effects on the aircraft, obtaining meteorological information from appropriate sources prior to and during flight will be especially critical for the safe operation of these aircraft.[382]

The Circular recognizes the Annex 3 to the Chicago Convention – *Meteorological Service for International Air Navigation* has a requirement for aircraft on its registry operating on international air routes to make automated routine observations, if so equipped. RPA may not be so equipped. Likewise, there is a requirement for all aircraft to make special observations whenever severe turbulence, severe icing, severe mountain wave, thunderstorms, hail, dust, stone and volcanic ash are encountered during a flight. However, RPA may not be able to comply with these provisions as the pilot is remote from the aircraft, and the aircraft may not have the sensors to detect these phenomena.[383]

It is also recognized that conversely, the RPA specifically equipped for such purposes may in fact be used to monitor meteorological conditions, relaying information back to ground sensors. These aircraft could potentially be used in

---

[380] *Id.* Paragraph 5.23.
[381] *Id.* Paragraph 5.27.
[382] Id. Paragraph 5.28.
[383] Id. Paragraph 5.29.

2.5 Unmanned Aircraft Systems

conditions and locations where manned aircraft cannot safely operate such as in hurricanes, convective weather or in the vicinity of volcanic ash/gases.[384]

One of the critical elements in UAS operations is the security of the system as security is a vital issue for RPA with aspects that are both similar and unique when compared with manned aircraft. As a remote pilot station is similar in purpose and design to a cockpit, it must likewise be secure from sabotage or unlawful malicious interference. Chapter 13 of Annex 6 to the Chicago Convention – Part I – *International Commercial Air Transport – Aeroplanes* contains SARPs to secure the flight crew compartment. However, due to the fixed and exposed nature of the remote pilot station (as opposed to the restricted nature of a commercial aeroplane where the intrusion and use of heavier weapons is less likely) further consideration should be given to the potential vulnerability of the premises against unlawful interference.[385] Similarly, the aircraft itself must be stored and prepared for flight in a manner that will prevent and detect tampering and ensure the integrity of vital components.

2.5.2 Operations over the High Seas

The Circular prescribes that operators must have approval from the State of the operator before conducting operations in high seas airspace. They must likewise coordinate their operations with the ATS provider responsible for the airspace concerned.[386] Article 12 of the Chicago Convention unambiguously states that over the high seas, the rules in force shall be those under the Convention and each Contracting State undertakes to insure the prosecution of all persons violating the regulations applicable. This peremptory principle,[387] of adherence by States and aircraft bearing their nationality to any Standards and Recommended Practices (SARPs) adopted in regard to the high seas, effectively precludes any possible reliance by States on Article 38 of the Convention which allows States to deviate from SARPs in general. In other words, Annex 2 on Rules of the Air, which contains provisions relating to the operation of aircraft over the high seas, is sacrosanct and inviolable. The first legal issue that would emerge from this clear principle is the question of applicability of Annexes (other than Annex 2) to the high seas and whether their provisions, if directly related to the principles of manoeuvre and navigation of aircraft over the high seas, would be binding with no flexibility offered by Article 38 of the Convention. Kaiser offers the opinion:

---

[384] Id. Paragraph 5.30.

[385] Id. Paragraph 5.32.

[386] *Id.* at paragraph 3.19.

[387] Bin Cheng confirms that over the high seas there is absolutely no option for States to deviate from rules established under the Chicago Convention for the manoeuvre and operations of aircraft. See Cheng (1962), at 148.

Over the high seas, the rules of the air have binding effect under Article 12, Sentence 3 of the Chicago Convention. It should be clarified that rules of the air have a broader meaning than Annex 2 and encompass the Standards and Recommended Practices of all other Annexes as far as their application makes sense over the high seas.[388]

Kaiser is of course referring mainly to Annexes 10 and 11 to the Chicago Convention relating to air traffic services and air traffic management, while at the same time drawing the example of Annex 16 (on environmental) protection being applicable in a future date if extended beyond noise and engine emissions to the high seas under Article 12 of the Chicago Convention.[389] This argument, which would ascribe to the ICAO Council wider control over larger spans of the world's air space, would be acceptable only if provisions of other Annexes (other than those of Annex 2) would directly have a bearing on the manoeuvre and navigation of aircraft over the high seas, as exclusively provided for by Article 12 of the Chicago Convention.

The provision of air navigation services are mainly regulated by three Annexes to the Chicago Convention, namely Annex 2 (Rules of the Air), Annex 3 (Meteorological Service for International Air Navigation) and Annex 11 (Air Traffic Services).[390] Of these, compliance with Annex 2 is mandatory[391] and does not

---

[388] Kaiser (1995), at 455. Bin Cheng states that contracting States are expected to be able to exercise control over all that takes place within their territories, but outside their respective territories only over aircraft bearing their nationality. Bin Cheng, *supra*, note 408, at 110.

[389] *Ibid.*

[390] Article 54 l) of the Chicago Convention stipulates as a mandatory function of the Council the act of adopting, in accordance with Chapter VI of the Convention, international standards and recommended practices (SARPs) and for convenience designate them as Annexes to the Convention. Article 37 of the Convention reflects the areas in which SARPs should be developed and Annexes formed. Article 38 obliges contracting States to notify ICAO of any differences between their own regulations and practices and those established by international standards or procedures. The notification of differences however, does not absolve States from their continuing obligation under Article 37 to collaborate in securing the highest practicable degree of uniformity in international regulations, standards, and procedures.

[391] In October 1945, the Rules of the Air and Air Traffic Control (RAC) Division at its first session made recommendations for Standards, Practices and Procedures for the Rules of the Air. These were reviewed by the then Air Navigation Committee and approved by the Council on 25 February 1946. They were published as *Recommendations for Standards, Practices and Procedures – Rules of the Air* in the first part of Doc 2010, published in February 1946. The RAC Division, at its second session in December 1946–January 1947, reviewed Doc 2010 and proposed Standards and Recommended Practices for the Rules of the Air. These were adopted by the Council as Standards and Recommended Practices relating to Rules of the Air on 15 April 1948, pursuant to Article 37 of the Convention on International Civil Aviation (Chicago, 1944) and designated as Annex 2 to the Convention with the title *International Standards and Recommended Practices – Rules of the Air*. They became effective on 15 September 1948. On 27 November 1951, the Council adopted a complete new text of the Annex, which no longer contained Recommended Practices. The Standards of the amended Annex 2 (Amendment 1) became effective on 1 April 1952 and applicable on 1 September 1952.

give the States the flexibility provided in Article 38 of the Chicago Convention to register differences from any provisions of the Annex.

With regard to maritime navigation, the *United Nations Convention on the Law of the Sea (UNCLOS)*, Article 39, lays down the duties of ships and aircraft involved in transit navigation to the effect that ships and aircraft, while exercising the right of transit passage, should : proceed without delay through or over the strait; refrain from any threat or use of force against the sovereignty, territorial integrity or political independence of States bordering the strait, or in any other manner in violation of the principles of international law embodied in the Charter of the United Nations; refrain from any activities other than those incident to their normal modes of continuous and expeditious transit unless rendered necessary by force majeure or by distress; and comply with the relevant provisions of the Convention. Article 39 (3) explicitly states that aircraft in transit passage shall observe the Rules of the Air established by ICAO as they apply to civil aircraft and that state aircraft will normally comply with such safety measures and will at all times operate with due regard for the safety of navigation. The provision further states that at all times aircraft shall monitor the radio frequency assigned by the competent internationally designated air traffic control authority or the appropriate international distress radio frequency.

Standard 2.1.1 of Annex 2 to the Chicago Convention provides that the rules of the air shall apply to aircraft bearing the nationality and registration marks of a Contracting State, wherever they may be, to the extent that they do not conflict with the rules published by the State having jurisdiction over the territory overflown.[392] The operation of an aircraft either in flight or on the movement area of an aerodrome shall be in compliance with the general rules and, in addition, when in flight, either with: visual flight rules (VFR); or the instrument flight rules (IFR).[393] Standard 2.3.1 further provides that the pilot-in-command of an aircraft shall, whether manipulating the controls or not, be responsible for the operation of the aircraft in accordance with the rules of the air, except that the pilot-in-command may depart from these rules in circumstances that render such departure absolutely necessary in the interests of safety.

---

[392] The Council of the International Civil Aviation Organization resolved, in adopting Annex 2 in April 1948 and Amendment 1 to the said Annex in November 1951, that the Annex constitutes Rules relating to the flight and manoeuvre of aircraft within the meaning of Article 12 of the Convention. Over the high seas, therefore, these rules apply without exception.

[393] Information relevant to the services provided to aircraft operating in accordance with both visual flight rules and instrument flight rules in the seven ATS airspace classes is contained in 2.6.1 and 2.6.3 of Annex 11. A pilot may elect to fly in accordance with instrument flight rules in visual meteorological conditions or may be required to do so by the appropriate ATS authority.

2.5.3 Air Traffic Services

The provision of air traffic services[394] is addressed in Annex 11 to the Chicago Convention which provides *in limine* that Contracting States shall determine, in accordance with the provisions of the Annex and for the territories over which they have jurisdiction, those portions of the airspace and those aerodromes where air traffic services will be provided. They shall thereafter arrange for such services to be established and provided in accordance with the provisions of this Annex, except that, by mutual agreement, a State may delegate to another State the responsibility for establishing and providing air traffic services in flight information regions, control areas or control zones extending over the territories of the former.[395]

The Standards and Recommended Practices in Annex 11, together with the Standards in Annex 2, govern the application of the *Procedures for Air Navigation Services – Air Traffic Management*[396] and the *Regional Supplementary Procedures –* Rules of the Air and Air Traffic Services, contained in Doc 7030, Annex 11 pertains to the establishment of airspace, units and services necessary to promote a safe, orderly and expeditious flow of air traffic. A clear distinction is made between air traffic control service, flight information service and alerting service. Its purpose, together with Annex 2, is to ensure that flying on international air routes is carried out under uniform conditions designed to improve the safety and efficiency of air operation.

2.5.4 UAS as State Aircraft

One of the legal issues that has to be considered is that UAS are usually State aircraft and as such might not come within the purview of the Chicago Convention in the

---

[394] According to Paragraph 2.2 of the Annex, The objectives of the air traffic services shall be to: a) prevent collisions between aircraft; b) prevent collisions between aircraft on the manoeuvring area and obstructions on that area; c) expedite and maintain an orderly flow of air traffic; d) provide advice and information useful for the safe and efficient conduct of flights; e) notify appropriate organizations regarding aircraft in need of search and rescue aid, and assist such organizations as required.

[395] 24 Standard 2.1.1. It is also provided in the Annex that if one State delegates to another State the responsibility for the provision of air traffic services over its territory, it does so without derogation of its national sovereignty. Similarly, the providing State's responsibility is limited to technical and operational considerations and does not extend beyond those pertaining to the safety and expedition of aircraft using the concerned airspace. Furthermore, the providing State in providing air traffic services within the territory of the delegating State will do so in accordance with the requirements of the latter which is expected to establish such facilities and services for the use of the providing State as are jointly agreed to be necessary. It is further expected that the delegating State would not withdraw or modify such facilities and services without prior consultation with the providing State. Both the delegating and providing States may terminate the agreement between them at any time.

[396] Doc 4444, PANS-ATM.

2.5 Unmanned Aircraft Systems

context of regulation through an Annex to the Convention unless such an Annex were to address issues affecting civil aircraft. Article 3 (a) of the Chicago Convention provides that the Convention will be applicable only to civil aircraft and not to state aircraft. It is an inclusionary provision which identifies military, customs and police service aircraft as being included in an undisclosed list of state aircraft. The Convention contradicts itself in Article 3 (c), where it says that no state aircraft of a contracting State shall fly over the territory of another State or land thereon without authorization by special agreement or otherwise, and in accordance with the terms thereof. The question arises as to how an international treaty, which on the one hand prescribes that it applies only to civil aircraft, turns around and prescribes a rule for state aircraft. Article 3 (c) effectively precludes relief flights over the territory of a State by state aircraft if the State flown over or landed upon does not give authorization for the aircraft to do so.

The distinction between civil and state aircraft is unclear as the Chicago Convention does not go to any length in defining or specifying as to how the two categories have to be distinguished. The ICAO Assembly, at its 14th Session held in Rome from 21 August to 15 September 1962, adopted Resolution A14-25 (Coordination of Civil and Military Air Traffic) which was on the subject addressed in Article 3(d) – that the Contracting States undertake, when issuing regulations for their state aircraft, that they will have due regard to the safety of navigation of civil aircraft. In A14-25, the Assembly directed the Council to develop guidance material for the joint civil and military use of airspace, taking into account the various policies, practices and means already employed by States to promote the satisfactory coordination or integration of their civil and military air traffic services.

At its 21st Session of the Assembly, Held in Montreal from 21 September to 15 October 1974, ICAO saw the adoption of Resolution A21-21 (Consolidated Statement of Continuing Policies and Associated Practices Related Specifically to Air Navigation) where, at Appendix O, on the subject of coordination of civil and military air traffic, the Assembly resolved that the common use by civil and military aviation of airspace and of certain facilities and services shall be arranged so as to ensure safety, regularity and efficiency of international civil air traffic, and that States would ensure that procedures and regulations pertaining to their state aircraft will not adversely affect or compromise the regularity and efficiency of international civil air traffic. In order to effectively implement the proposals of the Resolution, Contracting States were requested to initiate and improve the coordination between their civil and military air traffic services and the ICAO Council was required to ensure that the matter of civil and military coordination in the use of airspace is included, when appropriate, in the agenda of divisional and regional meetings

Resolutions adopted by the ICAO Assembly over the years show a sustained and robust evolution in global policy sharing of airspace by civil and military aviation. At its 35th Session, held in Montreal in September/October 2004, the ICAO Assembly adopted Resolution A 35-14[397] (Consolidated statement of continuing

---

[397] *Assembly Resolutions in Force (as of 8 October 2004)*, ICAO Doc 9848, ICAO Montreal, at II-2.

ICAO policies and associated practices related specifically to air navigation) which, in Appendix P (Coordination of civil and military traffic) recognized that the airspace as well as many facilities and services should be used in common by both civil and military aviation. The resolution also went on to note that full integration of the control of civil and military air traffic may be regarded as the ultimate goal, and that coordination between States in achieving this goal should be the ultimate aim in resolving current difficulties. The Assembly resolved that the common use by civil and military aviation of airspace and of certain facilities and services shall be arranged so as to ensure the safety, regularity and efficiency of international air traffic and that regulations and procedures established by contracting States to govern the operation of their State aircraft over the high seas shall ensure that such operations do not compromise the safety, regularity and efficiency of international civil air traffic. The Resolution concludes that, to the extent practicable, these operations should comply with provisions of Annex 2 to the Chicago Convention on Rules of the Air.[398]

At the 36th Session of the ICAO Assembly held in Montreal from 18 to 28 September 2007, the Assembly adopted Resolution A 36–13 (Consolidated statement of continuing ICAO policies and associated practices related specifically to air navigation) which superseded Resolution A 35–14) which noted *inter alia* that the airspace as well as many facilities and services should be used in common by civil aviation and military aviation and resolved that the common use by civil and military aviation of airspace and of certain facilities and services shall be arranged so as to ensure the safety, regularity and efficiency of international traffic.[399]

At the 37th Session of the Assembly held in Montreal in September October 2010 the Assembly adopted Resolution A 37–15 (Consolidated statement of continuing ICAO policies and associated practices related specifically to air navigation) which superseded Resolution A 36–13) Appendix O of which was on coordination and cooperation of civil and military air traffic. The Assembly recognized *inter alia* that the flexible use of airspace by both civil and military air traffic may be regarded as the ultimate goal and improvement in civil/military coordination and cooperation offers an immediate approach towards more effective airspace management. The Assembly went on to resolve that the common use by civil and military aviation of airspace and of certain facilities and services shall be arranged so as to ensure safety, regularity and efficiency of civil aviation as well as to ensure the requirements of military air traffic are met, concluding that ICAO serves as an international forum that plays a role in facilitating improved civil/military cooperation, collaboration and sharing of best practices, and to provide the necessary follow-up activities that build on the success of the Global Air Traffic Management Forum on Civil/Military Cooperation held in 2009 with the support of civil/military partners.[400]

---

[398] *Id.* II-12.

[399] Assembly Resolutions in Force (as of 28 September 2007), ICAO Doc 9902, *Appendix O* at II-17.

[400] The theme of the Forum was *"Time to take it global: Meeting each other's needs without compromising the Mission"*. The event was held as a follow up to recommendations of the

2.5 Unmanned Aircraft Systems

With regard to Conventions other than the Chicago Convention, one can see some provisions which are relevant to the discussion on the distinction between civil and military aircraft, the latter of which, by implication, includes UAS. The Convention on the International Recognition of Rights in Aircraft (Geneva. 1948),[401] the Convention on Offences and Certain Other Acts Committed on Board Aircraft (Tokyo, 1963),[402] the Convention for the Suppression of Unlawful Seizure of Aircraft (The Hague, 1970)[403] and the Convention for the Suppression of Unlawful Acts Against the Safety of Civil Aviation (Montreal. 1971),[404] all contain a provision that "this Convention shall not apply to aircraft used in military, customs or police services." This appears to be a more simple way to indicate the scope of applicability of these Conventions than the provisions of Article 3 (a) and (b) of the Chicago Convention, although the end result seems to be the same. Furthermore. the clear implication is that all aircraft not so used would be subject to the provisions of the respective Conventions

The Convention on Damage Caused by Foreign Aircraft to Third Parties on the Surface (Rome, 1952)[405] states in Article 26 that, "this Convention shall not apply to damage caused by military, customs or police aircraft." It should be noted that a "military, customs or police aircraft" is not necessarily the same thing as an "aircraft used in military, customs and police services" although again the expression "military, customs or police aircraft" was left undefined. Similarly, other "state" aircraft fall within the scope of the Convention. However, the 1978 Protocol to amend this Convention reverts to more familiar language; it would amend Article 26 by replacing it with, "this Convention shall not apply to damage caused by aircraft used in military, customs and police services."

The Convention for the Unification of Certain Rules Relating to the Precautionary Attachment of Aircraft (Rome, 1933) provides that certain categories of aircraft are exempt from precautionary attachment, including aircraft assigned exclusively to a government service, including postal services, but not commercial aircraft. On

---

Eleventh Air Navigation Conference (Doc 9828, Rec. 1/2) concerning coordination with military authorities with a view to achieving enhanced airspace organization and management and as an integral supporting mechanism of the successful series of civil/military air traffic management summits instituted by the Air Traffic Control Association (ATCA). It was also a follow up to ICAO Assembly ResolutionA36-13, Appendix O, *Coordination of civil and military air traffic* wherein States are asked to take appropriate action to coordinate with military authorities to implement a flexible and cooperative approach to airspace organization and management. The Forum was intended to create awareness among civil and military policy makers and regulators, civil and military air navigation service providers (ANSPs) and civil and military airspace users, on the need to improve civil/military cooperation and coordination in support of an optimum use of airspace by all users.

[401] The Convention entered into force on 17 September 1953. See http://www.mcgill.ca/files/iasl/geneva1948.pdf.
[402] Signed at Tokyo on 14 September 1963. See ICAO Doc. 8364.
[403] Signed at the Hague on 16 December 1970. See ICAO Doc. 8920.
[404] Signed at Montreal on 23 September 1971. See ICAO Doc. 8966.
[405] Signed at Rome on 7 October 1952. See ICAO Doc. 7364.

the other hand, the Convention for the Unification of Certain Rules Relating to Assistance and Salvage of Aircraft or by Aircraft at Sea (Brussels), 1938 "apply to government vessels and aircraft, with the exception of military, customs and police vessels or aircraft ..."

The Convention for the Unification of Certain Rules Relating to International Carriage By Air (Warsaw, 1929)[406] applies, *inter alia* to all international carriage of persons. luggage or goods performed by aircraft for reward, regardless of the classification of the aircraft. Article 2 specifically provides that the Convention applies to carriage performed by the State or by legally constituted public bodies, but by virtue of the Additional Protocol, Parties may make a declaration at the time of ratification or accession that Article 2 (1) shall not apply to international carriage performed directly by the State. The Hague Protocol of 1955 to amend this Convention, in Article XXVI allows a State to declare that the Convention as amended by the Protocol shall not apply to the carriage of persons, cargo and baggage for its military authorities on aircraft, registered in that State, the whole capacity of which has been reserved by or on behalf of such authorities. Identical provisions are contained, *mutatis mutandis*, in the Guatemala City Protocol of 1971 (Article XXIII) the *1975* Additional Protocol No. 2 (Montreal), the *1975* Additional Protocol No. 3 (Montreal) and in Montreal Protocol No. 4 of 1975. It is submitted that Article 3 (b) of the Chicago Convention has no bearing on the applicability of these instruments of the "Warsaw System" which specify their own scope of applicability.

The Montreal Convention of 1999[407] which replaced the Warsaw Convention of 1929 also stipulates in its Article 1 that the Convention applies to all international carriage of persons, baggage or cargo performed by aircraft for reward. Like its predecessor, the Montreal Convention does not distinguish between civil and military or other State aircraft.

This analysis of some international air law instruments illustrates that many post-Chicago air law instruments (Geneva 1948, Tokyo 1963, The Hague 1970, Montreal 1971 and Rome 1952 and as amended in 1978) all have broadly similar provisions to Article 3 (a) and (b) of the Chicago Convention. The private air law instruments of the Warsaw System on the other hand, because of their nature, have adopted different formulae.

The provisions of the Chicago Convention and Annexes would not apply in a case where a state aircraft is (mistakenly or otherwise) operated on the basis that it is a civil aircraft. Similarly, the Geneva Convention of 1948, the Tokyo Convention of 1963, The Hague Convention of 1970, the Montreal Convention of 1971 and the

---

[406] Signed at Warsaw on 12 October 1929. The authentic French text of this Convention can be referred to in II Conference Internationale de Droit Prive Aerian (4-12 Octobre 1929). The English translation is as the Schedule to the United Kingdom Carriage by Air Act, 1932; 22 & 23 Geo.5, ch.36.

[407] *Convention for the Unification of Certain Rules for International Carriage by Air*, signed at Montreal on 28 May 1999. ICAO Doc 9740.

Rome Convention (*1952*) as amended in 1978, will also not be applicable where it is determined that the aircraft was "used in military, customs or police services". The converse, of a civil aircraft being operated on the basis that it is a state aircraft, would theoretically raise the same problems (i.e. legal regimes thought to be inapplicable are in fact applicable). Concern is not often expressed in this regard.

Another frequently mentioned difficulty is claimed to be the loss of insurance coverage in respect of the aircraft (hull), operator, crew and passengers or other parties where the aircraft is in fact state aircraft. The question whether a particular insurance coverage is rendered invalid in such situations is primarily a private law matter of the construction and interpretation of the insurance contract. Unless the contract has an exclusion clause which specifically makes reference to the classification in Article 3 of the Chicago Convention (e.g. loss of coverage where the operation is of a state (or civil) aircraft as defined in the Chicago Convention), then the Convention will have no bearing on the contract, and this issue of the loss of insurance coverage is not germane to this study. Frequently, the policy will exclude usage of the aircraft for any purpose other than those stated" in a Schedule; among the exclusions would be any use involving abnormal hazards. Nearly every aviation hull and liability policy now excludes losses due to war, invasion, hostilities, rebellion. etc., although insurance to cover such losses can usually be obtained by the payment of a higher premium. However, the instances mentioned do not require a determination of whether the aircraft is considered to be state or civil under the Chicago Convention.

A question sometimes asked is whether national civil laws and regulations would apply to civilian flight crews operating what is a state aircraft under the Chicago Convention. Would civil or military investigative and judicial processes be applied, for example. in the case of an accident? The answer would depend largely on the domestic laws of the State concerned. The fundamental principle is stated in Article 1 of the Convention: every State has complete and exclusive sovereignty over the airspace above its territory. Furthermore, subject to the provisions of the Convention, the laws and regulations of a contracting State relating to the admission to or departure from its territory of aircraft engaged in international air navigation, or the operation and navigation of such aircraft within its territory, shall be complied with by (civil) aircraft of other contracting States, upon entering or departing from or while in the territory of that State. *A fortiori*, state aircraft are also subject to the laws of the subjacent State.

In the case of an accident involving state aircraft, States are not bound by Article 26 of the Chicago Convention and Annex 13. However, they can voluntarily (through their legislation) apply' these provisions. Sometimes, the legislation specifies a different procedure in relation to military aircraft only; all other aircraft, including those used in customs or police services, are treated as civilian in this regard. In the case of other incidents, where for example the requisite over-flight permission has not been obtained by a state aircraft, which is then forced to land and charges brought against the crew, again the answer would depend on the domestic laws of the over-flown State and the factual circumstances. The classification of an aircraft as "state" aircraft under the Convention does not necessarily mean that military laws and procedures of a State would apply to that aircraft or its crew. The current or any different classification of aircraft under the Convention would not be determinative whether a particular State, in the exercise of its sovereignty, would make that aircraft and/or its crew subject to civil or military laws and regulations. As a matter of practice States usually apply military rules and processes to military aircraft and personnel only. At the international level, attempts to arrive at a common, acceptable definition of military aircraft have met with a singular lack of success.

Circular 328-AN/190 makes it clear that there are certain rules that States are required to adhere to in order to ensure that UAS operated under their control do not adversely affect civil air transport. The various provisions of the Chicago Convention and its Annexes cited in this article as well as the numerous ICAO Assembly resolutions quoted leave no room for doubt that there is an existing regime that addresses the safety of de-segregated air space when it comes to the operation of civil and State aircraft. This regime derives its legal legitimacy from the principles of State responsibility which are now accepted as binding on States. Article 1 of the *Articles of Responsibility* of the International Law Commission (ILC) expressly stipulates that every internationally wrongful act entails the international responsibility of a State.[408]

Paul Stephen Dempsey[409] sums it up well, when he says that the issue of air traffic management has two critical considerations, one relating to legal issues and the other impacting public policy. Dempsey states correctly that the skies belong to the public and the sovereign is but the trustee in this regard. Therefore, inasmuch as States cannot abdicate or pass on their responsibility and accountability of their traditional function and fiduciary responsibility, ICAO too has responsibility under Chapter XV of the Chicago Convention to assist States needing help with regard to the provision of air navigation services.

However, the bottom line with regard to legal challenges posed by the operation of UAS lies in the issue of liability and the responsibility of States. It is also now recognized as a principle of international law that the breach of a duty involves an obligation to make reparation appropriately and adequately. This reparation is regarded as the indispensable complement of a failure to apply a convention and is applied as an inarticulate premise that need not be stated in the breached convention itself.[410] The ICJ affirmed this principle in 1949 in the *Corfu Channel Case*[411] by holding that Albania was responsible under international law to pay compensation to the United Kingdom for not warning that Albania had laid mines in Albanian waters which caused explosions, damaging ships belonging to the United Kingdom. Since the treaty law provisions of liability and the general

---

[408] See Crawford (2002), p.77
[409] See Dempsey (2003), at 118–119.
[410] *In Re. Chorzow Factory (Jurisdiction) Case, (1927) PCIJ, Ser. A, no. 9* at 21.
[411] *ICJ Reports (1949)*, 4 at 23.

principles of international law as discussed complement each other in endorsing the liability of States to compensate for damage caused by space objects, there is no contention as to whether in the use of nuclear power sources in outer space, damage caused by the uses of space objects or use thereof would not go uncompensated. Furthermore, under the principles of international law, moral damages based on pain, suffering and humiliation, as well as on other considerations, are considered recoverable.[412]

The sense of international responsibility that the United Nations ascribed to itself had reached a heady stage at this point, where the role of international law in international human conduct was perceived to be primary and above the authority of States. In its Report to the General Assembly, the International Law Commission recommended a draft provision which required:

> Every State has the duty to conduct its relations with other States in accordance with international law and with the principle that the sovereignty of each State is subject to the supremacy of international law.[413]

This principle, which forms a cornerstone of international conduct by States, provides the basis for strengthening international comity and regulating the conduct of States both internally – within their territories – and externally, towards other States. States are effectively precluded by this principle of pursuing their own interests untrammelled and with disregard to principles established by international law.

These obligations are *erga omnes* affecting all States and thus cannot be made inapplicable to a State or group of States by an exclusive clause in a treaty or other document reflecting legal obligations without the consent of the international community as a whole. Besides, holding governments responsible will ensure proper quality control in the provision of air navigation services.

2.6 Regional Safety in Air Transport

Regional governance under State responsibility is an undeniable necessity in ensuring safety in air transport. However, since international air transport crosses boundaries and all regions of the world, a certain stringent global regulatory system is required, and this objective cannot be reached unless the safety oversight system is de-centralized into safety agencies in each region. This is because although air transport is still considered the safest mode of travel, constant and diligent vigilance in ensuring the safety of aviation is required. The International Civil Aviation Organization (ICAO), which is the global forum for aviation, provides this

---

[412] Christol (1991), at 231.

[413] *Report of the International Law Commission to the General Assembly on the Work of the 1st Session, A/CN.4/13*, June 9 1949, at 21.

guidance and direction and encourages States to form regional civil aviation bodies in regions where they do not exist. Since almost all regional civil aviation bodies have their own Constitutions, these Constitutions should recognize the need for increased cooperation with ICAO and ICAO should actively seek such an inclusion. Cooperation could include assignment of tasks related to oversight, accident investigation, sharing of information and the conduct of training in collaboration with ICAO.

Some of the measures that are suggested in this article include, the initiation by ICAO where appropriate, of periodic meetings with regional civil aviation bodies and the setting up of panels that include members from ICAO and each regional body on a selective basis. This panel could recommend and monitor issues of cooperation, including cooperation between the various regional organizations.

There is also the compelling need to define the role that regional organizations and regional civil aviation bodies would play in working closely with ICAO; and the adoption of a policy with regard to cooperation with regional organizations and regional civil aviation bodies. ICAO should also ascertain the role to be played by the regional offices in coordinating ICAO cooperation with such bodies. Consideration should be given to providing appropriate resources in regional offices where necessary.

The role played by Regional Safety Oversight Organizations (RSOOs)[414] has sustained serious discussions in the International Civil Aviation Organization[415] and in particular in the Council[416] of that Organization over the past year. At its 187th Session on 10 June 2009, the Council accepted several principles with regard to the role of ICAO in regional governance[417] which were placed before it by a multidisciplinary group comprising some Representatives on the Council and members of the ICAO Secretariat, which had been tasked with producing a report on ICAO and Regional Governance, by the Council.

---

[414] Sometimes referred to as Regional Safety Oversight Agencies (RASAs).

[415] Supra n. 1.

[416] 3. The ICAO Council is a permanent body responsible to the Assembly. It is composed of 36 Member States elected by the Assembly. In electing the members of the Council, the Assembly gives adequate representation to States of chief importance to air transport; States not otherwise included which make the largest contribution to the provision of facilities for international air navigation; and States not otherwise included whose designation will ensure that all the major geographic areas of the world are represented on the Council. The mandatory and permissive functions of the Council are stipulated in Articles 54 and 55 of the Convention on International Civil Aviation respectively. The Council has its genesis in the Interim Council of the Provisional International Civil Aviation Organization (PICAO). PICAO occupied such legal capacity as may have been necessary for the performance of its functions and was recognized as having full juridical personality wherever compatible with the Constitution and the laws of the State concerned. See Interim Agreement on International Civil Aviation, opened for signature at Chicago, December 7 1944, Article 3. Also in Hudson, *International Legislation*, Vol 1X, New York: 1942-1945, at 159. For a detailed discussion on the functions of the Council See Abeyratne (1992).

[417] C-WP/13339, 12/05/09.

2.6 Regional Safety in Air Transport

At its meeting, the Council agreed ICAO should enhance its cooperation with regional organizations and regional civil aviation bodies and vice-versa, both in the technical and economic fields and should ensure that the interests of States which do not belong to regional civil aviation bodies should not be jeopardized or compromised in the above context. The Council also recognized that while ICAO encourages the activities of States and regional civil aviation bodies in facilitating the implementation of Standards and Recommended Practices of the Annexes to the Convention on International Civil Aviation (Chicago Convention)[418] (SARPs), States ultimately remained responsible for their obligations under the Convention on International Civil Aviation, notwithstanding whatever arrangements States may conclude with their regional civil aviation bodies. It was also agreed that ICAO should encourage States to form regional civil aviation bodies in regions where they do not exist, and that ICAO should define the role that regional organizations would play in working closely with ICAO.

This article examines the legal and regulatory basis of ICAO's role with regard to the establishment and functions of RSOOs with reference to selected regional oversight organizations around the world.

2.6.1 The Role of ICAO

The genesis of ICAO's role lies in the Chicago Convention[419], Article 55 a) of which states that the Council may, where appropriate and as experience may show to be desirable, create subordinate air transport commissions on a regional or other basis and define groups of States with or through which it may deal to facilitate the carrying out of aims of the Convention. Article 65 of the Convention allows the Council to enter into agreements with other international bodies for the maintenance of common services with and for common arrangements concerning personnel and, with the approval of the Assembly, allows the Council to enter into such other arrangements as may facilitate the work of the Organization.

This provision prompted the ICAO Assembly, at its First Assembly held on 6–27 May 1947 to adopt Resolution A 1–10, which recognized that there were a number of public organizations whose activities affect or are affected by ICAO and that the work of ICAO and the advancement of international civil aviation will be enhanced by close cooperation with such organizations. The Resolution authorized the Council to make appropriate arrangements with public international organizations whose activities affect international civil aviation and suggested that such arrangements be established through informal working arrangements rather than formal arrangements, wherever practicable.

---

[418] *Supra*, note 3.
[419] *Id.*

At its Tenth Session conducted from 19 June to 16 July in 1956, the ICAO Assembly adopted Resolution A10-5, which established a policy framework for ICAO and the European Civil aviation Conference (ECAC) which had its focus on regional cooperation. An agreement effective 12 July 1969 was signed by ICAO and ECAC which allowed the appointment of ICAO staff to the Secretariat of ECAC, and the provision of qualified and expert staff to the latter. A similar agreement effective 1 January 1978 was signed between ICAO and the African Civil Aviation Commission (AFCAC0 which provided *inter alia* that ICAO will provide, to the extent possible, secretariat services to AFCAC. There is also an agreement, which became effective on 1 January 1979 between ICAO and the Lating Civil Aviation Commission (LACAC) which called for close cooperation between the two Organizations and the provision of secretariat services to LACAC by ICAO.

There are numerous other instances of such regional cooperation which followed such as the signing of a memorandum of understanding by ICAO and LACAC on 1 October 1999 followed by an MoU between ICAO and ECAC on 11 November 1999. These were to be followed by a management service agreement between ICAO and AFCAC which came into effect on 1 January 2007.

These regional bodies are by themselves not RSOOs, although they could, and indeed should play an integral role in assisting States in their oversight functions and the RSOOs in their work. In this regard, it is critical to recognize that regional governance and regional safety oversight are prerogatives which devolve consequent responsibility on the State which cannot be delegated. The RSOO acts as an agent[420] of the State (which is the Principal) with delegated functions from the Principal. ICAO, to the extent of its empowerment by its member States, can assist in the oversight process, offer expertise, advice, technical support and even manage an RSOO if necessary.

Therefore, ICAO's position with regard to RSOOs is that it would give its full cooperation to regional governance as determined by States through carefully formulated policy which would in no way abdicate or delegate ICAO's responsibilities under the Chicago Convention. In the agency equation the State remains the Principal and the RSOO remains the Agent and ICAO remains the organization empowered by the States to carry out audits and report results thereof.

2.6.2 The Regional Safety Oversight Manual

ICAO's perspectives on regional safety oversight are reflected in its *Regional Safety Oversight Manual*[421] which provides guidance for States that wish to form

---

[420] An agency is a consensual relationship created by contract or law where the Principal grants authority to agents to act on behalf of or under the control of the Principal to deal with a third party.

[421] ICAO Doc 9734 AN/959 Part B (The Establishment and Management of a Regional Safety Oversight System).

2.6 Regional Safety in Air Transport

a regional safety oversight system. It formulates a regional strategy which attenuates and brings together the efforts, experiences and resources of contracting States. From ICAO's perspective the reasons for adopting a strategy to establish RSOOs are to: eliminate a duplication of effort by standardizing regulatory and enforcement provisions over a large area of aviation activities; achieve economies of scale leading to efficiency and effectiveness; pool human and financial resources; institute effective regional programmes through the joint action of States; address external factors and constraints more effectively; develop and implement a safety management system[422] that would allow for the implementation of similar standards and procedures to measure the safety performance of civil aviation organizations in the region; supplement shortfalls in the scope of domestic and bilateral interventions; prove organizational ability by testing activities before making important commitments under national programmes; meet industry expectations; demonstrate improved regional solidarity; improve the objectivity and independence of inspectors; and develop a capability for drafting and amending regulations and procedures.[423]

ICAO believes that RSOOs should have goals which reflect national priorities, and recommends that States consider providing adequate and efficient resources and ensure coordination between high level government officials who are responsible for aviation safety. The most incontrovertible principle ICAO believes in is, as mentioned earlier, that the responsibility for safety oversight within the State remains with the State and cannot be delegated.[424] The critical activities of a State are identified in the *Manual* as: the licensing of operational personnel; the certification of aircraft, air operators, aerodromes, and maintenance organizations; the control and supervision of licensed personnel, certified products and approved organizations; the provision of air navigation services[425]; and aircraft accident and incident investigation.[426]

ICAO's role, according to the *Manual*, is to provide assistance to States that are willing to enter into a cooperative agreement for the establishment and management of an RSOO on the basis of an agreement between ICAO and the interested parties. ICAO may also manage an RSOO until such time as the State concerned develops an ability to perform the task. The Organization may also provide technical and logistical support as well as necessary information and documents that may be needed by an RSOO and also, upon a request of a State monitor the effectiveness of an RSOO. Where necessary and as requested, ICAO could, according to the

---

[422] For a detailed discussion on safety management systems see Abeyratne (2007).

[423] *Id*. Para 2.2.4 at pp. 2–3.

[424] *Id*. Para. 2.2.5 at pp. 2–4.

[425] Including meteorological information, aeronautical telecommunications, search and rescue services, charts and the distribution of information.

[426] *Id*. Para. 2.4.5 at pp. 2–6.

Manual, provide advice or propose action required in the proper running of an RSOO.[427]

In this regard the Symposium was also advised that ICAO has conducted a series of safety audits throughout the world and information of these audits were available.

2.6.3 The ICAO AFI Comprehensive Implementation Programme (ACIP)

The ICAO Assembly, at its 36th Session held in September-October 2007, adopted Resolution A36-1 which recognized that it was essential that there be increased coordinated efforts under ICAO leadership to reduce serious deficiencies in the Africa-Indian Ocean (AFI) Region which are detrimental to the functioning an further development of international civil aviation. The resolution also noted that the Council of ICAO had already taken steps to address safety issues through the development of a Comprehensive Regional Implementation Plan for Aviation Safety in Africa (the AFI Plan)[428] and that many Contracting States in the AFI Region may not have the technical or financial resources to comply with the requirements of the Chicago Convention and its Annexes. It also recognized the need to coordinate, under the ICAO umbrella, activities of all stakeholders providing assistance to States in the AFI Region. Resolution A36-1 therefore urged Contracting States of the AFI Region to commit to the achievement of the goals and objectives of the AFI Plan and to ongoing transparency with regard to the progress accomplished. It instructed the Council to coordinate the contributions towards the implementation of the AFI Plan and to ensure a stronger ICAO leadership role in coordinating activities, initiatives and implementation strategies aimed specifically at meeting the goals and objectives of the Plan, in order to achieve sustained improvement of flight safety in the AFI Region and to allocate resources to the relevant Regional Offices accordingly; Furthermore the Council was instructed to ensure the continued development of new working relationships integrating the capabilities of the bureaux at Headquarters with the resources of Regional Offices.

The Council started with the initial premise that the problems facing the States in the AFI Region were not dissimilar to what other States around the world were facing. The one unique feature of the AFI region, however, was that the acute economic and political issues facing States of the AFI region were indeed complex in nature and therefore posed a challenge to aviation safety in the region which

---

[427] *Id.* Para 3.5 at pp. 3–5.

[428] The Comprehensive Regional Implementation Plan for Aviation Safety in Africa (AFI Plan) was developed by ICAO with a view to addressing the various concerns expressed by the ICAO Council on the status of safety of aircraft operating in the African and Indian Ocean Region.

clearly required a new approach. The AFI Plan was a response to this particular exigency and was developed with a view to addressing aviation safety concerns and supporting African States in their endeavour to meet their international obligations for safety oversight under the applicable Annexes to the Chicago Convention.

The AFI Plan was considered by a high-level conference convened in Montreal on September 17, 2007, which unreservedly endorsed it. Consequently, the AFI Plan was presented to the 36th Session of the ICAO Assembly, held from 18 to 28 September 2007 and the Assembly requested both the Council and the Secretary General to implement the AFI Plan within the shortest possible period. The Assembly's AFI Plan Resolution, A36-1, also emphasized a heightened leadership role and accountability by ICAO for the Plan's effective implementation, supported by strong programme management and coordination activities. Thus, under ICAO's leadership, the AFI Plan calls for optimal collaboration between regulatory agencies and industry in the implementation of initiatives aimed at rectifying safety deficiencies.

To this end, specific objectives have been developed requiring ICAO to: a) Increase compliance with ICAO SARPs and industry best practices; b.) Increase the number of qualified personnel at the industry and oversight levels. c) Improve the quality of inspectors and other civil aviation staff through training. d) Ensure impartial and unimpeachable investigation and reporting of serious accidents and incidents. e) Enhance regional cooperation. f) Enhance capacity of regional and sub-regional safety oversight systems. g) Improve assistance in oversight to least developed States. h) Provide expert aviation knowledge within the reach of the targeted States via the web.

The Secretary General established the AFI Comprehensive Implementation Programme (ACIP– which was dissolved in 2010 and reverted to the earlier AFI Plan) with a view to ensuring that the aforesaid objectives are met. For this purpose, he nominated members to the Steering Committee of ACIP to oversee its work. ACIP, which is aimed at implanting the AFI Plan recognizes the basic requirements of a safety oversight programme to be a regulatory framework; responsibility and accountability; accident and incident investigation; and enforcement policy, all of which were necessary for a State safety programme. He identified the essential elements for a safety risk management programme as an adequate oversight organization with trained personnel, equipment, tools, guidance material and processes and procedures for exchanging safety critical information.

The basic requirements for an effective safety culture are management commitment and safety accountability; appointment of key safety personnel; coordination of emergency response planning and safety management system documentation.

2.6.4 Safety Oversight in Africa

It is heartening that safety levels in Africa were improving although the accident rates in Africa were still above world average levels. The main instances of accidents in African States were mostly seen in conflict ridden States, particularly in the Democratic Republic of Congo, Angola and Sudan. The main reasons for accidents were ageing aircraft, lack of financing for fleet modernization and the lack of adequate skilled personnel. The reduction of accident rates in Africa could be attributed to the adoption of IOSA,[429] the ICAO Global Safety Roadmap[430] and the proactive approach taken towards safety by the key stakeholders. He concluded that RSOOs would enhance safety oversight and that toward this end there was a need for African States to have autonomous and regionalized civil aviation authorities. He was also of the view that there was a need to pool scarce resources in training and to harmonize standards on conditions of service and the operation of regional airlines. One of the key features of RSOOs was their ability to be better able to provide attractive conditions of service for their personnel and to mobilize resources for infrastructure.

2.6.4.1 The Banjul Accord Group (BAG)

The BAG States (Cape Verde, Gambia, Ghana, Guinea Conakry, Liberia, Nigeria and Sierra Leone) play a significant role in leadership in aviation safety oversight. The 10th Plenary Session of the BAG States held on 30–31 October 2008 reached the conclusion that a regional safety oversight agency and a regional accident investigation agency must be established. In this regard BAG States had sent a letter seeking assistance to ICAO. ICAO was requested for assistance in developing a framework for the establishment of the BAG Regional Safety Oversight Organization (BAGASOO) and the development of a framework for the establishment of a Regional Accident Investigation Agency (BAGAIA). ICAO was also requested to assist in developing operational regulations and guidance material for the implementation of BAGASOO and BAGAIA and the development of a guidance manual for the implementation of a regional safety programme.

It is also noteworthy that the Directors General of Civil Aviation of the BAG States met with ACIP management in Accra on 15–16 December 2008 and

---

[429] The IATA Operational Safety Audit (IOSA) programme is an internationally recognised and accepted evaluation system designed to assess the operational management and control systems of an airline. IOSA's quality audit principles are designed to conduct audits in a standardised manner. With the implementation and international acceptance of IOSA, airlines and regulators achieve the following benefits: a reduction of costs and audit resource requirements for airlines and regulators; continuous updating of standards to reflect regulatory revisions and the evolution of best practices within the industry; a quality audit programme under the continuing stewardship of IATA; accredited audit organisations with formally trained and qualified auditors; accredited training organisations with structured auditor training courses; a structured audit methodology, including standardised checklists; elimination of audit redundancy through mutual acceptance of audit reports; and development of auditor training courses for the airline industry.

[430] For a discussion on the Roadmap, see Abeyratne (2009b).

developed an action plan and determined the scope of cooperation, resulting in a letter of understanding.

Three events which had already been held and were significant in the evolution of BAGASOO were: The Global Aviation Roadmap Workshop (Abuja, April 2008); GAP Analysis with GASRs in All BAG States (May-June 2008); and the 10th Plenary Session of the BAG States (Banjul, October 2008). Priority action of BAGASOO, as already presented to the ICAO Council included: development of a framework for the accelerated establishment of BAGASOO; Development of a framework for the establishment of BAGAIA; development of operational regulations and guidance material for the implementation of BAGASOO and BAGAIA and the development of guidance material for the implementation of the regional programme.

The ICAO Council, in November 2008, had requested Contracting States, industry and donors to assist the BAG States in implementing priority projects and actions as determined during the 10th Plenary Session of the BAG States, and that ACIP take necessary measures to assist BAG States in the implementation of such projects.

2.6.5 Safety Oversight in Central America (ACSA)

ACSA, the Central American Agency for Aeronautical Safety, created in 2001, is committed to the regional capability of safety oversight, through reduced bureaucracy and the avoidance of duplication of efforts. ACSA concentrates on sharing expertise and resources among member States and was ISO 9001 compliant in terms of quality management. ACSA is an agency of COCESNA, which was tasked with implementing plans pertaining to COCESNAs Regional Information Aeronautical System (SIAR) on safety issues, continuous surveillance, licensing and qualified personnel.

2.6.6 Safety Oversight in the Caribbean

The Civil Aviation Authorities of Barbados, Belize, Guyana, Haiti, Jamaica, OECS (Anguilla, Antigua and Barbuda, Dominica, Grenada, Montserrat, St. Lucia, St. Kitts and Nevis, St. Vincent and the Grenadines), Suriname, and Trinidad and Tobago, Member States of the Caribbean Community, signed an agreement in late 2001, formalizing their participation in and support for a cooperative approach to aviation safety oversight. The Agreement provided for the establishment of an "Association of Civil Aviation Authorities of the Caribbean" (ACAAC) under the umbrella of the Caribbean Community (CARICOM) Secretariat and to form its operating arm, the Regional Aviation Safety Oversight System (RASOS). Each RASOS member authority was mandated to implement the provisions of all ICAO

Annexes. It was the RASOS mandate to assist them with specific regard to Annexes 1, 6, and 8 of the Chicago Convention. The mandate also involved aiding, facilitating, harmonizing and sharing resources for the provision of aviation safety oversight services in 13 small nations in the Caribbean region. Although all participating authorities belong to States that are members of CARICOM, membership in the CARICOM was never a pre-requisite for membership in the ACAAC.

RASOS Member States developed a formal agreement signed by Heads of State in 2008 to widen the regional organization's mandate to include all ICAO Annexes. This marked a major step forward in elevating the RASOS status by establishing it as a new entity, and renaming RASOS as the Caribbean Aviation Safety and Security Oversight System (CASSOS) and having it designated as an Institution of the Community by the Conference of Heads of Government pursuant to Article 21 of the *Revised Treaty of Chaguaramas Establishing the Caribbean Community Including the CARICOM Single Market and Economy*. The agreement subsumed RASOS into CASSOS and ACAAC no longer exists. CASSOS has full juridical personality, and its Board of Directors report to the Ministers who constitute the CARICOM Council for Trade and Economic Development (COTED).

Originally, the Members of ACAAC implemented the Regional Aviation Safety Oversight System (RASOS), to share resources and reduce the cost of providing the required airworthiness and flight operations oversight services to individual member states. The RASOS concentrated on the optimization of the use of the region's technical resources. Its focus was to strengthen the civil aviation authorities, promote the upgrading and harmonization of regulations, standards, procedural guidance material, inspector training and to enhance the ability of the individual states to fully discharge their safety oversight responsibilities. The sub-regional approach chosen is consistent with the global strategy promoted by ICAO to address safety oversight problems of contracting states. RASOS core operations were funded by equal annual member CAA contributions and it was self sufficient during its existence, and by its frugality was able to commence CASSOS with no additional expenditures.

Assistance from the FAA between the years 2003–2008 provided numerous training courses for RASOS Member CAA's inspectors and other technical staff, ICAO courses were delivered in the region with regard to aerodrome certification and dangerous goods and PEL. Transport Canada assisted with some compliance/enforcement training, and medical examiner training. The FAA provides extensive in-country assistance by providing technical experts in an effort that was aimed at advising and assisting Members to achieve IASA category one and compliance with Annexes 1, 6 and 8. This assistance included mentoring of inspectors and technical advice provided during re-certification of air operators. Other ongoing FAA technical assistance pursuant to Technical Assistance Agreements was aimed at full implementation in the first half of 2008 of a common and ICAO compliant computer based written knowledge testing system. A regionally developed, harmonized, common license format and production system has been installed in all member authorities and could be made be available commercially to any other authorities that are interested in such a system. Transport Canada continues to assist

2.6 Regional Safety in Air Transport

with training of civil aviation medical examiners and cabin safety inspectors and is offering ongoing training support in SMS and aviation compliance and enforcement. Common qualifications and training standards for inspectors have been enunciated to facilitate resource sharing, that is, the trans national use of inspectors, and procedures for designation and delegation of authority and for requesting, tasking and deploying trans national inspectors have been developed and approved. A Policy and Procedures Manual was developed to guide the management and operation of RASOS and in its latest version will now be used to guide and direct CASSOS activities. Inspector guidance material is shared freely between Members and is well harmonized already. It is anticipated that development of unitary common guidance material will follow the development of common "regional" civil aviation regulations while at present, all regulations are based on adaptations of the ICAO MCAR and are virtually identical. Other initiatives include harmonized enforcement procedures and inspection procedures.

In 2009 four of the original seven RASOS Member CAAs continued to meet the IASA Category One standards. CASSOS, in a manner similar to that used by the European Aviation Safety Agency (EASA), performs reviews of its Members using experienced inspectors from the region and checklists derived from IASA and ICAO USOAP checklists. Reports developed for the Member authorities are reviewed and the results are used to determine, prioritize and respond toregion-wide needs. The reviews assist with harmonization activities and have also assisted members to prepare for IASA and ICAO audits some Members find them useful to develop compliance action plans. There remains an ongoing need for on-site mentoring and training of technical personnel and for technical assistance in all Member CAAs, particularly as the CASSOS mandate has been significantly widened and new expectations arise resulting from changes to the Annexes and technology.

Using needs assessment methodology the regional body has identified the need for professional training and recurrent qualification training of airport operators' personnel. It has from its own resources and assisted by a member of the FAA airport standards staff, delivered a three day seminar on aerodrome manual preparation to some 33 aerodrome specialists from the region.

The RASOS web site contains public information and members' only sections. The inspectors' section contains downloadable inspection forms, some common guidance material, flight test forms and other data required by the region's technical staff. The site also hosts a safety newsletter, links to Member CAA sites, links contact to RASOS and provides a secure 128 bit encrypted e-mail service for the Directors, RASOS staff and all technical safety inspectors in the RASOS group and other selected officials who have been working with RASOS. It is a very strong tool for communication, information and data sharing and for providing a public identity for the organization as well as serving as a virtual office for RASOS personnel. The website is being updated and changed to reflect CASSOS operations and that should be completed by October, 2009.

CASSOS has adopted the European Center for Civil Aviation Incident Reporting System (ECCAIRS) for incident and accident reporting and, in a regional project,

CASSOS Members use a common, harmonized regional inspection planning, tracking and reporting system. While this system respects national security, confidentiality and sovereignty as required, it provides a valuable tool for analysis and tracking of trends and allow development of appropriate safety and regulatory interventions. CASSOS will share inspection data as in the European Safety Audit of Foreign Aircraft (SAFA) system. Seminars in ALAR CFIT accident reduction have been delivered and this will continue under CASSOS with a much widened safety promotion mandate. The regional organization has assisted members with accident and incident investigation and it is envisaged that this will grow into a truly regional service as the benefits of a centralized investigating office are beyond question. The foregoing summarizes the major efforts of the past seven years toward safety oversight harmonization within the CARICOM CSME framework.

All of the above initiatives have been aimed at building a strong regional regulatory and Safety Management System to enhance civil aviation safety in Member states and throughout the region and are continuing under CASSOS. Funding at this moment in time is limited to provision of two technical experts and one administration person. Future development and strengthening of the regional safety oversight capability may require increasing member contributions or new sources of funding or assistance.

The direct beneficiaries of the regional CASSOS institution activities are the participating States of CARICOM and will include any other States or Territories in the region that might become part of the regional aviation safety oversight mechanism during the next few years. Other direct beneficiaries of CASSOS activities are the owners and operators of aircraft and all who use the aviation industry infrastructure and services in the CASSOS States. One must not overlook the indirect beneficiaries of the air transport, aviation services and infrastructure that includes the tourism and business sectors of the economies. External benefits flow to the States from the improved aviation safety environment resulting from the upgraded aviation infrastructure and the increased surveillance and enforcement of the safety standards established by ICAO.

CASSOS has matured from its fairly humble beginnings during 7 years of hard work by all persons involved and now has 7 years of successful operating experience in coordinated, cooperative, harmonized, self-sufficient group efforts aimed at providing safety oversight services to the high economic value air transportation system in all participating States as well as to other States whose airlines operate into the region. This high level of achievement will continue as CASSOS continues its growth into a truly regional institution.

The immediate benefits of regional cooperation are evident from the constantly improving track record of results of the ICAO and FAA safety oversight audits of member CAAs. Benefits are also accruing to members from the mutual technical cooperation, mutual technical assistance, attainment of greater numbers of trained and qualified technical inspectors, and the valuable technical expertise contributions made by all Members in their efforts to achieve and sustain compliance with international aviation safety oversight standards at affordable costs.

A strong regional safety oversight partnership has been forged. Future activities are aimed at establishing a permanent Headquarters, undertaking new regional projects such as a single upper airspace control system, introduction of new ATM surveillance technology and air navigation technology, and managing safety initiatives and interventions to keep the aviation system loss rates as low as possible.

2.6.7 Safety Oversight in South Asia

South Asian Regional Initiative (SARI) comprises a group of authorities from South Asia including those of Bangladesh, Bhutan, India, Maldives, Nepal, Pakistan and Sri Lanka. SARI was created during the EU-South Asia Civil Aviation Cooperation Programme in 2006. SARIs objective is to create a forum for civil aviation authorities in South Asia to foster regulatory convergence. It was mentioned that EASA and the European aerospace industry support the activities of SARI with a view to reviving the Cooperation Programme.

SARI was created to update insufficient and outdated legislation, regulations and guidance on aviation safety and provide a legal foundation for regional cooperation in South Asia. It is also aimed at assisting airlines and regulators in interpreting rules. One of the main goals of SARI is to ensure that the region retains qualified technical personnel and to eliminate duplication of efforts and confusion of functions within civil aviation authorities.

2.6.8 Safety Oversight in Europe

Arguably the most developed regional implementation system of safety oversight lies in Europe, through the European Aviation Safety Agency (EASA). EASA has evolved through a gradual process of regulation in Europe where the roots of institutionalized regional regulation of aviation safety goes back to 1954 with the creating of the European Civil Aviation Conference (ECAC). ECAC is a consultative body whose resolutions and other recommendations for Europe are subject to approval of the European Union (EU) member States. The predecessors of EASA were the Joint Airworthiness Authorities (JAA) who began their work in 1970 with the initial mandate given by the EU to produce common certification codes for large aircraft and engines. The Authorities were intrinsically associated with ECAC and subsequently came to be renamed as the Joint Aviation Authorities, retaining the earlier acronym. The new JAA had an extended mandate reaching into operations, maintenance, licensing, certification and design standards for all types of aircraft. One of the products of JAA was

JARs (Joint Aviation Requirements) which carried regulatory thrust withinh the EU.[431]

EASA came into being on 15 July 2002,[432] replacing JAA in 2009. It is an agency of the European Union and is governed by EU public law. It stands separate from the EC institutions such as the European Council, European Parliament and the European Commission, and has its own legal *persona*. Its main areas of focus, in terms of ensuring the highest standards within Europe, are aviation safety and environmental protection (from an aviation perspective) in which it develops rules within Europe. EASA employs inspectors to monitor the implementation within EU States of ICAO SARPs and cooperates with national authorities on the issuance of certificates of airworthiness and the licensing of technical crew.

EASA's responsibilities include: giving expert advice to the EU for drafting new legislation; implementing and monitoring safety rules, including inspections in the Member States; type-certification of aircraft and components, as well as the approval of organisations involved in the design, manufacture and maintenance of aeronautical products; authorization of third-country (non EU) operators; and safety analysis and research. These responsibilities are growing to meet the challenges of the fast-developing aviation sector. It is envisioned that, in a few years, the Agency will also be responsible for safety regulations regarding airports and air traffic management systems.

On 28 September 2003, EASA took over responsibility for the airworthiness and environmental certification of all aeronautical products, parts, and appliances designed, manufactured, maintained or used by persons under the regulatory oversight of EU Member States.

The Agency's certification work also includes all post-certification activities, such as the approval of changes to, and repairs of, aeronautical products and their components, as well as the issuing of airworthiness directives to correct any potentially unsafe situation. All type-certificates are therefore now issued by the European Aviation Safety Agency and are valid throughout the European Union.

On the same date the Agency became the competent authority to approve and oversee the organisations involved in the design of aeronautical products, parts and appliances. It also carries out the same role for foreign organisations involved in the manufacture or maintenance of such products.

To execute its tasks within the present period of building up its resources, the Agency relies on national aviation authorities who have historically filled this role and concludes contractual arrangements to this effect.

If one were to compare the sharing of responsibilities between national and regional authorities involving ACSA, BAGASOO, CASSOA, CASSOS and EASA, on initial airworthiness, EASA, ACSA and BAGASOO share responsibility with national authorities on functions (not accountability) while CASSOA and CASSOS

---

[431] *As per European Regulation 3911/11.*

[432] *EASA was created by Regulation EC No 1592/2002 by the European Parliament and the Council of the EU.*

2.6 Regional Safety in Air Transport

do not show such sharing. On continuing airworthiness all agencies show shared responsibility at national and regional levels. This is also the case in operations and licensing. On air traffic management and airports there is consistency in all agencies in sharing responsibility of functions.

On regulatory aspects, all agencies in the comparison exercise share responsibilities on the development of regulations while none have a mandate to develop hard law (which is as should be, as legislation is the sole purview of sovereign States). CASSOS and EASA have the power to issue certificates while the other three do not have this power. In terms of judicial control, ACSA is under the jurisdiction of the American Court of Justice, while CASSOA is under the East African Court of Justice. CASSOS comes under the Caribbean Court of Justice while EASA falls under the jurisdiction of the European Court of Justice.

So where does this leave ICAO? To start with, it is incontrovertible that ICAO's involvement in RSOOs is both inevitable and pervasive. Whatever the regional or national initiatives may be in this area, since air transport involves the operation of foreign built aircraft in States, their certification and licensing has to be carried out according to standards set by ICAO in the relevant Annexes to the Chicago Convention.[433] This is particularly recognized by Article 33 of the Convention which provides for the acceptance by one State of certificates of airworthiness/competency and licenses issued by another, provided such document conform to the specifications set by the Convention. Therefore, whether it be the International Safety Assessment Programme (IASA) of the United States or EASA of the European Union, It is likely that any prudent State will refuse to recognize documents if they do not meet the Standards of the ICAO Annexes. In terms of empowerment and enforcement with regard to implementation, Article 87 of the Convention provides that the ICAO Council can ban the operation of an airline worldwide if such operation does not conform to a final decision of the Council. Therefore, ICAO is indispensable for the global operation of aircraft within the safety oversight equation.[434]

However, regional initiatives such as those discussed above can give ICAO support and an impetus in its work. In this regard it must be stated that there are many measures that ICAO could take in making its work easier and more effective. Firstly, ICAO should enhance its cooperation with regional organizations and regional civil aviation bodies and *vice-versa*, both in the technical and economic fields. In doing so, ICAO should ensure that the interests of States which do not belong to regional civil aviation bodies should not be jeopardized or compromised. It should be noted that, while ICAO encourages the activities of States and regional

---

[433] *Regulation EC No 2111/2005 requires that certificates and licences issued in the EC member States have to conform to international safety standards contained in the Chicago Convention and its Annexes.*

[434] *For a more in depth discussion on the powers of the ICAO Council in the field of Safety, See Abeyratne (2009c), at 196–206.*

civil aviation bodies in facilitating the implementation of SARPs, States ultimately remain responsible for their obligations under the Convention on International Civil Aviation, notwithstanding whatever arrangements States may conclude with their regional civil aviation bodies.

ICAO should also encourage States to form regional civil aviation bodies in regions where they do not exist. Since almost all regional civil aviation bodies have their own Constitutions, these Constitutions should recognize the need for increased cooperation with ICAO and ICAO should actively seek such an inclusion. Cooperation could include assignment of tasks related to oversight, accident investigation, sharing of information and the conduct of training in collaboration with ICAO.

ICAO should, where appropriate, initiate periodic meetings with regional civil aviation bodies and set up a panel that includes members from ICAO and each regional body on a selective basis. This panel could recommend and monitor issues of cooperation, including cooperation between the various regional organizations.

ICAO should define the role that regional organizations and regional civil aviation bodies would play in working closely with ICAO; and should adopt a policy with regard to cooperation with regional organizations and regional civil aviation bodies. ICAO should also ascertain the role to be played by the regional offices in coordinating ICAO cooperation with such bodies. Consideration should be given to providing appropriate resources in regional offices where necessary.

Regional governance under State responsibility is an undeniable necessity and ICAO has been unreserved in supporting it. ICAO's policy would be one driven by its ability to provide assistance and to implement this policy ICAO needs close cooperation with the regional civil aviation bodies and the political will of States. Any ICAO policy, however enthusiastically adopted would be destitute of effect if there is no endorsement and support of its member States.

2.7 Safety Oversight Audits

One of the significant results of the 36th Session of the ICAO Assembly, held in September 2007, was the adoption of two Resolutions calling for the sharing of information through the ICAO Council pertaining to safety and security audits conducted by ICAO. This brings to bear a certain shift of focus from the original confidentiality of the audits to one of limited transparency. It also raises the more compelling issue as to what the legal principles applicable are that would attribute to the Council the ability to divulge information and the limitations if any, on carrying out the instructions of the Assembly, which is one of the mandatory functions of the Council. The question also arises as to whether such a function could be sustained in the face of other overriding factors, one of which is the extent to which ICAO stands empowered by its constituent member States to divulge information pertaining to aviation activities in their territories.

2.7.1 Blacklisting of Airlines

A compelling issue involving aviation safety in the European Union concerns the EU's policy on blacklisting[435] of airlines considered unsafe. This process is predicated upon common criteria that take into consideration results of checks carried out in European airports and determination as to whether airlines flying into Europe are using poorly maintained, antiquated or obsolete aircraft. Other considerations are inability of airlines identified as unsafe to rectify shortcomings and the inability or unwillingness of the authorities charged with the safety of such airlines to enforce safety standards on airlines.

The Appendix to 473/2006 of March 2006 stipulates that the criteria for blacklisting would broadly be based on safety deficiencies reflected in reports that clearly show such deficiencies or persistent failure by the carrier concerned to address deficiencies identified by ramp inspectors performed under the Safety Assessment of Foreign Aircraft (SAFA) programme implemented by the Joint Aviation Authorities (JAA). The Joint Aviation Authorities is an associated body of the European Civil Aviation Conference (ECAC) representing the civil aviation regulatory authorities of a number of European States who have agreed to co-operate in developing and implementing common safety regulatory standards and procedures. This co-operation is intended to provide high and consistent standards of safety and a "level playing field" for competition in Europe. Much emphasis is also placed on harmonising the JAA regulations with those of the USA.

Another criterion is the effective enforcement of an operating ban imposed on a carrier by a third country predicated upon substantiated deficiencies related to international safety standards. Yet another is substantiated accident-related information or serious incident-related information indicating latent systemic safety deficiencies. The lack of ability and/or willingness of a carrier to address safety deficiencies that is demonstrated by lack of transparency or failure to communicate in a timely manner in response to a query by an EU member State regarding a carrier's safety practices and record and an inappropriate or insufficient corrective action plan are also criteria that would prompt inclusion in the black list.

Civil aviation authorities may also be subject to action under the blacklist if they do not pay sufficient attention to audits and related corrective action plans emerging from the ICAO's USOAP or any applicable Community law. The ban could be a complete one when shortcomings are due to action or inaction of the civil aviation authority concerned and a partial ban on an airline where it could operate into the Community only with its aircraft that meet the safety standards.

---

[435] Regulation 2111/2005 of 14 December 2005 of the European Parliament and Council required that a list be established within the European Community of air carriers that would be subject to an operating ban within the territory of the Community. This Regulation also provided for informing passengers of the identity of the carriers concerned. On 22 March 2006, by Regulation 473/2006 the European Commission laid down rules of implementation for the Community list of carriers which would be subject to an operating ban within the territory of the Community.

On the positive side, the flexing of the EU's muscles has brought to bear the awareness of the world of high risk airlines and gives the aviation community a sense of direction on the enormity of the issue. However, on the negative side is the lack of transparency in the process where criteria for blacklisting are not clearly spelt out. Furthermore, there is a perceived inequity in identifying an entire airline or aviation authority. There is also the seemingly impracticable requirement that a State or airline has only 10 days to respond to the European Community after being included in the blacklist.

2.7.2 Policy Aspects of Blacklisting

Arguably the most significant shortcoming of blacklisting is that it is a political measure rather than a solution toward improving aviation safety. For instance, in the African region where the accident rate is 6–7% worse than the global average, where 18.5% of the fatal airline accidents occurred in Africa during the year 2006 there are socio economic issues at the core and blacklisting would be a punitive measure rather than a corrective one. It must be noted that air transport in Africa generates 470,000 jobs and adds $11.3 billion to the GDP of the continent. Therefore, blacklisting could adversely affect trade and tourism in Africa and act as a trade sanction, harming both the populace and the airlines concerned.

The fact that there are no global standards for blacklisting although there is a global forum for aviation in the form of ICAO, sends confusing signals both to the consumer as well as to the industry. It has been argued that the blacklisting process of the EU is both unfair and counter-productive since it points a finger and brings to disrepute carriers operating outside the territorial boundaries of the EU. A concerted global audit programme, such as the audit programme under the ICAO Universal Safety Oversight Audit Programme (USOAP) within the Global Aviation Safety Plan (GASP)might well be a more effective tool as it has the potential to address core issues and global safety concerns. There is also the highly effective IATA Operational Safety Audit (IOSA) Programme[436] which is an internationally recognised and accepted evaluation system designed to assess the operational management and control systems of an airline. IOSA uses internationally recognised quality audit principles, and is designed so that audits are conducted in a standardised and consistent manner.

Inherent in the IOSA Programme is a degree of quality, integrity and security such that mutually interested airlines and regulators can all comfortably accept IOSA audit reports. As a result, the industry will be in a position to achieve the benefits of cost-efficiency through a significant reduction in audit redundancy.

---

[436] It must be noted that in March 2006, ICAO and IATA agreed to share information from their respective audit programmes.

2.7 Safety Oversight Audits

It can be mentioned that, when compared to the EU policy on blacklisting, a much more reasonable process obtains in the United States in the nature of the International Aviation Safety Assessments (IASA) Programme. The United States Federal Aviation Administration (FAA) established the IASA program through public policy in August of 1992. The FAA's foreign assessment program focuses on a country's ability, and not the individual air carrier's ability, to adhere to international standards and recommended practices for aircraft operations and maintenance established by the ICAO which is the United Nation's technical agency for aviation.

The IASA policy describes how the FAA would assess whether a foreign civil aviation authority (CAA) complied with the minimum international standards for aviation safety oversight established by ICAO. In obtaining information relevant to its assessment, the FAA meets with the foreign CAA responsible for providing the safety oversight to its carriers, reviews pertinent records and meets with officials of the subject foreign air carriers. The FAA then analyzes the collected information to determine whether the CAA complies with ICAO standards regarding the oversight provided to the air carriers under its authority. This determination is part of the basis for FAA recommended courses of action to the Department of Transportation on the initiation, continuation, or expansion of air services to the United States by the carriers overseen by that CAA. The IASA program applies to all foreign States with air carriers proposing or have existing air service to the United States under an economic authority issued by the Department.

In connection with the public disclosure policy, the FAA has established two categories of ratings for States to signify the status of a CAA's compliance with minimum international safety standards: Category 1, which stipulates that a State complies with ICAO Standards. This involves the issue as to whether a country's civil aviation authority has been assessed by FAA inspectors and has been found to license and oversee air carriers in accordance with ICAO aviation safety standards. Category 2, which includes States which do not comply with ICAO Standards. In this category are States whose civil aviation authority (CAA) have been assessed by the Federal Aviation Administration and determined that they do not provide safety oversight of its air carrier operators in accordance with the minimum safety oversight standards established by ICAO.

This rating is applied if one or more of the following deficiencies are identified:

- The country lacks laws or regulations necessary to support the certification and oversight of air carriers in accordance with minimum international standards;
- The CAA lacks the technical expertise, resources, and organization to license or oversee air carrier operations;
- The CAA does not have adequately trained and qualified technical personnel;
- The CAA does not provide adequate inspector guidance to ensure enforcement of, and compliance with, minimum international standards; and,
- The CAA has insufficient documentation and records of certification and inadequate continuing oversight and surveillance of air carrier operations.

This category consists of two groups of States. In one group are States that have air carriers with existing operations to the United States at the time of the assessment. While in Category 2 status, carriers from these States will be permitted to continue operations at current levels under heightened FAA surveillance. Expansion or changes in services to the United States by such carriers are not permitted while in category 2, although new services will be permitted if operated using aircraft wet-leased from a duly authorized and properly supervised U.S. carrier or a foreign air carrier from a category 1 country that is authorized to serve the United States using its own aircraft.

In the second group are States that do not have air carriers with existing operations to the United States at the time of the assessment. Carriers from these States will not be permitted to commence service to the United States while in Category 2 status, although they may conduct services if operated using aircraft wet-leased from a duly authorized and properly supervised U.S. carrier or a foreign air carrier from a Category 1 country that is authorized to serve the United States with its own aircraft.

No other difference is made between these two groups of States while in a category 2 status.

It must be noted that ICAO member States, at the 36th Session of the ICAO Assembly in September 2007, adopted Resolution A36-2 (Unified strategy to resolve safety-related deficiencies) whereby it is recognized that the establishment of regional and sub-regional safety oversight systems, including regional safety oversight organizations, has great potential to assist States in complying with their obligations under the Chicago Convention through economies of scale and harmonization on a larger scale. However, the Resolution calls for a unified strategy (thus implicitly rejecting a fragmented or regional approach) involving all ICAO member States, ICAO and other concerned parties in civil aviation operations that could render assistance to States experiencing difficulties in correcting safety related deficiencies identified through the ICAO safety oversight audits. The Resolution, while recognizing the safety enhancement contributions of regional organizations such as EASA of Europe, reminds ICAO member States of the need for surveillance of all aircraft operations, including foreign aircraft within their territory and to take action when necessary to preserve safety. Although this reminder seemingly pertains to surveillance regarding safety within the territory of a State or group of States, it is wide open, when it refers to taking action when necessary to preserve safety. Since safety is a global issue and affects the world across the globe, it would not be unjustified to argue that the resolution allows identification of unsafe airlines across the world particularly since nationals of one State or a group of States travel across the world and restricting the identification of unsafe aircraft to a geographic boundary would do a disservice to the efforts of ICAO and the international community in enhancing global aviation safety.

States have an overarching responsibility to ensure aviation safety not only within the parameters of State boundaries but wherever their reach would take them, in order to protect the travelling public. This calls clearly for international responsibility. From a legislative perspective, Article 33 of the Chicago Convention requires States to recognize as valid certificates of airworthiness and certificates of competency and licences issued or rendered valid by the Contracting State in which the aircraft in question is registered provided such certificates or licences are rendered valid and equal to the minimum standards prescribed by the Convention. This would mean that an exception to Article 33 would exist if States do not keep to such standards in issuing the documents. Since Annex 8 to the Chicago Convention prescribes these standards, it is fair to say that any document of a State referred to in Article 33 that does not stand up to the requirements of Annex 8 could be rejected by other States. The 1981 case of *British Caledonian Airways Ltd. v. Bond*[437], concerned the issuance by the Federal Aviation Administration of the United States of an Emergency Airworthiness Directive (EAD) grounding all domestic DC-10 aircraft based on an accident involving an American Airlines DC-10 on 25 May 1979 which crashed immediately after take-off killing all on board. Several foreign carriers operating to the United States with DC-10 aircraft filed an action in the US Court of Appeals in the District of Columbia Circuit which found that the legislative basis for action with regard to such instances was already covered in Article 33 of the Chicago Convention and that there was no need for implementing domestic legislation.

The above notwithstanding, the greatest damage that would be caused by outright and peremptory blacklisting would be to the development of the aviation industry and its contribution to the global economy. Airlines that have safety deficiencies need assistance, not censure and shame. Aviation has reached a juncture in its political history that requires investment in the industry to ensure safe, secure and environmentally progressive development. Investment in air transport will also have a beneficial effect on the robust synergy between aviation and tourism, particularly because of the overarching dependence by the tourism industry on aviation for the carriage of tourists to their destinations. The growing interdependence between the two industries has resulted in a significant increase in the combined contribution of aviation and tourism to the gross domestic product, generating employment and investment opportunities.

The pre-eminent goal of governments should be to create opportunities for the private sector for investment while at the same time creating expansion and employment within the State sector. In other words, the goal should be to crate a sound investment environment for everyone so that society as a whole would benefit. Of course, this is easier said than done, as developing nations have their own internal concerns and pressing needs brought to bear by both social and natural factors. It calls for a certain symbiosis between the developed and developing world as well as an enduring commitment from the international community to assist the developing world in three main areas: removing distortions in developed States that harm the investment climates of developing States; providing increased and effective assistance; and sharing knowledge and experience. These three areas of

---

[437] 665 F.3d.1153 (D.C. Cir.1981).

contribution from the international community have to be applied to the basic axiom that economic development requires adequate and effective transportation. Each country has a theoretically optimum amount of transport capacity. Transportation plays a multi faceted role in the pursuit of development objectives of a nation as well as the need to maintain international communication networks. Air transport enables goods and passengers to be transferred between and within production and consumption centres. Therefore, it could be argued that investment is vital to air transport.

2.7.3 Regulatory Oversight

Many of ICAO's 190 member States are already facing problems with respect to safety oversight. A glaring fact emerging from safety audits conducted by ICAO on States is that the findings of the initial safety oversight audit conducted by ICAO relating to the three Annexes to the Chicago Convention[438] – Annex 1– *Personnel Licensing*, Annex 6 – *Operation of Aircraft* and Annex 8 – *Airworthiness of Aircraft*, indicated that of the 181 Contracting States that were audited between March 1999 and July 2004, considerable numbers of States had deficiencies in respect of a number of requirements under these Annexes. Furthermore, audit follow-up missions have revealed that in many cases, significant deficiencies identified during the initial audits remain.

Several fatal accidents in August, October and December 2005, and some accidents in 2007 which clearly focused on safety deficiencies as the causative factor, have brought to bear the need for regulators and operators to re-focus on the safety of civil aviation. There is an onerous burden on States to ensure that, hand in hand with unfettered competition among carriers, rigid enforcement of technical regulation ensuring the safety of flight is necessary. This calls for the need to put into operation coordinated global safety management systems. This challenge calls for renewed thinking by all concerned in the best interests of the entire aviation industry.

Under ICAO's Universal Safety Oversight Audit Programme (USOAP), States are required to establish a safety programme where aircraft operators, maintenance organizations and services providers implement appropriate safety management systems. Safety Management Systems (SMS) are processes which proactively manage the projected increase in aircraft incidents and accidents brought about by the increase in air traffic movements. SMS require vigilance in the liberalization of air transport and the correspondent increase in capacity. At the Directors General of Civil Aviation Conference on a Global Strategy for Aviation Safety, convened by ICAO in Montreal from 20 to 22 March 2006, Canada defined a Safety

---

[438] Convention on International Civil Aviation, supra, note 3.

2.7 Safety Oversight Audits

Management System as a businesslike approach to safety. An SMS is a systematic, explicit and comprehensive process for the management of safety risks that integrates operations and technical systems with financial and human resource management, for all activities related to an air operator as an approved maintenance organization's certificate holder.[439]

At its 35th session, held in September-October 2004, the ICAO Assembly adopted Resolution A 35–7 which recognized that transparency and sharing of safety information is one of the fundamental tenets of a safe air transportation system and urged all Contracting States to share with other Contracting States critical safety information which may have an impact on the safety of international air navigation and to facilitate access to all relevant safety information. The Resolution also encouraged Contracting States to make full use of available safety information when performing their safety oversight functions, including during inspections as provided for in Article 16 of the Chicago Convention, which empowers State authorities to search aircraft of other States on landing or departure, and to inspect the certificates and other documents prescribed by the Convention. The Resolution also directed the Council to further develop practical means to facilitate the sharing of such safety information among Contracting States and reminds Contracting States of the need for surveillance of all aircraft operations, including foreign aircraft within their territory and to take appropriate action when necessary to preserve safety. A salient feature of Resolution A35-7 was that it touched on Article 54 j) of the Chicago Convention which makes it a mandatory function of the Council to report to Contracting States any infraction of the Convention, as well as any failure to carry out recommendations and determinations of the Council, and directed the Council to develop a procedure to inform all Contracting States, within the scope of Article 54 j) in the case of a State having significant compliance shortcomings with respect to ICAO safety-related SARPs.

At the 36th session of the ICAO Assembly, held in Montreal from 18 to 28 September 2007, ICAO member States adopted Resolution A36-2 (Unified Strategy to resolve safety-related deficiencies) which supersedes the earlier Resolution A35-7. The replacement of one Resolution by another is usually to keep the purpose, text and content of a Resolution current with some added revisions to accommodate new issues and exigencies. Resolution A36-2 urges all ICAO member States to share with other member States safety information which may have an impact on the safety of international air navigation and to facilitate access to all relevant safety information. The Resolution also, *inter alia*, requests the Council to strengthen the Implementation Support Development Programme, which, according to another Resolution adopted by the Assembly – Resolution A36-3 (Implementation Support and Development(ISD) Programme – Safety) *inter alia* directs the ICAO Council to implement the ISD Programme based on transparency and sharing of critical

---

[439] Management of Aviation Safety, Presented by Canada, DGCA/06-WP/15, 4/2/06, at 2.

information and requires the Council to continue the analysis of relevant safety-critical information for determining effective means of providing assistance to States in need and also for determining actions to be taken in relation to those States that have not rectified safety deficiencies.

The Assembly also adopted a third Resolution in this field – Resolution A 36–4 (Application of a continuous monitoring approach for the ICAO Universal Safety Oversight Audit Programme USOAP beyond 2010). This Resolution directs the Council *inter alia*, to examine the feasibility, among the various options that could be considered, of a new approach beyond the concept of continuous monitoring, to be implemented at the end of the current audit cycle in 2010, and to assess on an on going basis States' compliance with their oversight obligations and adapt the audit planning accordingly. The above measures adopted by the ICAO Assembly portends a shift of focus toward increased transparency and sharing of safety critical information between States and ICAO's role as the pre-eminent forum of civil aviation and its newly defined objectives that give ICAO the freedom to divulge certain information that is calculated to enhance aviation safety under USOAP.

References

Abeyratne RIR (1992) Law making and decision making powers of the ICAO council - a critical analysis. Zeitschrift Fur Luft-und Weltraumrecht 41(4):387–394

Abeyratne RIR (1998) Negligence of the airline pilot. Professional Negligence 14(4):219–231

Abeyratne RIR (2000) Access to financial statements of airline pilots and privacy issues - the silk air crash. Air and Space Law XXV(3):97–107, June 2000

Abeyratne RIR (2002) Funding an international financial facility for aviation safety. Ann Air Space Law XXVII:1–26

Abeyratne R (2007) State responsibility for safety management systems. J Aviat Manag 2:7–13

Abeyratne R (2007) Air Law and Policy. PublishAmerica, Baltimore

Abeyratne R (2009a) Regulating unmanned aerial vehicles- issues and challenges. Eur Transport Law XLIV(5):503–520

Abeyratne R (2009b) The role of ICAO in implementing the safety roadmap. Eur Transport Law XLIV(1):29–36

Abeyratne R (2009c) Aircraft registration, legal and regulatory issues. Ann Air Space Law XXXIV:173

Arnold K, de Leon PM (2010) Regulation (EC) 261/2004 in the Light of the Recent Decisions of the European Court of Justice: Time for a Change? Air Space Law XXXV(2):91–112, April 2010

Aust A (2000) Modern treaty law and practice. Cambridge University Press, Cambridge

Besco RE (2000) The myths of mental health screening. Air Safety Week 14(47):20, November 2000

Buergenthal T (1969) Law making in the international civil aviation organization. Syracuse University Press, Syracuse

Burridge AJ (1977) The dismissal of a pilot for poor airmanship — the employer's point of view. Aeronaut J:206, May 1977

Cheng B (1962) The law of international air transport. Oceania Publications, London

References

Christol CQ (1991) Space law past, present and future. Kluwer Law and Taxation Publishers, Deventer

Crawford WA (1989) Environmental tobacco smoke and airlines - health issues. Aerospace:12, July 1989

Crawford J (2002) The international law commission's articles on state responsibility: introduction, text and commentaries. Cambridge University Press, Cambridge, p 77

de Witte B (1998) Sovereignty and European integration: the weight of tradition. In: Slaughter A-M et al (eds) The European court and national courts: doctrine and jurisprudence. Hart, Oxford

Dempsey PS (2003) Privatization of the air: governmental liability for privatized air traffic services. Ann Air Space Law XXVII:95

Evans T (1995) How fit is the pilot doctor? Focus on Commercial Aviat Safety 18:8–9, Spring 1995

Groenewege AD (1999) Compendium of international civil aviation, 2nd edn. International Aviation Development Corporation, Canada

Hale M, Al-Seffar J (2008) Preliminary report on aerotoxic syndrome (AS) and the need for diagnostic neurophysiological tests. JAssoc Neurophysiol Sci 2:107–118

Holcomb LC (1988) Impact of environmental tobacco smoke on airline cabin air quality. Environ Tech. Lett:3, 11 June 1988

Kaiser SA (1995) Infrastructure, airspace and automation - air navigation issues for the 21st century. Ann Air Space Law XX-I:447

Klabbers J (2002) An introduction to international institutional law. Cambridge University Press, Cambridge

Koenig TH, Rustad ML (2005) Defence of Tort Law. New York University Press, New York, p 136

Lawrence C (2000) Film review, "Crusader in High Heels". The Mirror:20–23, April 5 2000

Milde M (2002) Legal aspects of airports constructed in the sea. In: Milde M, Khadjavi H (eds) Public international air law, vol 2. McGill University Faculty of Law, Montreal

Miyagi M (2005) Serious accidents and human factors. American Institute of Aeronautics and Astronautics, Virginia

Montgomery MR, Wier GT, Zieve FJ et al (1977) Human intoxication following inhalation exposure to synthetic jet lubricating oil. Clin Toxicol II(4):423–426

Phillips EH (1999) Accidents raise issue of pilot psychological testing. Aviat Week Space Technol 151(21):22, November 1999

Prosser WL (1964) Handbook of the law of Torts, 3rd edn. West Publishing, St. Paul

Rayman RB, McNaughten CB (1983) Smoke/fumes in the cockpit. Aviat Space Environ Med 54:738–740

Saba J (2003) The credibility of the international financial facility for aviation safety in an environment where "Security" and "Survival" are air transport priorities. Transport Law J 31(1):1–34, Fall 2003

Shaw MN (2003) International law, 5th edn. Cambridge University Press, Cambridge

Sinnett M (2007) 787 No-bleed systems: saving fuel and enhancing operational efficiencies. Aero Q:8, 04 April 2007

Slaughter A-M (2004) A new world order. Princeton University Press, New Jersey

Smith LB, Bhattacharya A, Lemasters G, Succop P, Puhula E, Medvelovic M, Joyce J (1997) Effect of chronic low level exposure to jet fuel on postural balance of US air force personnel. J Occup Environ Med 39:625–632

Snyder QC (1988) Psychological stress, anxiety and your airman certificate. Airline Pilot 67(8):16-20, September 1988

Speiser SM, Krause CF (1978) Aviat Tort Law 1:473

Steenblick JW (1995) Certifying pilots, to fly. Airline Pilot 64(2):30–33, February 1995

Street H (1953) Governmental liability. Cambridge University Press, Cambridge, Chap 2

Sudharshan HV (2003) Seamless sky. Ashgate, Aldershot

Thibeault C (1997) Special committee report: cabin air quality. Aviat Space Environ Med:68–80

Winder C, Balouet J-C (2001) Aircrew exposure to chemicals in aircraft: symptoms of irritation and toxicity. J Occup Health Safety – Aust-NZ 17(5):471–483

Winder C, Fonteyn P, Balouet J-C (2002) Aerotoxic syndrome: a descriptive epidemiological survey of aircrew exposed to in-cabin airborne contaminants. J Occup Health Safety – Aust-NZ 18(4):321–338

Chapter 3
Security Issues

Cyber Terrorism, attacks on the integrity of travel documents, the use of full body scanners and issues flowing therefrom, civil unrest as they threaten the security of airports and unlawful interference with civil aviation are issues that draw the attention of the aviation community in general and the air transport industry in particular.

3.1 Cyber Terrorism

Cyber crimes and cyber terrorism are becoming increasingly menacing and the latter has been identified as a distinct threat requiring attention. At the 21st Aviation Security Panel Meeting of ICAO (AVSECP/21, 22 to 26 March 2010) a new Recommended Practice related to cyber threats was proposed for adoption by the Council as part of amendment 12 to Annex 17 (Security) to the Convention on International Civil Aviation (Chicago Convention). It was adopted on 17 November 2010, will become effective on 26 March 2011 and applicable on 1 July 2011. This Recommended Practice suggests that each Contracting State develop measures in order to protect information and communication technology systems used for civil aviation purposes from interference that may jeopardize the safety of civil aviation. At the 22nd Meeting of the Panel, conducted by ICAO from 21 to 25 March 2011, the Panel noted the value of vulnerability assessments pertaining to cyber security in aviation whose objectives are to evaluate the efficiency of existing mitigation measures and identify any vulnerabilities from a threat-based perspective and further noted that better understanding of residual risks.

Cyber terrorism is an inescapable facet of today's world, and has led to substantial changes in the way we respond to terrorism. The international and national reliance placed on cyberspace for the development and maintenance of human interaction will further increase in the coming years and, in connection with this ongoing growth, dire threats and daunting challenges will proliferate from cyber terrorists. An advantage of cyber terrorism is its anonymous nature, permitting hackers to prevent any trace of their movements through checkpoints or physical evidence to be linked to them. Such terrorism can also be run on low budgets, as the only costs incurred are those related to interference with the computer programs of a State, through the purchase of the necessary computer equipment.[1] The unavoidable challenge posed by cyber terrorism is that our digital world, which allows us to create and share knowledge, also provides ample opportunity for terrorists to commit cyber crimes. The digital environment nurtures motivated offenders who are able to explore covert capabilities for exploiting vulnerabilities in that environment. Thus, limiting the opportunities for subterfuge is another challenge to be faced in the development of the cyber environment. Currently, the most ominous obstacle to be overcome is the lack of guards to prevent crimes against the digital world.[2]

In considering the above, we must first establish the difference, if any, between cyber crime and cyber terrorism and determine whether there are any common links. Cyber crime was termed "computer[3] crime" in the early stages of its evolution and has also been called "computer related crime" or "crime by computer".[4] At its most basic, cyber terrorism may be defined as "an assault on electronic

---

[1] Author Michael Hanlon envisions the consequences of a cyber attack as: "at first, it would be no more than a nuisance. No burning skyscrapers, no underground explosions, just a million electronic irritations up and down the land. Thousands of government web pages suddenly vanish... the disruption continues: thousands of popular websites, from eBay to YouTube, start malfunctioning or are replaced by malicious parodies. Tens of millions of pounds are wiped off the share price of companies like Amazon as fears grow that the whole Internet credit card payment network is now vulnerable and insecure... eventually, reports start to flood in that hundreds of thousands of personal bank accounts have been raided overnight". See Michael Hanlon, Attack on the Cyber Terrorists, Mail Online at http://www.dailymail.co.uk/sciencetech/article-457504/Attack-cyber-terrorists.html.

[2] Cohen and Felson (1979). James D. Zirin, writing to the *Washington Times* said: "It is an irony of the digital age that technology has aided the security forces in detecting and thwarting terrorist operations and has helped terrorists do their evil". See http://bit.ly/d41gsV.

[3] Computers have been defined as "systems of machines that process information in the form of letters, numbers, and other symbols, and that are self directing within predetermined limits". *Webster's New International Dictionary* defines a computer as "a mechanical or electronic apparatus capable of carrying out repetitious and highly complex mathematical operations at high speeds". Computers are used in business for the maintenance of inventories, the calculation and preparation of payrolls, etc.; in industry for the automatic operation of machinery, the control of refinery operations, etc.; and in research for the determination of flight characteristics of missiles and spacecraft, the prediction of the behaviour of substances acted upon by a number of variables, etc. These definitions were cited by the Canadian Supreme Court in *R. v. McLaughlin* [1980] 2 SCR 331 at 339.

[4] See House of Commons Standing Committee on Justice and Legal Affairs, Computer Crime, Final Report (1983), at 12. The *Oxford English Dictionary* defines cyber crime as: "crime or a crime committed using computers or the Internet". It is significant that, in 1998 an 18-year-old Israeli hacker Ehud Tenenbaum, popularly known as the "Analyzer," penetrated the computer systems of the Pentagon, National Aeronautics and Space Administration, Massachusetts Institute of Technology, Naval Undersea Warfare Center, and other highly protected computer systems in the United States. A United States Defense Department official called it "the most organized and

3.1 Cyber Terrorism

communication networks".[5] The Federal Bureau of Investigation provides a fuller definition: "the premeditated, politically motivated attack against information, computer systems, computer programs, and data which result in violence against non-combatant targets by sub-national groups or clandestine agents".[6] A commentator declares that cyber terrorism is terrorism in cyberspace, carried out through the use of computers, the Internet and technology-based networks or systems against infrastructures also supported by computers and such networks.[7] A further interpretation is that cyber terrorism involves employing computer networks in order to harm human life or sabotage critical States infrastructures in ways that may cause harm to human life.[8] Consideration of these definitions demonstrates that the activities related to both cyber crime and cyber terrorism are intended to sabotage infrastructure and disrupt computer systems. As can be seen, though the activities involved may be the same or similar in both categories, the motive behind cyber crime may differ to that in cyber terrorism. A 1999 report by the Centre for the Study of Terrorism and Irregular Warfare determined that the probability of significant cyber attacks in the future are in addition to the traditional physical attacks carried out by terrorists.[9]

Cyberspace, comprised of millions of fibre optic cables enabling servers, computers and routers, is the nervous system of any State's critically important infrastructures, significant among which is transportation. Attacks on cyberspace can cause immeasurable harm, particularly by disrupting education centres and religious places of worship, and essential services such as government, banking and finance, telecommunications, transportation, infrastructures, health and health care,

---

systematic attack the Pentagon has seen to date". See *Master hacker 'Analyzer' held in Israel*, 18 March 1998, at http://www.cnn.com/TECH/computing/9803/18/analyzer/index.html.

[5] http://wordnetweb.princeton.edu/perl/webwn.

[6] http://www.crime-research.org/Cyber_Terrorism_new_kind_Terrorism. Two other definitions are worth noting: "A criminal act perpetrated by the use of computers and telecommunications capabilities, resulting in violence, destruction and/or disruption of services to create fear by causing confusion and uncertainty within a given population, with the goal of influencing a government or population to conform to a particular political, social or ideological agenda" (United States National Infrastructure Protection Center); and "the use of computer network tools to shut down critical national infrastructure (such as energy, transportation, government operations) or to coerce or intimidate a government or civilian population" (Center for Strategic and International Studies, author James Lewis).

[7] Dunnigan (2003). Cyber terrorism brings to the fore two significant modern fears, those of technology and terrorism (see Embar-Seddon 2002).

[8] Definition provided by Shlomo Harnoy, Founder, senior Vice President and Professional manager at SDEMA Group, and Yossi Or, Vice President Information Security at SDEMA Group. The SDEMA Group is an integrated homeland security solutions partnership specializing in risk mitigation. SDEMA also offers information security services including market forward protection against cyber terrorism. This definition is also accepted in academic literature. See Weimann (2006).

[9] Dhanashree Nagre and Priyanka Warade, Cyber Terrorism, Vulnerabilities and Policy Issues "Facts Behind the Myth", at http://www.andrew.cmu.edu/user/dnagre.

power and energy generation and distribution, manufacturing, agriculture and food, electricity and water supply, and military defence. Of these, aerospace activities[10] and air traffic control[11] are prominent targets.

Cyber terrorism may be seen as a corollary to a change in control in manufacturing utilities, banking and communications, moving from secured national control to globally networked computers.[12] The threat of cyber terrorism is all the more real for having already occurred and that future occurrences could be prodigious. Blaise Pascal states in *Ars Cogitandi* that fear of harm ought to be proportional not merely to the gravity of the harm but also to the probability of an event.[13] Fundamentals of risk management define that, given similar conditions, the occurrence of an event in the future will follow the same pattern as in the past.[14] It seems a given, then, that we may face the daunting possibility of a nuclear 9/11 in the future,[15] possibly aided and abetted by cyber terrorism.

The events of 11 September 2001 demonstrated that the three most vulnerable targets for a terrorist attack are people, infrastructure and technology, since they are the foremost components of a successful economy. The incident also emphasised the inextricable interdependencies between physical and cyber infrastructures. Cyber terrorism is thus a "clear and present danger"[16] and the question has even been raised as to whether 9/11 was a result of cyber terrorism.[17]

---

[10] In March 1998, the website of the National Aeronautics and Space Administration NASA received a 'denial of service' attack, calculated to affect Microsoft Windows NT and Windows 95 operating systems. These attacks prevented servers from answering network connections and crashed computers, causing a blue screen to appear. The attacked systems were revived, but this attack was a follow-up to one perpetrated in February of the same year when, the United States Defense Department had unclassified networks penetrated for two weeks, with hackers accessing personnel and payroll information.

[11] On 18 March 1998, federal criminal charges were unsealed against a computer hacker who had disabled a key telephone company computer servicing Worcester airport. As a result of a series of commands sent from the hacker's personal computer, vital services to the Federal Aviation Administration control tower were disabled for six hours in March of 1997, see http://www.justice.gov/criminal/cybercrime/juvenilepld.htm. In April 2002, it was reported that hackers were able to penetrate a Federal Aviation Administration system and download unpublished information on airport passenger screening activities. See Poulsen (2002).

[12] *The White House, The National Strategy to Secure Cyberspace* (2003), at 5, at http://www.dhs.gov/xlibrary/assets/National_Cyberspace_Strategy.pdf.

[13] Ferguson (2008).

[14] *Ibid*.

[15] Bobbitt (2008).

[16] In the 1919 decision of *Schenk* v. *US* [249 US 47 (1919)], Justice Oliver Wendell Holmes used the words *clear and present danger* when the United States Supreme Court adjudicated the case of Charles Schenk, who had distributed leaflets allegedly calculated to incite and cause insubordination and obstruction in recruits of the American Socialist Party. The actions of Schenk were considered to constitute an offence under the *Espionage Act* of 1917. See also Stohl (2006).

[17] James Corbett, *The Corbett Report*, 9/11 and Cyberterrorism: Did the real "cyber 9/11" happen on 9/11? 17 July 2009, see http://www.corbettreport.com/articles/20090717_cyber_911.htm. For an informative discussion on cyber terrorism post-9/11 see Cortes (2004).

3.1 Cyber Terrorism

In taking action against cyber crimes, then President Bill Clinton, in a 1999 speech to the National Academy of Sciences said: "open borders and revolutions in technology have spread the message and the gifts of freedom, but have also given new opportunities to freedom's enemies... we must be ready... ready if our adversaries try to use computers to disable power grids, banking, communications and transportation networks, police, fire, and health services – or military assets.[18] In order to achieve this objective", President Clinton outlined a strategy based on a new programme "Cyber Corps" that would be in addition to and augment government efforts already in place to counter cyber terrorism, at the same time initiating new strategies calculated to strengthen the protection of critical systems. The President stated that he would seek the support of Congress to allocate $1.46 billion in the next federal budget for this strategy,[19] involving a 40% increase from previous spending on related efforts. This proposed measure was opportune as, according to a 2002 survey conducted by the Federal Bureau of Investigation and a San Francisco-based computer security institute, 90% of large corporations and government agencies in the United States had experienced unauthorized computer breaches in 2001.[20]

Under then President George Bush, the United States adopted, in 2003, a *National Strategy to Secure Cyberspace*, aimed at preventing cyber attacks against critical American infrastructures, reducing national vulnerability to cyber attacks and minimizing damage and recovery time from actual cyber attacks.[21] The Strategy defines the national priority as securing the government's cyberspace and national security and initiating international cooperation on cyberspace security. The strategy would be supported by a response system, threat and vulnerability reduction programme, and awareness and training programme, for national cyberspace security. A significant principle of this strategy was its recognition that efforts to counter cyber terrorism should involve strong, proactive collaboration between those providing essential services in the United States, since the federal government could not—and should not—secure nor interfere with the computer networks of banks, energy companies, transportation firms, and other activities of the private sector. Similarly, the federal government should not intrude into the computer networks of homes and small businesses, universities, and State and local agencies and departments. The *Strategy* therefore exhorted each American who depends on cyberspace and information networks to secure the part that they own or for which they are responsible.

---

[18] http://news.cnet.com/2100-1023-220532.html.

[19] http://news.cnet.com/2100-1023-220532.html#ixzz1HWoNI8fW.

[20] Misra (2003).

[21] *The National Strategy to Secure Cyberspace*, February 2003, Washington DC, Executive Summary, see http://www.dhs.gov/files/publications/editorial_0329.shtm. The Cyberspace Strategy is an implementing component of the *National Strategy for Homeland Security* and is complemented by a *National Strategy for the Physical Protection of Critical Infrastructures and Key Assets*.

The annual *Threat Assessment of the United States Intelligence Community for the Senate Select Committee on Intelligence*[22] of 2010 shows the extent of the cyber terrorism threat in its statement that the agility and technological innovation demonstrated by the cyber criminal sector far exceeds the response capability of network defenders. The *Threat Assessment* identified two components as particularly vulnerable to cyber attack: *Network Convergence*, the merging of distinct voice and data technologies to the point where all communications are relayed over a common network structure; and *Channel Consolidation*, the concentration of data on individual users captured by service providers through e-mails or instant messages. The *Threat Assessment* drew an implicit parallel between cyber terrorism and international organized crime, extrapolating the theory that international criminal organizations will increasingly hinder the ability of legitimate businesses to compete and may even push legitimate players out of the market.[23]

Further, whether conducted by an individual, a corporation or a State, cyber terrorism has the potential to target the electronic systems of companies that design and develop hardware and software used at airports and in air traffic control systems. Such terrorism may also target industries involved in the construction of aircraft and components, whether they are used for civil or military purposes. One commentator says:

> [h]ere, the objective is that of manipulating, in the design phase, software orhardware which will eventually come to be used in critical environments. The eventslinked to the theft of designs relating to the American F-35 project are an exampleof this kind of act.[24]

A review conducted in 2010 by the United States Government reported that the Federal Aviation Administration (FAA) computer systems remained vulnerable to cyber attacks, since most air traffic control facilities had not been enhanced to adequately respond to cyber intrusions.[25] The threat of cyber terrorism was

---

[22] *Threat Assessment of the United States Intelligence Community for the Senate Select Committee on Intelligence*, 2 February 2010, ATA February 2010 – Intelligence Community Statement for the Record, at 3, see http://www.cfr.org/intelligence/annual-threat-assessment-intelligence-community-senate-select-committee-intelligence-2010/p21369.

[23] It is disturbing that a recent study from the Department of Homeland Security highlighted how the information technology systems of the United States Computer Emergency Readiness Team, which are used by the National Cyber Security Division in its mission to be the focal point in terms of cyber security, both at the public and private levels, suffer from numerous and dangerous vulnerabilities linked above all to the problem of a poor information technology security culture amongst its employees, see Department of Homeland Security (DHS) – Office of Inspector General, *"DHS Needs to Improve the Security Posture of Its Cybersecurity Program Systems"*, at http://www.dhs.gov/xoig/assets/mgmtrpts/OIG_10-111_Aug10.pdf.

[24] Stefano Male, *Cyber Warfare and its Damaging Effects on Citizens*, September 2010, at https://www.infosecisland.com/download/index/id/30.html.

[25] Lolita C. Baldor, Cyber Security Still Issue for FAA, *Boston Globe*, 13 August 2010, at http://www.boston.com/news/nation/washington/articles/2010/08/13/cyber_security_still_issue_for_faa. A Department of Energy report released in May 2009 documented successful attacks that have affected FAA networks. In 2006, the FAA shut down a "portion of its air traffic control systems in

considered with regard to the Boeing *Dreamliner* 787, whereby the FAA reportedly claimed that "the plane may be at risk for hacking on to its on-board computer system, with disastrous consequences".[26]

In an aviation context, therefore, cyber terrorism has multiple facets that can disrupt air transport in many ways. Acts of cyber terrorism could be used to spread disinformation or engage in psychological warfare, with media attention being manipulated regarding possible threats, leading to disruptions in airport and aircraft operations. The end result would be a "fear factor", such as was seen in the immediate aftermath of 9/11, where individuals displayed increased reluctance to travel. This could, in turn, result in economic losses, particularly in States that are dependent on the tourism industry to boost their Gross National Products. At the most serious level, cyber terrorism could lead to fatalities, injuries and major damage at airports and to aircraft in flight.[27]

3.1.1 International Efforts

Offences against civil aviation, particularly with regard to unlawful interference with civil aviation related to aircraft have been addressed on three major occasions, though the *Tokyo Convention* of 1963, *The Hague Convention* of 1970 and the *Montréal Convention* of 1971.[28] Yet none of these conventions referred, whether directly or indirectly, to cyber terrorism.

The first such convention to do so, the 2010 *Convention on the Suppression of Unlawful Acts Relating to International Civil Aviation* adopted in Beijing,[29] states in Article 1d) that an offence is committed when a person destroys or damages air navigation facilities or interferes with their operation, if any such act is likely to endanger the safety of aircraft in flight. This clearly refers, inter alia, to cyber terrorism, yet links the offence exclusively to the safety of aircraft in flight. Article 2a) of the Convention provides that an aircraft is considered to be in flight at any time from the moment when all its external doors are closed following embarkation until the moment when any such door is opened for disembarkation. In the event of

---

Alaska" due to a "viral attack", and in 2008, FAA computers, again in Alaska, were compromised and 40,000 usernames and passwords were stolen. In 2009, an FAA "public-facing web application computer" was compromised, leading to the theft of "PII on 48,000 current and former FAA employees", see Nart Villeneuve, Thoughts on "Critical Infrastructure" 13 December 2009, at http://www.nartv.org/2009/12/13/thoughts-on-critical-infrastructure-protection/.

[26] Kim Zetter, FAA: New Boeing 787 Dreamliner may be Vulnerable to Hacker Attack, http://www.wired.com/politics/security/news/2008/01/dreamliner_security.

[27] See Guill (2000).

[28] See generally, Abeyratne (1998), which extensively discusses the treaties. See also, Abeyratne (2010a).

[29] *Convention on the Suppression of Unlawful Acts Relating to International Civil Aviation*, done at Beijing on 10 September 2010.

a forced landing, the flight would be deemed to continue until the competent authorities take over responsibility for the aircraft and for persons and property on board. For instance if, as a result of an act of cyber terrorism, a taxiing aircraft collided with an aircraft that had opened its doors for disembarkation, but passengers were still on board, such an act would not be considered an offence in terms of the passengers in the process of disembarkation. That is, the offender(s) would not be committing an offence under the convention either against the second aircraft or its disembarking passengers. Nonetheless, the Beijing Convention of 2010 is an initial step toward countering the threat of cyber terrorism, a threat directed often toward the target of air transport.

More generally, yet with relevance to the field of aviation, are the activities conducted since the 1980s by international organizations such as the United Nations, Council of Europe, INTERPOL, and the Organization for Economic Co-operation and Development[30] in response to the challenges posed by cyber crime. A significant result of such collective efforts was the publication of the *United Nations Manual on Cybercrime*[31] and 2001 United Nations Resolution[32] exhorting States, in the context of an earlier United Nations Resolution on Millennium Goals,[33] which recognized that the benefits of new technologies, especially information and communication-related technologies, are available to all, to ensure that their laws and practices eliminate safe havens for those who criminally misuse information technology. The Resolution also urged States to ensure the cooperation of law enforcement authorities in the investigation and prosecution of international cases of the criminal misuse of information technology, and that this should be coordinated among all concerned States. The Resolution further required information to be exchanged between States regarding the challenges faced in combating such criminal misuse and stated that law enforcement personnel should be trained and equipped to address any criminal misuse of information technology.

Further, the Resolution recognized that legal systems should protect the confidentiality, integrity and availability of data and computer systems from unauthorized impairment, ensure that criminal abuse is penalized, and that such systems should permit preservation of and quick access to electronic data pertaining to specific criminal investigations. The Resolution called upon mutual assistance regimes to ensure the timely investigation of the criminal misuse of information

---

[30] The mission of the Organization for Economic Co-operation and Development is to promote policies that will improve the economic and social well being of people around the world. The Organization provides a forum in which governments can work together to share experiences and seek solutions to common problems, and works with governments to understand what drives economic, social and environmental change.

[31] *United Nations Manual on the Prevention and Control of Computer Related Crime*, International Review of Criminal Policy, 43 and 44 (1999).

[32] *United Nations Resolution on Combating the Criminal Misuse of Information Technologies* General Assembly Resolution 55/63, United Nations General Assembly 55th Session, 81st Plenary Meeting, UN Doc A/RES/55/63 (2001).

[33] A/RES/55/2.

3.1 Cyber Terrorism

technology and the timely gathering and exchange of evidence in such cases. States were requested to make the public aware of the need to prevent and combat such criminal misuse. Finally, the Resolution called for information technology to be designed to help prevent and detect criminal misuse, trace criminals and collect evidence to the extent practicable, recognizing that countering the criminal misuse of information technology requires the development of solutions that take into account the protection of individual freedoms at the same time as their privacy and the preservation of the capacity of governments to fight such misuse.

A second significant activity borne of international collaborative responses to cyber crime was the adoption of the *Cybercrime Convention*[34] of the Council of Europe, opened for signature in November 2001, and which came into force on 1 July 2004. In the United States, this Convention was ratified by then President Bush on 22 September 2006 and entered into force on 1 January 2007. The main focus of this Convention is the risk that computer networks and electronic information might be used for committing criminal offences and that evidence relating to such offences may be stored in and transferred over these networks. States Parties to the Convention therefore expressed their view in a Preambular Clause that cooperation between States and private industry in combating cyber crime was necessary and that there was a need to protect legitimate interests in the use and development of information technology. The intent of the Convention falls under three goals:

a) Harmonizing domestic criminal substantive law elements of offences and connected provisions in the area of cyber crime;
b) Providing for domestic criminal procedural law powers necessary for the investigation and prosecution of such offences as well as other offences committed by means of a computer system or evidence in relation to which is in electronic form; and
c) Setting up a fast and effective regime of international cooperation.

Article 2 of the Convention requires each Party to adopt the legislative and other measures that may be necessary to establish access to the whole or any part of a computer system, when committed intentionally, without right, as a criminal offence under domestic law.[35] Additionally, a Party may require that the offence

---

[34] European Treaty Series no. 185. Forty-two European States, the United States, Canada and many other States were signatories to the Convention.

[35] A computer system under the Convention is a device consisting of hardware and software developed for the automatic processing of digital data and may include input, output, and storage facilities. It may stand alone or be connected in a network with other similar devices. "Automatic" means without direct human intervention, "processing of data" means that data in the computer system is operated by executing a computer programme, which is a set of instructions that can be executed by the computer to achieve an intended result. A computer can run different programmes and a computer system usually consists of different devices, such as a processor, central processing unit, and peripherals, which are devices that perform specific functions in interaction with the processing unit, such as printers, video screens, compact disc readers and writers, or other storage devices. See *Cybercrime Convention*, Explanatory Report, paragraph 23.

be committed by infringing security measures, with the intent of obtaining computer data or with other dishonest intent, or in relation to a computer system that is connected to another computer system. Other provisions call for States Parties to adopt legislative or other measures to counter illegal inception of transmission of computer data, data interception and exchange interception.[36] Of particular significance to aviation is Article 7 on the alteration of data and forgery, which requires each Party to adopt such legislative and other measures as may be necessary to establish as criminal offences under its domestic law, when committed intentionally and without right, the input, alteration, deletion, or suppression of computer data, resulting in inauthentic data, with the intent that such data be considered or acted upon for legal purposes as if it were authentic, regardless of whether or not the data is directly readable and intelligible. The Provision concludes that a Party may require an intent to defraud, or similar dishonest intent, before criminal liability may be imposed.

Article 7 also calls for the protection of certain measures adopted by the aviation community to ensure the integrity of passports and other machine readable travel documents using technology such as the Public Key Directory (PKD). The PKD has been developed using the quantum cryptography technique, intended to eradicate vulnerabilities to fraud in the use of digitally stored data. Quantum cryptography transmits information along cables by polarized photons rather than electronic signals. These photons are tiny particles of light sensitive enough that they immediately become corrupted when intercepted, thus rendering their message unintelligible and alerting both sender and recipient to the attempt at fraud or spying. The use of the PKD technique in passports provides a good example. In this case, the PKD is designed and proposed for use by customs and immigration authorities who verify biometric details in an electronic passport. The PKD is based on cryptography, an already viable tool that is now considered by the aviation community as a fail-safe method for ensuring the accuracy and integrity of passport information.

Biometric information in the identification of persons is another method that counters cyber terrorism and unlawful interference with computer imagery. Biometrics involve measuring the distinguishing physiological or behavioural traits of individuals and storing them in an automated database such as machine-encoded representations created by computer software algorithms, which compare these with the actual features. Biometrics that have been successfully used and are the most appropriate for this scientific process are facial recognition, fingerprinting and iris-recognition. Identification through biometrics involves four steps: first, the capture or acquisition of a biometric sample; second, the extraction or conversion of the sample into an intermediate form; and third, the creation of templates of this data for storage; and fourth, the comparison of the information offered in a travel document with that which is stored in the template.

---

[36] *Cybercrime Convention*, Articles 3, 4 and 5 respectively.

3.1.2 National Efforts

Interception of data is a significant offense that is a precursor to cyber crime and cyber terrorism. The *Cybercrime Convention* defines interception as:

> Listening to, monitoring or surveillance of the content of communications, to the procuring of the content of data either directly, through access and use of the computer system, or indirectly, through the use of electronic eavesdropping or tapping devices.[37]

Australia adopted the *Telecommunications (Interception and Access) Act* in 1979. Section 7(1) provides that a person must not intercept, authorize, suffer or permit another person to intercept or do any act or thing that will enable him or her or another person to intercept a communication passing over a telecommunications system.[38] An important provision is Section 108 (1), which provides that an offence is committed if a person, with intent and knowledge, accesses a stored communication or authorizes, suffers or permits another person to access a stored communication or does any act or thing that will enable them or another person to access a stored communication, where the intended recipient of the stored communication or the person who sent the stored communication had no knowledge of the offender's act.

In Canada, a Bill[39] was introduced in Parliament in 2005 aimed at introducing reforms to the legislative structure concerned with the unlawful interception of documents and communications. In the absence of specific legislation, parallels may be found in Canada's criminal legislation. For example, Section 184(1) of the *Canadian Criminal Code* provides that an agent of the State[40] may intercept, by means of any electromagnetic, acoustic, mechanical or other device, a private communication if either the originator of the private communication or the person intended by the originator to receive it has consented to the interception, or the agent of the State believes on reasonable grounds that there is a risk of bodily harm

---

[37] *Cybercrimes Convention* Explanatory Report, paragraph 53.

[38] The Act defines a telecommunications system as a service for carrying communications, by means of guided or unguided electromagnetic energy or both, the use of which enables communications to be carried over a telecommunications system operated by a carrier but not being a service for carrying communications solely by means of radio communication.

[39] Bill C-74. This Bill was introduced in the House of Commons by the Minister of Public Safety and Emergency Preparedness on 15 November 2005. The Bill refers to specific aspects of the rules governing lawful access, an investigative technique used by law enforcement agencies and national security agencies. The Bill was aimed at protecting victims against new technologies such as wireless data networks and voice over Internet protocol, which often present obstacles to the lawful interception of communications. Since such technologies can create "intercept safe havens" where criminal groups are able to operate without being detected, and in light of factors such as the deregulation of the telecommunications market, the Bill was intended to respond to the growing complexity of telephone networks that makes investigators' work more difficult and may result in delays in identifying suspects.

[40] An agent of the State is defined as a peace officer, and a person acting under the authority of, or in cooperation with, a peace officer.

to the person who consented to the interception and the purpose of the interception is to prevent such bodily harm. The provision goes on to require the agent of the State who intercepts a private communication to, as soon as practicable in the circumstances, destroy any recording of the private communication obtained during an interception, any full or partial transcript of the recording and any corresponding notes, if nothing in the private communication suggests that bodily harm, attempted bodily harm or threatened bodily harm has occurred or is likely to occur. Further, Section 287(1)(b) provides that anyone commits theft who fraudulently, maliciously, or without colour of right uses any telecommunications facility or obtains any telecommunications service.[41]

In the United Kingdom, the 2000 *Regulation of Investigatory Powers Act* was a legislative attempt by Parliament to unify in a single legal framework provisions countering the interception of information and communications. This Act does not discriminate between types of communications or the location at which communications are intercepted. Section 1.1. of the Act states that it is an offence to intercept, intentionally and without lawful authority, at any place in the United Kingdom, any communication in the course of its transmission by means of a public postal service or a public telecommunications system. Further, the Act prescribes it an offence to intercept, intentionally and without lawful authority, at any location in the United Kingdom, any communication while it is being transmitted via a public or private telecommunications system. Significantly, Section 4.1 deems conduct by an interceptor lawful if the interception of a communication in the course of its transmission by means of a telecommunications system constitutes interception carried out for the purpose of obtaining information about the communications of a person who is, or whom the interceptor has reasonable grounds for believing may be, in a State or territory outside of the United Kingdom. Such interception would relate to the use of a telecommunications service provided to persons in that State or territory, which is either a public telecommunications service or a telecommunications service that would be a public service if the persons to whom it was offered or provided were members of the public in a part of the United Kingdom.

In the United States, surveillance laws against interception were not fully defined until they were reformed in the 1986 *Electronic Communications Privacy Act*, adopted prior to the widespread introduction of the Internet and World Wide Web. Courts have referred to such laws as convoluted,[42] confusing and uncertain. In the decision of *Konop* v *Hawaiian Airlines*,[43] handed down by the United States Court of Appeal 9th Circuit in 2002, the court noted inter alia that the Act defines "electronic communication" as a "transfer" of signals, and that "unlike the definition of 'wire communication,' the definition of 'electronic communication' does

---

[41] Section 287 defines telecommunication as "any transmission, emission or reception of signs, signals, writing, images, sounds or intelligence of any nature by radio, visual, electronic or other electromagnetic system".

[42] *US.* v. *Smith* 155 F 3d 1051, at 1055 (9th Cir. 1998).

[43] 302 F 3d 868.

not include electronic storage of such communications", which led the Court to conclude that the Act was not equipped to handle modern forms of electronic communication.[44]

A particular feature of cyber terrorism is that the threat is enhanced by globalization and the ubiquity of the Internet. Given such a global problem, requiring a global solution, the one forum that can provide a global framework against cyber terrorism is the International Civil Aviation Organization (ICAO). A sustained global process of security risk assessment[45] is the first necessary step.

At the 21st Meeting of the ICAO Aviation Security Panel, conducted in Montréal, from 22 to 26 March 2010, a new Recommended Practice related to cyber threats was proposed for adoption by the Council as part of Amendment 12 to Annex 17 – *Security* to the *Convention on International Civil Aviation* (Chicago Convention). This Amendment was adopted on 17 November 2010, and will become effective on 26 March 2011 and applicable on 1 July 2011. The new Recommended Practice suggests that each Contracting State develop measures in order to protect information and communication technology systems used for civil aviation purposes from interference that may jeopardize the safety of civil aviation. The 22nd Meeting of the ICAO Aviation Security Panel, held in Montréal from 21 to 25 March 2011, noted the value of vulnerability assessments pertaining to cyber security in aviation, whose objectives are to evaluate the efficiency of existing mitigation measures and identify any vulnerabilities from a threat-based perspective. The Panel further noted that better understanding of residual risks will support a State's efforts to refine its risk response.

In pursuance of these objectives, ICAO, in collaboration with its Member States, could undertake a study to identify critical aviation information systems; review the effectiveness of existing mitigation measures established for such systems; identify any vulnerabilities in current security arrangements; analyse best practices on how to address these vulnerabilities; and determine how to better manage identified residual risks.

3.2 Attacks on the Integrity of Travel Documents

A passport is the property of the State which issues it. As such no State or other legal entity has a right to alter an existing passport of a holder or forge false documentation purporting to have been issued by a State. The fundamental issue

---

[44] *Id.*, 461.

[45] One definition of security risk assessment considered by the ICAO Aviation Security Panel at its Twenty-second Meeting: "an outcome based process, coordinated by the Appropriate Authority utilising all appropriate resources, consisting of an analysis of prevailing threat factors compared against current mitigation measures, with a view to determining levels of risk that result in the application of appropriate mitigation measures".

therefore, is whether a State or instrumentality of that State can use a forged passport with impunity, particularly in the course of criminal activity. On 19 January 2010, Mahmoud al-Mabhouh, considered to be a senior commander of Hamas, a radical Palestinian group, was assassinated at a hotel in Dubai in a manner usually employed by professionally trained military and secret service agencies. The killing was attributed to Mossad, the Israeli secret service. Quite apart from the criminality of the assassination, the issue of the use of forged passport was brought to bear by some countries which alleged that the perpetrators of the crime had allegedly forged passports that seemingly belonged to those countries. This international incident sparked off a legal and diplomatic discussion in the media as to the responsibility and liability of States – which condone or allow the use of these forged documents – towards those States that are affected.

The passport is a basic document in the transport by air of persons. Its use therefore is of fundamental importance as a travel document, not only because it reflects the importance of the sovereignty of a State and the nationality of its citizens but also because it stands for the inviolability of relations between States that are linked through air transport. The assassination of a leader of Hamas on 19 January 2010 by a group of individuals in Dubai who used forged passports belonging to various nations, raised a diplomatic outcry and brought to bear an important facet of air transport that is vulnerable to abuse and contention among States.

The fundamental issue that emerges is one that is critical to air law in the context of the integrity and ownership of the passport and its abuse in the course of criminal activity. There is also the issue, from a legal and diplomatic perspective as to whether a State or instrumentality of State, can, with impunity, use forged passports for travel of its staff on missions of espionage or assassination. A fortiori, an additional issue is whether a State could be complicit or condone or be seen to condone (in the absence of any action taken by the State to punish the miscreants) such abuse of travel documents belonging to other nations. In order to determine these issues, this article addresses two basic discussions: the first on complicity and condonation of a State and the second on the nature and integrity of the passport. Finally, it discusses issues of State responsibility, diplomacy and criminality.

On 19 January 2010, Mahmoud al-Mabhouh, considered to be a senior commander of Hamas, a radical Palestinian group, was assassinated at a hotel in Dubai in a manner usually employed by professionally trained military and secret service agencies. The killing was attributed to Mossad[46] The European Union, which considers Hamas a terrorist organization, nonetheless condemned the assassination of the Hamas leader and showed particular concern over the fact that the killers had used passports from Ireland, France, Germany and the UK – to coordinate their travel into Dubai from various parts of the world, synchronizing their arrival time

---

[46] Mossad is responsible for the collection of intelligence and other covert activities including military operations. It is one of the most integral parts of the Israeli intelligence community and reports directly to the Prime Minister of Israel. See http://en.wikipedia.org/wiki/Mossad.

3.2 Attacks on the Integrity of Travel Documents

from various flights into Dubai International Airport and checking into the hotel of the victim contemporaneously. The EU strongly condemned the fact that those involved in this action used fraudulent EU member states' passports and credit cards acquired through the theft of EU citizens' identities.[47]

Australia was another complainant who warned Israel that its friendly relations with Israel would be jeopardised if it were found to have condoned the suspected theft of three Australian citizens' identities which Mossad used to carry out its political assassination. The diplomatic impasse occurred when three Australians from Victoria living in Israel at the time were confirmed among 26 people from four nations whose tampered passports were allegedly used by a team of suspected Israeli Mossad agents who assassinated al-Mabhouh. Australian Prime Minister Kevin Rudd is reported to have stated that Australia would be vocal in its contempt of any State if it were found that it "... has been complicit in the use or abuse of the Australian passport system, let alone for the conduct of an assassination, and has treated Australia with contempt and there will therefore be action by the Australian government in response".[48] Dubai authorities are reported to have said that they were virtually certain Israeli agents carried out the killing and had released the identities of 11 people who travelled on forged British, Irish, French and German passports to kill al-Mabhouh in a hotel.[49]

There is seemingly a history behind alleged Mossad involvement in the use of fake foreign passports in its activities. Reportedly, in 2004 New Zealand's prime minister imposed diplomatic sanctions – restricting visas and cancelling high level visits – after two Mossad agents were caught trying to acquire passports fraudulently – one in the name of a tetraplegic man. Seven years earlier, Mossad assassins carrying Canadian passports with assumed names attempted to murder the Hamas leader Khaled Meshaal by spraying nerve agent into his ear as he entered his office in Amman.[50]

---

[47] Toby Vogel, EU Condemns Use of False Passports inn Hamas Killing, http://www.europeanvoice.com/article/2010/02/eu-condemns-use-of-false-passports-in-hamas-killing/67225.aspx.

[48] http://www.theaustralian.com.au/news/world/australians-caught-in-hit-on-hamas/story-e6frg6so-1225834538825 It is reported that in 1997, Mossad bungled the assassination of top Hamas leader Khalid Mishal, who was injected while in Jordan with a poison by Israeli agents travelling on Canadian documents. He survived after his assailants were captured by his bodyguards and Israel provided the antidote. In 2004, two Mossad agents were jailed in New Zealand after trying to obtain fake passports, one in the name of a cerebral palsy sufferer. *Ibid.*

[49] http://www.euractiv.com/en/foreign-affairs/eu-unhappy-israel-over-fake-passports-james-bond-killings-news-278602.

[50] David Sapsted, and Loveday Morris, Israel in the Dock Over Fake Passports, http://www.thenational.ae/apps/pbcs.dll/article?AID=/20100218/NATIONAL/702179796/1133/sport.
Hamas, which won 2006 legislative elections in the Palestinian territories, is shunned by the West for rejecting its calls to recognise Israel and renounce violence. Hit squads dispatched by Mossad have used foreign passports in the past, notably in 1997 when agents entered Jordan on Canadian passports and bungled an attempt to kill Meshaal with poison. In 1987, Britain protested to Israel about what London called the misuse by Israeli authorities of forged British passports and said it

The fundamental issue that emerges is one that is critical to air law in the context of the integrity and ownership of the passport and its abuse in criminal activity. There is also the issue, from a legal and diplomatic perspective is whether a State or instrumentality of State such as Mossad, can, with impunity, use forged passports for travel of its staff on missions of espionage or assassination. A fortiori, an additional issue is whether a State could be complicit or condone or be seen to condone (in the absence of any action taken by the State to punish the miscreants) such abuse of travel documents belonging to other nations.

3.2.1 Complicity

The fundamental issue in the context of State responsibility for the purposes of this article is to consider whether a State should be considered responsible for its own failure or non-feasance to prevent a private act that is a violation of its international responsibility towards a third State or whether the conduct of the State itself can be impugned by identifying a nexus between the perpetrator's conduct and the State. One view is that an agency paradigm, which may in some circumstances impute to a state reprehensibility on the ground that a principal–agent relationship between the State and the perpetrator existed, can obfuscate the issue and preclude one from conducting a meaningful legal study of the State's conduct.[51]

At the core of the principal–agent dilemma is the theory of complicity, which attributes liability to a State that was complicit in a private act. Hugo Grotius (1583–1645), founder of the modern natural law theory, first formulated this theory based on State responsibility that was not absolute. Grotius' theory was that although a State did not have absolute responsibility for a private offence, it could be considered complicit through the notion of *patienta* or *receptus*.[52] While the concept of *patienta* refers to a State's inability to prevent a wrongdoing, *receptus* pertains to the refusal to punish the offender.

The eighteenth Century philosopher Emerich de Vattel was of similar view as Grotius, holding that responsibility could only be attributed to the State if a sovereign refuses to repair the evil done by its subjects or punish an offender or deliver him to justice whether by subjecting him to local justice or by extraditing him.[53] This view was to be followed and extended by the British jurist Blackstone a

---

received assurances steps had been taken to prevent future occurrences. In 2003, the offices of several EU member countries in the Council's Justus Lipsius building, including France, Germany and the UK, were found to be bugged. Although the Union has been discrete over the incident, many consider Mossad to have been responsible for the wiretapping. *Ibid.*

[51] Caron (1998) cited in Becker (2006a).

[52] Grotius and Scott (1646).

[53] De Vattel and Fenwick (1916).

few years later who went on to say that a sovereign who failed to punish an offender could be considered as abetting the offence or of being an accomplice.[54]

A different view was put forward in an instance of adjudication involving a seminal instance where the Theory of Complicity and the responsibility of states for private acts of violence was tested in 1925. The case[55] involved the Mexico-United States General Claims Commission which considered the claim of the United States on behalf of the family of a United States national who was killed in a Mexican mining company where the deceased was working. The United States argued that the Mexican authorities had failed to exercise due care and diligence in apprehending and prosecuting the offender. The decision handed down by the Commission distinguished between complicity and the responsibility to punish and the Commission was of the view that Mexico could not be considered an accomplice in this case.

The Complicity Theory, particularly from a Vattellian and Blackstonian point of view is merely assumptive unless put to the test through a judicial process of extradition. In this Context it becomes relevant to address the issue through a discussion of the remedy.

3.2.2 Condonation

The emergence of the Condonation Theory was almost concurrent with the *Jane* case[56] decided in 1925 which emerged through the opinions of scholars who belonged to a school of thought that believed that States became responsible for private acts of violence not through complicity as such but more so because their refusal or failure to bring offenders to justice, which was tantamount to ratification of the acts in question or their condonation.[57] The theory was based on the fact that it is not illogical or arbitrary to suggest that a State must be held liable for its failure to take appropriate steps to punish persons who cause injury or harm to others for the reason that such States can be considered guilty of condoning the criminal acts and therefore become responsible for them.[58] Another reason attributed by scholars in support of the theory is that during that time, arbitral tribunals were ordering States to award pecuniary damages to claimants harmed by private offenders, on the basis that the States were being considered responsible for the offences.[59]

---

[54] Blackstone and Morrison (2001).

[55] *Laura M.B. Janes (USA)* v. *United Mexican States* (1925) 4 R Intl Arb Awards 82.

[56] *Ibid.*

[57] *Black's Law Dictionary* defines condonation as "pardon of offense, voluntary overlooking implied forgiveness by treating offender as if offense had not been committed."

[58] Jane's case, *Supra*, note 11, at 92.

[59] Hyde (1928).

The responsibility of governments in acting against offences committed by private individuals may sometimes involve condonation or ineptitude in taking effective action against terrorist acts, in particular with regard to the financing of terrorist acts. The United Nations General Assembly, on 9 December 1999, adopted the International Convention for the Suppression of the Financing of Terrorism,[60] aimed at enhancing international co-operation among States in devising and adopting effective measures for the prevention of the financing of terrorism, as well as for its suppression through the prosecution and punishment of its perpetrators.

The Convention, in its Article 2 recognizes that any person who by any means directly or indirectly, unlawfully or willfully, provides or collects funds with the intention that they should be used or in the knowledge that they are to be used, in full or in part, in order to carry out any act which constitutes an offence under certain named treaties, commits an offence. One of the treaties cited by the Convention is the International Convention for the Suppression of Terrorist Bombings, adopted by the General Assembly of the United Nations on 15 December 1997.[61]

The Convention for the Suppression of the Financing of Terrorism also provides that, over and above the acts mentioned, providing or collecting funds toward any other act intended to cause death or serious bodily injury to a civilian, or to any other person not taking an active part in the hostilities in the situation of armed conflict, when the purpose of such act, by its nature or context, is to intimidate a population, or to compel a government or an international organization to do or to abstain from doing any act, would be deemed an offence under the Convention.

The United Nations has given effect to this principle in 1970 when it proclaimed that:

> Every State has the duty to refrain from organizing or encouraging the organization of irregular forces or armed bands, including mercenaries, for incursion into the territory of another State. Every State has the duty to refrain from organizing, instigating, assisting or participating in acts of civil strife or terrorist acts in another State or acquiescing in organized activities within its territory directed towards the commission of such acts, when the acts referred to in the present paragraph involve a threat or use of force.[62]

Here, the words *encouraging* and *acquiescing in organized activities within its territory directed towards the commission of such acts* have a direct bearing on the concept of condonation and would call for a discussion about how States could

---

[60] International Convention for the Suppression of the Financing of Terrorism, adopted by the General Assembly of the United Nations in resolution 54/109 of 9 December 1999.

[61] A/52/653, 25 November 1997.

[62] Declaration on Principles of International Law Concerning Friendly Relations and Co-operation Among States in Accordance with the Charter of the United Nations, UN General Assembly Resolution 2625 (XXV) 24 October 1970.

3.2 Attacks on the Integrity of Travel Documents

overtly or covertly encourage the commission of such acts. One commentator[63] identifies three categories of such support: *Category I* support entails protection, logistics, training, intelligence, or equipment provided terrorists as a part of national policy or strategy; *Category II* support is not backing terrorism as an element of national policy but is the toleration of it; *Category III* support provides some terrorists a hospitable environment, growing from the presence of legal protections on privacy and freedom of movement, limits on internal surveillance and security organizations, well-developed infrastructure, and émigré communities.

Another commentator[64] discusses what he calls the '*separate delict theory*' in State responsibility, whereby the only direct responsibility of the State is when it is responsible for its own wrongful conduct in the context of private acts, and not for the private acts themselves. He also contends that indirect State responsibility is occasioned by the State's own wrongdoing in reference to the private terrorist conduct. The State is not held responsible for the act of terrorism itself, but rather for its failure to prevent and/or punish such acts, or for its active support for or acquiescence in terrorism.[65] Arguably the most provocative and plausible feature in this approach is the introduction by the commentator of the desirability of determining State liability on the theory of causation. He emphasizes that:

> The principal benefit of the causality based approach is that it avoids the automatic rejection of direct State responsibility merely because of the absence of an agency relationship. As a result, it potentially exposes the wrongdoing State to a greater range and intensity of remedies, as well as a higher degree of international attention and opprobrium for its contribution to the private terrorist activity.[66]

The causality principle is tied in with the rules of State Responsibility enunciated by the International Law Commission and Article 51 of the United Nations Charter which states that nothing in the Charter will impair the inherent right of individual or collective self-defense if an armed attack occurs against a Member of the United Nations, until the Security Council has taken measures necessary to maintain international peace and security. The provision goes on to say that measures taken by Members in the exercise of this right of self-defense will be immediately reported to the Security Council and will not in any way affect the authority and responsibility of the Security Council under the present Charter to take at any time such action as it deems necessary in order to maintain or restore international peace and security.

The International Law Commission has established that a crime against the peace and security of mankind entails individual responsibility, and is a crime of

---

[63] Steven Metz, State Support for Terrorism, Defeating Terrorism, Strategic Issue Analysis, at http://www.911investigations.net/IMG/pdf/doc-140.pdf.
[64] Becker (2006b).
[65] *Id*. Chapter 2, 67.
[66] Becker, *supra*, note 64 at 335.

aggression.[67] A further link drawing civil aviation to the realm of international peace and security lies in the Rome Statute of the International Criminal court, which defines a war crime, inter alia, as intentionally directing attacks against civilian objects; attacking or bombarding, by whatever means, towns, villages, dwellings or buildings which are undefended and which are not military objects; employing weapons, projectiles, and material and methods of warfare that cause injury.[68] The Statute also defines as a war crime, any act which is intentionally directed at buildings, material, medical units and transport, and personnel using the distinctive emblems of the Geneva Conventions in conformity with international law.[69]

3.2.3 Knowledge

Another method of determining State responsibility lies in the determination whether a State had actual or presumed knowledge of acts of its instrumentalities, agents or private parties which could have alerted the State to take preventive action. International responsibility of a State cannot be denied merely on the strength of the claim of that State to sovereignty. Although the Chicago Convention in Article 1 stipulates that the contracting States recognize that every State has complete and exclusive sovereignty over the airspace above its territory, the effect of this provision cannot be extended to apply to State immunity from responsibility to other States. Professor Huber in the *Island of Palmas* case[70] was of the view:

> Sovereignty in the relations between States signifies independence. Independence in regard to a portion of the globe is the right to exercise therein, to the exclusion of any other State, the functions of a State...Territorial sovereignty...involves the exclusive right to display the activities of a State.[71]

Professor Huber's definition, which is a simple statement of a State's rights, has been qualified by Starke as the residuum of power which a State possesses within the confines of international law.[72] Responsibility would devolve upon a State in whose territory an act of unlawful interference against civil aviation might occur, to other States that are threatened by such acts. The International Court of Justice (ICJ) recognised in the *Corfu Channel* Case:

---

[67] Draft Code of Crimes Against the Peace and Security of Mankind, International Law Commission Report, 1996, Chapter II Article 2.

[68] Rome Statute of the International Criminal Court, Article 8.2 (b) (ii), (V) and (XX).

[69] Id. Article 8.2 (b) (XXIV).

[70] The *Island of Palmas* Case (1928) 11 U.N.R. I.A.A. at 829.

[71] Ibid.

[72] Starke (1989).

every State's obligation not to allow knowingly its territory to be used for acts contrary to the rights of other States.[73]

In the famous *Corfu Channel* case, the International Court of Justice applied the subjective test and applied the fault theory. The Court was of the view that:

> It cannot be concluded from the mere fact of the control exercised by a State over its territory and waters that the State necessarily knew, or ought to have known, of any unlawful act perpetrated therein, nor yet that it necessarily knew, or should have known the authors. This fact, by itself and apart from other circumstances, neither involves prima facie responsibility nor shifts the burden of proof.[74]

The Court, however, pointed out that exclusive control of its territory by a State had a bearing upon the methods of proof available to establish the involvement or knowledge of that State as to the events in question.

Apart from the direct attribution of responsibility to a State, particularly in instances where a State might be guilty of a breach of treaty provisions, or violate the territorial sovereignty of another State, there are instances where an act could be imputed to a State.[75] Imputability or attribution depends upon the link that exists between the State and the legal person or persons actually responsible for the act in question. The legal possibility of imposing liability upon a State wherever an official could be linked to that State encourages a State to be more cautious of its responsibility in controlling those responsible for carrying out tasks for which the State could be ultimately held responsible. In the same context, the responsibility of placing mines was attributed to Albania in the *Corfu Channel* case where the court attributed to Albania the responsibility, since Albania was known to have knowledge of the placement of mines although it did not know who exactly carried out the act. It is arguable that, in view of the responsibility imposed upon a State by the Chicago Convention on the provision of air navigation services, the principles of immutability in State responsibility could be applied to an instance of an act or omission of a public or private official providing air navigation services.

The sense of international responsibility that the United Nations ascribed to itself had reached a heady stage at this point, where the role of international law in

---

[73] (1949) *I.C.J.R.*1, 22.

[74] The *Corfu Channel* Case, ICJ Reports, 1949, p. 4.

[75] There are some examples of imputability, for example the incident in 1955 when an Israeli civil aircraft belonging to the national carrier El Al was shot down by Bulgarian fighter planes, and the consequent acceptance of liability by the USSR for death and injury caused which resulted in the payment of compensation to the victims and their families. See 91 *ILR* 287. Another example concerns the finding of the International Court of Justice that responsibility could have been be imputed to the United States in the *Nicaragua* case, where mines were laid in Nicaraguan waters and attacks were perpetrated on Nicaraguan ports, oil installations and a naval base by persons identified as agents of the United States. See *Nicaragua* v. *the United States*, ICJ Reports 1986, 14. Also, 76 *ILR* 349. There was also the instance when the Secretary General of the United Nations mediated a settlement in which a sum of $ 7 million was awarded to New Zealand for the violation of its sovereignty when a New Zealand vessel was destroyed by French agents in New Zealand. See the *Rainbow Warrior* case, 81 *AJIL*, 1987 at 325. Als in 74 *ILR* at 241.

international human conduct was perceived to be primary and above the authority of States. In its Report to the General Assembly, the International Law Commission recommended a draft provision which required:

> Every State has the duty to conduct its relations with other States in accordance with international law and with the principle that the sovereignty of each State is subject to the supremacy of international law.[76]

This principle, which forms a cornerstone of international conduct by States, provides the basis for strengthening international comity and regulating the conduct of States both internally – within their territories – and externally, towards other States. States are effectively precluded by this principle of pursuing their own interests untrammelled and with disregard to principles established by international law.

The United Nations General Assembly, in its Resolution 56/83,[77] adopted as its Annex the International Law Commission's *Responsibility of States for Internationally Wrongful Acts* which recognizes that every internationally wrongful act of a State entails the international responsibility of that State[78] and that there is an internationally wrongful act of a State when conduct consisting of an action or omission is attributable to the State under international law and constitutes a breach of an international obligation of the State.[79] Article 5 of the ILC document provides that the conduct of a person or entity which is not an organ of State but which is empowered by the law of that State to exercise elements of the governmental authority shall be considered an act of State under international law, provided the person or entity is acting in that capacity in the particular instance.

In the *Pan Am* case,[80] where an aircraft was destroyed over Lockerbie the British allegation against Libya's involvement in the act of terrorism was that the accused individuals (Libyan nationals) had acted as part of a conspiracy to further the purposes of the Libyan Intelligence Services using criminal means that amounted to terrorism. The United Kingdom appeared to stress the point in the UN Security Council that Libya had failed to respond to the request for extradition of the implicated Libyan nationals, and arguably as a consequence, the Security Council adopted Resolution 731 on 21 January 1992 which expressed concerns over certain investigations which imputed reprehensibility to officials of the Libyan Government.[81]

ICAO has been working on the development of passports since 1968. The Seventh Session of the ICAO Facilitation Division in 1968 recommended that a

---

[76] Report of the International Law Commission to the General Assembly on the Work of the 1st Session, A/CN.4/13, June 9 1949, at 21.

[77] A/RES/56/83 56th Session, 28 January 2002

[78] *Id*. Article 1

[79] Id. Article 2.

[80] *Infra*, note 130.

[81] For a discussion on this point see Jorgensen (2000).

3.2 Attacks on the Integrity of Travel Documents

small panel of qualified experts including representatives of the passports and/or other border control authorities, be established: to determine the establishment of an appropriate document such as a passport card, a normal passport or an identity document with electronically or mechanically readable inscriptions that meet the requirements of document control; the best type of procedures, systems (electronic or mechanical) and equipment for use with the above documents that are within the resources and ability of Member States; the feasibility of standardizing the requisite control information and methods of providing this information through automated processes, provided that these processes would meet the requirements of security, speed of handling and economy of operation.[82]

A passport asserts that the person holding the passport is a citizen of the issuing State while a visa confirms that the State issuing the visa has granted the visa holder the non-citizen privilege of entering and remaining in the territory of the issuing State for a specified time and purpose. An ePassport is a type of Machine Readable Passport (MRP)[83] with an embedded microchip that contains data printed on the data page of the passport, including biographic and biometric information of the holder, and passport data. The chip also contains security features for preventing passport fraud and forgery and misuse of data stored on the chip. ePassports are easily recognised by the international ePassport symbol on the front cover.[84]

The techniques of biometrics employed in a machine readable travel document (MRTD), be it a visa or passport,[85] enable the user to uniquely encode a particular physical characteristic of a person into a biometric identifier or biometric template which can be verified by machine to confirm or deny a claim regarding a person's identity. Accordingly, biometric identification of a person either correctly establishes his identity as being consistent with what is claimed in the passport he is holding or brings to bear the possibility that the person carrying a particular passport is an imposter. A biometric is a measurable, physical characteristic or

---

[82] See Facilitation Division, Report of the Seventh Session, 14–30 May 1968, ICAO Doc 8750-FAL/564, Agenda Item 2.3, at 2.3-4. See also *AT-WP/1079, 1/12/70*, Attachment A, which sets out the Terms of Reference of the Panel.

[83] The machine readable passport (MRP) is a passport that has both a machine readable zone and a visual zone in the page that has descriptive details of the owner. The machine readable zone enables rapid machine clearance, quick verification and instantaneous recording of personal data. Besides these advantages, the MRP also has decided security benefits, such as the possibility of matching very quickly the identity of the MRP owner against the identities of undesirable persons, whilst at the same time offering strong safeguards against alteration, counterfeit or forgery. See Abeyratne (1992).

[84] http://www.dhs.gov/xtrvlsec/programs/content_multi_image_0021.shtm.

[85] The machine readable passport (MRP) is a passport that has both a machine readable zone and a visual zone in the page that has descriptive details of the owner. The machine readable zone enables rapid machine clearance, quick verification and instantaneous recording of personal data. Besides these advantages, the MRP also has decided security benefits, such as the possibility of matching very quickly the identity of the MRP owner against the identities of undesirable persons, whilst at the same time offering strong safeguards against alteration, counterfeit or forgery (Abeyratne 1992).

personal behavioral trait used to recognize the identity, or verify[86] the claimed identity of a person. In the modern context, biometrics are usually incorporated in an MRTD with a view to achieving five goals, the first of which is global interoperability[87] enabling the specifications of biometrics deployed in travel documents across the world to be applied and used in a universally operable manner. This is a critical need if the smooth application of biometric technology were to be ensured across borders. The second goal is to ensure uniformity within States in specific standard setting by States authorities who deploy biometrics in travel documents issued by them. The third is technical reliability, where States are required to ensure that technologies used in deploying biometrics are largely failure-proof and of sufficient quality and standard to ensure a State immigration authority reading documents issued by other States can determine that the details in the document do provide accurate verification of facts. Fourthly, the technology used has to be practical and not give rise to the need for applying disparate types of support technology at unnecessary cost and inconvenience to the user. The final goal is to ensure that the technology used will be sufficiently up to date for at least 10 years and also be backwardly compatible with new techniques to be introduced in the future.

Biometrics target the distinguishing physiological or behavioral traits of the individual by measuring them and placing them in an automated repository such as machine encoded representations created by computer software algorithms that could make comparisons with the actual features. Physiological biometrics that have been found to successfully accommodate this scientific process are facial recognition, fingerprinting and iris-recognition which have been selected by ICAO as being the most appropriate. The biometric identification process is fourfold: firstly involving the capture or acquisition of the biometric sample; secondly extracting or converting the raw biometric sample obtained into an intermediate form; and thirdly creating templates of the intermediate data that is converted into a template for storage; and finally the comparison stage where the information offered by the travel document is compared with that which is stored in the reference template.

Biometric identification gets into gear each time an MRTD holder (traveler) enters or exists the territory[88] of a State and when the State verifies his identity

---

[86] To "verify" means to perform a one-to-one match between proffered biometric data obtained from the holder of the travel document at the time of inquiry with the details of a biometric template created when the holder enrolled in the system.

[87] "Global interoperability" means the capability of inspection systems (either manual or automated) in different States throughout the world to exchange data, to process data received from systems in other States, and to utilize that data in inspection operations in their respective states. Global interoperability is a major objective of the standardized specifications for placement of both eye-readable and machine-readable data in all MRTDs.

[88] The Chicago Convention, *supra*, note3, defines, in Article 2, "territory of a State" as the land areas and territorial waters adjacent to the State under the sovereignty, suzerainty, protection and mandate of such State.

3.2 Attacks on the Integrity of Travel Documents

against the images or templates created at the time his travel document was issued. This measure not only ensures that the holder of the document is the legitimate claimant to that document and to whom it was issued, but also enhances the efficacy of any advance passenger information (API)[89] system used by the State to predetermine the arrivals to its territory. Furthermore, matching biometric data presented in the form of the traveler with the data contained in the template accurately ascertains as to whether the travel document has been tampered with or not. A three way check, which matches the traveler's biometrics with those stored in the template carried in the document and a central database, is an even more efficacious way of determining the genuineness of a travel document. The final and most efficient biometric check is when a four way determine is effected, were the digitized photograph is visually matched (non electronically) with the three way check described above.[90] In this context, it is always recommended that the traveler's facial image (conventional photograph) should be incorporated in the travel document along with the biometric templates in order to ensure that his identity could be verified at locations where there is no direct access to a central database or where the biometric identification process has not entered into the legal process of that location.

3.2.4 Security of the Passport

Production of passport books and travel documents, including the personalization processes, should be undertaken in a secure, controlled environment with appropriate security measures in place to protect the premises against unauthorized access. If the personalization process is decentralized, or if personalization is carried out in a location geographically separated from where the travel document blanks are made, appropriate precautions should be taken when transporting the blank

---

[89] API involves exchange of data information between airlines and customs authorities, where an incoming passenger's essential details are notified electronically by the airline carrying that passenger prior to his arrival. The data for API would be stored in the passenger's machine readable passport, in its machine readable zone. This process enables customs authorities to process passengers quickly, thus ensuring a smoother and faster clearance at the customs barriers at airports. One of the drawbacks of this system, which generally works well and has proven to be effective, is that it is quite demanding in terms of the high level of accuracy required. One of the major advantages, on the other hand, is the potential carried by the API process in enhancing aviation security at airports and during flight. See Abeyratne (2002a).

[90] Issuing States must ensure the accuracy of the biometric matching technology used and functions of the systems employed if the integrity of the conducted checks are to be maintained. They must also have realistic and efficient criteria regarding the number of travel documents checked per minute in a border control situation and follow a regular biometric identification approach such as facial recognition, fingerprint examination or iris identification system.

documents and any associated security materials to safeguard their security in transit.

There should be full accountability over all the security materials used in the production of good and spoiled travel documents and a full reconciliation at each stage of the production process with records maintained to account for all material usage. The audit trail should be to a sufficient level of detail to account for every unit of material used in the production and should be independently audited by persons who are not directly involved in the production. Certified records should be kept of the destruction of all security waste material and spoiled documents.

Materials used in the production of travel documents should be of controlled varieties and obtained only from bona fide security materials suppliers. Materials whose use is restricted to high security applications should be used, and materials that are available to the public on the open market should be avoided.

Sole dependence upon the use of publicly available graphics design software packages for originating the security backgrounds should be avoided. These software packages may however be used in conjunction with specialist security design software.

Security features and/or techniques should be included in travel documents to protect against unauthorized reproduction, alteration and other forms of tampering, including the removal and substitution of pages in the passport book, especially the biographical data page. In addition to those features included to protect blank documents from counterfeiting and forgery, special attention must be given to protect the biographical data from removal or alteration. A travel document should include adequate security features and/or techniques to make evident any attempt to tamper with it.

The combination of security features, materials and techniques must be well chosen to ensure full compatibility and protection for the lifetime of the document. There is another class of security features comprised of covert (secret) features, designed to be authenticated either by forensic examination or by specialist verification equipment. It is evident that knowledge of the precise substance and structure of such features should be restricted to very few people on a "need to know" basis. The purpose of these features is not to prevent counterfeiting but to enable authentication of documents where unequivocal proof of authenticity is a requirement (e.g. in a court of law). All travel documents should contain at least one covert security feature as a basic feature.

3.2.4.1 Threats to the Security of Passports

There are many threats to the security of passports such as: counterfeiting a complete travel document; photo-substitution; deletion/alteration of text in the visual or machine readable zone of the MRP data page; construction of a fraudulent document, or parts thereof, using materials from legitimate documents; removal and substitution of entire page(s) or visas; deletion of entries on visa pages and the

3.2 Attacks on the Integrity of Travel Documents

observations page; theft of genuine document blanks; and impostors (assumed identity; altered appearance).

To provide protection against these threats and others, a travel document requires a range of security features and techniques combined in an appropriate way within the document. Although some features can offer protection against more than one type of threat, no single feature can offer protection against them all. Likewise, no security feature is 100% effective in eliminating any one category of threat. The best protection is obtained from a balanced set of features and techniques providing multiple layers of security in the document that combine to deter or defeat fraudulent attack.[91]

Annex 9[92] to the Convention on International Civil Aviation, in Standard 3.7 requires ICAO member States to regularly update security features in new versions of their travel documents, to guard against their misuse and to facilitate detection of cases where such documents have been unlawfully altered, replicated or issued. Recommended Practice 3.9 suggests that member States incorporate biometric data in their machine readable passports, visas and other official travel documents, using one or more optional data storage technologies to supplement the machine readable zone, as specified in Doc 9303, Machine Readable Travel Documents. The required data stored on the integrated circuit chip is the same as that printed on the data page, that is, the data contained in the machine-readable zone plus the digitized photographic image. Fingerprint image(s) and/or iris image(s) are optional biometrics for member States wishing to supplement the facial image with another biometric in the passport. Member States incorporating biometric data in their Machine Readable Passports are to store the data in a contactless integrated circuit chip complying with ISO/IEC 14443 and programmed according to the Logical Data Structure as specified by ICAO.

3.2.4.2 The Diplomatic Fallout

Any diplomatic action in the context of the issues raised in this article must primarily be based on State responsibility. In turn, and as already discussed, the issue of responsibility hinges on knowledge, complicity and condonation of a State. The responsibility of a State is determined by the quantum of proof available that could establish intent or negligence of the State, which in turn would establish complicity or condonation on the part of the State concerned. One way to determine complicity or condonation is to establish the extent to which the State adhered to the obligation imposed upon it by international law and whether it breached its duty to others. In order to exculpate itself, the State concerned will have to demonstrate that either it did not tolerate the offence or that it ensured the punishment of the

---

[91] Machine Readable Travel Documents, ICAO Doc 9303 Part 1, Machine Readable Passports, Sixth Edition, 2006, III-4.

[92] Annex 9 to the Convention on International Civil Aviation, 12th Edition, 2006.

offender. *Brownlie* is of the view that proof of such breach would lie in the causal connection between the offender and the State.[93] In this context, the act or omission on the part of a State is a critical determinant particularly if there is no specific intent.[94] Generally, it is not the intent of the offender that is the determinant but the failure of a State to perform its legal duty in either preventing the offence (if such was within the purview of the State) or in taking necessary action with regard to punitive action or redress.[95]

There are a few principles that have to be taken into account when determining State responsibility. Firstly, there has to be either intent on the part of the State towards complicit or negligence reflected by act or omission. Secondly, where condonation is concerned, there has to be evidence of inaction on the part of the State in prosecuting the offender. Thirdly, since the State as an abstract entity cannot perform an act in itself, the imputability or attribution of State responsibility for acts of its agents has to be established through a causal nexus that points the finger at the State as being responsible. For example, The International Law Commission, in Article 4 of its Articles of State Responsibility states that the conduct of any State organ which exercises judicial, legislative or executive functions could be considered an act of State and as such the acts of such organ or instrumentality can be construed as being imputable to the State. This principle was endorsed in 1999 by the ICJ which said that according to well established principles of international law, the conduct of any organ of a state must be regarded as an act of State.[96]

The law of State responsibility has evolved through the years, from being a straightforward determination of liability of the State and its agents to a rapidly widening gap between the State and non State parties. In today's world private entities and persons could wield power similar to that of a State, bringing to bear the compelling significance and modern relevance of the agency nexus between the State and such parties. This must indeed make States more aware of their own susceptibility.

The United Nations General Assembly, in 2002 adopted Resolution A 56/83[97] on the subject of Responsibility of States for internationally wrongful acts. The Resolution, which was the result of work of the International Law Commission on the subject, provides that every internationally wrongful act of a State entails the international responsibility of that State[98] and that such an act is attributable to that State under international law and constitutes a breach of an international obligation of that State. Article 5 to the Annex to the Resolution states that the conduct of a

---

[93] Brownlie (1983).

[94] Report of the International Law Commission to the United Nations General Assembly, UNGOAR 56th Session, Supp. No. 10, *UN DOC A/56/10*, 2001 at 73.

[95] de Arechaga (1968).

[96] Differences Relating to Immunity from Legal Process of a Special Rapporteur, ICJ Reports 1999, 62 at 87.

[97] A/RES/56/83 Fifty sixty Session 28 January 2002.

[98] *Id.* Annex, Article 1.

3.2 Attacks on the Integrity of Travel Documents

person or entity which is not an organ of a State but which is empowered by the law of that State to exercise elements of the governmental authority shall be considered an act of the State under international law, provided the person or entity is acting in that capacity in the particular instance. If, as alleged, the assassination of Mahmoud al-Mabhouh was carried out by Mossad, which reports to the Isreali Prime Minister, Article 8 of the Annex to the Resolution is particularly relevant as it provides that the conduct of a person or group of persons would be considered an act of a State under international law if the person or group of persons is in fact acting on the instructions of, or under the direction or control of that State in carrying out the conduct. The Resolution also recognizes that there is a breach of an international obligation by a State when an act of that State is not in conformity with what is required of it by that obligation, regardless of its origin or character.[99]

Diplomatic relations between States are intrinsically linked to State responsibility and are based on relations between States dependent on comity.[100] The fundamental fact in this context is that international society is not an unchanging entity, but is subject to the ebb and flow of political life and activity.[101] Analogies to diplomatic relations between States arise often in the context of terrorism as in the 1985 *Rainbow Warrior* case. The sinking of the ship *Rainbow Warrior* in Auckland Harbour in New Zealand as a result of an officially organized undercover French military operation calculated to obstruct Greenpeace protests against French nuclear operations in the South Pacific is a good example. The destruction of the ship, which resulted in the death of a Dutch seaman, was directly attributable to the placing of explosives by an agency reporting to the French Ministry of Defence. This was construed by the Government of New Zealand and by the High Court of New Zealand as a violation of the principles of international law against the sovereignty of the country.[102] The Secretary General of the United Nations ruled that France should offer New Zealand an apology and that compensation should follow to the amount of $ 7,000,000. The *Rainbow Warrior* incident goes down in the annals of diplomatic relations as an act of international delinquency resulting in the criminal responsibility of a State.

Another analogous diplomatic incident occurred in 1981 when, on May 6, the US Department of State announced at a special press briefing that the United States Government had decided to require the Socialist People's Libyan Arab Jamahiriya

---

[99] *Id*, Article 12.

[100] In law, comity specifically refers to legal reciprocity - the principle that one jurisdiction will extend certain courtesies to other nations (or other jurisdictions within the same nation), particularly by recognizing the validity and effect of their executive, legislative, and judicial acts. The term refers to the idea that courts should not act in a way that demeans the jurisdiction, laws, or judicial decisions of another jurisdiction. Part of the presumption of comity is that other jurisdictions will reciprocate the courtesy shown to them.

[101] See generally Jennings and Watts (1992), Lauterpacht (1947), Chen (1951), Shaw (2003).

[102] *R*. v. *Mafart and Prieur*, New Zealand High Court, Auckland Registry, 22 November 1985, Per Davison CJ reported in 74 *ILR* 241.

to close its People's Bureau at Washington immediately and to withdraw all personnel within five working days. The Department's official statement was that from the first days of the Administration, both the President and the Secretary of State had made known their very real concern about a wide range of Libyan provocations and misconduct, including support for international terrorism. The United States made it officially clear that it had been concerned by a general pattern of unacceptable conduct by the People's Bureau in Washington, which was contrary to internationally accepted standards of diplomatic behaviour. The United States therefore asked the Libyans to close their People's Bureau in Washington and have given them five working days starting today to withdraw their personnel. This action reduced US relations with Libya to the lowest level consistent with maintenance of diplomatic relations and was followed with a travel advisory which stated that due to unsettled relations between the United States and the Government of Libya, the Department of State warned American citizens against any travel to or residence in Libya. Travellers were also informed that the US Embassy in Tripoli was closed and the US was not in a position to provide consular protection and assistance to Americans presently in Libya.[103]

In 1999, the Clinton administration warned Russia to voluntarily reduce the large number of intelligence officers operating in the United States or face cutbacks in diplomatic positions or expulsions. U.S. Ambassador James Collins delivered the message in Moscow during a meeting with Vladimir Putin, the former KGB domestic spying chief and currently Russia's top Security Council adviser, according to administration officials familiar with the issue. The warning followed two expulsions of Russian intelligence officers from the United States and the ouster of a U.S. Army attaché from Moscow a month earlier.[104]

The international treaty regulating diplomatic relations is the *Vienna Convention on Diplomatic Relations* of 1961.[105] The Convention does not explicitly make provision for the right to break diplomatic relations. It follows by implication from Article 2 which provides that the establishment of diplomatic relations takes place by mutual consent that if either State withdraws that consent diplomatic relations are broken. Breach therefore takes place normally in consequence of a unilateral act – even though it frequently follows a sequence of reciprocal or retaliatory moves between two States to downgrade their relations or a collective political decision by a number of States directed against another State whose conduct is regarded as unacceptable. Relations are broken from the moment of the initial action.[106] The other State has no option in the matter. There are no legal

---

[103] Dept. of State File No. P81 0101–1084.

[104] Bill Gertz, *The Washington Times*, 26 July 1999.

[105] Done at Vienna on 18 April 1961 and entered into force on 24 April 1964. United Nations Treaty Series, Vol. 5000 at p. 95.

[106] For an account of the series of incidents and complaints between France and Iran which led France to break diplomatic relations in July 1987, see *1987 AFDI 1000*. See also do Nascimento e Silva, *Diplomacy in International Law* p. 173–4.

3.2 Attacks on the Integrity of Travel Documents

limitations on the right of a State to break diplomatic relations with another, but the action is now invariably taken for political reasons. Practical considerations will almost always favour the continuation of relations, though not necessarily the retention of a permanent mission. This has become more obvious in the light of some recent cases where diplomatic relations subsisted even while armed conflict was taking place between sending and receiving States – as between India and Pakistan in 1965 and 1971.

A breach of diplomatic relations generally precludes direct contact between sending and receiving States other than what is needed to effect orderly departure and some form of interim regime. It does not, however, preclude the sending and receiving of special missions (which may later herald a resumption of normal diplomatic relations), meetings between diplomatic representatives of the two States in a third State (for example the regular meetings in Warsaw over many years of representatives of the United States and of the People's Republic of China) or contacts between representatives of the two States to an international organization. Detailed rules on permissible contacts are usually provided in the internal diplomatic service regulations of each State. It is often a feature of modern diplomacy that those on occasion a much-advertised breach of relations may turn out to be only partially real. This occurs when two States, having broken off diplomatic relations, usually on the initiative of one of them, continue an active, if quiet, direct relationship despite the appointment of third States to protect the interests of each in the territory of the other State.[107]

Whatever unilateral diplomatic action an aggrieved State might take, be it on grounds of sovereignty or the violation of its national property (passports) and the rights of its citizens who held the passports, there are certain legal nuances in the Mahmoud al-Mabhouh case which are incontrovertible. Falsification of passports and identity theft are serious criminal offences under most national laws. It could well be that these are also offences under the laws of Israel. Falsification of a national passport, whatever its country or nationality might be, by a member of the Israeli intelligence services brings to bear issues that are much more serious than mere breaches of diplomatic courtesy or relations. Under the theory of condonation any government involved would be seriously implicated, were it to turn out that it was aware that falsified travel documents were being used by its security agency as has been suggested by some.

The international community should therefore condemn the extra judicial killing of Mahmoud al-Mabhouh as a breach of international law and those involved must unequivocally declare as to whether they were aware that falsified travel documents were being used by Mossad in relation to this operation and/or any other. If there is cogent evidence implicating the Israeli Government, the international community must also require the former to confirm whether its intelligence services were involved in the murder of Mahmoud al-Mabhouh and demand that the Israeli

---

[107] D. James, "Diplomatic Relations and Contacts", *1991 BYIL 375*.

government confirm whether or not their intelligence services used falsified passports for this or any other operation or whether they have done since any assurance that they would not do so. Furthermore the international community should seek an assurance from the Israeli government that their intelligence operatives will never falsify passports for use in operations and require the Israeli government to condemn the killing of Mahmoud al-Mabhouh as a breach of international law. Finally an assurance must be sought from the Israeli government that they will extradite to Dubai any of those identified by the Dubai authorities as having been involved in the killing to face trial for murder and to Ireland, Britain, France and/or Germany to face trial for offences arising out of the abuse of passports issued by those countries.

At present, the issue of extradition could be settled through the United Nations and its Organs such as the Security Council[108] and the International Court of Justice (ICJ).[109] Of noteworthy practical relevance with regard to the complicity theory, particularly on the issue of extradition and whether one State can demand the extradition of offenders harbored in another State is the opinion given by the ICJ[110] on the explosion over Lockerbie, Scotland on 21 December 1988 of PAN AM Flight 103. The explosion is believed to have been caused by the detonation of a plastic explosive concealed in a portable cassette player/radio. The ICJ noted that it was a general principle of international law that no State could be compelled to extradite its nationals and that the State concerned held the prerogative of trying the accused of a crime in its own territory. The ICJ was encumbered with the discussion as to whether the Court had jurisdiction over a United Nations Security Council Resolution on the issue. The essence of the views of the learned judges of the ICJ was that the complimentary roles played by the United Nations Security Council and the ICJ would devolve responsibility on States to respect both these organs on the subject of extradition of private offenders.

It appears that the question in The ICJ's was whether the Security Council, by its Resolution 748 (1992) which required Libya to extradite its nationals either to the United States or to the United Kingdom, had the authority to override an established

---

[108] The Security Council is the branch of the United Nations charged with the maintenance of international peace and security. Its powers, outlined in the Charter of the United Nations, include the establishment of peacekeeping operations, the establishment of international sanctions, and the authorization for military action. The Security Council's power are exercised through its Resolutions. The Permanent members of the Security Council are the United States of America, United Kingdom, France, the Russian Federation and the Republic of China.

[109] The International Court of Justice (ICJ) is the principal judicial organ of the United Nations (UN). It was established in June 1945 by the Charter of the United Nations and began work in April 1946.The Court's role is to settle, in accordance with international law, legal disputes submitted to it by States and to give advisory opinions on legal questions referred to it by authorized United Nations organs and specialized agencies. The Court is composed of 15 judges, who are elected for terms of office of 9 years by the United Nations General Assembly and the Security Council. It is assisted by a Registry, its administrative organ. Its official languages are English and French.

[110] I.C.J. Reports 1980, 116.

principle of international law. The answer to this question was, in the view of one judge, in the affirmative.

If a State found reprehensible is unable or unwilling to make reparations as requested, there is nothing to prevent a State from unilaterally terminating diplomatic relations with any that State if the former wishes to do so.

3.3 Full Body Scanners and Emergent Issues

Aviation is an important global business and a significant driver of the global economy. It is vital, therefore, that stringent measures are taken to counter acts of unlawful interference with civil aviation. The *Convention on International Civil Aviation* signed at Chicago on 7 December 1944, states in its *Preamble* that whereas the development of civil aviation may help preserve friendship and understanding among the people of the world, yet, its abuse could become a threat to general security.

The genealogy of the term *"Terrorism"* lies in Latin terminology meaning "to cause to tremble"*(terrere)*. Since the catastrophic events of 11 September 2012, we have seen stringent legal measures taken by the United States to attack terrorism, not just curb it. The famous phrase "war on terror" denotes pre-emptive and preventive strikes carried out through applicable provisions of legitimately adopted provisions of legislation. The earliest example is the *Air Transportation Safety and System Stabilization Act* (ATSAA) enacted by President Bush less than 2 months after the 9/11 attacks. Then, 2 months after the attacks, in November 2001, Congress passed the *Aviation and Transportation Security Act* (ATSA) with a view to improving security and closing the security loopholes which existed on that fateful day in September 2001. The legislation paved the way for a huge federal body called the Transportation Security Administration (TSA) which was established within the Department of Transportation. The Homeland Security Act of 2002 which followed effected a significant reorganization of the Federal Government.

Since the events of 11 September 2001, there have been several attempts against the security of aircraft in flight. These threats have ranged from shoe bombs to dirty bombs to explosives that can be assembled in flight with liquids, aerosols and gels. In every instance the global community has reacted with pre emptive and preventive measures which prohibit any material on board which might seemingly endanger the safety of flight. Some jurisdictions have even gone to extremes in prohibiting human breast milk and prescriptive medications on board.

New and emerging threats to civil aviation are a constant cause for concern to the aviation community. Grave threats such as those posed by the carriage of dangerous pathogens on board, the use of cyber technology calculated to interfere with air navigation systems, and the misuse of man portable air defence systems are real and have to be addressed with vigour and regularity. The International Civil Aviation

Organization has been addressing these threats for some time and continues to do so on a global basis.

It is a platitude to say that aviation security is a largely reactive process. It will be recalled that after the spate of hijackings in the late 1960s and 1970s, States rushed to install detectors with X-Ray capability at the entrance to the aircraft gate. Then, as the *displacement theory*[111] demonstrated, terrorists moved their attention towards attacking airports, which prompted States to install screening equipment at centralized points in the terminal itself. In similar vein, in the aftermath of the attempted bombing of an aircraft on 25 December 2009 by a person who is alleged to have carried explosives in his undergarments, some States began to look seriously into tightening airport security, particularly through a more stringent body scanning process. While the United States toughened screening measures on US bound flights, particularly with regard to passengers arriving from 14 targeted nations,[112] airports in the United Kingdom began the use of full body scanners at both Heathrow and Manchester airports.[113] In Canada, Rob Merrifield, Minister of State for Transport is reported to have stated that 44 scanners have been ordered to be used on passengers selected for secondary screening at Canadian airports. The machines, which can scan through clothing, will be installed in Vancouver, Calgary, Edmonton, Winnipeg, Toronto, Ottawa, Montreal and Halifax.[114] This measure is partly due to the fact that the Christmas day incident was later classified as having occurred due to a serious lapse in security.

Full body scanners, costing about $ 250,000 each and claimed by some security experts as an effective tool in detecting hidden explosives, show the contours of the human body as well as body parts in some detail, prompting some to question the legality and ethical justification of their use. In the United States, passengers handpicked for a full-body scan can opt out of the screening, but if they do, they must submit to full-body pat-downs by an officer of the Transport Security

---

[111] The Displacement Theory suggests that removing opportunity for crime or seeking to prevent a crime by changing the situation in which it occurs (see Situational Crime Prevention) does not actually prevent crime but merely moves it around. There are five main ways in which this theory suggests crime is moved around: crime can be moved from one location to another (geographical displacement); crime can be moved from one time to another (temporal displacement); crime can be directed away from one target to another (target displacement); one method of committing crime can be substituted for another (tactical displacement); and one kind of crime can be substituted for another (crime type displacement).

[112] Afghanistan, Algeria, Cuba, Iran, Iraq, Lebanon, Libya, Nigeria, Pakistan, Saudi Arabia, Somalia, Sudan, Syria and Yemen. See US Toughens Screening for US- Bound Flights, *Air Letter*, No. 16,896, Monday 04 January 2010, at 1.

[113] UK Airports Commence Use of Full Body Scanners, *Air Letter*, No. 16,918, Wednesday 03 February 2010 at 2. According to this report, scanning equipment were scheduled to be installed in Birmingham in late February 2010. *Ibid.*

[114] CBC News, January 5 2010. See: http://www.cbc.ca/canada/story/2010/01/05/security-canada-us-airport.html#ixzz0eVT3wBNY.

3.3 Full Body Scanners and Emergent Issues

Administration (TSA).[115] The technology was introduced a couple of years ago, but U.S. airports have been slow to install the machines, partly because of privacy concerns raised by some members of Congress and civil liberties groups.

It must be noted that in the United States, the Fourth Amendments states:

> The right of the people to be secure in their persons, houses, papers and effects, against unreasonable searches and seizures, shall not be violated, and no Warrants shall issue, but upon probable cause, supported by Oath or affirmation, and particularly describing the place to be searched, and the persons or things to be seized.[116]

The significance of this provision lies in the fact that the prohibition is against unreasonable searches, and that too by agents of the governments, the latter fact being borne out by a strong *cursus curiae* in the United States.[117] It can therefore be assumed that the Fourth Amendment may not be applicable in instances where scanning is carried out by airport security officers who are not government agents.[118] If, however, the scanning at the airport is conducted by officers of the government, by law, the consent of the passenger has to be obtained before such scanning is carried out.[119]

States which are installing full body scanners are fully aware that their use could bring to bear issues of privacy. However, it should be noted that this is just one more reactive step – to ensure that no person enters an aircraft with explosives hidden in his underwear – and the only known way to respond to this new threat is to use full body scanner. The question then arises as to whether the responsibility of the State toward its constituents and those using aircraft for transport from and to their territory, to prevent private acts of terrorism overrides the right of privacy of the individual. This article will address the balance between the two interests.

3.3.1 *The Right of Privacy of the Passenger*

The *Convention on International Civil Aviation* of 1944 (Chicago Convention),[120] which established the regulatory framework for international civil aviation, underscores the fundamental aim of States in the context of civil aviation to exchange privileges which friendly nations have a right to expect from each other. In his message to the Conference in Chicago, President Roosevelt said:

---

[115] Philip Rucker, TSA Tries to Assuage Concerns About Full Body Scans, *Washington Post*, Monday January 4 2010 at 1.

[116] *Us Constitution*, Article 1 Sec. 4 Clause 6.

[117] See Kathleen Sweet, Aviation Security and Passenger Rights, *Aviation Security Management*, Volume Two, Andrew R. Thomas ed, Praeger Security International: Westport Connecticut, 2008, at 45.

[118] *Ibid*.

[119] See *US* v. *Favela* 247 F.3d.838, 2001 and *U.S* v. *Eustaquio* 198 R.3d 1068 (8th Cir.1999).

[120] Convention on International Civil Aviation, signed at Chicago on 7 December 1944. See ICAO Doc 7300/9 Ninth Edition, 2008.

"the Conference is a great attempt to build enduring institutions of peace, which cannot be endangered by petty considerations or weakened by groundless fears".[121]

The Chicago Convention embodies in its *Preamble* the need to create and preserve friendship and understanding among the nations and peoples of the world, and cautions Contracting States that the abuse of this friendship and understanding can become a threat to general security. Article 13 of the Convention provides that the laws and regulations of a Contracting State as to the admission to and departure from its territory of passengers, crew or cargo of aircraft, such as regulations relating to entry, clearance, immigration, passports, customs and quarantine shall be complied with by or on behalf of such passengers, crew or cargo upon entrance into or departure from, or while within the territory of that State. This provision ensures that a Contracting State has the right to prescribe its own internal laws with regard to passenger clearance and leaves room for a State to enact laws, rules and regulations to ensure the security of that State and its people at the airport. However, this absolute right is qualified so as to preclude unfettered and arbitrary power of a State, by Article 22 which makes each Contracting State agree to adopt all practicable measures, through the issuance of special regulations or otherwise, to facilitate and expedite navigation of aircraft between the countries.

The above notwithstanding, there are three rights of privacy relating to the display and storage and use of personal data:

1. The right of an individual to determine what information about oneself to share with others, and to control the disclosure of personal data;
2. The right of an individual to know what data is disclosed, and what data is collected and where such is stored when the data in question pertains to that individual; the right to dispute incomplete or inaccurate data; and
3. The right of people who have a legitimate right to know in order to maintain the health and safety of society and to monitor and evaluate the activities of government.[122]

It is incontrovertible that the data subject has a right to decide what information about oneself to share with others and more importantly, to know what data is collected about him. This right is balanced by the right of a society to collect data about individuals that belong to it so that the orderly running of government is ensured.

The data subject, like any other person, has an inherent right to his privacy.[123] The subject of privacy has been identified as an intriguing and emotive one.[124] The right to privacy is inherent in the right to liberty, and is the most comprehensive of rights and the right most valued by civilized man.[125] This right is susceptible to

---

[121] Proceedings of the International Civil Aviation Conference, Chicago, Ikllinois, November 1–December 7 1944 The Department of State, Vol. 1 at p. 43.

[122] Hoffman (1980).

[123] Abeyratne (2001, 2002b).

[124] Young (1978).

[125] Warren and Brandies (1890–1891).

3.3 Full Body Scanners and Emergent Issues

being eroded, as modern technology is capable of easily recording and storing dossiers on every man, woman and child in the world.[126] The data subject's right to privacy, when applied to the context of the full body scanner is brought into focus by Alan Westin who says:

> Privacy is the claim of individuals, groups or institutions to determine for themselves when, how, and to what extent information is communicated to others.[127]

The role played by technology in modern day commercial transactions has affected a large number of activities pertaining to human interaction. The emergence of the information superhighway and the concomitant evolution of automation have inevitably transformed the social and personal life styles and value systems of individuals, created unexpected business opportunities, reduced operating costs, accelerated transaction times, facilitated accessibility to communications, shortened distances, and removed bureaucratic formalities.[128] Progress notwithstanding, technology has bestowed on humanity its corollaries in the nature of automated mechanisms, devices, features, and procedures which intrude into personal lives of individuals. For instance, when a credit card is used, it is possible to track purchases, discovering numerous aspects about that particular individual, including, food inclination, leisure activities, and consumer credit behaviour.[129] In similar vein, computer records of an air carrier's reservation system may give out details of the passenger's travel preferences, inter alia, seat selection, destination fondness, ticket purchasing dossier, lodging keenness, temporary address and telephone contacts, attendance at theatres and sport activities, and whether the passenger travels alone or with someone else.[130] In similar vein, does it follow that a full body scanning exercise would reveal imperfections of the human body which person would desire to keep private? This scheme of things may well give the outward perception of surveillance attributable to computer devices

---

[126] As far back as in 1973 it was claimed that ten reels, each containing 1,500 m of tape 2.5 cm wide, could store a 20 page dossier on every man, woman, and child in the world. See Jones (1973).

[127] Westin (1970).

[128] Orwell (1978).

[129] For a detailed analysis of the implications of credit cards with respect to the right of privacy see Nock (1993).

[130] The paramount importance of airline computer reservation system records is reflected in the world-renowned cases *Libyan Arab Jamahiriya* v. *United Kingdom* and *Libyan Arab Jamahiriya* v. *United States of America* regarding the PANAM 103 accident at Lockerbie, Scotland in 1988, where the International Court of Justice requested air carriers to submit to the Court the defendants' flight information and reservation details. See International Court of Justice. News Release 99/36, "Questions of Interpretation and Application of the 1971 Montreal Convention arising from the Aerial Incident at Lockerbie" (1 July 1999), online: http://www.icj-cij.org/icjwww/idocket/iluk/iluk2frame.html (date accessed: 14 July 2000). In a similar vein, Arthur R. Miller describes the significance of airline computer reservation system records when dealing with federal, state, local, and other types of investigations where these dossiers could provide valuable information. See also Miller(1971).

monitoring individuals' most intimate activities, preferences and physical attributes, leading to the formation of a genuine "traceable society".[131]

The main feature of this complex web of technological activity is that an enormous amount of personal information handled by such varied players from the public and private sector, may bring about concerns of possible "data leaks" in the system, a risk that could have drastic legal consequences affecting an individual's rights to privacy.

At the international level, privacy was first recognized as a fundamental freedom in the *Universal Declaration of Human Rights*.[132] Thereafter, several other human rights conventions followed the same trend, granting to individuals the fundamental right of privacy.[133] The pre-eminent concern of these international instruments was to establish a necessary legal framework to protect the individual and his rights inherent to the enjoyment of a private life.

Privacy represents different things for different people.[134] The concept per se has evolved throughout the history of mankind, from the original non-intrusion approach, which defended an individual's property and physical body against unwanted invasions and intrusions, then manifesting in whom to associate with, later enlarging its scope to include privacy as the individual's decision-making right,[135] and culminating in the control over one's personal information.[136] Thus, the conceptual evolution of privacy is directly related to the technological advancement of each particular period in history.

---

[131] See Scott (1995), Burnham (1983). *A contrario* to the argument supported in this thesis that the advancement of technology directly affects the intimacy of individuals. U.S. Circuit Judge Richard Posner favours the idea that other factors, such as urbanisation, income, and mobility development have particularly weakened the information control that, for instance, the government has over individuals: this denotes that individuals' privacy has increased. See Posner (1978).

[132] The text reads: "No one shall be subjected to arbitrary interference with his privacy, family, home or correspondence, nor to attacks upon his honor and reputation. Everyone has the right to the protection of the law against such interference or attacks". See *Universal Declaration of Human Rights*. GA Res. 217(III), 10 December 1948, Art. 12.

[133] See International Covenant on Civil and Political Rights, GA Res. 2200 (XXI), 16 December 1966, Art. 17; American Declaration on the Rights and Duties of the Man (1948), Art. 5; American Convention on Human Rights, 22 November 1969, San Jose, Costa Rica, Art. 11; Convention for the Protection of Human Nations Convention on Migrant Workers, A/RES/45/158, 25 February 1991, Art. 14; United Nations Convention on Protection of the Child, GA Res. 44/25, 12 December 1989, Art. 16.

[134] See Regan (1995), Freund (1971).

[135] In this case, the US Supreme Court acknowledged the right of women to have abortions based on the grounds that the federal government could not interfere within her "decisional privacy" sphere. See *Roe v. Wade*, 410 U.S. 113 (1973). See also Cate (1997), Zelermyer (1959).

[136] In a remarkable case concerning the legality of a national census scheduled by the authorities, the German Constitutional court connected the individual's liberty and the personal data processing of the intended census, to rule that if the individuals do not know for what purposes and who is collecting the data, that situation will eventually create an abdication of the individual's rights to the processor's command, "which cannot be tolerated in a democratic society". See Simitis (1995). See also Hoffer (2000), Gavison (1980).

3.3 Full Body Scanners and Emergent Issues

The right of privacy, as enunciated by the United States Judge Thomas M. Cooley, was the right "to be let alone" as a part of a more general right to one's personality. This idea was given further impetus by two prominent young lawyers, Samuel D. Warren and Louis D. Brandeis,[137] in 1890.[138] Before this idea was introduced, the concept of privacy reflected primarily a somewhat physical property or life. The foundations of "information privacy", whereby the individuals would determine when, how, and to what extent information about themselves would be communicated to others, inextricably drawing the right of control of information about oneself,[139] is a cornerstone of privacy. With the development of computer capabilities to handle large amounts of data, privacy has been enlarged to include the collection, storage, use, and disclosure of personal information.[140] The notion of informational privacy protection, a typically American usage, has been particularly popular both in the United States and Europe, where the term "data protection" is used.[141]

Self-determination in the right to protect one's privacy was first judicially embraced by the German Bundesverfassungsgericht in 1983. The US Supreme court followed this trend by adopting the principle of privacy self-determination in *DOJ v. Reporters Comm. for Freedom of the Press*.[142]

It must be borne in mind that privacy is not an absolute, unlimited right that operates and applies in isolation.[143] It is not an absolute right, applied unreservedly, to the exclusion of other rights. Hence there is frequently the necessity to balance privacy rights with other conflictive rights, such as the freedom of speech and the right to access information when examining individuals' rights *vis-à-vis* the interest of society.[144] This multiplicity of interests will prompt courts to adopt a balanced approach when adjudicating on a person's rights, particularly whose interests of a State are involved.

---

[137] See Cooley (1888), as cited in Warren and Brandeis (1980).

[138] The definition of privacy as the "Right to be Alone" is often erroneously attributed to Warren and Brandeis. See Warren & Brandeis. See Cooley (1888) as cited in Warren and Brandeis (1980). Additionally the concept of privacy as "the right to be let alone", and "the right most valued by civilized man: was embraced by US courts in the landmark dissenting opinion of Justice Louis D. Brandeis in *Olmsted* v. *United States*. See *Olmsted* v. *United States*, 277 U.S. 438, 478 (1928) [hereinafter *Olmstead*.]

[139] See Westin (1967). For a similar conceptualisation of privacy, see Fried (1978).

[140] See Reidenberg (1995).

[141] The former Privacy Commissioner of British Columbia, Canada, has asserted that privacy was originally a "non-legal concept". See Flaherty (1991). The term "data protection" has been translated from the German word *Datenschutz*, referring to a set of policies seeking to regulate the collection, storage, use, and transfer of personal information. See Bennet (1992).

[142] See DOJ v. Reporters Comm. for Freedom of the Press, 489 U.S. 749 AT 763 (1988).

[143] See Simmel (1971).

[144] See Halpin (1997). See also Foschio (1990). For a comprehensive study on the conflictive interest on privacy and the mass media and the Freedom of Speech, see Pember (1972), Prowda (1995). See also J. Montgomery Curtis Memorial Seminar (1992).

Since the data contained in equipment such as body scanners may be subject to trans-border storage, there is a compelling need to consider the introduction of uniform privacy laws in order that the interests of the data subject and the data seeker are protected. Although complete uniformity in privacy legislation may be a difficult objective to attain[145] (as has been the attempt to make other aspects of legislation uniform), it will be well worth the while of the international community to at least formulate international Standards and Recommend Practices (in the lines of the various ICAO Annexes) to serve as guidelines of State conduct. After all, as Collin Mellors pointed out:

> Under international agreements... privacy is now well established as a universal, natural, moral and human right. Article 12 of the Universal Declaration of Human Rights, Article 17 of the United Nations Covenant on Civil and Political Rights and Article 8 of the European Convention for the Protection of Human Rights and Fundamental Freedoms, all specify this basic right to privacy. Man everywhere has occasion to seek temporary "seclusion or withdrawal from society" and such arrangements cannot define the precise area of the right to privacy.[146]

It is such a definition that is now needed so that the two requirements of ensuring respect for information about individuals and their privacy on the one hand, and the encouragement of free and open dissemination of trans-border data flows on the other, are reconciled.

As for the use of information resulting in a full body scan, such information is purely biological and should be used only for purposes of identifying weapons or dangerous objects on the person with an explicit undertaking by the authorities concerned who use the information that it will not be used for any purpose other than for purposes of scanning. Before a process for the collection of such information is formally put into practice, legal issues pertaining to privacy, cultural sensitivity and ethical justification should be carefully thought out, and given foremost consideration.

In the provision of biometric data, the provider of the information and the receiver thereof are both under obligation to ensure that the data is not used for any purpose other than clearance of the owner of the information through customs barriers. This information may not later be used for commercial or other gain for instance for advertising purposes (such as using the physical profile of a prominent actor or actress whose biometric information originally given for customs clearance).[147]

In the body scanning process, there is an implicit link between ownership rights and privacy. Data protection legislation, including data privacy laws have been enacted by many countries throughout the world for two main reasons: protection of privacy; and ensuring of access by the owner to his information stored in a

---

[145] *Computers and Privacy in the Next Decade*, Lance J. Hoffman ed. op. cit at 146.

[146] Collin Mellors, *Governments and the Individual – Their Secrecy and His Privacy*, cited in, *A Look at Privacy*, John B.Young ed., Supra, note 15, at 94.

[147] See *Gould Estate* v. *Stoddart Publishing Company* (1996) O.J. No. 3288 (Gen. Div)

computer. Although the exact nature can vary from State to State, there is a common thread that weaves itself into the fabric of legislation in general, to ensure that: data is obtained by lawful means and processed in a fair manner; data is stored for the legitimate purpose intended and not used for any purpose incompatible with the original purpose; the collection of data is done in a reasonable manner and not excessively in order to store data over and above what is necessary; the accuracy of the data should be ensured; and the time of preservation of data is limited to the period during which such data is used.

The protection of human rights is the most significant and important task for a modern State, particularly since multi ethnic States are the norm in today's world. Globalization and increased migration across borders is gradually putting an end to the concept of the nation State, although resistance to reality can be still seen in instances where majority or dominant cultures impose their identity and interests on groups with whom they share a territory. In such instances, minorities frequently intensify their efforts to preserve and protect their identity, in order to avoid marginalization. Polarization between the opposite forces of assimilation on the one hand and protection of minority identity on the other inevitably causes increased intolerance and eventual armed ethnic conflict. In such a scenario, the first duty of governance is to ensure that the rights of a minority society are protected.

3.3.2 Security of the State

The foregoing discussion addressed the right of privacy of the individual which is paramount over most legal considerations. The only factor that would override this would be the security of State. Inherent to the concept of security of State is State responsibility to its citizens and others who are in its territory. The fundamental issue in the context of State responsibility for the purposes of this article is to consider whether a State should be considered responsible for its own failure or non-feasance to prevent a private act of terrorism against civil aviation or whether the conduct of the State itself can be impugned by identifying a nexus between the perpetrator's conduct and the State. One view is that an agency paradigm, which may in some circumstances impute to a state reprehensibility on the ground that a principal–agent relationship between the State and the perpetrator existed, can obfuscate the issue and preclude one from conducting a meaningful legal study of the State's conduct.[148]

At the core of the principal–agent dilemma is the theory of complicity, which attributes liability to a State that was complicit in a private act. Hugo Grotius (1583–1645), founder of the modern natural law theory, first formulated this theory

---

[148] Caron (1998) cited in Becker (2006a).

based on State responsibility that was not absolute. Grotius' theory was that although a State did not have absolute responsibility for a private offence, it could be considered complicit through the notion of *patienta* or *receptus*.[149] While the concept of *patienta* refers to a State's inability to prevent a wrongdoing, *receptus* pertains to the refusal to punish the offender.

The eighteenth Century philosopher Emerich de Vattel was of similar view as Grotius, holding that responsibility could only be attributed to the State if a sovereign refuses to repair the evil done by its subjects or punish an offender or deliver him to justice whether by subjecting him to local justice or by extraditing him.[150] This view was to be followed and extended by the British jurist Blackstone a few years later who went on to say that a sovereign who failed to punish an offender could be considered as abetting the offence or of being an accomplice.[151]

A different view was put forward in an instance of adjudication involving a seminal instance where the Theory of Complicity and the responsibility of states for private acts of violence was tested in 1925. The case[152] involved the Mexico-United States General Claims Commission which considered the claim of the United States on behalf of the family of a United States national who was killed in a Mexican mining company where the deceased was working. The United States argued that the Mexican authorities had failed to exercise due care and diligence in apprehending and prosecuting the offender. The decision handed down by the Commission distinguished between complicity and the responsibility to punish and the Commission was of the view that Mexico could not be considered an accomplice in this case.

The Complicity Theory, particularly from a Vattellian and Blackstonian point of view is merely assumptive unless put to the test through a judicial process of extradition. In this Context it becomes relevant to address the issue through a discussion of the remedy.

The emergence of the Condonation Theory was almost concurrent with the *Jane* case[153] decided in 1925 which emerged through the opinions of scholars who belonged to a school of thought that believed that States became responsible for private acts of violence not through complicity as such but more so because their refusal or failure to bring offenders to justice, which was tantamount to ratification of the acts in question or their condonation.[154] The theory was based on the fact that it is not illogical or arbitrary to suggest that a State must be held liable for its failure to take appropriate steps to punish persons who cause injury or harm to others for the reason that such States can be considered guilty of condoning the criminal acts

---

[149] Grotius and Scott (1646).

[150] De Vattel and Fenwick (1916).

[151] Blackstone and Morrison (2001).

[152] *Laura M.B. Janes (USA)* v. *United Mexican States* (1925) 4 R Intl Arb Awards 82.

[153] *Ibid*.

[154] *Black's Law Dictionary* defines condonation as "pardon of offense, voluntary overlooking implied forgiveness by treating offender as if offense had not been committed."

and therefore become responsible for them.[155] Another reason attributed by scholars in support of the theory is that during that time, arbitral tribunals were ordering States to award pecuniary damages to claimants harmed by private offenders, on the basis that the States were being considered responsible for the offences.[156]

The responsibility of governments in acting against offences committed by private individuals may sometimes involve condonation or ineptitude in taking effective action against terrorist acts, in particular with regard to the financing of terrorist acts. The United Nations General Assembly, on 9 December 1999, adopted the International Convention for the Suppression of the Financing of Terrorism,[157] aimed at enhancing international co-operation among States in devising and adopting effective measures for the prevention of the financing of terrorism, as well as for its suppression through the prosecution and punishment of its perpetrators.

The Convention, in its Article 2 recognizes that any person who by any means directly or indirectly, unlawfully or willfully, provides or collects funds with the intention that they should be used or in the knowledge that they are to be used, in full or in part, in order to carry out any act which constitutes an offence under certain named treaties, commits an offence. One of the treaties cited by the Convention is the International Convention for the Suppression of Terrorist Bombings, adopted by the General Assembly of the United Nations on 15 December 1997.[158]

The Convention for the Suppression of the Financing of Terrorism also provides that, over and above the acts mentioned, providing or collecting funds toward any other act intended to cause death or serious bodily injury to a civilian, or to any other person not taking an active part in the hostilities in the situation of armed conflict, when the purpose of such act, by its nature or context, is to intimidate a population, or to compel a government or an international organization to do or to abstain from doing any act, would be deemed an offence under the Convention.

The United Nations has given effect to this principle in 1970 when it proclaimed that:

> Every State has the duty to refrain from organizing or encouraging the organization of irregular forces or armed bands, including mercenaries, for incursion into the territory of another State. Every State has the duty to refrain from organizing, instigating, assisting or participating in acts of civil strife or terrorist acts in another State or acquiescing in organized activities within its territory directed towards the commission of such acts, when the acts referred to in the present paragraph involve a threat or use of force.[159]

---

[155] Jane's case, *Supra*, note 55, at 92.

[156] Hyde (1928).

[157] International Convention for the Suppression of the Financing of Terrorism, adopted by the General Assembly of the United Nations in resolution 54/109 of 9 December 1999.

[158] A/52/653, 25 November 1997.

[159] Declaration on Principles of International Law Concerning Friendly Relations and Co-operation Among States in Accordance with the Charter of the United Nations, UN General Assembly Resolution 2625 (XXV) 24 October 1970.

Here, the words *encouraging* and *acquiescing in organized activities within its territory directed towards the commission of such acts* have a direct bearing on the concept of condonation and would call for a discussion about how States could overtly or covertly encourage the commission of such acts. One commentator[160] identifies three categories of such support: *Category I* support entails protection, logistics, training, intelligence, or equipment provided terrorists as a part of national policy or strategy; *Category II* support is not backing terrorism as an element of national policy but is the toleration of it; *Category III* support provides some terrorists a hospitable environment, growing from the presence of legal protections on privacy and freedom of movement, limits on internal surveillance and security organizations, well-developed infrastructure, and émigré communities.

The Convention, in its Article 2 recognizes that any person who by any means directly or indirectly, unlawfully or wilfully, provides or collects funds with the intention that they should be used or in the knowledge that they are to be used, in full or in part, in order to carry out any act which constitutes an offence under certain named treaties, commits an offence. The treaties listed are those that are already adopted and in force and which address acts of unlawful interference with such activities as deal with air transport and maritime transport. Also cited is the International Convention for the Suppression of Terrorist Bombings, adopted by the General Assembly of the United Nations on 15 December 1997.

The *Convention for the Suppression of the Financing of Terrorism* also provides that, over and above the acts mentioned, providing or collecting funds toward any other act intended to cause death or serious bodily injury to a civilian, or to any other person not taking an active part in the hostilities in the situation of armed conflict, when the purpose of such act, by its nature or context, is to intimidate a population, or to compel a government or an international organization to do or to abstain from doing any act, would be deemed an offence under the Convention.

The above notwithstanding, one cannot ignore the incontrovertible fact that security is a systemic process. The mere use of full body scanners by no means ensures total security against acts of unlawful interference with civil aviation. As the Christmas day incident showed, intelligence gathering, sharing of information, and more importantly, integration and analysis of such information[161] are critical to the security chain. What is most important is to establish a global security culture that ensures the cooperation of the 190 ICAO member States by working in harmony. A perceived inadequacy of the global framework of aviation security is the lack of an implementation arm. ICAO has taken extensive measures to

---

[160] Steven Metz, State Support for Terrorism, Defeating Terrorism, Strategic Issue Analysis, at http://www.911investigations.net/IMG/pdf/doc-140.pdf.

[161] As stated by President Obama, "this was not a failure to collect intelligence...it was a failure to integrate and understand the intelligence that we already had..." Airline Bomber could Have Been Stopped, *The Air Letter*, Tuesday 05 January 2010, No. 16,897, at 1.

introduce relevant international conventions as well as Standards and Recommended Practices (SARPs) in Annex 17 to the Chicago Convention. There is also a highly classified *Aviation Security Manual* developed by ICAO which is provided to States. Additionally, the Organization provides focused security training courses to its member States. However, ICAO's role is largely confined to rule making and the provision of guidance, bringing to bear the need for an aviation security crisis management team on a global scale that could work towards effectively precluding acts of terrorism.

3.3.3 Flight NW 253

The attempted bombing of Northwest Airlines flight 253 on Christmas day of 2009, which was arguably the turning point in the initiative across the globe to use full body scanners, once again issued the ominous reminder that aviation is vulnerable and that there is a compelling need for a global security culture that recognizes and applies uniform global security standards. The first step toward achieving this objective is the realization by the aviation community of new and emerging threats to aviation and the need to share information. Since the event, several countries have imposed stringent security standards at airports and initiated inter-governmental meetings to discuss closer cooperation among States in the hope that such measures will effectively respond to the threat. Both ICAO and IATA took immediate measures after the event in issuing statements and called for a closer look by States at their security measures and a re-assessment of available responses to threats posed to aviation.

It is now abundantly clear that collection and sharing of information, improvement of technology and the strengthening of security standards are critical in addressing the problem of unlawful interference with civil aviation. Other aspects such as curbing the financing of terrorism, taking a closer look at the responsibility of States, and increasing vigilance on a global scale are also significant considerations.

On 25 December 2009, a person on board Northwest Airlines flight 253, flying from Amsterdam to Detroit, attempted to detonate an explosive device[162] which failed to explode, but ignited injuring the offender and two other passengers. The flight crew and some passengers restrained the offender, who was arrested when the A330 airliner landed safely in Detroit. The Summary of the United States White House Review which was issued subsequent to the event reflected preliminary findings which were that several opportunities that might have allowed the counter-terrorism community to take cohesive action by "connecting the dots" of information available prior to the attempted bombing of the airliner, brought to bear

---

[162] The device used contained the explosive PETN (Pentaerytritol) which was stitched into his underwear.

a need to assist counter-terrorism analysts in their preventive work.[163] The White House Review highlighted human errors which resulted in failure to identify, correlate and piece together an emerging terrorist plot against the United States. Also identified were weak points in assigning responsibility and accountability and shortcomings of the watch-list system that would have enabled the offender to board the flight.

Reportedly, as a result of the event, US officials reacted by tightening security measures for all US bound passengers, advising persons flying from 14 handpicked nations[164] that they would be subject to stringent screening.[165] Dutch authorities are known to have stated that they planned to put full body scanners into use within 3 weeks and British authorities had similar comments.[166]

The International Civil Aviation Organization (ICAO) responded on 30 December 2009 with a communiqué to its points of contacts in member States to the effect inter alia that ICAO Member States were encouraged to conduct a risk assessment, taking into consideration all the relevant factors, and implement appropriate screening measures. Where additional screening measures are considered necessary, these may include the application of explosives trace detection technology, physical searches and randomly-deployed explosives detection canine teams, among others.

The communiqué reminded ICAO Member States of the need for cooperation in all matters related to aviation security, particularly with regard to the requirements of Standard 2.4.1 of Annex 17 – *Security*, which required each member State to ensure that requests from other Contracting States for additional security measures in respect of a specific flight(s) by operators of such other States are met, as far as may be practicable. The requesting State was required to give consideration to alternative measures of the other State that are equivalent to those requested.

Finally ICAO stated that it was committed to provide all possible assistance in this matter, and will seek to establish the necessary communication and

---

[163] The *Aviation Safety Journal* records that there was a similar finding five years before the flight 253 incident where "Information was not shared ... analysis was not pooled ... often the handoffs of information were lost across the divide separating the foreign and domestic agencies of the government. The finding recommended that "improved use of 'no-fly' and 'automatic selectee' lists should not be delayed .... This screening function should be performed by the TSA [Transportation Security Administration], and it should utilize the larger set of watch-lists maintained by the federal government"... the TSA ... must give priority attention to improving the ability of screening checkpoints to detect explosives on passengers". See *http://asj.nolan-law.com/2009/12/red-flags-ignored-in-underpants-bomber-caper-same-as-in-911/*.

[164] Afghanistan, Algeria, Cuba, Iran, Iraq, Lebanon, Libya, Nigeria, Pakistan, Saudi Arabia, Somalia, Sudan, Syria and Yemen.

[165] US Toughens Screening for US Bound Fights, The Air Letter, Monday 4 January 2010, No. 10,896, at 1. According to the same report the Transportation Security Administration advised all passengers flying into the United States from abroad that they will be subject to random screening or so called "threat based screens".

[166] *Ibid.*

3.3 Full Body Scanners and Emergent Issues

coordination mechanisms to ensure that any response to this latest threat is provided with the greatest possible degree of harmonization. In this respect, States were reminded of the need to register and provide up-to-date information to ICAO for the Aviation Security Point of Contact Network. Giovanni Bisignani, Director General of the International Air Transport Association (IATA), in a letter to the Secretary of the US Department of Homeland Security, noted inter alia that security was a government responsibility which was a shared priority with industry, urging that a comprehensive review of security systems followed.[167]

While it is quite true that, as stated by President Obama, "this was not a failure to collect intelligence…it was a failure to integrate and understand the intelligence that we already had…"[168] the event resurfaces three fundamental truths about aviation security: it is a global issue and there are new and emergent threats to civil aviation; there needs to be a global security culture; and information sharing between parties is essential. This article will address these issues.

At its 33rd session held in Montreal from 25 September – 5 October 2001, the ICAO Assembly adopted Resolution A33-1[169] which was a direct response to the terrorist acts of 9/11. The Resolution recognized that a new type of threat was posed to civil aviation which required new concerted efforts and policies of cooperation on the part of States. The Resolution also urges all ICAO member States to ensure, in accordance with Article 4 of the Chicago Convention,[170] that civil aviation is not used for any purpose inconsistent with the aims of the Convention, and to hold accountable and punish severely those who misuse civil aircraft as weapons of destruction, including those responsible for planning and organizing such acts or for aiding, supporting or harbouring perpetrators. It also called upon States to cooperate with each other in this endeavour and to ensure that ICAO Standards and Recommended Practices (SARPs) relating to aviation security are adhered to. Finally the Resolution directed the Council of ICAO and the Secretary General to act urgently to address new and emerging threats to civil aviation, in particular to review the adequacy of existing aviation conventions on security.

A Special Sub Committee of the Legal Committee of ICAO met in Montreal from 3 to 6 July 2007 to discuss the preparation of one or more instruments addressing new and emerging threats. One of the issues addressed at this meeting was the unlawful transport of biological, chemical, nuclear weapons and other dangerous substances on board aircraft.

Earlier, the Secretary General of ICAO, Dr. Taieb Cherif, addressing the China Civil Aviation Development Forum on 9 May 2007, stated that although the global

---

[167] http://iata.org/pressroom/pr/2009-12-30-02.htm.

[168] Airline Bomber could Have Been Stopped, *The Air Letter*, Tuesday 05 January 2010, No. 16,897, at 1.

[169] Resolution A33-1, Declaration on misuse of civil aircraft as weapons of destruction and other terrorist acts involving civil aviation, Assembly Resolutions in Force (as of 8 October 2004) ICAO Doc. 9848. at VII-1.

[170] Supra n. 3

air transport system remains as secure as ever, yet events such as the illegal terrorist plot in the United Kingdom in the Summer of 2006, potentially involving liquids used as explosives, reminds us how vulnerable the system is. On another aviation platform, Giovanni Bisignani, Director General of IATA stressed at its Annual General Meeting held in Vancouver from 3 to 5 June 2007 that the industry had changed tremendously in 5 years since 9/11. Bisignani stated that, 6 years after the tragic events of 2001, air travel was much more secure but there were unlimited ways to attack the aircraft integrity. He added that there was no perfect security system and terrorists change tactics and weapons. Bisignani rightly pointed out that terrorists are studying what measures the industry is adopting; and that all the air industry can do is make the system strong enough to constitute sufficient deterrent and make aircraft a harder target to hit.

3.3.4 The AVSEC Panel

The Aviation Security Panel of ICAO met at its 20th Meeting in Montreal from 30 March to 3 April 2009. One of the key areas of discussion at this meeting concerned new and emerging threats to civil aviation. The Panel worked through the Working Group on New and Emerging Threats and noted that significant progress in efforts to proactively identify vulnerabilities and potential gaps in existing measures had been made, that would strengthen *Annex 17* (Aviation Security)to the Convention on International Civil Aviation (Chicago Convention).[171] At this meeting, the European Civil Aviation Conference (ECAC) stressed the importance of the challenge posed by cyber threats in light of the current lack of related provisions in Annex 17.

Consequently, the Panel considered the threat of cyber attacks, and some members stressed that this threat is significant. With reference to a proposal to include a Recommended Practice in Annex 17 to ensure that information and communication technology systems used for civil aviation purposes be protected from cyber attacks, the Panel agreed that, given the complexity of this issue, which involves air traffic management systems, aircraft design and operations, the matter requires further analysis by the Working Group on New and Emerging Threats prior to inclusion in Annex 17 or any guidance material. This analysis will be disseminated over the secure website by the end of June 2009 and, depending on the results of the analysis, the Working Group on Amendment 12 to Annex 17 will develop a proposal for amending the Annex, to be presented to the Panel at its 21st meeting.

The Panel also considered the merits of building unpredictability into the aviation security regime. While concern was expressed regarding the impact of

---

[171] Supra n 3

3.3 Full Body Scanners and Emergent Issues

unpredictable security measures on passenger confidence in aviation security, many Panel members supported implementation of the concept because of its value as a deterrent. It was suggested that States adopt an approach providing for a baseline regime, but with the addition of unpredictable measures, thus achieving a balance between certainty and unpredictability. With regard to an amendment to Annex 17 in this regard, the need for introducing unpredictability into the aviation security regime was considered, and it was agreed that unpredictability should be promoted in principle but not prescribed. The Panel suggested that if an Annex 17 specification related to unpredictability were to be developed, it would be necessary to ensure that the introduction of this concept by States does not diminish the level of security or result in delays for passengers. Further, the Panel noted that appropriate guidance material may be required to address the potential negative impact of introducing the concept of unpredictability, and proposed the development of guidance material related to unpredictability prior to the introduction of an amendment to Annex 17.

A Conclusion of the Panel was, inter alia, that the threat of cyber attacks is real and cannot be ignored, and that further analysis by the Working Group on New and Emerging Threats would be appropriate. Another Conclusion was that the ICAO focal point of contact (PoC) Network is an important tool for sharing critical threat information and should be used more effectively, and that the Secretariat should consider the establishment of a web-based community page. Yet another was that the concept of building unpredictability into the aviation security regime is in principle a useful tool, however, concerns expressed regarding the possible impact on the level of security and the impact on passenger confidence should be resolved before its inclusion as a Recommended Practice in Annex 17.

The Recommendations of the Panel were that:

a) The Working Group on New and Emerging Threats propose its new name, terms of reference and composition, including suggestions on how observers might participate in the Working Group, as well as details of its evolving collaboration with the G8 Group, at the 21st Panel meeting;
b) The Working Group evaluate the threat of cyber attacks and disseminate the results of its analysis on the secure website by the end of June 2009 and that, depending on the results of this analysis, the Working Group on Amendment 12 to Annex 17 consider developing an amendment to Annex 17 for presentation at the 21st Panel meeting;
c) The ICAO Secretariat issue an electronic bulletin reminding States of the importance of subscribing to the PoC Network and providing information on its usage; and
d) the concept of building unpredictability into the aviation security regime be further considered.[172]

---

[172] See Report of the Aviation Security (AVSEC) Panel, Twentieth Meeting, AVSECP/20 at 2.1.

3.3.4.1 The Need for a Security Culture

Since the events of 11 September 2001, there have been several attempts against the security of aircraft in flight. These threats have ranged from shoe bombs to dirty bombs to explosives that can be assembled in flight with liquids, aerosols and gels. In every instance the global community has reacted with pre emptive and preventive measures which prohibit any material on board which might seemingly endanger the safety of flight. Some jurisdictions have even gone to extremes in prohibiting human breast milk and prescriptive medications on board.

New and emerging threats to civil aviation are a constant cause for concern to the aviation community. Grave threats such as those posed by the carriage of dangerous pathogens on board, the use of cyber technology calculated to interfere with air navigation systems, and the misuse of man portable air defence systems are real and have to be addressed with vigour and regularity. The International Civil Aviation Organization has been addressing these threats for some time and continues to do so on a global basis.

Since the events of 11 September 2001 took place, the most critical challenge facing international civil aviation remains to be the compelling need to ensure that the air transport industry remains continuous and its consumer is assured of sustained regular, safe and secure air transport services. The Air Transport Association (ATA), in its 2002 State of the United States Airline Industry Statement, advised that, in the United States, the combined impact of the 2001 economic downturn and the precipitous decline in air travel following the 11 September 2001 attacks on the United States resulted in devastating losses for the airline industry which are likely to exceed $7 billion and continue through 2002.[173] Of course, the overall picture, which portended a certain inevitable gloom for the air transport industry, was not the exclusive legacy of United States' carriers. It applied worldwide, as was seen in the abrupt downfall of air traffic globally during 2001. The retaliation by the world community against terrorism, which is an ongoing feature in world affairs, increased the airline passenger's fear and reluctance to use air transport. In most instances in commercial aircraft purchasing, air carriers cancelled or postponed their new aircraft requisition orders. Many carriers, particularly in developing countries, were seen revisiting their cost structures and downsizing their human resource bases. It is incontrovertible that another similar event or series of events will inevitably plunge the aviation industry into similar despair and destitution.

ICAO has a security oversight programme called the Universal Security Audit Programme (UASP). The ICAO Universal Security Audit Programme (USAP), launched in June 2002, represents an important initiative in ICAO's strategy for

---

[173] State of the United States Airline Industry, *A Report on Recent Trends for United States Carriers*, Air Transport Association: 2002, Statement by Carol B. Hallett, President and CEO, ATA.

3.3 Full Body Scanners and Emergent Issues

strengthening aviation security worldwide and for attaining commitment from States in a collaborative effort to establish a global aviation security system.

The programme, which is part of the Aviation Security Plan of Action, provides for the conduct of universal, mandatory and regular audits of the aviation security systems in all ICAO member States. The objective of the USAP is to promote global aviation security through the auditing of States on a regular basis to assist States in their efforts to fulfil their aviation security responsibilities. The audits identify deficiencies in each State's aviation security system, and provide recommendations for their mitigation or resolution.

Implementation of the programme commenced with the first aviation security audit taking place in November 2002 and between three and four audits continue to be conducted around the world each month. The 35th Session of the Assembly held from 28 September to 8 October 2004 mandated ICAO to maintain strict confidentiality of all State-specific information derived from audits conducted under the Universal Security Audit Programme (USAP). However, in order to promote mutual confidence in the level of aviation security between States, the Assembly urged all Contracting States to "share, as appropriate and consistent with their sovereignty, the results of the audit carried out by ICAO and the corrective actions taken by the audited State, if requested by another State".[174]

While noting the importance of continuing bilateral exchanges of information between States, the 36th Session of the Assembly, held from 18 to 28 September 2007, also recognized the value of proposals presented by the Council and Contracting States for the introduction of a limited level of transparency with respect to ICAO aviation security audit results[175] The Assembly directed the Council to consider such an introduction of a limited level of transparency, balancing the need for States to be aware of unresolved security concerns with the need to keep sensitive security information out of the public realm. In doing so, the Assembly emphasized that it was essential that any methodology developed to provide for increased transparency also ensure the appropriate safeguarding of a State's security information in order to prevent specific information that could be used to exploit existing vulnerabilities from being divulged.

The 36th Session of the ICAO Assembly adopted Resolution A 36–20,[176] *Appendix E* of which addresses the USAP. As mentioned earlier, it must be emphasized that the Resolution inter alia directs the Council to consider the introduction of a limited level of transparency with respect to ICAO aviation security audits, balancing the need for States to be aware of unresolved security

---

[174] A35-9, Appendix E, Resolving Clause 4; and Recommended Practice 2.4.5 of Annex 17 — *Security*).

[175] Resolution A36-20, A36-WP/336 and Plenary Action Sheet No. 3.

[176] Resolution A 36–20, Consolidated statement on the continuing CA policies related to the safeguarding of international civil aviation against acts of unlawful interference, Report of the Executive Committee (Report Folder) Assembly, Thirty –sixth Session, A36 – WP/336, p/46, at 16–2.

concerns with the need to keep sensitive security information out of the public realm and requests the Council to report to the next ordinary session of the Assembly (in 2010) on the overall implementation of the USAP.

Since the launch of the USAP in 2002, 169 aviation security audits and 77 follow-up missions have been conducted.[177] The audits have proven to be instrumental in the ongoing identification and resolution of aviation security concerns, and analysis reveals that the average implementation rate of Annex 17 Standards in most States has increased markedly between the period of the initial audit and the follow-up mission.

A critical part of the audit process is the requirement that all audited States submit a corrective action plan to address deficiencies identified during an audit. As directed by the Council, all States are notified (by State letter and on the USAP secure website) of those states that are more than 60 days late in submitting a corrective action plan. As of 31 July 2007, there were seven States that were more than 60 days late. In the case of late corrective action plans, repeated reminders are sent to States, including at the level of the Secretary General and with the involvement of the applicable Regional Office, and ICAO assistance is offered should the State require advice or support in the preparation of its action plan. Extensive feedback is provided to each audited State on the adequacy of its corrective action plan, and an ongoing dialogue is maintained where necessary to provide support in the implementation of proposed actions.

ICAO performs comprehensive analyses of audit results on levels of compliance with Annex 17 – *Security* Standards on an ongoing basis (globally, by region and by subject matter). This statistical data is made available to authorized users on the USAP secure website and is shared with other relevant ICAO offices as a basis for prioritizing training and remedial assistance projects. As of 31 July 2007, 77 follow-up missions had been conducted. These missions take place 2 years after the initial audit with the purpose of validating the implementation of State corrective action plans and providing support to States in remedying deficiencies. These missions are normally conducted by the applicable Regional Office, with close coordination through Headquarters. The results of the follow-up visits indicate that the majority of States have made significant progress in the implementation of their corrective action plans.

A high-level ICAO Secretariat Audit Results Review Board (ARRB) has been established as part of an overall coordinated strategy for working with States that are found to have significant compliance shortcomings with respect to ICAO Standards and Recommended Practices (SARPs). The ARRB examines both the

---

[177] The 36th Session of the ICAO Assembly was informed that there are some 150 certified auditors on the USAP roster, from 59 States in all ICAO regions. The participation of certified national experts in the audits under the guidance of an ICAO team leader has permitted the programme to be implemented in a cost-effective manner while allowing for a valuable interchange of expertise.

safety and security histories of specific States and provides an internal advisory forum for coordination among ICAO's safety, security and assistance programmes.

A security culture, if such were to exist among ICAO's member States, would mean that the States would be aware of their rights and duties, and, more importantly, assert them. Those who belong to a security culture also know which conduct would compromises security and they are quick to educate and caution those who, out of ignorance, forgetfulness, or personal weakness, partake in insecure conduct. This security consciousness becomes a "culture" when all the 190 member States as a whole makes security violations socially and morally unacceptable within the group.

A significant issue pertaining to the infusion of a security culture in States is the compelling need to globally curb the financing of terrorism. The United Nations General Assembly, on 9 December 1999, adopted the International Convention for the Suppression of the Financing of Terrorism, aimed at enhancing international cooperation among States in devising and adopting effective measures for the prevention of the financing of terrorism, as well as for its suppression through the prosecution and punishment of its perpetrators.

The Convention, in its Article 2 recognizes that any person who by any means directly or indirectly, unlawfully or wilfully, provides or collects funds with the intention that they should be used or in the knowledge that they are to be used, in full or in part, in order to carry out any act which constitutes an offence under certain named treaties, commits an offence. The treaties listed are those that are already adopted and in force and which address acts of unlawful interference with such activities as deal with air transport and maritime transport. Also cited is the International Convention for the Suppression of Terrorist Bombings, adopted by the General Assembly of the United Nations on 15 December 1997.

The *Convention for the Suppression of the Financing of Terrorism* also provides that, over and above the acts mentioned, providing or collecting funds toward any other act intended to cause death or serious bodily injury to a civilian, or to any other person not taking an active part in the hostilities in the situation of armed conflict, when the purpose of such act, by its nature or context, is to intimidate a population, or to compel a government or an international organization to do or to abstain from doing any act, would be deemed an offence under the Convention.

Acts of international terrorism that have been committed over the past two decades are too numerous to mention. Suffice it to say, that the most deleterious effect of the offense is that it exacerbates international relations and endangers international security. From the isolated incidents of the 1960s, international terrorism has progressed to becoming a concentrated assault on nations and organizations that are usually susceptible to political conflict, although politics is not always the motivation of the international terrorist. International terrorism has been recognized to engulf acts of aggression by one State on another as well as by an individual or a group of individuals of one State on another State. The former typifies such acts as invasion, while the latter relates to such individual acts of violence as hijacking and the murder of civilians in isolated instances. In both instances, the duties of the offender-State have been emphatically recognized. Such duties are to condemn such acts and take necessary action.

The United Nations has given effect to this principle in 1970 when it proclaimed that:

> Every State has the duty to refrain from organizing or encouraging the organization of irregular forces or armed bands, including mercenaries, for incursion into the territory of another State. Every State has the duty to refrain from organizing, instigating, assisting or participating in acts of civil strife or terrorist acts in another State or acquiescing in organized activities within its territory directed towards the commission of such acts, when the acts referred to in the present paragraph involve a threat or use of force.

The most pragmatic approach to the problem lies in identifying the parameters of the offense of international terrorism and seeking a solution to the various categories of the offense. To obtain a precise definition would be unwise, if not impossible. Once the offense and its parasitic qualities are clearly identified, it would become necessary to discuss briefly its harmful effects on the international community. It is only then that a solution can be discussed that would obviate the fear and apprehension we suffer in the face of this threat.

Finally on this subject, it must be said that in building a security culture within ICAO member States it is imperative that consideration should also be given to the development of a process for ensuring that all Member States are notified when deficiencies identified during the course of a USAP audit remain unaddressed for a sustained period of time. A notification process could involve the use of information which does not divulge specific vulnerabilities but enables States to initiate consultations with the State of interest to ensure the continued protection of aviation assets on a bilateral basis. Such a notification process may result in a strengthened ability on the part of ICAO to ensure that States unwilling to meet basic security standards will be held accountable and allow for a limited amount of transparency in the security audit programme without divulging specific potential security vulnerabilities.

3.3.4.2 Sharing Information

Assembly Resolution A 35–1 (adopted at the 35th Session of the ICAO Assembly in Montreal in September/October 2004),[178] calls upon ICAO member States to study the ways and means to reinforce the prevention of terrorist attacks by means of explosives, in particular by enhancing international cooperation and information exchange in developing technical means of detection of explosives, giving increased attention to the detection of explosive devices on the human body.

---

[178] Resolving Clause 4. See Assembly Resolutions in Force (as of 28 September 2007), Doc 9902, at 1–44.

3.3 Full Body Scanners and Emergent Issues   219

The Passenger Name Record

One category of information in this regard is the Passenger Name Record (PNR).[179] The Passenger Name Record (PNR) is a subject that has been under intense scrutiny by the Council of ICAO, which has developed PNR Data Guidelines that have been transmitted to Contracting States for their comments[180] This exercise was carried out on the understanding that, in the present context of the compelling need for the enhancement of aviation security, the global aviation community has shown an increased interest[181] in adding the PNR data as a security measure in addition to the already existing Advanced Passenger Information (API)[182] and the Machine Readable Travel Document (MRTD), which, although primarily are facilitation tools, greatly assist States authorities in ensuring border security.

A new Recommended Practice concerning the PNR data was included in Annex 9 to the Chicago Convention (Facilitation) after being adopted by the ICAO Council in March 2005.[183] This Recommended Practice, which supplements an already existing Recommended Practice[184] provides that Contracting States requiring

---

[179] The air transport industry regards a *Passenger Name Record* (PNR), as a generic term applicable to records created by aircraft operators or their authorized agents for each journey booked by or on behalf of any passenger. The data is used by operators for their own commercial and operational purposes in providing air transportation services. The definition applicable in the United States identifies a PNR as a repository of information that air carriers would need to make available upon request under existing regulations and refers to reservation information contained in a carrier's electronic computer reservation system.

[180] See Attachment to State Letter EC 6/2-05/70, Passenger Name Record (PNR) data, 9 June 2005.

[181] The advantage of collection by States of PNR Data was first discussed by the global aviation community at the Twelfth Session of the ICAO Facilitation Division that was held in Cairo, Egypt from 22 March to 1 April 2004. Consequently, the Division adopted Recommendation B/5, that reads as follows:

"It is recommended that ICAO develop guidance material for those States that may require access to Passenger Name Record (PNR) data to supplement identification data received through an API system, including guidelines for distribution, use and storage of data and a composite list of data elements [that] may be transferred between the operator and the receiving State."

Pursuant to this recommendation, In June 2004, the Air Transport Committee of the ICAO Council requested the Secretary General to establish a Secretariat Study Group to develop Guidelines on PNR data transfer. The Council, in endorsing Recommendation B/5, directed that these Guidelines were to be submitted early in 2005.

[182] See, Abeyratne (2002a). Also by the same author, Abeyratne (2001), and also by Abeyratne (2003).

[183] Recommended Practice 3.48 which provides: "Contracting States requiring Passenger Name Record (PNR) access should conform their data requirements and their handling of such data to guidelines developed by ICAO".

[184] Recommended Practice 3.47, which provides inter alia that Contracting States should, where appropriate, should introduce a system of advance passenger information which capture certain passport and visa information prior to departure, for onward transmission to relevant public authorities by electronic means.

Passenger Name Record (PNR) access should conform their data requirements and their handling of such data to guidelines developed by ICAO. It is worthy of note that Article 13 of the Chicago Convention provides that the laws and regulations of a Contracting State as to the admission to or departure from its territory of passengers, crew or cargo of aircraft, such as regulations relating to entry, clearance, immigration, passports, customs, and quarantine shall be complied with, by or on behalf of such passengers, crew or cargo upon entrance into or departure from, or while within the territory of that State. This provision gives a State the discretion to specify the information it requires relating to persons wishing to gain entry into its territory. Accordingly, a State may require aircraft operators operating flights to, from or in transit through airports within its territory to provide its public authorities, upon request, with information on passengers such as PNR data.

The philosophy underlying the importance of PNR data and their efficient use by States for enhanced expediency in border crossing by persons is embodied in the General Principles set out in Chapter 1 of Annex 9 which require Contracting States to take necessary measures to ensure that: the time required for the accomplishment of border controls in respect of persons is kept to the minimum[185]; the application of administrative and control requirements causes minimum inconvenience; exchange of relevant information between Contracting States, operators and airports is fostered and promoted to the greatest extent possible; and, optimal levels of security, and compliance with the law, are attained.

Contracting States are also required to develop effective information technology to increase the efficiency and effectiveness of their procedures at airports.[186]

Advance Passenger Information (API)

The API process requires the carrier to capture passport details prior to departure and the transmit the details by electronic means to the authorities at destination. By this process, authorities can screen the passengers through their databases with a view to identifying potentially high-risk individuals. In addition to he security

---

[185] There is an abiding symbiosis between security and facilitation in the field of air transport. While security is of paramount interest to the global aviation community, it must not unduly disrupt or in any adversely affect the expediency of air transport. To this end, Recommended Practice 2.2 of Annex 9 – Facilitation – to the Chicago Convention suggests that Each Contracting State should whenever possible arrange for security controls and procedures to cause a minimum of interference with, or delay to the activities of civil aviation provided the effectiveness of these controls and procedures is not compromised. See McMunn (1996).

[186] It must be noted that Annex 9 specifies that the provisions of the Annex shall not preclude the application of national legislation with regard to aviation security measures or other necessary controls.

3.3 Full Body Scanners and Emergent Issues

advantage, this process also assists in reducing congestion at airports and consequently decreases delays in border control processing.[187]

Some States consider API to be a compelling information tool that enables public authorities to manage a potential threat of unlawful interference with civil aviation and also to expedite clearance on arrival.[188] The implementation of such a system requires a great deal of regulation as it involves data capturing and processing. It has also to be noted that under the Chicago Convention, a State has the right to request information in order for proper border control to be established.[189] Therefore, this is a matter of national policy which is further buttressed by the concept of sovereignty of nations.[190]

The United States is arguably one of the most active jurisdictions in adopting recent legislation with regard to airline passengers carried into the territory of a State. On 19 November 2001, The President signed into law the *Aviation and Transportation Security Act* which added on a new requirement that each carrier, foreign and domestic, operating a passenger flight involving foreign air transportation into the United States, must transmit to the US Customs electronically and in advance of the arrival of the flight, a related passenger manifest and crew manifest containing certain specifically required information of such persons. Following this law, the Customs authorities of the United States published an interim rule[191] in the Federal Register on 31 December 2001 which requires air carriers, for each flight subject to the *Aviation and Transportation Security Act*, to transmit to Customs, by means of an electronic data interchange system approved by Customs, a passenger manifest and, by way of a separate transmission, using the same system, a crew manifest.

---

[187] "[...] This technique is beginning to be used by Border Control Agencies and it has the potential to reduce considerably the inconvenience and delay experienced by some travelers due to border controls." Facilitation Division-Eleventh Session, (1995) ICAO Doc FAL/11-IP/2.

[188] Refer to Recommendation Practice 3.34 of Annex 9: "Where appropriate Contracting States should introduce a system of advanced passenger information which involves the capture of certain passport or visa details prior to departure, the transmission of the details by electronic means to public authorities, and the analysis of such data for risk management purposes prior to arrival in order to expedite clearance. To minimize handling time during check-in, document reading devices should be used to capture the information in machine readable travel documents. When specifying the identifying information on passengers to be transmitted, Contracting States should only require information that is found in the machine readable zones of passports and visas that comply with the specifications contained in Doc 9303 (series), Machine Readable Travel Documents. All information required should conform to specifications for UN/EDIFACT PAXLST message formats.

[189] Chicago Convention-Art. 13: "The laws and regulations of a contracting State as to the admission to or departure from its territory of passengers, crew or cargo of aircraft, such as the regulations relating to entry, clearance, immigration, passports, customs, and quarantine shall be complied with by or on behalf of such passengers, crew or cargo upon entrance into or departure from, or while within the territory of that State."

[190] Chicago Convention-Art. 1: "The contracting States recognize that every State has complete and exclusive sovereignty over the airspace above its territory."

[191] 66 FR 67482.

The Passenger Name Record (PNR) information so required must electronically provide Customs with access to any and all PNR data elements concerning the identity and travel plans of the passenger to any flight in foreign air transportation to and from the United States, to the extent that the carrier in fact possesses the required data elements in its reservation system and/or departure control system.

On Section 402 of The United States *Enhanced Border Security and Visa Entry Reform Act of 2002* amends section 231 of the Immigration and Nationality Act by providing that for each commercial vessel or aircraft transporting any person to any seaport or airport of the United States...it shall be the duty of an appropriate official...to provide...manifest information about each passenger, crew member, and other occupant transported on such vessel or aircraft prior to arrival at that port. This new provision admits of the valid use of advance passenger information (API) to determine the admissibility of a person to the United States as well as the admissibility of a person as a passenger in an aircraft. The provision details the type of information that may be required. Section 231 is amended in (f) which states that no operator of any private or public carrier that is under a duty to provide information shall be granted papers until the requirements of the provision are complied with. Sub section (g) prescribes penalties to be imposed on carriers who do not comply with the requirement of providing information to the authorities.

The significance of these requirements to the carrier's right in refusing a passenger boarding is that such requirements may impose upon the carrier the added responsibility of being doubly vigilant as to the safety of flights performed by the carrier into the United States. It is therefore evident that the above legislation imposes an obligation on air carriers in the United States to be vigilant and aware of persons they have contracted to carry.

The No Fly List

A vexed issue that flight 253 brought to bear is the effectiveness of "no fly" lists which contain names of persons who are considered a threat to aviation should they travel by air. These lists are maintained by individual States (and not by Organizations such as ICAO) where such States effectively preclude the travel of a potentially dangerous passenger. This is a matter purely for the State concerned as the prerogative of admitting a person to its territory lies exclusively on the State. However, the maintenance of a no fly list is not a fool proof measure as there have been occasions (such as when an 8 year old boy who shared the same name with a suspect on a no fly list was patted down by Customs authorities every time he travelled).[192]

---

[192] http://www.cbsnews.com/stories/2006/10/05/60minutes/main2066624.shtml. A "false positive" occurs when a passenger who is not on the No Fly List has a name that matches or is similar to a name on the list. Such a passenger will not be allowed to board a flight unless they can differentiate themselves from the actual person on the list – usually by showing a middle name or date of birth. In some cases, false positive passengers have been denied boarding or have missed flights because they could not easily prove that they were not the person on the No Fly List.

3.3 Full Body Scanners and Emergent Issues

A potentially dangerous passenger can be identified either by spontaneous offensive conduct at the pre-boarding stage, a name-match or by criminal profiling, which is an investigative technique based on knowledge of the human personality and various psychological disorders that afflict the human being. However, a no fly list in particular may well raise the issue of privacy, particularly in instances of the false positive. The data subject, like any other person, has an inherent right to his privacy. The subject of privacy has been identified as an intriguing and emotive one.[193] The right to privacy is inherent in the right to liberty, and is the most comprehensive of rights and the right most valued by civilized man.[194] This right is susceptible to being eroded, as modern technology is capable of easily recording and storing dossiers on every man, woman and child in the world.[195] The data subject's right to privacy, when applied to the context of the machine readable travel document (MRTD) is brought into focus by Alan Westin who says:

> Privacy is the claim of individuals, groups or institutions to determine for themselves when, how, and to what extent information is communicated to others.[196]

Legally speaking, there are three rights of privacy relating to the storage and use of personal data:

1. The right of an individual to determine what information about oneself to share with others, and to control the disclosure of personal data;
2. The right of an individual to know what data is disclosed, and what data is collected and where such is stored when the data in question pertains to that individual; the right to dispute incomplete or inaccurate data; and
3. The right of people who have a legitimate right to know in order to maintain the health and safety of society and to monitor and evaluate the activities of government.[197]

It is incontrovertible therefore that the data subject has a right to decide what information about oneself to share with others and more importantly, to know what data is collected about him. This right is balanced by the right of a society to collect data about individuals that belong to it so that the orderly running of government is ensured.

The role played by technology in modern day commercial transactions has affected a large number of activities pertaining to human interaction. The emergence of the information superhighway and the concomitant evolution of automation have inevitably transformed the social and personal life styles and value systems of individuals,

---

[193] Young (1978).

[194] Warren and Brandies (1890–1891).

[195] As far back as in 1973 it was claimed that ten reels, each containing 1,500 m of tape 2.5 cm wide, could store a twenty page dossier on every man, woman, and child in the world. See Jones (1973).

[196] Westin (1970).

[197] Hoffman (1980).

created unexpected business opportunities, reduced operating costs, accelerated transaction times, facilitated accessibility to communications, shortened distances, and removed bureaucratic formalities.[198] Progress notwithstanding, technology has bestowed on humanity its corollaries in the nature of automated mechanisms, devices, features, and procedures which intrude into personal lives of individuals. For instance, when a credit card is used, it is possible to track purchases, discovering numerous aspects about that particular individual, including, food inclination, leisure activities, and consumer credit behaviour.[199] In similar vein, computer records of an air carrier's reservation system may give out details of the passenger's travel preferences, inter alia, seat selection, destination fondness, ticket purchasing dossier, lodging keenness, temporary address and telephone contacts, attendance at theatres and sport activities, and whether the passenger travels alone or with someone else.[200] This scheme of things may well give the outward perception of surveillance attributable to computer devices monitoring individuals' most intimate activities and preferences, leading to the formation of a genuine "traceable society".[201]

The above notwithstanding, it must be borne in mind that privacy is not an absolute, unlimited right that operates and applies in isolation.[202] It is not an absolute right, applied unreservedly, to the exclusion of other rights. Hence there is frequently the necessity to balance privacy rights with other conflictive rights, such as the freedom of speech and the right to access information when examining individuals' rights *vis-à-vis* the interest of society.[203] This multiplicity of interests

---

[198] See generally Orwell (1984).

[199] For a detailed analysis of the implications of credit cards with respect to the right of privacy see Nock (1993).

[200] The paramount importance of airline computer reservation system records is reflected in the world-renowned cases *Libyan Arab Jamahiriya* v. *United Kingdom* and *Libyan Arab Jamahiriya* v. *United States of America* regarding the PANAM 103 accident at Lockerbie, Scotland in 1988, where the International Court of Justice requested air carriers to submit to the Court the defendants' flight information and reservation details. See International Court of Justice. News Release 99/36, "Questions of Interpretation and Application of the 1971 Montreal Convention arising from the Aerial Incident at Lockerbie" (1 July 1999), online: http://www.icj-cij.org/icjwww/idocket/iluk/iluk2frame.html (date accessed: 14 July 2000). In a similar vein, Arthur R. Miller describes the significance of airline computer reservation system records when dealing with federal, state, local, and other types of investigations where these dossiers could provide valuable information. See also Miller (1971).

[201] See Scott (1995), Burnham (1983). *A contrario* to the argument supported in this thesis that the advancement of technology directly affects the intimacy of individuals. U.S. Circuit Judge Richard Posner favours the idea that other factors, such as urbanisation, income, and mobility development have particularly weakened the information control that, for instance, the government has over individuals: this denotes that individuals' privacy has increased. See Posner (1978).

[202] See Simmel (1971).

[203] See Halpin (1997). See also Foschio (1990). For a comprehensive study on the conflictive interest on privacy and the mass media and the Freedom of Speech, see Pember (1972), Prowda (1995). See also J. Montgomery Curtis Memorial Seminar (1992).

will prompt courts to adopt a balanced approach when adjudicating on a person's rights, particularly when the interests of his own State are involved.

A perceived inadequacy of the global framework of aviation security is the lack of an implementation arm. ICAO has taken extensive measures to introduce relevant international conventions as well as Standards and Recommended Practices (SARPs) in Annex 17 to the Chicago Convention. There is also a highly classified *Aviation Security Manual* developed by ICAO which is provided to States. Additionally, the Organization provides focused security training courses to its member States. However, ICAO's role is largely confined to rule making and the provision of guidance, bringing to bear the need for an aviation security crisis management team on a global scale that could work towards effectively precluding acts of terrorism.

3.4 Suppressing the Financing of Terrorism

Suppressing the financing of terrorism is a major measure against terrorism. However it is just one of the many tools in a conglomeration of connected measures that goes toward combating terrorism. Therefore, any study of the financing of terrorism and the fight against it would not be complete with an analysis of the collective counter-terrorism measures to be used in combating the problem. This article discusses the subject of suppressing the financing of terrorism against the backdrop of the offence of terrorism and its related issues.

The United Nations General Assembly, on 9 December 1999, adopted the *International Convention for the Suppression of the Financing of Terrorism*,[204] aimed at enhancing international co-operation among States in devising and adopting effective measures for the prevention of the financing of terrorism, as well as for its suppression through the prosecution and punishment of its perpetrators.

---

[204] http://www.un.org/law/cod/finterr.htm. On 26 October 2001 President George W. Bush signed the USA Patriot Act (the contrived acronym for "PATRIOT" being *Providing Appropriate Tools Required to Intercept and Obstruct Terrorism*) and the full title being *Uniting and Strengthening America by Providing Appropriate Tools Required to Intercept and Obstruct Terrorism Act of 2001*. The Act empowered law enforcement agencies to search telephone, e-mail communications, medical, financial, and other records; eases restrictions on foreign intelligence gathering within the United States. It also extended the powers of the Secretary of State to regulate financial transactions, particularly those involving foreign individuals and entities; and the discretion of law enforcement and immigration authorities in detaining and deporting immigrants suspected of terrorism-related acts. A significant feature of this legislation was its extended coverage that included domestic terrorism.

The Convention, in its Article 2 recognizes that any person who by any means directly or indirectly, unlawfully or wilfully, provides or collects funds with the intention that they should be used or in the knowledge that they are to be used, in full or in part, in order to carry out any act which constitutes an offence under certain named treaties, commits an offence. The treaties[205] listed are those that are already adopted and in force and which address acts of unlawful interference with such activities as deal with air transport and maritime transport. Also cited is the International Convention for the Suppression of Terrorist Bombings,[206] adopted by the General Assembly of the United Nations on 15 December 1997.

The Convention for the Suppression of the Financing of Terrorism also provides that, over and above the acts mentioned, providing or collecting funds toward any other act intended to cause death or serious bodily injury to a civilian, or to any other person not taking an active part in the hostilities in the situation of armed conflict, when the purpose of such act, by its nature or context, is to intimidate a population, or to compel a government or an international organization to do or to abstain from doing any act, would be deemed an offence under the Convention.[207]

The use of the word "terrorism" in the title of the Convention brings to bear the need to examine in greater detail both the etymology and the connotations of the word in modern parlance. The term "terrorism" is seemingly of French origin and is believed to have been first used in 1798. "Terrorism", which originally had connotations of criminality to one's conduct, is now generally considered a system of coercive intimidation brought about by the infliction of terror or fear. The most frustrating obstacle to the control of unlawful acts against international peace is the paucity of clear definition of the offence itself. Many attempts at defining the offence have often resulted in the offence being shrouded in political or national barriers.

---

[205] Convention for the Suppression of Unlawful Seizure of Aircraft, done at The Hague on 16 December 1970; Convention for the Suppression of Unlawful Acts against the Safety of Civil Aviation, done at Montreal on 23 September 1971; Convention on the Prevention and Punishment of Crimes against Internationally Protected Persons, including Diplomatic Agents, adopted by the General Assembly of the United Nations on 14 December 1973; International Convention against the Taking of Hostages, adopted by the General Assembly of the United Nations on 17 December 1979; Convention on the Physical Protection of Nuclear Material, adopted at Vienna on 3 March 1980; Protocol for the Suppression of Unlawful Acts of Violence at Airports Serving International Civil Aviation, supplementary to the Convention for the Suppression of Unlawful Acts against the Safety of Civil Aviation, done at Montreal on 24 February 1988; Convention for the Suppression of Unlawful Acts against the Safety of Maritime Navigation, done at Rome on 10 March 1988; Protocol for the Suppression of Unlawful Acts against the Safety of Fixed Platforms located on the Continental Shelf, done at Rome on 10 March 1988; and International Convention for the Suppression of Terrorist Bombings, adopted by the General Assembly of the United Nations on 15 December 1997

[206] United Nations General Assembly A/52/653, 25 November 1997. See http://www.un.org/law/cod/terroris.htm.

[207] Article 2 b).

In 1980 the Central Intelligence Agency of the United States of America adopted a definition of terrorism which read:

> Terrorism is the threat or use of violence for political purposes by individuals or groups, whether acting for or in opposition to established governmental authority, when such actions are intended to shock, stun or intimidate victims. Terrorism, has involved groups seeking to overthrow specific regimes, to rectify perceived national or group grievances, or to undermine international order as an end in itself.[208]

This all embracing definition underscores the misapprehension that certain groups which are etched in history such as the French Resistance of Nazi occupied France during World War 11 and the Contras in Nicaragua would broadly fall within the definitive parameters of terrorism. In fact, this formula labels every act of violence as being "terrorist" engulfing in its broad spectrum such diverse groups as the Seikigunha of Japan and the Mujahedeen of Afghanistan, although their aims, modus operandi and ideologies are different. James Adams prefers a narrower definition which reads:

> a terrorist is an individual or member of a group that wishes to achieve political ends using violent means, often at the cost of casualties to innocent civilians and with the support of only a minority of the people they claim to represent.[209]

Even this definition although narrower than the 1980 definition cited above is not sufficiently comprehensive to cover for instance the terrorist who hijacks an air plane for his own personal gain. The difficulty in defining the term seems to lie in its association with political aims of the terrorist as is found in the definition that terrorism is really:

> terror inspired by violence, containing an international element that is committed by individuals or groups against non-combatants, civilians, States or internationally protected persons or entities in order to achieve political ends.[210]

The offence of terrorism has also been defined as one caused by:

> ...any serious act of violence or threat thereof by an individual. Whether acting alone or in association with other persons which is directed against internationally protected persons, organizations, places, transportation or communication systems or against members of the general public for the purpose of intimidating such persons, causing injury to or the death of such persons, disrupting the activities of such international organizations, of causing loss,

---

[208] The CIA website states: The Intelligence Community is guided by the definition of terrorism contained in Title 22 of the US Code, Section 2656f(d):

- The term "terrorism" means premeditated, politically motivated violence perpetrated against noncombatant targets by subnational groups or clandestine agents.
- The term "international terrorism" means terrorism involving the territory or the citizens of more than one country.
- The term "terrorist group" means any group that practices, or has significant subgroups that practice, international terrorism. See https://www.cia.gov/news-information/cia-the-war-on-terrorism/terrorism-faqs.html.

[209] Adams (1989).

[210] Silets (1987); see also Lee (2005).

detriment or damage to such places or property, or of interfering with such transportation and communications systems in order to undermine friendly relations among States or among the nationals of different States or to extort concessions from States.[211]

It is time that terrorism is recognised as an offence that is *sui generis* and one that is not always international in nature and motivated by the political aims of the perpetrator. For the moment, if terrorism were to be regarded as the use of fear, subjugation and intimidation to disrupt the normal operations of humanity, a more specific and accurate definition could be sought, once more analysis is carried out on the subject. One must always be mindful however, that without a proper and universally acceptable definition, international cooperation in combating terrorism would be impossible.

A terrorist act is one which is *mala in se* or evil by nature and has been associated with the political repression of the French Revolution era where, it is said, the word terrorism was coined. A terrorist is a *hostis humani generis* or common enemy of humanity.

International terrorism has so far not been defined comprehensively largely due to the fact that owing to its diversity of nature the concept itself has defied precise definition. However, this does not preclude the conclusion that international terrorism involves two factors. They are:

1. The commission of a terrorist act by a terrorist or terrorists; and
2. The "international" element involved in the act or acts in question i.e., that the motivation for the commission of such act or acts or the eventual goal of the terrorist should inextricably be linked with a country other than that in which the act or acts are committed.

Perhaps the oldest paradigm of international terrorism is piracy which has been recognized as an offense against the law of nations and which is seen commonly today in the offense of aerial piracy or hijacking.

Acts of international terrorism that have been committed over the past two decades are too numerous to mention. Suffice it to say, that the most deleterious effect of the offense is that it exacerbates international relations and endangers

---

[211] Nechayev, Serge, Revolutionary Catechism, cited in (Rapoport 1971). Another noteworthy definition was the one that was adopted at the Conference of the International Law Association in Belgrade, 1980 which states:

The definition of "international terrorist offence" presented here is more comprehensive than the definitions which appear in the multilateral convention relating to the control of international terrorism which has been concluded in the past two decades. The term comprehends serious criminal acts, such as murder, assault, arson, kidnapping, extortion, sabotage and the use of explosives devices which are directed towards selected targets. These targets include internationally protected persons, places and international civil aircraft which are already protected under the conventional or customary international law. See. Delaney (1979). See also, The Draft Convention of the International Law Association, Belgrade Conference (Committee on International Terrorism), August 1980, at 9, for definitions of "terrorism" proposed by the Haitian and French delegations at the Conference.

international security. From the isolated incidents of the 1960s, international terrorism has progressed to becoming a concentrated assault on nations and organizations that are usually susceptible to political conflict, although politics is not always the motivation of the international terrorist. International terrorism has been recognized to engulf acts of aggression by one State on another as well as by an individual or a group of individuals of one State on another State. The former typifies such acts as invasion, while the latter relates to such individual acts of violence as hijacking and the murder of civilians in isolated instances. In both instances, the duties of the offender-State have been emphatically recognized. Such duties are to condemn such acts and take necessary action.

The United Nations has given effect to this principle in 1970 when it proclaimed that:

> Every State has the duty to refrain from organizing or encouraging the organization of irregular forces or armed bands, including mercenaries, for incursion into the territory of another State. Every State has the duty to refrain from organizing, instigating, assisting or participating in acts of civil strife or terrorist acts in another State or acquiescing in organized activities within its territory directed towards the commission of such acts, when the acts referred to in the present paragraph involve a threat or use of force.[212]

The most pragmatic approach to the problem lies in identifying the parameters of the offense of international terrorism and seeking a solution to the various categories of the offense. To obtain a precise definition would be unwise, if not impossible. Once the offense and its parasitic qualities are clearly identified, it would become necessary to discuss briefly its harmful effects on the international community. It is only then that a solution can be discussed that would obviate the fear and apprehension we suffer in the face of this threat.

3.4.1 Acts of International Terrorism

It is said that terrorism is a selective use of fear, subjugation and intimidation to disrupt the normal operations of society. Beyond this statement which stands both for national and international terrorism any attempt at a working definition of the words "international terrorism" would entail complications. However, in seeking a solution which would lead to the control of international terrorism it is imperative that contemporaneous instances of the infliction of terror be identified in order that they may be classified either as acts of international terrorism or as mere innocuous acts of self defence. Broadly acts of international terrorism may be categorized into two distinct groups. In the first category may be included what are termed as acts of oppression such as the invasion of one state by another. In the second category are acts which are deviously claimed to be acts of defence. While the former is self

---

[212] A/52/653, 25 November 1997.

explanatory, the latter – by far the more prolific in modern society – can be identified in four separate forms of manifestation. They are:

(a) Acts claimed to be committed in self defence and in pursuance of self-determination to circumvent oppression;
(b) Nonviolent acts committed internationally which are calculated to sabotage and destroy an established regime;
(c) Random acts of violence committed internationally by an individual or groups of individuals to pressurize a State or a group of individuals to succumb to the demands of terrorists; and
(d) Acts committed internationally which aid and abet national terrorism.

With the exception of the first category of invasion, the others are *prima facie* acts of international terrorism which are essentially extensions of national terrorism. That is to say that most acts of international terrorism are a species of the genus national terrorism.

3.4.1.1 Acts of Defence

Some States claim that internal oppression either by foreign invasion or by an internal totalitarian regime necessitates guerrilla warfare for the achievement of freedom. With more emphasis, it has been claimed that one state must not be allowed to exploit and harass another and that the physical manifestation of desire to attain freedom should not be construed as terrorism. Often, such acts of self defence prove to take extreme violent forms and manifest themselves overseas, thus giving rise to international terrorism. Acts of defence, as they are called, are common forms of international terrorism and are categorized as political violence. These acts take the form of acts of disruption, destruction, injury whose purpose, choice of targets or victims, surrounding circumstances, implementation and/or effects have political significance.

Organized political groups plan strikes and acts of violence internally while extensions of these groups carry out brutal assassinations, kidnapping and cause severe damage to property overseas. The retaliatory process which commences as a token of self defence transcends itself to terroristic violence which is totally ruthless and devoid of moral scruples. Usually, a cause which originates as dedicated to self defence and self determination aligns itself to gaining the support of the people, disarming the military strength of the regime against which it rebels and above all seeks to strengthen itself in order that the terrorist movement attains stability. In this instance terrorist acts seek primarily to carry out a massive propaganda campaign in the international community while at the same time concentrating more on individual instances of terrorism in populated urban areas which attract attention more than those committed in isolated areas. Advertising a cause in the international community becomes an integral part of political terrorism of this nature.

Both the international community and the governments concerned should be mindful that acts of defence can be treated as such only in instances where people

3.4 Suppressing the Financing of Terrorism   231

defend themselves when they are attacked and not when retaliatory measures are taken in isolation to instil fear in the international community. To that extent, acts of defence can be differentiated from acts of terrorism.

3.4.1.2 Nonviolent Acts

There are instances where terrorism extends to destabilizing an established regime or a group of persons by the use of threats which are often calculated to instil fear in the international community. Typical examples of this kind of terrorism are the spreading of false propaganda and the invocation of threats which unhinge both the nation or a group of persons against whom the threats are carried out and the nations in which such acts are said to be committed. There have been instances in the past where export consumer commodities of a nation such as food items have been claimed to be poisoned in order that foreign trade between nations be precluded. Although such acts are devoid of actual physical violence, they tend to unhinge the economic stability of a nation particularly if such nation depends solely on the export of the item in question. In such instances, international terrorism assumes proportions of great complexity and succeeds at least temporarily to disrupt the infrastructural equilibrium of the nation against which such threat is aimed. The government concerned is immediately placed on the defensive and attempts counter-propaganda. In spreading propaganda of this nature, the media is the terrorist's best friend. He uses the media of television and radio as a symbolic weapon to instil fear in the public and to cripple the persons or government against which his attack has been aimed. The effect of publicity on people is truly tangible, whether it pertains to the statement of facts or whether it relates to the issuance of threats. Primarily, media terrorism creates an emotional state of apprehension and fear in threatened groups and secondly, draws world attention to the existence of the terrorists and their cause. In both instances, the terrorist succeeds in creating a credibility gap between his target and the rest of the world. Psychological terrorism of this nature is perhaps the most insidious of its kind. It is certainly the most devious.

3.4.1.3 Random Acts of Violence

A random act of violence is normally a corollary to a threat though not necessarily so. Often as it happens, the international community is shocked by a despicable act of mass murder and destruction of property which takes the world completely by surprise. Responsibility for the act is acknowledged later though in many instances no responsibility is claimed. In the latter instance when no responsibility is claimed, the offended nation and the world at large are rendered destitute of an immediate remedy against the offense. Even if motive is imputed to a particular terrorist group, the exercise of sanction becomes difficult as the international community would not condone sanction in the absence of concrete and cogent evidence.

The difficulty lies largely on the fact that any terrorist act is usually carefully planned and executed. Often one observes that the terrorist cautiously retracts his steps obscuring all evidence unless he seeks publicity. The average terrorist is a militant who employs tactics aimed at instilling fear in the minds of the international community. His acts are calculated to instil fright and paralyse the infrastructure of a state by totally exhausting the strength of his target. He further disarms his target by introducing the element of surprise to his attack. Perhaps the most outstanding element of a random attack is the psychological element where excessive and sporadic acts of violence instil both fear and psychological disorientation in a society. This in turn contributes to undermining and weakening a government's authority and control. The disruptive influence that terrorism of this kind exercises over society often creates disharmony within the political circles of a nation and unhinges the psychological behavioural pattern of an organized society. Most often the gap between the citizen and the established government both in the State in which the act is committed and in the state against which the act is committed is widened as the average citizen tends to regard his personal security as the most inviolate of rights that has to be protected by his government.

3.4.1.4 Acts Which Aid and Abet National Terrorism

The fourth facet of international terrorism pertains to acts which promote national terrorism and which are committed outside the State against which the terrorist cause exists. These acts manifest themselves in the maintenance of overseas training camps for terrorists where guerilla warfare, techniques of assassination, destruction and sabotage are taught to terrorist groups who, after sustained training, return to their country and practice what they learn overseas. Such training camps are conducted usually by revolutionary groups and mercenaries on the request of terrorist organizations. A natural corollary to this trend is the collection of funds overseas for the financing of such training programmes, the purchase of arms, ammunition and explosives and the collection of monies involved in meeting the costs incurred by foreign propaganda.

Indirect acts of international terrorism such as those which aid and abet national terrorism indicate clearly that although there is no identifiable definition of the word "terrorism", the word itself can no longer be associated only with violent acts of aggression. In fact, recent studies reflect that any organized campaign of international terrorism involves both direct and indirect acts in equal proportion.

Broadly, international terrorism embitters humanity and antagonizes one nation against another, one human being against another. The eventual consequence of the problem is aggression and even war. The main aim of use of the psychological element by the international terrorist which is by far the most obnoxious and objectionable ambition is to polarize humans and society. However, its immediate manifestation and future development are not without features sufficient to cause grave concern to the world.

3.4 Suppressing the Financing of Terrorism

Acts of international terrorism, whether in the form of violent or non violent acts have clear and immediate international consequences. They are numerous in nature and warrant a separate study. However, in effect they obtain for the miscreant the same result of creating disharmony and disruption in society. The concept has grown in recent times to portend more serious problems to the international community. Those problems are worthy of comment.

Terrorism has so far not reached the proportions of being an international conspiracy although one group identifies its objectives and purpose with another. We have not had the misfortune of seeing all terrorist groups band together to work as a composite element. This has not happened for the reason that diverse ideologies and religions have kept each group separate. Nevertheless, there is a strong identity bond between groups and even evidence that one helps the other with training and military aid, even though their causes are quite different. The link between terrorist groups is an important consideration for the world as close association between groups could strengthen a weak force and nurture it to maturity. In addition, strong and established terrorist organizations, under cover of burgeoning groups, could carry out campaigns which would cover their tracks and make identification difficult. In most instances, this was found to be true and investigation reveals that a small group, not too significant at that time to take account of, has been responsible for an act or acts whereas later it is revealed that a much stronger group had masterminded the offenses for its benefit. Another important feature of the growing incidence of international terrorism is the assistance the terrorists receive from the advancement of technology in communication, the manufacture of sophisticated weaponry and the proliferation of nuclear armament. In today's context, terrorism has blown to unmanageable proportions with the use of advanced weapons of destruction. Arms control plays a vital role in the control of aggression and it naturally follows that terrorism too benefits from the availability of new modes of aggression. The vulnerability of the international community has been mainly brought upon by the paucity of adequate security measures to prevent nuclear theft. With the growth of the nuclear power industry, developed nations exposed themselves to the vulnerability of theft by power groups, in whose hands nuclear weapons act as threats of destruction. The most effective counter measure that can be taken in this instance against the threat of nuclear theft is to take such effective measures as are necessary to protect the stored items and to make known to the terrorist the high risk involved in an attempt to steal such material. Ideally, any hope of theft must be obviated. This can be achieved by strengthening governmental security.

3.4.1.5 Problems of Deterrence

The only deterrence that would be effective against terrorism of any nature is broadly based on the success of convincing the terrorist that the risk he takes outweighs the benefits which may accrue to his cause by his act. The futility of attempting to wipe out terrorism by the use of military force or the threat of general sanction on an international level is apparent. The terrorist has to be shown that any attempt at terrorist activity would cause him and his cause more harm than good. Deterrence in this context attains fruition when effective punitive sanctions are prescribed and carried out whilst simultaneously denying the terrorist his demands. In both instances the measures taken should be imperatively effective. It is not sufficient if such measures are merely entered into the statute books of a State or incorporated into international treaty. The international community has to be convinced that such measures are forceful and capable of being carried out.

However, deterrence does not stop at the mere imposition of effective sanction nor does it complete its task by the denial of terrorist demands. Perhaps the most effective method of countering terrorism is psychological warfare. The terrorist himself depends heavily on psychology. His main task is to polarize the people and the establishment. He wants popular support and a sympathetic ear. He wants a lot of people listening and watching, not a lot of people dead. Counter measures taken against a terrorist attack, be it hostage taking, kidnapping or a threat of murder, should essentially include an effective campaign to destroy the terrorist's credibility and sincerity in the eyes of the public. Always, the loyalty of the public should be won over by the target and not by the terrorist. It is only then that the terrorist's risk outweighs the benefits he obtains. To achieve this objective it must be ensured that the terrorist receives publicity detrimental to him, showing the public that if the threatened person, group of persons or State comes to harm, the terrorist alone is responsible. Therefore, the most practical measures that could be adopted to deter the spread of terrorism can be accommodated in two chronological stages:

1. Measures taken before the commission of an offense such as the effective imposition and carrying out of sanctions and the refusal to readily comply with the demands of the terrorist;
2. Measures taken after the commission of the act such as the skilful use of the media to destroy the credibility of the terrorist cause and to convince the people that the responsibility for the act devolves at all stages solely upon the terrorist.

One difficulty in exercising deterrence against terrorism in general and international terrorism in particular is that often, the measures taken are not effective enough to convince the terrorist that in the end, more harm would be caused to him than good. Negotiation with the terrorist in particular has to be done by professionals specially trained for the task. A fortiori, the media has to be handled by specialists with experience. Things would be much more difficult for the terrorist if these were done. The greatest problem of deterrence is the pusillanimity of the international community in the face of terrorism and the feeble response offered by States as a composite body. The reasons for this hesitation on the part of the international community to adopt effective measures against international terrorism are by no means inexplicable. When one state supports a revolutionary cause which is aimed against another, it is quite natural that the terrorist is aware of the support he is capable of obtaining from at least one part of the already polarized world. Therein lies the problem.

3.4.1.6 The Practical Solution

The primary objective of international peace and security is the endeavour to preserve the right to life and liberty. This right is entrenched in Article 3 of the Universal Declaration of Human Rights of 1948 and is accepted today as constituting an obligation on all member States to recognize the legally and morally binding nature of the Declaration. Therefore the destruction of human life and the restriction of liberty are acts committed against international law and order. International terrorism destroys both life and liberty. Indeed there need be no doubt in our minds that international terrorism is illegal. To begin with, there should be more awareness in the world today that every human being has the inherent right to life and that the right is protected by law. Any act of terrorism being illegal, becomes subject to law and its punitive sanctions. However, in this instance, unlike in a simple instance of murder where sanction itself may act as a deterrent, the two forces of law and sanction are not sufficient to curb terrorism. The international community should realize that the solution to terrorism lies rather in its prevention than in its cure. Therefore the problem has to be approached solely on the basis that the terrorist on the one hand has to be dissuaded that his act may not succeed while on the other he has to be persuaded that even if he succeeded in committing the act of terrorism, it would not achieve for him the desired results. The philosophy of warfare against terrorism is therefore based on one single fact – that of convincing the terrorist that any attempt at committing a terrorist act would be fruitless and would entail for him unnecessary harm. This simple philosophy should be adopted gradually in stages with the sustained realization that each measure taken is as important as the next and that all measures should be adopted as a composite element and not as those that are mutually exclusive.

A potential terrorist can therefore be attacked in two ways:

1. By the adoption of practical measures to discourage the commission of the act;
2. By the adoption of such effective measures as would impose severe punitive sanctions if the act is committed.

In the first instance measures of self help are imperative. They should be adopted with careful planning and the terrorist should be made aware that the community at large are afforded the full protection of these measures. They are:

a) The establishment of a system of intelligence which would inform the state concerned of an impending terrorist attack;
b) The establishment of counter–terrorism mechanisms which would effectively preclude such catalysts as the collection of arms, ammunition and weaponry;
c) The adoption of such practical measures of self-help and attack as are necessary in an instance of an attack;
d) The existence of the necessary machinery to retain the confidence and sympathy of the public at all times;
e) The persuasion necessary to convince the public that terrorism of any kind is evil and should not be condoned, whatever its cause is.

The second instance is concerned with measures taken in the event a terrorist act is committed. If strongly enforced with unanimity, such measures as the imposition of laws which bind all nations to view terrorist acts as crimes against humanity can be an effective deterrent. A fortiori, sanctions would further discourage the terrorist.

3.4.2 Money Laundering

In theory, the financing of terrorism and money laundering are antithetical in that while the former involves the support of terrorism through the injection of funds the latter involves covering money acquired through acts of criminality in a cloak of legitimacy. In other words money laundering is "the process by which the source and ownership of criminally derived wealth and property is changed to confer on it a perception of legitimacy".[213] From the point of view of the criminal, there seem to be three basic requirements: (a) the need to conceal the true ownership and origin of the proceeds; (b) the need to maintain control of the proceeds; and (c) the need to change the form of the proceeds.[214] However, it is incontrovertible that there is a certain synergy between the two, as one feeds off the other and often the money gained from acts of terrorism are laundered and put back into funding acts of terrorism The enormity and wide-spread nature of money laundering on a global scale is reflected by the fact that the International Monetary Fund has estimated that, during the decade 1999–2009 the aggregate size of money-laundering was anywhere between 2% and 5% of the world's gross domestic product.[215] Just as an example, it is reported that in 2007, the money generated by organized crime in the United Kingdom was 15 billion Pounds Sterling, of which 10 billion was laundered.[216]

The most fundamental measure taken against money laundering is to criminalize it and to adopt legislative and other measures to identify, trace, freeze, seize and confiscate the proceeds of crime. To this the international effort is quite significant, particular concerning concerted action taken by the Financial Action Task Force (FATF), which is an inter-governmental body whose purpose is the development and promotion of national and international policies to combat money laundering

---

[213] *Money Laundering and the Financing of Terrorism*, Report published by the authority of the House of Lords of England, Volume 1, 22 July 2009. See http://www.coe.int/t/dghl/monitoring/moneyval/activities/UK_Parlrep.pdf.

[214] *Id* at 7. The offences associated with money laundering are: the conversion or transfer of property for the purpose of concealing or disguising its illicit origin or of assisting any person to evade the legal consequences of his actions; the concealment or disguise of the true nature, source, location, disposition, movement, rights with respect to, or ownership of criminally derived property; and the acquisition, possession or use of criminally derived property.

[215] Camdessus, Michel, (Managing Director of the IMF), "Money Laundering: the Importance of International Countermeasures", an address to the Plenary Meeting of the Financial Action Task Force on Money Laundering in Paris, 10 February 1998.

[216] Treasury (2007).

3.4 Suppressing the Financing of Terrorism

and terrorist financing. The FATF is therefore a "policy-making body" created in 1989 that works to generate the necessary political will to bring about legislative and regulatory reforms in these areas.[217] The FATF has recommended that: the 1999 United Nations Convention for the Suppression of the Financing of Terrorism and relevant UN Security Council Resolutions be ratified and implemented; the financing of terrorism, terrorist acts and terrorist organisations be criminalized; terrorist assets be frozen and confiscated; the widest possible assistance be extended to other countries' law enforcement and regulatory authorities for terrorist financing investigations; suspicious transactions related to terrorism be reported; anti-money laundering requirements on alternative remittance systems be imposed; ensure that entities, in particular non-profit organisations, cannot be misused to finance terrorism; and customer identification measures in international and domestic wire transfers be strengthened.

The Council of Europe Convention on Laundering, Search, Seizure and Confiscation of the Proceeds from Crime and on the Financing of Terrorism,[218] signed at Warsaw in December 2005 provides, in Article 2 that each Party shall adopt such legislative and other measures as may be necessary to enable it to confiscate instrumentalities and proceeds or property the value of which corresponds to such proceeds and laundered property. The Convention goes on to say that Parties may provide for mandatory confiscation in respect of offences which are subject to the confiscation regime which may include money laundering, drug trafficking, trafficking in human beings and any other serious offence.[219] Furthermore it is stated in the Convention that each Party shall adopt such legislative or other measures as may be necessary to require that, in respect of a serious offence or offences as defined by national law, an offender demonstrates the origin of alleged proceeds or other property liable to confiscation to the extent that such a requirement is consistent with the principles of its domestic law.[220] There is also provision for each Party to adopt such legislative and other measures as may be necessary to enable it to identify, trace, freeze or seize rapidly property which is liable to confiscation, in order in particular to facilitate the enforcement of a later confiscation.[221] A significant provision of the Convention lies ain Article 5 which states that each Party shall adopt such legislative and other measures as may be necessary to ensure that the measures to freeze, seize and confiscate also encompass: the

---

[217] http://www.fatf-gafi.org/pages/0,2987,en_32250379_32235720_1_1_1_1_1,00.html. The 34 members of the FATF are: Argentina; Australia; Austria; Belgium; Brazil; Canada; China; Denmark; the European Commission; Finland; France; Germany; Greece; the Gulf Co-operation Council; Hong Kong; Iceland; Ireland; Italy; Japan; Luxembourg; Mexico; the Kingdom of the Netherlands; New Zealand; Norway; Portugal; Russia; Singapore; South Africa; Spain; Sweden; Switzerland; Turkey; the United Kingdom; and the United States.

[218] *http://conventions.coe.int/Treaty/EN/Treaties/Html/198.htm*

[219] *Article 3.3.*

[220] *Article 3.4.*

[221] *Article 4.*

property into which the proceeds have been transformed or converted; property acquired from legitimate sources, if proceeds have been intermingled, in whole or in part, with such property, up to the assessed value of the intermingled proceeds; and income or other benefits derived from proceeds, from property into which proceeds of crime have been transformed or converted or from property with which proceeds of crime have been intermingled, up to the assessed value of the intermingled proceeds, in the same manner and to the same extent as proceeds.

Cutting off funding and enforcing rigid legislation against the financing of terrorism and implementing such is indeed an effective international measure against terrorism. While some countries have taken effective measures in this regard others are developing legislation.[222] However, for any measure against terrorism, the fundamental and compelling factor is the need to infuse a security culture among nations. A security culture, if such were to exist among States, would mean that the States would be aware of their rights and duties, and, more importantly, assert them. Those who belong to a security culture also know which conduct would compromises security and they are quick to educate and caution those who, out of ignorance, forgetfulness, or personal weakness, partake in insecure conduct. This security consciousness becomes a "culture" when all the States of the world as a whole make security violations socially and morally unacceptable within the group.

With regard to other measures against terrorism, the first step that should be taken to deter terrorism is to be equipped with the expertise to detect a potential threat beforehand and to be prepared for an attack. The next is to intensify security in all susceptible areas, particularly in such places as airports, subway terminals, etc. Surveillance of all people who are seen in such areas as are revealed to be targets of terrorist acts is imperative. There should be more awareness of the threat of terrorist activity particularly in international airports and international bus and train terminals where travel documents should be checked and passengers double checked. Electronic surveillance of passports and other documents have proved to be effective methods of deterrence in this context. Perhaps the most important facet of surveillance is the use of personnel who do not reveal their identity to the public but unobtrusively mingle with the crowds. This category of person can easily detect an irregularity without arousing suspicion and without alarming the common man. It is recommended that together with the armed personnel there should also be trained personnel who in all informality may work together with the security forces in such instances. Another significant requirement is the support of the people. The media should be made maximum use of to educate the common man as to how to react in an emergency and also to be totally distrustful of the terrorist whose acts are

---

[222] For a list of countries and a discussion see http://www.iss.co.za/pubs/Other/ahsi/Goredema_Botha/pt2chap12.pdf.

calculated to evoke sympathy. The State or persons against whom the terrorist attack is launched should, at all times, use the media to convince the public that responsibility for any destruction or harm resulting from a terrorist act devolves totally on the terrorist.

3.5 Civil Unrest and Aviation

What started on 17 December 2010 with an act of self immolation by Mohammed Bouazizi, a 26-year-old man trying to support his family by selling fruits and vegetables in the central town of Sidi Bouzid in Tunisia, led to massive protests in the country, resulting in the overthrow of Zine El Abidine Ben Ali, the country's president on 14 January 2011. On 25 January 2011, protests, at least partly inspired by the toppling of the authoritarian government in Tunisia, erupted in Egypt[223] and grew increasingly worse. As a result, Hosni Mubarak, President of Egypt, was deposed within weeks of a virulent peoples' uprising. Contemporaneous protests went on other States such as in Algeria, Yemen, and Bahrain, the last of which held a "day of rage" on February 14, instigated by youths, and inspired by events in Egypt and Tunisia. Furthermore, at the time of writing, there was acute unrest in Libya as a result of mass civil unrest.[224]

In the context of the Libyan crisis, many airlines adopted a cautious approach in planning their flights to Libya while others cancelled scheduled flights.[225] Stocks of European airlines rapidly declined and airlines such as British Airways and KLM cancelled their flight to Tripoli.[226] An inevitable corollary to intensifying violence in Libya, which is a large oil supplier to Europe, would be that airlines will be forced to charge higher fares. IATA observed that if the unrest were to continue in the various countries in the Middle East and North Africa, airlines would be forced

---

[223] Tourism and transport combined forms the largest industry in the world. Air transport is a significant driver of tourism and visitors arriving by air directly support approximately 6.7 million jobs worldwide in the tourism industry with the foreign exchange they spend during their travels. Both the tourism industry and air transport industry depend on the policies of governments and the individual stability of States for their sustenance and development. The unrest wrought by mass protests in North Africa and the Middle East in 2011 seriously disrupted tourism and air transport. Tourism earned Egypt more than 11 billion dollars in the last fiscal year. In the third quarter of 2010, Egypt was receiving about 280 million US dollars a week from tourism. See http://www.suite101.com/content/tourism-crisis-as-foreign-visitors-desert-egypt-a342840.

[224] Wikipedia identifies civil unrest with synonyms such as civil disorder, or civil strife, which are broad terms typically used by law enforcement to describe one or more forms of disturbance caused by a group of people.. Examples of civil disorder include, but are not necessarily limited to: illegal parades; sit ins; and other forms of obstructions; and other forms of crime. http://en.wikipedia.org/wiki/Civil_disorder.

[225] Airlines wary on operating to Libya, *Air Letter*, No. 17,180, Thursday 24 February 2011at p. 3.

[226] *Id.* at p.4.

to stop operating flights into those States, which would definitely result in significant losses to the airlines.[227]

In Libya, the runway at Benghazi airport was destroyed as a result of the continuing clashes between anti-government protesters and security forces.[228] It is reported that protesters against the government of Libya had surrounded the airport and the government of the United Kingdom, among others, was "urgently seeking landing permission from the Libyan Government" for a charter aircraft to airlift stranded British citizens out of the country.[229] The first point of contact of a tourist is the airport and if the airport premises is under severe civil unrest and attack, there will be no tourists visiting that country.

The security of a State is entirely dependent on the level of peace prevailing in its territory and any breach of that security, starting at the entry points to its territory, will also impact on loss of income for the State as the case is with tourism. Most, if not all countries affected by the civil unrest in the Middle East and North Africa are tourist intensive and their income will suffer immensely. Many States issued travel advisories on Tunisia, Egypt and Libya. At the time of the unrest in early 2011, Tunisia was recovering from the devastating effects on its tourism industry brought about by the terrorist attacks of 2001 and 2002 when the country lost a substantial number of tourists from its traditional markets of France, Germany, Italy and the United Kingdom. The 500,000 German tourists lost in the process was a big blow to Tunisia's tourism.[230] With regard to Egypt, hotel capacity increased by approximately 7,000 rooms between 2009 and 2010 to a total of 220,000 hotel rooms. In December 2010 The United Nations World Tourism Organization (UNWTO) increased its collaboration with Egypt in enhancing the country's tourist intake worked closely with Egypt in enhancing its capacity to measure the economic impact of tourism and provide consistent, internationally benchmarked tourism statistics.[231] With such an upsurge in tourism promotion, It is therefore heartening that tourism in Tunisia and Egypt, States that carried out a successful revolution in overthrowing their existing regimes, did not suffer for too long and recovered quickly. UNWTO has expressed its appreciation of proactive efforts by national authorities to restore confidence among tourists and by foreign governments to update travel advisories accordingly. Tourism is a significant contributor to both

---

[227] Airlines set for losses as mid-east unrest continues, *Air Letter*, No. 17,181, Friday, 25 February 2011at p.3.

[228] Kelly Reals, Runway at Benghazi Airport Destroyed: Capita Symonds, *Air Transport Intelligence News*, 22 February 2011. Seehttp://www.flightglobal.com/articles/2011/02/22/353498/runway-at-libyas-benghazi-airport-destroyed-capita.html.

[229] *Ibid.*

[230] This loss gradually balanced from the new European markets and especially from Poland, Czech Republic and Hungary. See http://www.traveldailynews.com/pages/show_page/23601-Tunisia-unveils-new-tourism-plan.

[231] http://www.ameinfo.com/252453.html.

3.5 Civil Unrest and Aviation 241

countries' economies and, as tourism returns to normalcy, overall economic recovery can be stimulated.[232]

As the situation in both Egypt and Tunisian returns to normal, tourism stakeholders from the private and public sectors have reacted accordingly. Major tourism sites are open to the public, airlines have resumed flights, tour operators in many of the main source markets have restarted selling holidays and governments have updated their travel advisories to reflect the unfolding situation.

From an aviation and tourism perspective the unrest in these regions has impelled the markets to respond with oil prices shooting skywards to $119 a barrel for Brent crude. these higher oil prices is highly worrying for airlines. Having retrenched and cut back, airlines were hoping for a return to profitability in 2011 as growth returns following the downturn. However, the latest rise in oil prices could, as IATA forecasts extinguish any airline gains this year, causing a global domino effect on aviation. leaving carriers with heavy losses. Airlines were hoping for a return to profitability in 2011 as growth returns following the downturn.[233]

3.5.1 Keeping Airports Open

To begin with, The 37th Session of the ICAO Assembly of the International Civil Aviation Organization (ICAO)[234] held from 28 September to 10 October 2010 officially recognized that ICAO has three Strategic Objectives: safety, security and environmental protection and sustainability of air transport. The last strategic objective, although relevant to the consequences of civil unrest on air transport by no means impels ICAO to intervene in the internal affairs of States or to ensure that amidst the clash of arms air transport carries on regardless. However, what it does is to draw a nexus between ICAO and the Convention on International Civil

---

[232] http://www.traveldailynews.com/pages/show_page/41810-UNWTO-welcomes-signs-of-tourism-recuperation-in-Egypt-and-Tunisia.

[233] http://www.aerosocietychannel.com/aerospace-insight/2011/02/shifting-sands/.

[234] The International Civil Aviation Organization, a specialized agency of the United Nations, was established by Article 44 of the *Convention on International Civil Aviation* (Chicago Convention), signed at Chicago on 7 December 1944 (*infra*, note 12). The main objectives of ICAO are to develop the principles and techniques of international air navigation and to foster the planning and development of air transport. ICAO has 190 Contracting States. ICAO's Mission and Vision Statement is "to achieve its mission of safe, secure and sustainable development of civil aviation through cooperation amongst its member States". In December 2004, following a decision by the 35th Session of the ICAO Assembly, the Council of ICAO approved six Strategic Objectives for 2005–2010: They were: safety; security; environmental protection; efficiency; continuity; and rule of law. From 2011, ICAO's Strategic Objectives will be based on safety; security; environmental protection and the sustainable development of air transport. However, during the 37th Session of the Assembly, the Strategic Objectives were reduced to three: safety; security; environmental protection and sustainable development of air transport.

Aviation (Chicago Convention)[235] which provides inter alia that an aim of ICAO is to foster the planning and development of international air transport so as to meet the needs of the peoples of the world for safe, regular, efficient and economical air transport.[236]

The Chicago Convention requires States to keep their airports open to all airlines operating into and out of their territories and provide meteorological, radio and other information as well as facilities such as ground services. Of course, one might argue that Article 89 of the Chicago Convention enables Contracting States to have freedom of action irrespective of the provisions of the Convention in case of war, whether belligerents or neutrals. It also allows a State which has declared a state of national emergency (and notifies the ICAO Council of such) to have the same freedom of action notwithstanding the provisions of the Convention. Therefore, unless a State is at war (which the Convention does not define)[237] or has declared a state of national emergency, it would be bound by the provisions of the Convention.

The first duty of a Contracting State not falling within the purview of Article 89 of the Chicago Convention is to keep its airport open to all incoming aircraft. Article 15 of the Convention requires inter alia that, uniform conditions shall apply to the use, by aircraft of every contracting State, of all air navigation facilities, including radio and meteorological services, which may be provided for public use for the safety and expedition of air navigation. This condition is subject to Article 9 which stipulates that each contracting State may, for reasons of military necessity or public safety, restrict or prohibit uniformly the aircraft of other States from flying over certain areas of its territory, provided that no distinction in this respect is made between the aircraft of the State whose territory is involved, engaged in international scheduled airline services, and the aircraft of the other contracting States likewise engaged. The provision goes on to say that Each contracting State reserves also the right, in exceptional circumstances or during a period of emergency, or in the interest of public safety, and with immediate effect, temporarily to restrict or prohibit flying over the whole or any part of its territory, on condition that such restriction or prohibition will be applicable without distinction of nationality to aircraft of all other States.

The question arises as to whether a State in which there is acute civil unrest is bound to follow the abovementioned principles of the Chicago Convention. States or international organizations which are parties to such treaties have to apply the

---

[235] Signed at Chicago on 7 December 1944. See ICAO Doc 7300 9th Edition: 2006.

[236] *Id.* Article 44 d).

[237] Article 31.1 of the *Vienna Convention on the Law of Treatie* provides that "a treaty shall be interpreted in good faith in accordance with the ordinary meaning to be given to the terms of the treaty in their context and in the light of its object and purpose". See *Vienna Convention on the Law of Treaties 1969*, done at Vienna on 23 May 1969. The Convention entered into force on 27 January 1980. United Nations, *Treaty Series*, vol. 1155, p. 331. The ordinary meaning of war can be considered as a behavior pattern of organized violent conflict typified by extreme aggression, societal disruption, and high mortality. This behavior pattern involves two or more organized groups. http://en.wikipedia.org/wiki/War.

3.5 Civil Unrest and Aviation 243

treaties they have signed and therefore have to interpret them. Although the conclusion of a treaty is generally governed by international customary law to accord with accepted rules and practices of national constitutional law of the signatory States, the application of treaties are governed by principles of international law. If however, the application or performance of a requirement in an international treaty poses problems to a State, the constitutional law of that State would be applied by courts of that State to settle the problem. Although Article 27 of the *Vienna Convention*[238] requires States not to invoke provisions of their internal laws as justification for failure to comply with the provisions of a treaty, States are free to choose the means of implementation they see fit according to their traditions and political organization.[239] The overriding rule is that treaties are juristic acts and have to be performed.

3.5.2 Airport and Aviation Security

The biggest threat to security in the vicinity of the airport, where aircraft landing and takeoff are at their lowest altitude, is Man Portable Air Defence Systems (MANPADS). Since the events of 11 September 2001, there have been several attempts against the security of aircraft in flight through the misuse of Man Portable Air Defense Systems (MANPADS).[240] The threat of MANPADS to aviation security is by far the most ominous and the international aviation community has made some efforts through ICAO. MANPADS have posed a serious threat to aviation security. On 5 January 1974, 220 soldiers and 200 police sealed off five square miles around Heathrow International airport in London after receiving reports that terrorists had smuggled SA-7s into Britain in the diplomatic pouches of Middle-Eastern embassies and were planning to shoot down an El Al airliner.[241]

Another significant incident occurred on 13 January 1975 when an attempt by terrorists to shoot down an El Al plane with a missile was believed to have brought civil aviation to the brink of disaster. Two terrorists drove their car onto the apron at Orly airport, where they set up a rocket launcher and fired at an El Al airliner which was about to take off for New York with 136 passengers. The first round missed the target thanks to the pilot's evasive action and hit the fuselage of a Yugoslav DC-9 aeroplane waiting nearby to embark passengers for Zagreb. The rocket failed to

---

[238] *Id.*

[239] Reuter (1989).

[240] The use of SAMs and anti-tank rockets by terrorists goes back to 1973. On 5 September 1973 Italian police arrested five Middle-Eastern terrorists armed with SA-7s. The terrorists had rented an apartment under the flight path to Rome Fumicino Airport and were planning to shoot down an El Al airliner coming in to land at the airport. See Dobson and Payne (1987).

[241] Mickolus (1980).

explode and no serious casualties were reported. After firing again and hitting an administration building, which caused some damage, the terrorists escaped by car? A phone call from an individual claiming responsibility for the attack was received at Reuters. The caller clearly implied that there would be another such operation, saying 'Next time we will hit the target'.

In fact, 6 days later another dramatic though unsuccessful attempt did occur at Orly airport. The French authorities traced the attack to the PFLP Venezuelan terrorist, and leader of the PFLP group in Europe, Carlos.[242] It is also known that once again an El Al airliner had been deliberately chosen as a target by Gadafi in an attempt to avenge the loss of the Libyan airliner shot down by Israel over the Sinai Desert.[243]

MANPADS are extremely effective weapons which are prolific in their availability worldwide. The significance of the abuse of MANPADS as a threat to civil aviation in the airport context is that MANPADS could be used in the vicinity of the perimeter of the airport or in the airport premises itself in view of the short range needed to hit an aircraft approaching an airport or departing from one. Introduced in the 1950s and originally meant to deter terror attacks from air to ground to be used by State authorities and other protection agencies, these weapons have got into the wrong hands and are being used against civil and military aviation. The surface to air MANPAD is a light weapon which offers very little warning before impact, and is often destructive and lethal.[244] They are cheap, easily carried, handled and concealed. It is claimed that there are at least 100,000 and possibly in excess of 500,000 systems in inventories around the world and several thousands of these are vulnerable to theft from State authorities.[245] It is also claimed that there is a 70% chance that a civil aircraft will be destroyed if hit by a MANPAD.[246] A study conducted and published in early 2005 by the Rand Corporation concludes that, based on the effects of the attacks of September 11 2001, it is plausible for air travel in the United States to fall by 15–20% after a successful MANPADS attack on a commercial airliner in the United States.[247] The international aviation community is aware that civil aircraft are particularly vulnerable to hand held ground to air missiles and that susceptibility avoidance techniques (calculated to avoid being hit) and vulnerability avoidance (survival after being hit) systems must be in place. This is particularly so since tracking the

---

[242] Christopher Dobson and Ronald Payne, *supra*, p. 53.

[243] . *Ibid*.

[244] The lethality of the weapon can be reflected by the 340 MANPADS used by Afghan Mujahedeen rebels to successfully hit 269 soviet aircraft. See http://www.janes.com/security/international_security/news/.

[245] MANPADS, *Ploughshares Monitor* Autumn 2004, at 83.

[246] *Ibid*. The deadly accuracy and ease of handling of MANPADS were demonstrated when Somali gunmen shot down two US MH-60 Black Hawk helicopters in October 1993.

[247] Infrastructure Safety and the Environment, *Protecting Commercial Aviation against the Shoulder-Fired Missile Threat*, Rand Corporation, 2005, at 9.

3.5 Civil Unrest and Aviation

proliferation of MANPADS is difficult since any intelligence gathered on this particular threat is usually ex post facto, through the recovery of launchers or fragments from expended missiles. Contrary to popular belief, the MANPAD is considerably durable and can be used several years after inactivity, with recharged batteries.

The World's attention to the deadly threat posed by MANPADS was further drawn in November 2002 when there was an unsuccessful attempt to bring down a civilian aircraft leaving Mombasa, Kenya. Over the past 35 years, significant developments have taken place in dangerous weapons systems creating more opportunities for terrorists. The ready acceptance of new modern technologies by the international community and our growing dependence on them have created many targets, such as nuclear and civil aircraft in flight. Similarly, developments in electronics and microelectronics, and the trend towards miniaturization and simplification have resulted in a greater availability of tactical weapons with longer ranges and more accuracy that are also simpler to operate. One of the most effective developments in individual weaponry is portable, precision-guided munitions (PGMs), which are lightweight and easy to operate. They can usually be carried and operated by a single person. The United States-made Stinger, the British-made Blowpipe and the Russian-made SA-7 missiles are examples of these smaller weapons. These are shoulder-fired, anti-aircraft missiles that have infra-red, heat-seeking sensors in the projectile that guide it to the heat emitted from an aircraft engine. It is known that more than 60 States possess SA-7 missiles and there is no doubt that most of them maintain strict security measures to prevent the outflow of the weapons. However, it has been alleged that some States, including Libya, have supplied PGMs to terrorist organizations. It is incontrovertible that in the hands of terrorists these missiles are not likely to be used against conventional targets such as tanks and military fighter aircraft. Of particular concern is the prospect of civilian airliners being shot at by SAMs and anti-tank rockets as they land at or take off from airports[248] Dr. Richard Clutterbuck subsumes the great threat of missile attacks:

> Recent years have seen increasing use of expensive and sophisticated surface-to-surface and surface-to-air missiles (SSM and SAM) by terrorists, generally of Russian or East European origin and redirected by Arab Governments, notably Colonel Gadafi's. Continuing development of these weapons for use by regular armies will ensure that new and more efficient versions will become available for terrorists.[249]

With increased airport security, the possibility of placing explosive devices on civil aircraft is becoming more difficult, but now the same destructive result can be achieved far more easily by using modern missiles or rockets.

Perimeter security at the airport is a vital element in ensuring security of the airport itself as well as the security of incoming and outgoing aircraft. For a successful missile attack against aircraft, the firing position has to be located within range of the

---

[248] Hanle (1989), Ofri (1984), Pierre (1975–1976), Dorey (1983).
[249] Clutterbuck (1991).

flight path. A missile's guidance system is such that the weapon has to be fired within a few degrees of the flight path if the infra-red guidance is to locate the target. Accordingly, a possible preventive measure would be to prevent terrorists from getting into a firing position with their missiles. However, it would be very difficult to cut off areas of up to 6 km wide that lie in the paths of aircraft as they land and take off. This measure is therefore impracticable if not impossible.[250] This difficulty can be overcome to an extent by patrolling the outer areas of airports in times of stringent security conditions might prevent such attacks. Even in times when no specific threat has been received, it is within the capacity of most States to monitor those strips of land from which a SAM could be launched and thus minimize the risk. At the same time, these security operations would deter terrorists from spending vital resources on buying SAMs given the limited possibilities for their use.

Although the success rate so far of Western States in preventing terrorist missile attacks against civil aviation is satisfactory, and security forces, with the help of good intelligence, have been successful in tracking down and capturing missiles before they could be used, it is not unlikely that there will be attempts to use surface-to-air missiles to attack civil aviation in the near future. As some targets are becoming more difficult for terrorists to attack it can be anticipated that they will make efforts to overcome the enhanced security systems as well as redirecting their efforts towards less secure targets. The displacement of the increasingly ineffective system of hijacking by missile attacks against civil aviation is a real threat.

Another aspect in securing aviation in times of civil unrest is diplomacy and the meaning and purpose of aviation as interpreted by the founding fathers of the Chicago Convention. Given its strategic objective on sustanbility of air transport and its compelling diplomatic role which ICAO has played over the past 66 years with aplomb and competence, member States of ICAO could well consider the role of aviation in bringing about peace. The importance of aviation toward maintaining peace has been accepted since World War 2 and is aptly reflected in the Statement of the British at that time, that civil aviation holds the key to power and importance of a nation and therefore it must be regulated or controlled by international authority. Lord Beaverbrook for the British Government of that time stated in Parliament:

> Our first concern will be to gain general acceptance of certain broad principles whereby civil aviation can be made into a benign influence for welding the nations of the world together into a closer cooperation...it will be our aim to make civil aviation a guarantee of international solidarity, a mainstay of world peace.[251]

The intensely political overtones that moulded the incipient civil aviation system of the world immediately after the War, thereby incontrovertibly establishing the relevance of diplomacy, international politics and international relations in civil aviation, is borne out by the statement of the first President of the ICAO Council, Edward Warner, when he said:

---

[250] Dorey (1983).

[251] *Flight*, Vol. XLV No. 1331, January 27, 1944, at pp. 97–98.

3.5 Civil Unrest and Aviation

> It is well that we should be reminded...if the extent of the part which diplomatic and military considerations have played in international air transport, even in periods of undisturbed peace. We shall have a false idea of air transport's history, and a very false view of the problems of planning its future, if we think of it purely as a commercial enterprise, or neglect the extent to which political considerations have been controlling in shaping its course.[252]

In retrospect, it must be noted that this statement is a true reflection of what civil aviation stood for at that time, and, more importantly, that the statement has weathered the passage of time and is true even in the present context. A more recent commentator correctly observes that over the past decades, civil aviation has had to serve the political and economic interests of States and that, in this regard, ICAO has alternated between two positions, in its unobtrusive diplomatic role and its more pronounced regulatory role.[253]

> An inherent characteristic of aviation is its ability to forge inroads into human affairs and promote international discourse. It also promotes international goodwill and develops "a feeling of brotherhood among the peoples of the world".[254] Therefore, it has been claimed that problems of international civil aviation constitute an integral part of the universal political problems of world organization and therefore aviation problems cannot be solved without involving the world political and diplomatic machinery.[255] It is at these crossroads that one encounters the profound involvement of the United Nations mechanism in general and ICAO in particular.

As for ICAO, the Organization has acted in the past on non-aviation issues which were related to the need to ensure peace. At its various sessions the ICAO Assembly has addressed instances of social injustice such as racial discrimination as well as threats to commercial expediency achieved through civil aviation. The 15th session of the ICAO Assembly adopted Resolution A15-7 (Condemnation of the Policies of Apartheid and Racial Discrimination of South Africa) where the Resolution urged South Africa to comply with the aims and objectives of the Chicago Convention, on the basis that the apartheid policies constitute a permanent source of conflict between the nations and peoples of the world and that the policies of apartheid and racial discrimination are a flagrant violation of the principles enshrined in the Preamble to the Chicago Convention.[256]

The Preamble[257] was also quoted in Resolution A17-1 (Declaration by the Assembly) which requested concerted action on the part of States towards

---

[252] Warner (1942).

[253] Sochor (1991).

[254] Schenkman (1955).

[255] *Id*. Vi.

[256] See *Repertory Guide to the Convention on International Civil Aviation*, Second Edition, 1977, Preamble – 1.This subject was also addressed at a later session of the Assembly when the Assembly, at its 18th Session adopted Resolution A18-4 (Measures to be taken in pursuance of Resolutions 2555 and 2704 of the United Nations General Assembly in Relation to South Africa).

[257] The Preamble to the Chicago Convention states., inter alia, that the future development of international civil aviation can greatly help to create and preserve friendship and understanding

suppressing all acts which jeopardize safety and orderly development of international civil aviation. In Resolution A20-2 (Acts of Unlawful Interference with Civil Aviation) the Assembly reiterated its confidence that the development of international civil aviation can be an effective tool in bringing about friendship and understanding among the peoples of the world.

These discussions would suggest that civil unrest, as it affects aviation, and aviation, as it affects the sustenance of peace are two sides of the same coin.

3.5.2.1 The Beijing Convention of 2010

Aviation is an important global business and a significant driver of the global economy. It is vital, therefore, that stringent measures are taken to counter acts of unlawful interference with civil aviation. Following a diplomatic conference, held in Beijing from 30 August to 10 September 2010 under the auspices of the International Civil Aviation Organization, representatives from more than 80 States adopted two international air law instruments for the suppression of unlawful acts relating to civil aviation. The two instruments are the *Convention on the Suppression of Unlawful Acts Relating to International Civil Aviation* and the *Protocol Supplementary to the Convention for the Suppression of Unlawful Seizure of Aircraft*. This discussion will only be on the Beijing Convention.

The Beijing Convention (See the full text in the **APPENDIX**) serves international civil aviation well, by requiring parties to criminalize a number of new and emerging threats to the safety of civil aviation, including using aircraft as a weapon and organizing, directing and financing acts of terrorism. These new treaties reflect the international community's shared effort to prevent acts of terrorism against civil aviation and to prosecute and punish those who would commit them. The treaties promote cooperation between States while emphasizing the human rights and fair treatment of terrorist suspects. The Convention also obligates States to criminalize the transport of biological, chemical, nuclear weapons and related material.

Many provisions of the Convention, which is a newcomer to aviation security in the context of some new provisions it introduces, may need reflection, particularly in interpreting the intent of its founding fathers.

Under the auspices of ICAO, a diplomatic conference, held in Beijing from 30 August to 10 September 2010, composed of representatives from more than 80 States,[258] adopted two international air law instruments for the suppression of unlawful acts relating to civil aviation.

---

among the nations and peoples of the world, yet its abuse can become a threat to the general security; and It is desirable to avoid friction and to promote that cooperation between nations and peoples upon which the peace of the world depends.

[258] Some 400 participants from more than 80 States and international organizations attended the Conference. The Conference unanimously elected Mr. XIA Xinghua from China as the President, and Mr. Terry Olson from France as the First-Vice President.

3.5 Civil Unrest and Aviation

The two instruments adopted by the Diplomatic Conference are the *Convention on the Suppression of Unlawful Acts Relating to International Civil Aviation* and the *Protocol Supplementary to the Convention for the Suppression of Unlawful Seizure of Aircraft*. This article will only discuss the former instrument.

Since the 1960s, a number of treaties on aviation security have been concluded under the auspices of ICAO.[259] These legal instruments criminalize acts against international civil aviation, such as hijacking and sabotage, and facilitate the cooperation between States to make sure that such acts do not go unpunished. The treaties adopted in Beijing further criminalize the act of using civil aircraft as a weapon, and of using dangerous materials to attack aircraft or other targets on the ground. The unlawful transport of biological, chemical and nuclear weapons and their related material[260] becomes now punishable under the treaties. Moreover, directors and organizers of attacks against aircraft and airports will have no safe haven. Making a threat against civil aviation may also trigger criminal liability. The Convention also implicitly addresses the threat of cyber attacks on aviation, as the discussion to follow will illustrate.

Aviation is an important global business and a significant driver of the global economy. It is vital, therefore, that stringent measures are taken to counter acts of unlawful interference with civil aviation. The *Convention on International Civil Aviation* signed at Chicago on 7 December 1944,[261] states in its *Preamble* that whereas the development of civil aviation may help preserve friendship and understanding among the people of the world, yet, its abuse could become a threat to general security. Therefore, air transport is intrinsically linked to peace and is far

---

[259] See generally Abeyratne (1998) which discusses extensively the treaties. See also, Abeyratne (2010a).

[260] Abeyratne (2007).

[261] The Convention on International Civil Aviation, signed at Chicago on 7 December 1944 (*supra*, note 1), which is the founding document of commercial aviation, in its Preamble, recognizes that the future development of international civil aviation can greatly help to create and preserve friendship and international understanding among the nations and peoples of the world, yet its abuse can become a threat to the general security; and it is desirable to avoid friction and to promote that cooperation between nations and peoples upon which the entire peace of the world depends. See ICAO Doc 7300/9 Ninth Edition: 2006, *Preamble*. The most significant modernist construction of the role of civil aviation in securing world peace and security comes from language used in the letters of invitation issued by the United States to the participant States to the Chicago Conference that, consequent to the war, the restorative processes of prompt communication may greatly facilitate the return to the processes of peace. However, the conscious awareness of the parties to the Convention, that in securing this peace, prudent economic and business principles must not be compromised, should not be forgotten. See Proceedings *of the International Civil Aviation Conference*, Chicago, Illinois, November 1–7 December 1944, US Department of State Volume 1 at 7.

removed from its antithesis – terrorism – which is usually linked with acts of unlawful interference with civil aviation.

The genealogy of the term "*Terrorism*" lies in Latin terminology meaning "to cause to tremble"(*terrere*). Since the catastrophic events of 11 September 2012, we have seen stringent legal measures taken by the United States to attack terrorism, not just curb it. The famous phrase "war on terror" denotes pre-emptive and preventive strikes carried out through applicable provisions of legitimately adopted provisions of legislation. The earliest example is the *Air Transportation Safety and System Stabilization Act* (ATSAA) enacted by President Bush less than 2 months after the 9/11 attacks. Then, 2 months after the attacks, in November 2001, Congress passed the *Aviation and Transportation Security Act* (ATSA) with a view to improving security and closing the security loopholes which existed on that fateful day in September 2001. The legislation paved the way for a huge federal body called the Transportation Security Administration (TSA) which was established within the Department of Transportation. The *Homeland Security Act of 2002* which followed effected a significant reorganization of the Federal Government.

All this goes to show that the law plays a significant role in ensuring aviation security, and the Beijing instruments play a crucial role in furthering this objective. Since the events of 11 September 2001, there have been several attempts against the security of aircraft in flight. These threats have ranged from shoe bombs to dirty bombs to explosives that can be assembled in flight with liquids, aerosols and gels. In every instance the global community has reacted with preemptive and preventive measures which prohibit any material on board which might seemingly endanger the safety of flight. Some jurisdictions have even gone to extremes in prohibiting human breast milk and prescriptive medications on board.

New and emerging threats to civil aviation are a constant cause for concern to the aviation community. Grave threats such as those posed by the carriage of dangerous pathogens on board, the use of cyber technology calculated to interfere with air navigation systems, and the misuse of man portable air defence systems are real and have to be addressed with vigour and regularity. ICAO has been addressing these threats for some time and continues to do so on a global basis.

The Aviation Security Panel of ICAO met at its 20th Meeting in Montreal from 30 March to 3 April 2009. One of the key areas of discussion at this meeting concerned new and emerging threats to civil aviation. The Panel worked through the Working Group on New and Emerging Threats and noted that significant progress in efforts to proactively identify vulnerabilities and potential gaps in existing measures had been made, that would strengthen *Annex 17* (Aviation Security) to the Convention on International Civil Aviation. At this meeting, the European Civil Aviation Conference (ECAC) stressed the importance of the challenge posed by cyber threats in light of the current lack of related provisions in Annex 17.

Consequently, the Panel considered the threat of cyber attacks, and some members stressed that this threat is significant. With reference to a proposal to include a Recommended Practice in Annex 17 to ensure that information and communication technology systems used for civil aviation purposes be protected from cyber attacks, the Panel agreed that, given the complexity of this issue, which involves air traffic management systems, aircraft design and operations, the matter requires further analysis by the Working Group on New and Emerging Threats prior to inclusion in Annex 17 or any guidance material. Depending on the results of the analysis, the Working Group on Amendment 12 to Annex 17 will develop a proposal for amending the Annex, to be presented to the Panel at its 21st meeting.

The Panel also considered the merits of building unpredictability into the aviation security regime. While concern was expressed regarding the impact of unpredictable security measures on passenger confidence in aviation security, many Panel members supported implementation of the concept because of its value as a deterrent. It was suggested that States adopt an approach providing for a baseline regime, but with the addition of unpredictable measures, thus achieving a balance between certainty and unpredictability. With regard to an amendment to Annex 17 in this regard, the need for introducing unpredictability into the aviation security regime was considered, and it was agreed that unpredictability should be promoted in principle but not prescribed. The Panel suggested that if an Annex 17 specification related to unpredictability were to be developed, it would be necessary to ensure that the introduction of this concept by States does not diminish the level of security or result in delays for passengers. Further, the Panel noted that appropriate guidance material may be required to address the potential negative impact of introducing the concept of unpredictability, and proposed the development of guidance material related to unpredictability prior to the introduction of an amendment to Annex 17.

A Conclusion of the Panel was, inter alia, that the threat of cyber attacks is real and cannot be ignored, and that further analysis by the Working Group on New and Emerging Threats would be appropriate. Another Conclusion was that the ICAO focal point of contact (PoC) Network is an important tool for sharing critical threat information and should be used more effectively, and that the Secretariat should consider the establishment of a web-based community page. Yet another was that the concept of building unpredictability into the aviation security regime is in principle a useful tool, however, concerns expressed regarding the possible impact on the level of security and the impact on passenger confidence should be resolved before its inclusion as a Recommended Practice in Annex 17.

The 37th Session of the ICAO Assembly, which was held from 28 September to 8 October 2010 at ICAO Headquarters in Montréal, built on the achievements of the diplomatic Conference in Beijing in September 2010 by recognizing the need to strengthen aviation security worldwide. In a Declaration on Aviation Security, unanimously adopted by participants, international commitment was reaffirmed to enhance aviation security collaboratively and proactively through screening technologies to detect prohibited articles, strengthening of international standards, improvement of security information-sharing and provision of capacity-building assistance to States in need.

The Assembly also put its full support behind the new ICAO Comprehensive Aviation Security Strategy.

It must be underscored that the following discussion is not meant to criticize a fine treaty that will serve as a landmark against new and emerging threats to civil aviation. Rather, the discourse is meant to be creative and attenuate the logicality behind certain key provisions of the Convention. The author's comments in the Conclusion are mere observations which the drafters of the Convention would undoubtedly have answers to.

Convention on the Suppression of Unlawful Acts Relating to International Civil Aviation

The Beijing Convention has a short Preamble, which sets the tone and theme of the treaty. It recognizes *in limine* that the State Parties to the instrument are deeply concerned that unlawful acts against civil aviation jeopardize the safety and security of persons and property; seriously affect the operation of air services, airports and air navigation; and undermine the confidence of the peoples of the world in the safe and orderly conduct of civil aviation for all States. States Parties also recognize that new types of threats against civil aviation require new concerted efforts and policies of cooperation on the part of States. As such they are convinced that in order to better address these threats, there is an urgent need to strengthen the legal framework for international cooperation in preventing and suppressing unlawful acts against civil aviation.

New Types of Threat

The distinctive feature of this treaty, which makes it stand out from its predecessors, is that it bases itself on responding to new and emergent threats to security. As already mentioned, This subject has its genesis in the ICAO Aviation Security Panel which met at its 20th Meeting in Montreal from 30 March to 3 April 2009. One of the key areas of discussion at this meeting concerned new and emerging threats to civil aviation.

Cyber-terrorism has the advantage of anonymity, which enables the hacker to obviate checkpoints or any physical evidence being traceable to him or her. It is a low budget form of terrorism where the only costs entailed in interfering with the computer programs of an air transport system would be those pertaining to the right computer equipment.

Any interference with air transport, which would be inextricably linked to the purpose of international civil aviation as enunciated in the *Preamble* to the Chicago Convention, which states that the future development of international civil aviation can greatly help to create and preserve friendship and understanding among the nations and people of the world, yet, its abuse can become a threat to the general security.

3.5 Civil Unrest and Aviation   253

The leakage of dangerous pathogens[262] from laboratories also presents an ominous analogy to the aviation sector in that the same could well occur in the carriage of such dangerous goods by air. Although past instances of the escape of dangerous pathogens are small in number, nonetheless their occurrence and the threat posed to the wellbeing of humanity cannot be underestimated. In 2002 when Anthrax spores escaped from two military laboratories in the United States, the authorities agreed that the leakage was due to a security lapse.[263] In 2003 a string of such leakages occurred in Asia, this time of the SARS virus.[264]

Offences Under the Convention

The first offence identified by the Convention relates to any person committing an offence if that person unlawfully and intentionally performs an act of violence against a person on board an aircraft in flight if that act is likely to endanger the safety of that aircraft.[265] This offence has three salient elements: the offence has to be committed by a person "on board"[266] an aircraft; the aircraft has to be "in flight"; and the act perpetrated should endanger the safety of the aircraft. According to the Convention, the aircraft is considered to be in flight at any time from the moment when all its external doors are closed following embarkation until the moment when any such door is opened for disembarkation; in the case of a forced landing, the flight shall be deemed to continue until the competent authorities take over the responsibility for the aircraft and for persons and property on board.[267]

---

[262] Pathogens are microorganisms (including bacteria, viruses, rickettsia, parasites, fungi) or recombinant microorganisms (hybrid or mutant) that are known or are reasonably expected to cause infectious disease in humans or animals.

[263] An year earlier, a covert event occurred in October 2001 when anthrax spores were sent through the mail exposing persons in the eastern USA to contaminated mail resulting in deaths, illnesses and identified exposures to Anthrax. Overt, announced events, in which persons are warned that an exposure has occurred, have taken place in the United States, although most of these were determined to have been hoaxes, that is, there were no true exposures to infectious agents.

[264] The leakages occurred in China, Taiwan and Singapore. See Air-Tight Security, *Intersec*, June 2007 33–35 at 34.

[265] *Convention on the Suppression of Unlawful Acts Relating to International Civil Aviation*, done at Beijing on 10 September 2010, Article 1. (a).

[266] The offender has to be physically inside the aircraft. This offence therefore does not *ex facie* apply to an offence committed outside the aircraft. The Convention does not define "on board". However, it must be noted that the term "on board" has been judicially defined in absolute terms to mean that as long as a person is physically in the aircraft, it matters not whether the flight had been terminated or not. See *Herman* v. *Trans World Airlines*, 330 N.Y.S.D. 2nd 829 (Sup.Ct. 1972) where the Court held that although the aircraft in which the passenger was travelling had been hijacked and flown to the desert, and the passenger was kept in the aircraft for several days, he was nonetheless considered to have been on board, irrespective of whether the purpose of the flight had been fulfilled or not. See also *Pfug* v. *Egyptair*, 961F. 2d. 26 (2nd Cir.1992).

[267] *Supra*, note 269, Article 2 (a).

The next consideration within this specific offence is that the act perpetrated should endanger the safety of the aircraft. This seemingly excludes acts of air rage which in many instances only affect the safety of the person against whom the offence is committed. By restricting the offence to safety of the aircraft in flight, the Convention has ensured that every offence under this provision must essentially endanger the safety of the aircraft in which the offence is committed. Undoubtedly the ultimate arbiters of the Convention, which were the ICAO member States, had a reason for adopting this approach and it would be interesting to learn of the rationale of this approach, particularly in the context of a recommendation offered by the ICAO Aviation Security Panel at its 21st Meeting held from 22 to 26 March 2010, where the Panel suggested that the *Secretariat Study Group on Unruly Passengers* be reconvened in order to study the issue of unruly passengers and consider whether the existing international legal regime should be re-examined.

The second offence under the Convention is committed when a person destroys an aircraft in service or causes damage to such an aircraft which renders it incapable of flight or which is likely to endanger its safety in flight.[268] An aircraft is considered to be in service from the beginning of the pre-flight preparation of the aircraft by ground personnel or by the crew for a specific flight until 24 h after any landing; furthermore, the period of service will, in any event, extend for the entire period during which the aircraft is in flight as defined in the Convention.[269] This provision does not seem to cover an act which causes damage to an aircraft but which does not affect the safety of the flight. Therefore, a wilful or wanton act committed by a member of a technical team (for example a maintenance engineer) at pre flight stage, if it damages the aircraft but does not affect the safety of a flight would not, under this provision, be considered an unlawful act relating to international aviation.

The third offence identified by the Convention relates to a person who places or causes to be placed on an aircraft in service, by any means whatsoever, a device or substance which is likely to destroy that aircraft, or to cause damage to it which renders it incapable of flight, or to cause damage to it which is likely to endanger its safety in flight.[270] Here again, the offence must relate to the destruction of the aircraft or damage which renders the aircraft unserviceable, or adversely affects the safety of the aircraft. It is interesting that the Convention does not define the words

---

[268] Id. Article 1 (b).
[269] *Id*, Article 2 (b).
[270] *Id*, Article 1 (c).

3.5 Civil Unrest and Aviation

"device"[271] or "substance".[272] It is even more interesting that the Convention did not include the word "weapon" as it has done in a following provision.[273]

The fourth offence is a first for any treaty on unlawful interference with civil aviation. It provides that an offence is committed when a person destroys or damages air navigation facilities or interferes with their operation, if any such act is likely to endanger the safety of aircraft in flight.[274] This undoubtedly refers, inter alia to cyber terrorism, but links the offence exclusively to the safety of aircraft in flight. If therefore as a result of an act of cyber terrorism, a taxiing aircraft collides with an aircraft which has opened its doors for disembarkation but the passengers are still on board awaiting disembarkation, that act would not be considered an offence in terms of the passengers in the process of disembarkation. In other words, the offender would not be committing an offence under the Treaty either against the second aircraft or its disembarking passengers.

The Beijing Treaty of 2010 is a step forward in the right direction with the threat of cyber terrorism looming, affecting the peace of nations. Air transport could well be a target towards the erosion of that peace. The maintenance of international peace and security is an important objective of the United Nations,[275] which recognizes one of its purposes as being inter alia:

> To maintain international peace and security, and to that end: take effective collective measures for the prevention and removal of threats to the peace, and for the suppression of acts of aggression or other breaches of peace, and to bring about by peaceful means, and in conformity with the principles of justice and international law, adjustment or settlement of international disputes or situations which might lead to a breach of the peace.[276]

It is clear that the United Nations has recognized the application of the principles of international law as an integral part of maintaining international peace and security and avoiding situations which may lead to a breach of the peace.

---

[271] The free Online Dictionary defines "device" as inter alia a contrivance or an invention serving a particular purpose, especially a machine used to perform one or more relatively simple tasks, or a technique or means, or a plan or scheme, especially a malign one. See http://www.thefreedictionary.com/device.

[272] The free Online Dictionary defines "substance" as that which has mass and occupies space; matter or a material of a particular kind or constitution. See http://www.thefreedictionary.com/substance.

[273] Article 1.2 of the Convention provides that A person commits an offence if that person unlawfully and intentionally, using any device, substance or weapon: (a) performs an act of violence against a person at an airport serving international civil aviation which causes or is likely to cause serious injury or death; or (b) destroys or seriously damages the facilities of an airport serving international civil aviation or aircraft not in service located thereon or disrupts the services of the airport, if such an act endangers or is likely to endanger safety at that airport.

[274] Beijing Convention, *supra*, Article 1 (d).

[275] Charter of the United Nations and Statute of the International Court of Justice, Department of Public Information, United Nations, New York, DPI/511 – 40108 (3–90), 100 M at 1.

[276] *Id.* at 3.

Cyber terrorism would not only affect the security of air transport. One commentator says:

> Cyber-terrorism can be used in many ways. In its simplest form, it can be used as a means of disinformation or psychological warfare by manipulating media attention regarding possible threats, thus causing disruption to airport and aircraft operations. This could result in the reluctance of persons to travel which, in turn, could affect the economies of nations dependent on the movement of air passengers. In its most serious form, cyber-terrorism could lead to fatalities, injuries and major damage at airports and to aircraft in flight.[277]

The particularity of cyber-terrorism is that the threat is enhanced by globalization and the ubiquity of the Internet. It is a global problem in search of a global solution.

The fifth offence identified in the Convention covers an instance where a person communicates information which that person knows to be false, thereby endangering the safety of an aircraft in flight.[278] This seemingly rules out a message communicated negligently, where the purveyor of the message did not bother to find out the veracity of the information he was providing. Furthermore, this provision raises an important issue. One could argue that the exclusivity of "safety in flight" may unduly restrict the scope of this provision. For instance, if a phony telephone call claims that there would be a bomb on board a flight that would be operated the next day, and the air operator cancels that flight incurring an economic loss, there would be no offence as the aircraft in question was not "in flight" as defined in the Convention. This consideration may be particularly relevant in the context of the title of the Treaty which is *"Convention on the Suppression of Unlawful Acts Relating to International Civil Aviation"* which obviously does not restrict itself to safety or security issues. Another consideration is that if such a communication were to come just as the doors of an arriving aircraft are opened for disembarkation and passengers are injured or killed in a stampeded, this provision would not apply.

The sixth offence is a throwback on 9/11 and provides that any person who uses an aircraft in service for the purpose of causing death, serious bodily injury, or serious damage to property or the environment commits an offence.[279] The interesting feature of this provision is that it has included environmental damage that could be caused by such an act. In the maritime context, there are analogous provisions[280] where important safeguards are prescribed when a State Party takes measures against a ship, including boarding. The safeguards include: not

---

[277] Guill (2000).

[278] Beijing Convention, *supra*, Article 1 (e).

[279] *Id*, Article 1. (f).

[280] Convention for the Suppression of Unlawful Acts Against the Safety of Maritime Navigation, Protocol for the Suppression of Unlawful Acts Against the Safety of Fixed Platforms Located on the Continental Shelf adopted 10 March 1988; Entry into force 1 March 1992; 2005 Protocols: Adopted 14 October 2005; Entry into force 28 July 2010.

3.5 Civil Unrest and Aviation

endangering the safety of life at sea; ensuring that all persons on board are treated in a manner which preserves human dignity and in keeping with human rights law; taking due account of safety and security of the ship and its cargo; ensuring that measures taken are environmentally sound; and taking reasonable efforts to avoid a ship being unduly detained or delayed.[281] *The European Convention on the Protection of the Environment through Criminal Law*[282] calls upon each State Party to adopt such appropriate measures as may be necessary to establish as criminal offences or administrative offences, liable to sanctions or other measures under its domestic law, when committed intentionally or with negligence the unlawful discharge, emission or introduction of a quantity of substances or ionising radiation into air, soil or water.[283]

The next provision[284] makes it an offence to release or discharge from an aircraft in service any BCN[285] weapon or explosive, radioactive, or similar substances in a manner that causes or is likely to cause death, serious bodily injury or serious damage to property or the environment. This provision provides, inter alia a response to bio terrorism, which is a new and emerging threat to civil aviation.[286] A bioterrorism attack is the deliberate release of viruses, bacteria, or other germs (agents) used to cause illness or death in people, animals, or plants. These agents are typically found in nature, but it is possible that they could be changed to increase their ability to cause disease, make them resistant to current medicines, or to increase their ability to be

---

[281] *Id.* Article 8 bis.

[282] Strasbourg, XI.1998.

[283] *Id.* Article 4.

[284] Beijing Convention, *supra*, Article 1 (g).

[285] According to Article 2 (h) BCN weapons are (a) biological weapons, which are: (i) microbial or other biological agents, or toxins whatever their origin or method of production, of types and in quantities that have no justification for prophylactic, protective or other peaceful purposes; or (ii) weapons, equipment or means of delivery designed to use such agents or toxins for hostile purposes or in armed conflict. (b) "chemical weapons", which are, together or separately: toxic chemicals and their precursors, except where intended for: (A) industrial, agricultural, research, medical, pharmaceutical or other peaceful purposes; or (B) protective purposes, namely those purposes directly related to protection against toxic chemicals and to protection against chemical weapons; or (C) military purposes not connected with the use of chemical weapons and not dependent on the use of the toxic properties of chemicals as a method of warfare; or (D) law enforcement including domestic riot control purposes, as long as the types and quantities are consistent with such purposes; munitions and devices specifically designed to cause death or other harm through the toxic properties of those toxic chemicals which would be released as a result of the employment of such munitions and devices; any equipment specifically designed for use directly in connection with the employment of munitions and devices and nuclear weapons and other nuclear explosive devices.

[286] A Special Sub Committee of the Legal Committee of ICAO met in Montreal from 3 to 6 July 2007 to discuss the preparation of one or more instruments addressing new and emerging threats. One of the issues addressed at this meeting was the unlawful transport of biological, chemical, nuclear weapons and other dangerous substances on board aircraft.

spread into the environment. Biological agents can be spread through the air, through water, or in food. Terrorists may use biological agents because they can be extremely difficult to detect and do not cause illness for several hours to several days. While some bioterrorism agents, such as the smallpox virus, can be spread from person to person some agents such as anthrax are incapable of doing so.

There have been several noteworthy instances of bioterrorism in the past[287] as early as 1915,[288] which send an ominous message that it is a distinct possibility in the aviation context. Until recently in the United States of America, most biological defence strategies have been geared to protecting soldiers on the battlefield rather than looking after ordinary people in cities. In 1999, the University of Pittsburgh's Center for Biomedical Informatics deployed the first automated bioterrorism detection system, called RODS (Real-Time Outbreak Disease Surveillance). RODS is designed to draw collect data from many data sources and use them to perform signal detection, that is, to detect a possible bioterrorism event at the earliest possible moment. RODS, and other similar systems, collect data from sources including clinical data, laboratory data, and data from over-the-counter drug sales. In 2000, Michael Wagner, the co director of the RODS laboratory, and Ron Aryel, a subcontractor, conceived of the idea of obtaining live data feeds from "non-traditional" (non-health-care) data sources. The RODS laboratory's first efforts eventually led to the establishment of the National Retail Data Monitor, a system which collects data from 20,000 retail locations nation-wide.

Another noteworthy provision[289] to follow states that where it is an offence to perform an act of violence against a person at an airport serving international civil aviation which causes or is likely to cause serious injury or death; or to destroy or seriously damage the facilities of an airport serving international civil aviation or aircraft not in service located thereon or disrupts the services of the airport, if such an act endangers or is likely to endanger safety at that airport. It is quite curious that an attack against an airport, to be classified under this Convention has to endanger safety *at* the airport and does not include safety *of* the airport. Therefore the

---

[287] In 1984 followers of the Bhagwan Shree Rajneesh attempted to control a local election by incapacitating the local population by infecting salad bars in eleven restaurants, doorknobs, produce in grocery stores and other public domains with Salmonellas typhimurium in the city of The Dalles, Oregon. The attack caused about 751 people to get sick (there were no fatalities). This incident was the first known bioterrorist attack in the United States in the twentieth century. In September and October of 2001, several cases of anthrax broke out in the United States which were reportedly caused deliberately. This was a well-publicized act of bioterrorism. It motivated efforts to define biodefense and biosecurity.

[288] In 1915 and 1916, Dr Anton Dilger, a German-American physician used cultures of anthrax and glanders with the intention of committing biological sabotage on behalf of the German government. Other German agents are known to have undertaken similar sabotage efforts during World War I in Norway, Spain, Romania and Argentina.

[289] Beijing Convention, *supra*, Article 1.2.

3.5 Civil Unrest and Aviation

provision seems to imply that any wanton damage to an airport or its infrastructure; insofar as it does not affect the safety of persons would not be an offence.

Other provisions follow, which address threats to commit the offences discussed above[290] and attempts to commit such offences[291] and make them offences under the Convention. An interesting provision is contained in Article 3 of the Convention which states that each State Party undertakes to make the offences discussed above punishable by severe penalties. Here, the key word is "undertakes". It is worthy of note that the drafters have not used the word "shall" which would have made the requirement peremptory. In regular parlance "undertake" would mean to "accept as a challenge or promise to do or accomplish or enter upon an activity or enterprise"[292] Another definition of "undertake" is "to agree to be responsible for a job or project and do it".[293] Therefore, logically, one could argue that Article 3 makes States Parties promise that they would make offences under the Convention punishable. On the other hand, the word "shall" would have made the requirement obligatory.

This logicality notwithstanding, one should consider this conundrum in its legal perspective. Article 31.1 of the *Vienna Convention on the Law of Treaties*[294] provides that "a treaty shall be interpreted in good faith in accordance with the ordinary meaning to be given to the terms of the treaty in their context and in the light of its object and purpose". One has therefore to inquire as to what the ordinary meaning is that was given to the word "undertake" by the drafters of the Beijing Convention.

The Convention does not apply to military, customs or police services. In the absence of a definition of these services, one could seek guidance from the Chicago Convention which denies its application to State aircraft and goes onto to say that aircraft used in military, customs and police services shall be deemed to be State aircraft.[295]

In the face of the use of the word "undertakes" in Article 3 as discussed, one notices that Article 8.1 on the issue of jurisdiction provides that each State Party *shall* (author's emphasis) take such measures as may be necessary to establish its jurisdiction over the offences discussed above: (a) when the offence is committed in the territory of that State; (b) when the offence is committed against or on board an aircraft registered in that State; (c) when the aircraft on board which the offence is

---

[290] *Id*, Article 1. 3.

[291] *Id*. Article 1.4.

[292] http://www.audioenglish.net/dictionary/undertake.htm.

[293] http://www.macmillandictionary.com/dictionary/american/undertake.

[294] *Vienna Convention on the Law of Treaties 1969*, done at Vienna on 23 May 1969. The Convention entered into force on 27 January 1980. United Nations, *Treaty Series*, vol. 1155, p. 331.

[295] Chicago Convention, *Supra*, note 3, Article 3. a) and b). For a clear and compelling discussion on the interpretation of the terms state aircraft and military, customs and police aircraft see Michael Milde, *International Air Law and ICAO*, Eleven Publishing: Utrecht, 2008 at 69–71.

committed lands in its territory with the alleged offender still on board; (d) when the offence is committed against or on board an aircraft leased without crew to a lessee whose principal place of business or, if the lessee has no such place of business, whose permanent residence is in that State; (e) when the offence is committed by a national of that State. Each State Party may also establish its jurisdiction over any such offence when the offence is committed against a national of that State; or when the offence is committed by a stateless person whose habitual residence is in the territory of that State.[296]

With regard to extradition of offenders, the Convention obligates (again with the word "shall") the State Party in the territory of which the alleged offender is found if it does not extradite that person, without exception whatsoever and whether or not the offence was committed in its territory, to submit the case to its competent authorities for the purpose of prosecution. Those authorities are required to take their decision in the same manner as in the case of any ordinary offence of a serious nature under the law of that State.[297]

It is incontrovertible that, given the various innovative terrorist acts perpetrated against civil aviation, the Beijing Treaty of 2010 is a proactive and timely initiative of both ICAO and the international civil aviation community. In this regard it must be noted that this treaty was adopted, as are other treaties, by State Parties to the Beijing Conference and ICAO was the initiator and facilitator of the Conference. Therefore, one could assume that whatever the treaty provides is in accord with and responds to the needs of the member States of ICAO.

Firstly, one is struck by the title of the Convention, which is *Convention on the Suppression of Unlawful Acts Relating to International Civil Aviation*. Therefore, one is to take it that the purpose of the treaty is to suppress unlawful acts relating to civil aviation. Yet, it is incontrovertible that this is an instrument exclusively addressing aviation security, although the word "security" is not used in the document. It is mainly concerned with the safety of aircraft in service or aircraft in flight, with one provision on safety pertaining to airports. Excluded from its purview are such unlawful acts as negligent entrustment which is now an unlawful and criminal act in the United Kingdom and Scotland which many common law countries may follow.[298] Also excluded from the purview of the Convention is any unlawful act calculated to cancel flights causing economic loss to air carriers. The Convention, in all practicality therefore remains one that is adopted to suppress unlawful acts relating to the safety of international civil aviation.

Secondly, one would observe that the Convention does not cover all instances of air rage although all such acts are unlawful acts relating to international civil aviation. Furthermore, the almost exclusive insistence on safety of "aircraft in flight" and the narrow definition of "in flight" as contained in the Convention

---

[296] Beijing Convention, *supra*, Article 8.2.
[297] *Id*. Article 10.
[298] Abeyratne (2010b).

may not cover every instance of air transport operated by a carrier when passengers are still on board an aircraft.

The above notwithstanding, the Beijing Convention serves international civil aviation well, by requiring parties to criminalize a number of new and emerging threats to the safety of civil aviation, including using aircraft as a weapon and organizing, directing and financing acts of terrorism. This new treaty reflects the international community's shared effort to prevent acts of terrorism against civil aviation and to prosecute and punish those who would commit them. The treaty promotes cooperation between States while emphasizing the human rights and fair treatment of terrorist suspects.

The Convention also obligates States to criminalize the transport of biological, chemical, nuclear weapons and related material. The provisions in the treaty reflect the nexus between non-proliferation and terrorism and ensure that the international community will act to combat both. This treaty also strengthens global efforts to ensure that these extraordinarily dangerous materials will not be transported via civil aircraft for illicit purposes and, if such attempts are made, those responsible will be held accountable under the law.

Under the circumstances, this landmark treaty leaves no room for doubt that it is a valuable contribution towards the enhancement of collaboration between States to curb unlawful acts against international civil aviation. The abovementioned features of the Beijing Treaty undoubtedly makes it a timely and proactive initiative of ICAO and the international aviation community.

Finally, it must be mentioned that treaties of this nature, although essential and innovative, merely offer an ex post facto response to aviation security. They should be accommodated by proactive measures such as one suggested by the ICAO AVSEC Panel at its 21st Meeting in March 2010: that there should be more innovative identification of passengers. Undoubtedly, this recommendation could be extended to employees and others coming into contact with aircraft and airports.

3.5.2.2 Extraordinary Rendition

Extraordinary rendition is the handing over of a person from one jurisdiction to another, without initial determination as to the possibility of that person's guilt.[299] A stronger definition is:

> The alleged acts of the U.S. authorities in the wake of "911" when suspected high-value terrorists were – without trial-abducted/arrested ("snatched" in the media jargon) in one territory and transported by aircraft to another territory for interrogation by US agents or

---

[299] In contrast, a rendition flight is a flight which takes a felon to a place which has jurisdiction to adjudicate the crime in question. rendition is therefore a legal measure and has its genesis in the need for the US to recapture fugitive slaves. See Article 4, Section 2, Clause 2 of the United States Constitution.

delivered to security forces in other countries where they would not enjoy the protection against torture or other abuses.[300]

It is reported that the application and significance of this practice had a change of focus in the 1980s when foreign delinquents were transferred by United States authorities to be interrogated in countries with which the United States did not have extradition treaties, which in turn carried the connotation that such delinquents would be treated differently than they would have been in countries of the west.[301] From a conceptual standpoint, extraordinary rendition is diametrically at variance with principles of international law, which has certain safeguards against the transportation of a person against his will, unless a proper judicial determination has been made in favour of such a transfer. Protocol No. 7 to the *Convention for the Protection of Human Rights and Fundamental Freedoms*,[302] provides that an alien lawfully resident in the territory of a State shall not be expelled therefrom except in pursuance of a decision reached in accordance with law and shall be allowed: (a) to submit reasons against his expulsion, (b) to have his case reviewed, and (c) to be represented before the competent authority or a person or persons designated by that authority.[303] An alien may be expelled before the exercise of his rights, when such expulsion is necessary in the interests of public order or is grounded on reasons of national security.[304] Furthermore, Article 13 of the *International Covenant on Civil and Political Rights* (ICCPR)[305] provides that an alien lawfully in the territory of a State Party may be expelled therefrom only in pursuance of a decision reached in accordance with law and shall, except where compelling reasons of national security otherwise require, be allowed to submit the reasons against his expulsion and to have his case reviewed by, and be represented before the competent authority or a person or persons especially designated by the competent authority.[306]

---

[300] Milde (2008). Consistent with this definition, Wikipedia states that some journalists have called extraordinary rendition "*torture by proxy*". See http://en.wikipedia.org/wiki/Extraordinary_rendition.

[301] Ingrid Detter Francopan, Extraordinary Rendition and the Law of War, North Carolina Journal of International Law and Commercial Regulation, Summer 2008 (33 N.C.J. Int'lL. &Com. Reg). 657 at 659.

[302] *Convention for the Protection of Human Rights and Fundamental Freedoms as amended by Protocol No. 1Rome, 4.XI.1950. The text of the Convention had been amended according to the provisions of Protocol No. 3 (ETS No. 45), which entered into force on 21 September 1970, of Protocol No. 5 (ETS No. 55), which entered into force on 20 December 1971 and of Protocol No. 8 (ETS No. 118), which entered into force on 1 January 1990, and comprised also the text of Protocol No. 2 (ETS No. 44) which, in accordance with Article 5, paragraph 3 thereof, had been an integral part of the Convention since its entry into force on 21 September 1970.*

[303] Art. 1, Nov. 22, 1984, Eur. T.S. No. 117. Article 1, titled "*Procedural safeguards relating to expulsion of aliens.*

[304] *Ibid.*

[305] International Covenant on Civil and Political Rights, adopted and opened for signature, ratification and accession by General Assembly resolution 2200A (XXI) of 16 December 1966, (*entered into force* on 23 March 1976, in accordance with Article 49).

[306] International Covenant on Civil and Political Rights Art. 13, opened for signature Dec. 16, 1966, 999 U.N.T.S. 171.

3.5 Civil Unrest and Aviation

The *Third Geneva Convention on Prisoners of War of 1949*[307] stipulates that, in the case of armed conflict not of an international character occurring in the territory of one of the High Contracting Parties, each Party to the conflict shall be bound to apply, as a minimum, that persons taking no active part in the hostilities, including members of armed forces who have laid down their arms and those placed hors de combat by sickness, wounds, detention, or any other cause, shall in all circumstances be treated humanely, without any adverse distinction founded on race, colour, religion or faith, sex, birth or wealth, or any other similar criteria. To this end the Convention prohibits: violence to life and person, in particular murder of all kinds, mutilation, cruel treatment and torture; taking of hostages; outrages upon personal dignity, in particular, humiliating and degrading treatment; the passing of sentences and the carrying out of executions without previous judgment pronounced by a regularly constituted court affording all the judicial guarantees which are recognized as indispensable by civilized peoples. The Convention also provides that the wounded and sick shall be collected and cared for. It also provides that an impartial humanitarian body, such as the International Committee of the Red Cross, may offer its services to the Parties to the conflict. In the case of *Hamdan* v. *Rumsfeld*[308] the decision of the Supreme Court suggests that all activities carried out in the "war on terror" of the United States must pass the minimum standards set out in the Convention.[309]

The above notwithstanding, there are at least three controversial cases where the rendition of the victims by air have been questionable.[310] This brings to bear two areas of concern: false arrest and imprisonment; and the significance of the role of air transport and its meaning and purpose in the backdrop of circumstances of war.

False Arrest and Imprisonment

False arrest is unlawful or unjustifiable arrest and is committed where a person unlawfully, intentionally or recklessly restrains another's freedom of movement from a particular place.[311] Physical detention is an essential ingredient for grounding an action in false imprisonment. Thus if a person agrees to go to a police station voluntarily, he has not been arrested even though the person taking him would have arrested him on refusal to go.[312]

The *Fourth Amendment* to the U.S. Constitution stipulates that unless there is probable cause, a person could not be subject to a search warrant or arrest. A good

---

[307] Convention (III) relative to the Treatment of Prisoners of War. Geneva, 12 August 1949.
[308] 126S.Ct.2749 (2006).
[309] *Id.* 2794–97.
[310] Sattertwaite (2007).
[311] Smith and Hogan (1992). See also *Rahman* v. *Queen* (1985) 81 Cr App Rep 349 at 353.
[312] *Campbell* v. *Tormey* (1969) 1 All E.R. 961, cited in Smith & Hogan, *Criminal Law, op.cit.* at 432.

example of probable cause is the 1967 case of *Terry* v. *Ohio*,[313] which arose from an arrest stemming from a policeman becoming suspicious of two men when one of them walked up the street, peered into a store, walked on, started back, looked into the same store, and then conferred with his companion. The other suspect repeated this ritual, and between them the two men went through this performance about a dozen times before following a third man up the street. The officer, thinking they were preparing to commit a misdemeanour and might therefore be armed, confronted the men, asked their names and patted them down, thereby discovering pistols the plaintiff and his companion. In affirming Terry's conviction for carrying a concealed weapon, the Supreme Court concluded that where a police officer observes unusual conduct which leads him reasonably to conclude in light of his experience that criminal activity may ensue and that the person with whom he is dealing may be armed and presently dangerous, where in the course of investigating this behavior he identifies himself as a policeman and makes reasonable inquiries, he is entitled for the protection of himself and others in the area to conduct a carefully limited search of the outer clothing of such persons in an attempt to discover weapons which might be used to assault him. Physical detention is an essential ingredient for grounding an action in false imprisonment. Thus if a person agrees to go to a police station voluntarily, he has not been arrested even though the person taking him would have arrested him on refusal to go.[314]

The law provides that where an arresting officer has reasonable grounds for suspecting that an arrestable offence has been committed, he may arrest without a warrant anyone whom he has reasonable grounds for suspecting to be guilty of that offence.[315] The offence of false imprisonment is one of "basic intent" and despite the paucity of authority as to whether the element of *mens rea* is necessary to constitute the offence of false imprisonment,[316] at least one decision[317] has recognized the requirement.

In the early case of *Christy* v. *Leachinsky*[318] Lord Simonds, while observing that it was the right of every citizen to be free from arrest and that he should be entitled to resist arrest unless that arrest is lawful, concluded that a person cannot be arrested unless he knows why he is being arrested.[319] This principle has however, since been replaced by Section 28 of the *Police and Criminal Evidence Act* 1984 which provides that where a person is arrested, otherwise than being informed that he is under arrest, he must be so informed as soon as practicable afterwards. While this

---

[313] 392 U.S. 1 (1968), argued 12 Dec. 1967, decided 10 June 1968

[314] *Campbell* v. *Tormey* (1969) 1 All E.R. 961, cited in Smith & Hogan, *Criminal Law, op.cit.* at 432.

[315] *Police and Criminal Evidence Act* 1984 Section 24 (6).

[316] False imprisonment is generally considered under civil actions where the element of mens rea is not relevant.

[317] *Re Hutchins* (1988) *Crim L R* 379.

[318] (1947) 1 *All E.R.* 567.

[319] *Id.* at 575.

provision holds incontrovertible the fact that a person who is arrested has to be informed of the grounds for his arrest, it dispenses with the exclusive need to inform the person at the time of arrest.[320]

In the more recent case of *Murray* v. *Ministry of Defence*[321] the plaintiff sued the Crown for false imprisonment on the ground that she had been detained and questioned by members of the armed forces for 30 min before they indicated to her that she was under arrest. She claimed that her arrest took place only when she was informed that she was under arrest and that the preceding detention was therefore unlawful. The House of Lords at appeal held that where a person was detained or restrained by a police officer and knew that he was being detained or restrained, such detention amounted to an arrest even though no formal words of arrest were spoken by the officer. Since the plaintiff had been under restraint from the moment she was identified, and must have realised that she was under restraint, she was deemed to have been under arrest from that moment, notwithstanding that the arrest took place formally, a half hour later.

Lord Griffiths, quoting an earlier decision,[322] endorsed the principle that arrest did not depend merely on the legality of the act but on the fact whether the person arrested had been deprived of his liberty. His Lordship went on address the decision in *Christy* v. *Leachinsky*[323] and noted:

> There can be no doubt that in ordinary circumstances, police should tell a person the reason for his arrest at the time they make the arrest. If a person's liberty is being restrained he is entitled to know the reason. If the police fail to inform him, the arrest will be unlawful with the consequence that if police are assaulted as the suspect resists arrest, he commits no offence. Therefore, if he is taken to custody, he will have action for wrongful imprisonment.

However, *Christy* v. *Leachinsky* made it clear that there are exceptions to this rule.[324]

The exceptions that Lord Griffiths referred to were those expressed by Viscount Simon where, when circumstances were such, that the person detained knew the general nature of the alleged offence, the requirement for informing him of the fact and grounds for his arrest did not arise. Viscount Simon held that technical or precise language need not be used and since any person is entitled to his freedom, if restraint was used and he knew the reason for such restraint, that was enough.[325]

There is however, no need anymore to rely on this aspect of the *Christy* decision since, as discussed earlier in this paper, statute has now explicitly laid down the law, leaving no room for ambivalence on the subject.

---

[320] Smith & Hogan, *Criminal Law*, op. cit. at 438.
[321] (1988) 2 All E.R. 521.
[322] *Spicer* v. *Holt* (1976) 3 All E. R. 71 at 79.
[323] *Supra*, note, 315.
[324] *Id*. 526.
[325] *Christy* v. *Leachinsky, supra*, note 315, at 572–573.

The Use of Air Transport for Peaceful Purposes

The attacks of 11 September 2001 inevitably highlighted the strategic position of civil aviation both as an industry vulnerable to attack and as an integral tool in ensuring peace and security in the world. The modernist view of civil aviation, as it prevailed when the Chicago Convention was signed on 7 December 1944, was centred on State sovereignty[326] and the widely accepted post-war view that the development of international civil aviation can greatly help to create and preserve friendship and understanding among the nations and peoples of the world, yet its abuse can become a threat to general security.[327] This essentially modernist philosophy focussed on the importance of the State as the ultimate sovereign authority which can overrule considerations of international community welfare if they clashed with the domestic interests of the State. It gave way, in the 1960s and 1970s to a post-modernist era of recognition of the individual as a global citizen whose interests at public international law were considered paramount over considerations of individual State interests.

The 11 September 2001 events led to a new era that now calls for a neo-post modernist approach which admits of social elements and corporate interests being involved with States in an overall effort at securing world peace and security. The role of civil aviation in this process is critical, since it is an integral element of commercial and social interactivity and a tool that could be used by the world community to forge closer interactivity between the people of the world.

The Chicago Convention was signed at the height of the modernist era of social justice and commercial interaction. As *Milde* says: "It is in the first place a comprehensive codification/unification of public international law, and, in the second, a constitutional instrument of an international inter-governmental organization of universal character".[328] Be that as it may, the real significance of the Convention, particularly as a tool for ensuring political will of individual States, lies in the fundamental philosophy contained in its Preamble. In its Preamble, the Convention enunciates a message of peace through aviation. It makes mention of the future development of international civil aviation being able to help preserve friendship and understanding among the nations of the world, while its abuse (i.e. abuse of future development of international civil aviation) can become a threat to "the general security". By "general security" the Chicago Conference meant the prevention of threats to peace. These words have been interpreted in the widest possible sense by the Assembly of the ICAO[329] at its various sessions to cover instances of social injustice such as racial discrimination as well as threats to

---

[326] Article 1 of the Chicago Convention provides that the Contracting States recognize that every State has complete and exclusive sovereignty over airspace above its territory.

[327] Preamble to the Chicago Convention, *supra* note 3.

[328] Milde (1994).

[329] ICAO is the specialized agency of the United Nations dealing with international civil aviation. t has 190 member States, all of whom signed or ratified the Chicago Convention.

commercial expediency achieved through civil aviation. The 15th session of the ICAO Assembly adopted Resolution A15-7 (Condemnation of the Policies of Apartheid and Racial Discrimination of South Africa) where the Resolution urged South Africa to comply with the aims and objectives of the Chicago Convention, on the basis that the apartheid policies constitute a permanent source of conflict between the nations and peoples of the world and that the policies of apartheid and racial discrimination are a flagrant violation of the principles enshrined in the Preamble to the Chicago Convention.[330]

The Preamble was also quoted in Resolution A17-1 (Declaration by the Assembly) which requested concerted action on the part of States towards suppressing all acts which jeopardize safety and orderly development of international civil aviation. In Resolution A20-2 (Acts of Unlawful Interference with Civil Aviation) the Assembly reiterated its confidence that the development of international civil aviation can be an effective tool in bringing about friendship and understanding among the peoples of the world.

The general discussions which took place during the Chicago Conference gives one an overall view of the perspectives of each State, particularly in terms of what they expected out of the Convention with regard to the role to be played by civil aviation in ensuring peace, security and economic development in the world in the years to come. However, in some cases of extraordinary rendition, unfortunately, the converse has happened and in the instances already mentioned the attendant circumstances have even caused strife among nations. Arguably, the most dangerous risk of arrest of an innocent airline passenger suspected of terrorist activity would arise from profiling. It is an incontrovertible fact that profiling is a useful tool in the pursuit of the science of criminology. Profiling is also a key instrument in a sociological context and therefore remains a sustained social science constructed through a contrived process of accumulation of single assumptions and propositions that flow to an eventual empirical conclusion. However, profiling raises well reasoned latent fears when based on a racial platform. Jonathan Turley, Professor of Constitutional Law at George Washington University, in his testimony before a United States House of Representatives Committee on Airport Security regarding the use of racial profiling to identify potentially dangerous travellers observed:

> [R]acial profiling is to the science of profiling as forced confessions are to the art of interrogation. Like forced confessions, racial profiling achieves only the appearance of effective police work. Racial profiling uses the concept of profiling to shield or obscure a racist and unscientific bias against a particular class or group. It is the antithesis of profiling in that it elevates stereotypes over statistics in law enforcement.[331]

---

[330] See *Repertory Guide to the Convention on International Civil Aviation*, Second Edition, 1977, Preamble – 1.This subject was also addressed at a later session of the Assembly when the Assembly, at its 18th Session adopted Resolution A18-4 (Measures to be taken in pursuance of Resolutions 2555 and 2704 of the United Nations General Assembly in Relation to South Africa).
[331] Turley (2002).

Notwithstanding this telling analogy, and the apprehensions one might have against racial profiling, it would be imprudent to conclude that racial profiling is per se undesirable and unduly discriminatory, particularly in relation to profiling at airports which should essentially include some considerations of ethnic and national criteria. This article will examine the necessary elements that would go into effective and expedient airport profiling of potential undesirable passengers. It will also discuss legal issues concerned with the rights of the individual with regard to customs and immigration procedures. The rights of such persons are increasingly relevant from the perspective of ensuring air transport security and refusing carriage to embarking passengers who might show profiles of criminality and unruly behaviour on board.

A legitimate profiling process should be based on statistically established indicators of criminality which are identified through a contrived aggregation of reliable factors. The application of this criterion to airport profiling would immediately bring to bear the need to apply nationality and ethnic factors to passenger profiles. Although one might validly argue that racial profiling would entail considerable social and political costs for any nation, while at the same time establishing and entrenching criminal stereotypes in a society, such an argument would be destitute of effect when applied to airport security which integrally involves trans boundary travel of persons of disparate ethnic and national origins. This by no means implies that racial profiling is a desirable practice. On the contrary, it is a demeaning experience to the person subjected to the process and a de facto travel restriction and barrier. It is also a drain on law enforcement resources that effectively preclude the use of proven and conventional uses of enforcement.

The sensitive conflict of interests between racial profiling per se, which at best is undesirable in a socio-political context, and airport profiling, raises interesting legal and practical distinctions between the two. Among these the most important distinction is that airport profiling is very serious business that may concern lives of hundreds if not thousands in any given instance or event. Profiling should therefore be considered justifiable if all its aspects are used in screening passengers at airports. Nationality and ethnicity are valid baseline indicators of suspect travellers together with other indicators which may raise a 'flag' such as the type of ticket a passenger holds (one way instead of return) and a passenger who travels without any luggage.

Racial profiling, if used at airports, must not be assumptive or subjective. It must be used in an objective and non discriminatory manner alongside random examinations of non-targeted passengers. All aspects of profiling, including racial and criminal profiling, should as a matter of course be included in the Computer Assisted Passenger Screening System (CAPS)[332] without isolating one from the

---

[332] The CAPS system was adopted in 1994 by Northwest Airlines to single out high risk passengers. After the TWA flight 800 disaster in July 1996, the Clinton Administration appointed the Al Gore Commission to study aviation security. The Commission recommended that all airlines use the CAPS system provided profiling did not rely on material of a constitutionally suspect nature such as race, religion or national origin of United States citizens.

3.5 Civil Unrest and Aviation

other. In this context the now popular system of compliance examination (COMPEX) is a non threatening, non discriminatory process which transcends the threshold debate on "profiling" by ensuring a balanced and proper use of profiling in all its aspects by examining "non targeted" passengers as well as on a random basis.

Another critical distinction to be drawn between discriminatory and subjective racial profiling on the one hand and prudent airport profiling on the other is the blatant difference between racism and racial profiling. The former is built upon the notion that there is a causal link between inherent physical traits and certain traits of personality, intellect or culture and, combined with it, the idea that some races are inherently superior to others.[333] The latter is the use of statistics and scientific reasoning that identify a set of characteristics based on historical and empirical data. This brings to bear the clear difference between "hard profiling", which uses race as the only factor in assessing criminal suspiciousness and "soft profiling" which uses race as just one factor among others in gauging criminal suspiciousness.

The nature of Air Transport in Extraordinary Rendition

Extraordinary rendition is an act of state and therefore the question arises as to whether aircraft used in this activity are military aircraft. Article 3 (a) of the Chicago Convention provides that the Convention will be applicable only to civil aircraft and not to state aircraft. It is an inclusionary provision which identifies military, customs and police service aircraft as being included in an undisclosed list of state aircraft. The Convention contradicts itself in Article 3 (c), where it says that no state aircraft of a contracting State shall fly over the territory of another State or land thereon without authorization by special agreement or otherwise, and in accordance with the terms thereof. The question arises as to how an international treaty, which on the one hand prescribes that it applies only to civil aircraft, turns around and prescribes a rule for state aircraft. Article 3 (c) effectively precludes relief flights over the territory of a State by state aircraft if the State flown over or landed upon does not give authorization for the aircraft to do so. *Milde* cites several instances of different types of aircraft being used in rendition flights[334] and concludes that a State aircraft may be identified by the design of the aircraft and its technical characteristics; registration marks; ownership; and type of operation.[335]

The distinction between civil and state aircraft is unclear as the Chicago Convention does not go to any length in defining or specifying as to how the two categories have to be distinguished. The ICAO Assembly, at its 14th Session held in Rome from 21 August to 15 September 1962, adopted Resolution A14-25

---

[333] *Britannica Macropedia*, 15 Ed. Vol. 9. at p. 880.
[334] Milde. *supra* at 478.
[335] *Id.* 481–482.

(Coordination of Civil and Military Air Traffic) which was on the subject addressed in Article 3(d) – that the Contracting States undertake, when issuing regulations for their state aircraft, that they will have due regard to the safety of navigation of civil aircraft. In A14-25, the Assembly directed the Council to develop guidance material for the joint civil and military use of airspace, taking into account the various policies, practices and means already employed by States to promote the satisfactory coordination or integration of their civil and military air traffic services.

At its 21st Session of the Assembly, Held in Montreal from 21 September to 15 October 1974, ICAO saw the adoption of Resolution A21-21 (Consolidated Statement of Continuing Policies and Associated Practices Related Specifically to Air Navigation) where, at Appendix O, on the subject of coordination of civil and military air traffic, the Assembly resolved that the common use by civil and military aviation of airspace and of certain facilities and services shall be arranged so as to ensure safety, regularity and efficiency of international civil air traffic, and that States would ensure that procedures and regulations pertaining to their state aircraft will not adversely affect or compromise the regularity and efficiency of international civil air traffic. In order to effectively implement the proposals of the Resolution, Contracting States were requested to initiate and improve the coordination between their civil and military air traffic services and the ICAO Council was required to ensure that the matter of civil and military coordination in the use of airspace is included, when appropriate, in the agenda of divisional and regional meetings.[336]

The ICAO Assembly, at its 36th Session (Montreal, 18–27 September, 2007) adopted Resolution A36-13 (Consolidated statement of continuing ICAO policies and associated practices related specifically to air navigation), Appendix O of which reiterates for the most part the text of Appendix O of Resolution A21-21, adding that the ICAO Council should endeavour to support States in the establishment of civil/military agreements by providing advice and guidance.[337]

One of the fundamental issues in the determination of aircraft category under Article 3 of the Chicago Convention is the use of civil aircraft in some instances for military purposes.[338] In emergency situations, States may acquire or in any other manner use civil aircraft for the transport of military personnel or goods meant for official use. In such circumstances any determination of the category of the aircraft concerned must be made taking into account all pertinent circumstances of the flight. Perhaps the most fundamental difference between the operation of civil and military aircraft lay in the fact that, although they were expected to share the same

---

[336] It will be recalled that the ICAO Council, at the Sixth Meeting of its 37th Session, held on 15 May 1959, noted the need for the Secretary General to pursue as effectively as possible the problem of accommodation of civil and military traffic in the available airspace. The efforts of the Secretary General were primarily meant to focus on the prevention of mid air collisions by the proper coordination of civil and military air traffic.

[337] See Assembly Resolutions in Force (as of 28 September 2007) Doc. 9902, at II-I7 to II-18.

[338] See Abeyratne (1997).

3.5 Civil Unrest and Aviation

skies, the procedures by which they did this varied greatly. Civil aircraft depended entirely on predetermined flight paths and codes of commercial conduct which varied depending on aircraft type and types of traffic carried, whereas military aircraft operated in line with the exigency of a situation and were not necessarily always guided by predetermined flight paths. This dichotomy led to the adoption of Resolution A10-19 by the Tenth Session of the ICAO Assembly in 1956. The Assembly Resolution, while recognizing that the skies (airspace) as well as many other facilities and services are commonly shared between civil and military aviation, focused on ICAO's mandate to promote safety of flight[339] and reinforced the thrust of Article 3(d) of the Chicago Convention. The Resolution called for all Contracting States to co-ordinate between their various aeronautical activities in order that the common use of airspace inter alia be so arranged that safety, regularity and efficiency of international civil air navigation be safeguarded.

There is also the issue of military aircraft being used in some circumstances for civil aviation purposes. At the time of writing, there were no clear international rules, generally accepted, whether conventional or customary, as to what constitutes state aircraft and what constitutes civil aircraft in the field of air law.[340] Often, particular international air law instruments will in some way make reference to these or similar concepts, either without defining them, or at the most providing very broad general definitions which sometimes vary from one instrument to another. The situation also appears to be the same in the domestic legislation of States, with the meaning of terms such as "public aircraft", "state aircraft", "civil aircraft" or "private aircraft" varying according to the State in question and the object and purpose of the legislation.

Military aircraft, more than any other kind of aircraft including customs and police aircraft, personifies the public or sovereign power of a State, and several attempts have been made to arrive at an internationally acceptable definition thereof. The Treaty of Versailles of 1919 ending World War I, provided that the armed forces of Germany must not include any military or naval air forces, and several attempts were made to distinguish between military and commercial (or civil) aeronautical material. Between 1919 and 1922, the relationship between civil and military aviation was debated at length by three international committees of air experts which met, respectively, in Paris, Geneva and Washington. They concluded independently that no means could be devised to prevent the conversion of civil aviation to military purposes, which would not at the same time prejudice the

---

[339] As per Article 44 of the Chicago Convention.

[340] In the earliest days of this century, jurists divided aircraft into two categories, public and private, with differing applicable legal regimes. The majority of European powers which replied to a questionnaire submitted by the French Government in 1909 agreed that public and private aircraft should be distinguished. The first diplomatic conference on air navigation, which met in Paris in 1910, defined public aircraft as "aircraft employed in the service of a contracting State, and placed under the orders of a duly commissioned official of that State." More particularly, a very specific regime to govern military aircraft was outlined. The Conference did not formally adopt a convention, but provisions drafted heavily influenced the Paris Convention of 1919.

development of civil aviation. In 1920, the Supreme War Council of the Paris Peace Conference asked one of these committees, the Aeronautical Advisory Commission to the Peace Conference, which had given its opinion in 1919, to draw up rules to distinguish between civil aviation and the military and naval aviation forbidden by the Peace Treaties. The Commission, referring to its 1919 report, replied that the task was impossible. The Supreme Council insisted that the rules be drawn up; after several months of debates, the Commission submitted what is known as "The Nine Rules" of 1922, for differentiating between military and civil aircraft. The distinction was based on technical criteria such as engine size, speed, "useful load" etc. It soon became clear that many civil aircraft fulfilled these criteria and the Rules were later abandoned.

At the Chicago Conference of 1944, which paved the way for the adoption of the Chicago Convention, a United States proposal of a Convention on Air Navigation provided that the Convention 'shall be applicable only to civil aircraft.' "Civil aircraft" was defined as "any aircraft other than military, naval, customs and police aircraft of any State or any political subdivision thereof." A Canadian draft repeated, *mutatis mutandis*, the provisions of Chapter VII of the Paris Convention. Air Navigation principles were allocated to Subcommittee 2 of Committee I. The United States draft was used as the primary basis for discussion in Subcommittee 2. On 10 November 1944, the following suggestions were referred to the drafting Committee of Subcommittee 2. chaired by Mr. J.C. Cooper (United States): "(a) that the term 'civil aircraft' be used as suggested. in place of 'private aircraft' as used in the Paris Convention; (b) that a definition of 'military aircraft' be drafted which would cover military, naval, and air forces; (c) that the Status of military and state (customs and police) aircraft in relation to the requirements of the Convention be defined in a separate section." On that day, Sweden proposed an amendment which later emerged as Article 3 (d). The drafting Committee examined the issue and responded with what became Article 3 of the Chicago Convention in its final form. The metamorphosis of the relevant provisions of the United States draft into Article 3 (a) to (c) took place entirely in the drafting Committee, and no official record exists of the reasons behind the shift from the Paris Convention or even the U.S. draft. It should be noted that no definition of military aircraft was provided.

With regard to Conventions other than the Chicago Convention, one can see some provisions which are relevant to the discussion on the distinction between civil and military aircraft. The Convention on the International Recognition of Rights in Aircraft (Geneva, 1948), the Convention on Offences and Certain Other Acts Committed on Board Aircraft (Tokyo, 1963), the Convention for the Suppression of Unlawful Seizure of Aircraft (The Hague, 1970) and the Convention for the Suppression of Unlawful Acts Against the Safety of Civil Aviation (Montreal. 1971), all contain a provision that "this Convention shall not apply to aircraft used in military, customs or police services." This appears to be a more simple way to indicate the scope of applicability of these Conventions than the provisions of Article 3 (a) and (b) of the Chicago Convention, although the end result seems to be the same. Furthermore. the clear implication is that all aircraft not so used would be subject to the provisions of the respective Conventions (paragraph 1.4 above refers).

The Convention on Damage Caused by Foreign Aircraft to Third Parties on the Surface (Rome, 1952) states in Article 26 that, "this Convention shall not apply to damage caused by military, customs or police aircraft." It should be noted that a "military, customs or police aircraft" is not necessarily the same thing as an "aircraft used in military, customs and police services" although again the expression "military, customs or police aircraft" was left undefined. Similarly, other "state" aircraft fall within the scope of the Convention. However, the 1978 Protocol to amend this Convention reverts to more familiar language; it would amend Article 26 by replacing it with, "this Convention shall not apply to damage caused by aircraft used in military, customs and police services."

The Convention for the Unification of Certain Rules Relating to the Precautionary Attachment of Aircraft (Rome, 1933) provides that certain categories of aircraft are exempt from precautionary attachment, including aircraft assigned exclusively to a government service, including postal services, but not commercial aircraft. On the other hand, the Convention for the Unification of Certain Rules Relating to Assistance and Salvage of Aircraft or by Aircraft at Sea (Brussels, 1938 "apply to government vessels and aircraft, with the exception of military, customs and police vessels or aircraft ...".

The Convention for the Unification of Certain Rules Relating to International Carriage By Air Warsaw, 1929) applies, inter alia. to all international carriage of persons. luggage or goods performed by aircraft for reward, regardless of the classification of the aircraft. Article 2 specifically provides that the Convention applies to carriage performed by the State or by legally constituted public bodies, but by virtue of the Additional Protocol, Parties may make a declaration at the time of ratification or accession that Article 2 (1) shall not apply to international carriage performed directly by the State. The Hague Protocol of 1955 to amend this Convention, in Article XXVI allows a State to declare that the Convention as amended by the Protocol shall not apply to the carriage of persons, cargo and baggage for its military authorities on aircraft, registered in that State, the whole capacity of which has been reserved by or on behalf of such authorities. Identical provisions are contained, *mutatis mutandis*, in the Guatemala City Protocol of 1971 (Article XXIII) the *1975* Additional Protocol No. 2 (Montreal), the *1975* Additional Protocol No. 3 (Montreal) and in Montreal Protocol No. 4 of 1975. It is submitted that Article 3 (b) of the Chicago Convention has no bearing on the applicability of these instruments of the "Warsaw System" which specify their own scope of applicability.

This analysis of some international air law instruments illustrates that many post-Chicago air law instruments (Geneva 1948, Tokyo 1963, The Hague 1970, Montreal 1971 and Rome 1952 and as amended in 1978) all have broadly similar provisions to Article 3 (a) and (b) of the Chicago Convention. The private air law instruments of the Warsaw System on the other hand, because of their nature, have adopted different formulae.

The provisions of the Chicago Convention and Annexes would not apply in a case where a state aircraft is (mistakenly or otherwise) operated on the basis that it is a civil aircraft. Similarly, the Geneva Convention of 1948, the Tokyo Convention of 1963, The Hague Convention of 1970, the Montreal Convention of 1971 and the Rome Convention (*1952*) as amended in 1978, will also not be applicable where it is determined that the aircraft was "used in military, customs or police services". The converse, of a civil aircraft being operated on the basis that it is a state aircraft, would theoretically raise the same problems (i.e. legal regimes thought to be inapplicable are in fact applicable). Concern is not often expressed in this regard.

Another frequently mentioned difficulty is claimed to be the loss of insurance coverage in respect of the aircraft (hull), operator, crew and passengers or other parties where the aircraft is in fact state aircraft. The question whether a particular insurance coverage is rendered invalid in such situations is primarily a private law matter of the construction and interpretation of the insurance contract. Unless the contract has an exclusion clause which specifically makes reference to the classification in Article 3 of the Chicago Convention (e.g. loss of coverage where the operation is of a state (or civil) aircraft as defined in the Chicago Convention), where the Convention will have no bearing on the contract, and the issue of the loss of insurance coverage becomes moot, the Chicago Convention's application to the insurance contract would prevail. Frequently, the policy will exclude usage of the aircraft for any purpose other than those stated" in a Schedule; among the exclusions would be any use involving abnormal hazards. Nearly every aviation hull and liability policy now excludes losses due to war, invasion, hostilities, rebellion. etc., although insurance to cover such losses can usually be obtained by the payment of a higher premium. However, the instances mentioned do not require a determination of whether the aircraft is considered to be state or civil under the Chicago Convention.

A question sometimes asked is whether national civil laws and regulations would apply to civilian flight crews operating what is a state aircraft under the Chicago Convention. Would civil or military investigative and judicial processes be applied, for example. in the case of an accident? The answer would depend largely on the domestic laws of the State concerned. The fundamental principle is stated in Article 1 of the Convention: every State has complete and exclusive sovereignty over the airspace above its territory. Furthermore, subject to the provisions of the Convention, the laws and regulations of a contracting State relating to the admission to or departure from its territory of aircraft engaged in international air navigation, or the operation and navigation of such aircraft within its territory, shall be complied with by (civil) aircraft of other contracting States, upon entering or departing from or while in the territory of that A fortiori, state aircraft are also subject to the laws of the subjacent State.

In the case of an accident involving state aircraft, States are not bound by Article 26 of the Chicago Convention and Annex 13. They can voluntarily (through their legislation) so apply' these provisions. Sometimes, the legislation specifies a different procedure in relation to military aircraft only; all other aircraft, including those used in customs or police services, are treated as civilian in this regard. In the case of other incidents, where for example the requisite over-flight permission has not been obtained by a state aircraft, which is then forced to land and charges brought against the crew, again the answer would depend on the domestic laws of the over-flown State and the factual circumstances. It is impossible to give a definitive answer in a vacuum. but it is the view of the Secretariat that the classification of an aircraft as "state" aircraft under the Convention does not necessarily mean that military laws and procedures of a State would apply to that aircraft or its crew. The current or any different classification of aircraft under the Convention would not be determinative whether a particular State, in the exercise of its sovereignty, would make that aircraft and/or its crew subject to civil or military laws and regulations. As a matter of practice States usually apply military rules and processes to military aircraft and personnel only. Paragraph 2.1.3 above shows that at the international level, attempts to arrive at a common, acceptable definition of military aircraft have met with a singular lack of success.

The question may arise as to the status of airline pilots and other crew under the Geneva Conventions of August 12, 1949 for the protection of war victims (the Red Cross Conventions) which apply, inter alia, in all cases of declared war or armed conflict between two parties, even if the state of war is not recognized by one of them. The Convention for the Amelioration of the Condition of the Wounded and Sick in Armed Forces in the Field (No. I) applies to "persons who accompany the armed forces without actually being members thereof, such as civil members of military aircraft crews ... provided they have received authorization from the armed forces which they accompany" and to "... the crews of civil aircraft of the Parties to the conflict ..." The same provisions are found in Convention for the Amelioration of the Condition of the Wounded, Sick and Shipwrecked Members of the Armed Forces at Sea (No. 2) and the Convention relative to the Treatment of Prisoners of War (No. 3). The Geneva Conventions, and in particular the provisions quoted above, do not link their own scopes of applicability to the determination under the Chicago Convention of the status of an aircraft. The Conventions of 1949 refer to civil and military aircraft, but not to these terms as "defined" under Chicago. Consequently, the provisions of the Chicago Convention does not, and cannot, determine whether and to what extent the flight crew of an aircraft is given protection by these Conventions.

Apart from the significance of aviation in its role in transporting suspected delinquents under extraordinary rendition, particularly in the face of its intended role in securing world peace, avition could act as a useful evidentiary tool in tracing rendition flights. For example, flight logs, which are kept by aviation authorities for years have been used to trace the transportation of particular prisoners from one jurisdiction to another.[341] The most significant role played by aviation, however, is in assisting the world community in realizing that current political and diplomatic problems mostly emerge as a result of the inability of the world to veer from its self serving concentration on individual perspectives to collective societal focus. This distorted approach gives rise to undue emphasis being placed on rights rather than duties; on short-term benefits rather than long-term progress and advantage and on purely mercantile perspectives and values rather than higher human values.

---

[341] Solomon (2007).

Against this backdrop, the fundamental principle and the overriding theme of international civil aviation has been, and continues to be, the need to foster friendship and understanding among the people of the world with the ultimate objective of ensuring global peace. Toward this end both the principles of air navigation and aviation economics have to ensure that aviation is developed in a manner that would make sure the world has a safe, reliable, economical and efficient civil aviation system.

References

Abeyratne RIR (1992) The development of the machine readable passport and visa and the legal rights of the data subject. Ann Air Space Law/Annales de Droit Arien et Spatial XVII (Part II) 99:1–31
Abeyratne RIR (1997) The use of civil aircraft and crew for military purposes. Annals Air Space Law/Annales de Droit Arien et Spatial XXII(II):1–23
Abeyratne RIR (1998) Aviation security. Aldershot, Ashgate, pp 131–196
Abeyratne RIR (2001) The exchange of airline passenger information – issues of privacy. Commun Law 6(5):153–162
Abeyratne RIR (2002a) Intellectual property rights and privacy issues: the aviation experience in API and biometric identification. J World Intellectual Property 5(4):631–650
Abeyratne RIR (2002b) Attacks on America – privacy implications of heightened security measures in the United States, Europe, and Canada. J Air Law Commerce 67(1)
Abeyratne RIR (2003) Profiling of passengers at airports – imperatives and discretions. Eur Transport Law XXXVIII(3):297–311
Abeyratne RIR (2007) The safe carriage of dangerous pathogens by air: legal and regulatory issues. Eur Transport Law XLII(6):689–704
Abeyratne R (2010a) Aviation security law. Heidelberg, Springer, pp 205–264
Abeyratne RIR (2010b) Negligent entrustment of leased aircraft and crew: some legal issues. Air Space Law 35(1):33–44
Adams J (1989) The financing of terror. Simon & Schuster, New York, p 12
Becker T (2006) Terrorism and the state, hart monographs in transnational and international law. Hart Publishing, p 155
Becker T (2006b) Terrorism and the state; rethinking the rules of state responsibility. Hart Publishing, Portland
Bennet CJ (1992) Regulating privacy. Cornell University Press, Ithaca, NY, 13
Blackstone W, Morrison W (eds) (2001) 4 Commentaries on the laws of England (1765–69). Cavendish, London, p 68
Bobbitt P (2008) Terror and consent: the wars for the twenty first century. Knopf, New York, at 98–179
Brownlie I (1983) System of the law of nations: state responsibility, Part 1. Clarendon, Oxford, p 39
Burnham D (1983) The rise of the computer state. Random House, New York, p 20
Caron DD (1998) The basis of responsibility: attribution and other trans-substantive rules. In: Lillich RB, Magraw DB (eds) The Iran-United States claims tribunal: its conclusions to state responsibility, vol 109 (Irvington-on-Hudson) NY. Transnational Publishers, pp 153–54
Cate FH (1997) Privacy in the information age. Brookings Institution Press, Washington, DC, p 49
Chen TC (1951) The international law of recognition, London
Clutterbuck R (1991) Living with terrorism. Butterworths, London, p 175
Cohen and Felson (1979) Social change and crime rate trends: a routine activity approach. Am Sociol Rev 44:588–589

Cooley TM (1888) A treatise on the law of torts, 2nd edn. Callaghan, Chicago
Cortes WI (2004) Cyber terrorism post 9/11 in the Western Hemisphere. Monograph presented to the Inter American Defence College as a requisite for obtaining the diploma of completion for the course on defence and hemispheric security, Fort Leslie J. McNair, Washington DC, April 2004
de Arechaga EJ (1968) International responsibility. In: Sorenson M (ed) Manual of public international law. St. Martin's Press, New York, 531 at 535
De Vattel E, Fenwick CG (tr) (1916) 2, The law of nations or, the principles of natural law: applied to the conduct and to the affairs of nations and sovereigns. Legal Classics Library, New York, NY, p 72
Delaney RF (1979) World terrorism today. Calif Western Int Law J 9:454
Dobson C, Payne R (1987) Appendix B: the chronology of terror: 1968–1987. In: War without end: the terrorists: an intelligence dossier. Sphere Books, London, p 366
Dorey FC (1983) Aviation security. Granada, London, p 142
Dunnigan JF (2003) The next war zone: confronting the global threat of cyber terrorism. Citadel Press, New York, p 4
Embar-Seddon SA (2002) Cyber terrorism: are we under siege? Am Behav Sci 45(6):1033–1043, at 1034
Ferguson N (2008) The ascent of money. The Penguin Press, New York, at 188
Flaherty DH (1991) On the utility of constitutional rights to privacy and data protection. Case W Res 41:831 at 833–834
Foschio LG (1990) Motor vehicle records: balancing individual privacy and the public's legitimate need to know. In: Kuferman TR (ed) Privacy and publicity. Meckler, London, p 35
Freund PA (1971) Privacy: one concept or many. In: Pennnock JR, Chapman JW (eds) Privacy. Atherton Press, New York, p 182
Fried C (1978) Privacy: economics and ethics a comment on posner. Ga L Rev 12:423 at 425
Gavison R (1980) Privacy and the limit s of the law. Yale L J 89:421
Grotius H, Scott JB (tr) (1646) 2 De Jure Belli Ac Pacis, pp 523–26
Guill M (2000) Cyber-terrorism poses newest and perhaps elusive threat to civil aviation. ICAO J:18
Halpin A (1997) Rights & law analysis & theory. Hart Publishing, Oxford, p 111
Hanle DJ (1989) Terrorism: the newest face of warfare. Pergamon-Brassey's, New York, p 185
Hoffer S (2000) World cyberspace law. Juris Publishing, at 8.1
Hoffman LJ (ed) (1980) Computers and privacy in the next decade. Academic Press, New York, 142
Hyde C (1928) Concerning damages arising from neglect to prosecute. 22 Am J Int L 140:140–142
J. Montgomery Curtis Memorial Seminar (1992) The public, privacy and the press: have the media gone too far? American Press Institute, p 2
Jennings RY, Watts AD (eds) (1992) Oppenheim's international law, 9th edn. London
Jones RV (1973) Some threats of technology to privacy, privacy and human rights. In: Robertson AH (ed) Presented at the third colloquy about the European convention on human rights, Brussels, 30 Sept–3 Oct 1970. Manchester University Press
Jorgensen NHB (2000) The responsibility of states for international crimes. Oxford University Press, Oxford, pp 249–254
Lauterpacht H (1947) Recognition in international law, Cambridge
Lee D (2005) Why terrorism? An agent-based model of culture and violence. Paper presented at the annual meeting of the International Studies Association, Hilton Hawaiian Village, Honolulu, Hawaii, Mar 05:2005
McMunn MK (1996) Aviation security and facilitation programmes are distinct but closely intertwined. ICAO J51:9 at 7
Mickolus EF (1980) Transnational terrorism: a chronology of events, 1969–1979. Aldwych Press, London, p 428
Milde M (1994) The Chicago convention – are major amendments necessary or desirable 50 years later? Ann Air Space Law XIX (Part I):401–452 at p 403

Milde M (2008) "Rendition flights" and international air law. Zeitschrift Fur Luft-und Weltraumrecht 4:477–486 at 477
Miller AR (1971) The assault on privacy. The University of Michigan Press, Ann Arbor, MI, 42
Misra S (2003) High tech terror. The American City and County, at 118
Nock SL (1993) The costs of privacy. Aldine De Gryter, New York, 43
Ofri A (1984) Intelligence and counterterrorism. ORBIS:49
Orwell G (1978) Nineteen eighty-four. Clarendon Press, Oxford
Orwell G (1984) Nineteen eighty-four. Clarendon Press, Oxford
Pember DR (1972) Privacy and the press. University of Washington Press, Seattle, p 227
Pierre AJ (1975–1976) The politics of international terrorism. ORBIS 19:1256
Posner R (1978) The right of privacy. Ga L Rev 12(3):393 at 409
Poulsen K (2002) FAA confirms hack attack, security focus, p 4–25, at http://www.securityfocus.com/news/378
Prowda JB (1995) A layer's ramble down the information superhighway: privacy and security of data. Fordham L Rev 64:738 at 769
Rapoport DC (1971) Assassination and terrorism. Canadian Broadcasting Corporation, Toronto, p 79
Regan PM (1995) Legislating privacy. The University of North Caroline Press, Chapel Hill, NC, p 33
Reidenberg JR (1995) Data protection law and the European union's directive: the challenge for the United States: setting standards for fair information practice in the U.S. private sector. Iowa L Rev 80:497 at 498
Reuter P (1989) Introduction to the law of treaties. Pinter Publishers, London and New York, p 16
Sattertwaite ML (2007) Rendered meaningless: extraordinary rendition and the rule of law. George Washington Law Rev 75:1333, at 1335–1336
Schenkman J (1955) International civil aviation organization. Librairie E. Droz, Geneve, p 6
Scott GG (1995) Mind your own business – the battle for personal privacy. Insight Books, New York, p 307
Shaw MN (2003) International law, 5th edn., Cambridge, p 367
Silets HL (1987) Something special in the air and on the ground: the potential for unlimited liability of international air carriers for terrorist attacks under the warsaw convention and its revisions. JALC 53:321 at 358
Simitis S (1995) From the market to the polis: the ec directive on the protection for personal data. Iowa L Rev 80:445 at 447–448
Simmel A (1971) Privacy is not an isolated freedom. In: Pennnock JR, Chapman JW (eds) Privacy. Atherton Press, New York, p 71
Smith and Hogan (1992) Criminal law, 7th edn. Butterworths, London, p 431
Sochor E (1991) The politics of international aviation. Macmillan, London, p xvi
Solomon D (2007) Breaking Jeppessen. Metroactive June 13–17
Starke JG (1989) Introduction to international law, 10th edn. Butterworths, London, p 3
Stohl M (2006) Cyber terrorism: a clear and present danger, the sum of all fears, breaking point or patriot games? Crime Law Soc Change 46:223–238
Treasury HM (2007) The financial challenge to crime and terrorism, p 8
Turley J (2002) A useful mechanism. Transec:24
Warner E (1942) Foreword to international air transport and national policy by Lissitzyn O.J. New York, p V
Warren SD, Brandeis LD (1980) The right of privacy. Harv L Rev 4(5):193 at 195
Warren SD, Brandies LD (1890–1891) The right to privacy. Harvard Law Rev 4:193
Weimann G (2006) Terror on the internet: the new arena, the new challenges. United States Institute of Peace Press, Washington, DC, at 148
Westin A (1967) Privacy and freedom. Atheneum, New York, p 368
Westin AF (1970) Privacy and freedom. Bodley Head, at 124
Young JB (1978) A look at privacy. In: Young JB (ed) privacy. Willey, New York, p 1
Zelermyer W (1959) Invasion of privacy. Syracuse University Press, Syracuse, p 16

Chapter 4
Environmental Issues

Climate Change

4.1 The Reality of Climate Change

Climate change is no longer a theory and has fast become an unequivocal reality and a defining issue of our time. Its enormity can be identified numerically. For instance 2005 was the warmest year on record. There has been a 33% rise in global carbon dioxide emissions since 1987. The Inter-Governmental Panel on Climate Change (IPCC) records that 5 million extra people are at risk of hunger by the year 2020 if climate change continues unabated. The 2003 heat wave killed 35,000 people in Europe. Environmental campaigner Sheila Watt-Cloutier, in her article "A Human Issue" in the May 2007 issue of *"Our Planet"* – the magazine of the United Nations Environment Programme, says that there are palpable signs of drastic climate change in the Arctic, which she calls the health barometer for the planet. Whatever happens in the world occurs first in the Arctic – the home of Inuit. In 2004 certain conclusions were reached by the Arctic Climate Impact Assessment (ACIA) as a result of work carried out by 300 scientists from 15 countries. Among the results, according to Watt-Cloutier, is that for Inuit, warming is likely to disrupt or even destroy their hunting and food-sharing culture as reduced sea ice causes populations to decline or become extinct. The Inuit have lived in the arctic for thousands of years and their culture and economy reflect their homeland. Climate change in the arctic would therefore infringe the basic human right of the Inuit to life.

In the same issue of *Our Planet*, Basanta Shresta, in his article "Mountain Tsunamis" states that glaziers are retreating in the face of accelerating global warming, as human activities cause steadily increasing concentrations of greenhouse gases in the atmosphere and their melting is an important indicator of climate change. This is confirmed by the Fourth Assessment Report of the IPCC, released in the first half of 2007, which records that most of the observed increase in global averaged temperatures since the mid twentieth century is very likely due to the observed increase in anthropogenic greenhouse gas concentrations. According to the Report, rising temperatures in the Arctic have caused a decline of 2.7% of sea ice since 1978. A third of the glazier surface in Bolivia and Peru has disappeared since the 1970s. The Report concludes that climate change is one of the most critical global challenges of our time.

The former Secretary General of the United Nations, Kofi Annan, in his Report to the Fifty Ninth Session of the UN General Assembly in March 2005, observed that one of the greatest environmental and developmental challenges in the twenty-first century will be to control and cope with climate change. The Secretary General drew the attention of the General Assembly to the fact that entry into force in February 2005 of the 1997 Kyoto Protocol to the United Nations Framework Convention on Climate Change was an important step toward dealing with global warming. Since the Protocol extends only until 2012, the Secretary General called upon the international community to agree on stabilization targets for greenhouse gas concentrations beyond that date. To achieve this goal, scientific advancement and technological innovation must be mobilized for carbon management and energy efficiency, the responsibility for which lies with the countries that contribute to most of the problems.

The Third Conference of the Parties to the United Nations Framework Convention on Climate Change (Climate Change Convention) was held from to 11 December 1997 at Kyoto, Japan. Significantly the States parties to the Convention adopted a protocol (Kyoto Protocol) on 11 December 1997 under which industrialized countries have agreed to reduce their collective emissions of six greenhouse gases by at least 5% by 2008–2012. Ambassador Raul Estrada-Oyuela, who had chaired the Committee of the Whole established by the Conference to facilitate the negotiation of a Protocol text, expressed the view that the agreement will have a real impact on the problem of greenhouse gas emissions and that 11 December 1997 should be remembered as the "Day of the Atmosphere"

The Kyoto Protocol, in Article 1 (a) (v) calls each State Party to achieve progressive or phasing out of market imperfections, fiscal incentives, tax and duty exemptions and subsidies in all greenhouse gas emitting sectors that run counter to the objective of the Convention and application of market instruments. The subject of emissions leading to trading is addressed initially in Article 3 of the Protocol which requires States Parties to ensure that their aggregate anthropogenic carbon dioxide equivalent emissions of the greenhouse gases listed in Annex A do not exceed their assigned amounts, calculated pursuant to their quantified emission limitation and reduction commitments inscribed in Annex B. The provision also requires States parties to the Protocol to reduce their overall emissions of greenhouse gases by at least 5% below 1990 levels in the commitment period 2008–2012. Article 3 (6) goes further, in providing that States Parties shall be allowed a certain degree of flexibility in implementation of Article 3 and the reduction of their emission standards.

This approach has been endorsed by many learned and informed commentators, among whom is former US Vice President Al Gore, who in his movie "An Inconvenient Truth" highlighted the fact that the Earth's atmosphere is thin enough that we can change its composition with the emissions we produce through industrial activity and transportation. Gore, who has dedicated his career both as a

4.1 The Reality of Climate Change

distinguished politician in the United States where he served under President Clinton as Vice President and also as an erudite academic with impressive credentials, makes the frightening but accurate claim that the vastly increasing levels of Carbon Dioxide we produce can thicken the atmosphere so that the rays of the Sun which fall on Earth and bounce back as infra red rays beyond the atmosphere cannot escape the thick atmospheric layer at the rate they did before and are trapped within, making the world warmer. This phenomenon has resulted in Carbon Dioxide being termed a greenhouse gas as it causes a greenhouse effect by retaining heat within the atmosphere.

The environmental crisis that we face now is that the solar energy that we receive from the Sun, which under normal circumstances and for years has been retained in quantities that benefit the Earth in terms of the balance of its ecosystem and biodiversity, and the rest released by way of infra red rays, is not being released as it should as more infra red rays are being trapped due to the Carbon Dioxide induced layer surrounding the atmosphere. Popularly called "global warming" this phenomenon is the result of the observed increase in the average temperature of the Earth's atmosphere and oceans in recent decades. It is claimed that the Earth's average near-surface atmospheric temperature rose $0.6 \pm 0.2°C$ ($1.1 \pm 0.4°F$) in the twentieth century. The current scientific consensus is that most of the observed warming over the last 50 years is likely to have been attributable to human activities. As already stated, the main cause of the human-induced component of warming is the increase in atmospheric greenhouse gases (GHGs), especially Carbon Dioxide (CO_2), due to activities such as burning of fossil fuels, land clearing, and agriculture. Greenhouse gases are gases that contribute to the greenhouse effect. This effect was first described by Joseph Fourier in 1824, and was first investigated quantitatively in 1896 by the Swedish Chemist Svante Arhenius.

Another vocal commentator is George Monbiot, a radical thinker and visiting professor at Oxford Brookes University, who claims in his book "HEAT: How to Stop the Planet from Burning", that if in the year 2030, carbon dioxide concentrations in the atmosphere remain as high as they are today, the likely result is an increase of 2° warming above pre-industrial levels, which is the point beyond which certain major ecosystems begin collapsing. The collapse of one's environment would eventually bring about catastrophic results and threaten human existence. When this prospect is viewed in the context of human rights, climate change, and our apathy toward it would infringe our basic right to life, as in the case of Inuit cited above.

The basic protection of human rights starts at the least desirable level of existence. Although this is a minimalist approach, it is arguably the best starting point and perhaps the only one we have. A right is something due to a person by just claim, legal guarantee or moral principle. It is a power, privilege or immunity accrued to a person by law and is a legally enforceable claim that another will do or will not do a given act. It is also a recognized and protected interest, the violation of which is wrong. Therefore, the starting point should be in the words "just claim" "legal guarantee" and "moral principle". These claims and guarantees based on moral principles should be justiciable.

The question now is, how could we enforce the human right to life and ensure that life is not jeopardized by continued climate change? The first step of course would be to recognize the enormity of climate change. The second is to enforce measures. One of such measures would be the "greening" of human activity including industry. As an example, one can cite the admirable step taken by Virgin Atlantic Chairman, Sir Richard Branson who has formed a new company – Virgin Fuels – that would invest millions of dollars in alternate fuels. He has pledged to reinvest profits from his airline and rail transport businesses to alternative energy, explaining that his initiative was prompted by the concern he had regarding the welfare of the world affected by global warming. This move is but part of Branson's actions to coerce the aviation industry to cut fuel consumption, for which he has set aside an investment of US$3 billion over the next 10 years to fight global warming.

Branson's initiative, which is both laudable and timely, follows a secular trend where, since the time of the introduction of jet aircraft, civil aviation authorities have made inroads into the realm of fuel consumption and made sustained efforts to address the issue of environmental impacts caused by aircraft operations. This is primarily because of the inevitable corollary to the exclusive use of petroleum as industrial fuel. This in turn resulted in the depletion of global oil reserves, where, as far back as 1949 oil was recognized as a finite non renewable resource. Both factors – pollution caused by engine emissions as well as the limits of global oil resources – prompted wide ranging studies on the optimal use of fuel in the aviation industry and alternative fuel sources.

Being already aware of this trend, Ryanair, known to be Europe's greenest airline, is investing some E 17 billion on its fleet replacement and expansion programme which began in 1999. All of Ryanair's older Boeing 737-200 aircraft have already been replaced with next generation 737-800 aircraft, which has made Ryanair the airline with the youngest and most modern fleet in all of Europe. These measures put Ryanair in the forefront as an airline which minimises and continues to reduce fuel burn and carbon dioxide emissions per passenger kilometre. The low cost business model used by Rayanair (and other low cost carriers) which involves the use of secondary airports with no holding patterns as in busier airports and point to point services (which eliminate multiple landings) help increase fuel efficiency and restrict emissions. It is reported that Ryanair's fleet replacement has resulted in an overall reduction in fuel consumption of 52% between 1998 and 2006.

On the other side of the Atlantic, American Airlines, with one of the largest fleets in the world, is also involved in a comprehensive aircraft weight-reduction program. The airline has removed ovens and galleys from aircraft on which hot food is not served. It also carries less potable water on flights. Although earlier it was routine for maintenance personnel to simply fill the water tanks prior to flight, a study conducted by the airline revealed that usually less than half the water was being consumed. Following this realization, the maintenance department designed a $1 valve that shuts off the filler hose when the tank reaches 75% capacity rather than change the water-tank filling procedure, which resulted in a reduction of aircraft weight by about 100 pounds and a saving of approximately $2.8 million in fuel annually.

Carbon offsetting is one way of mitigating the effects of climate change. This act of mitigating ("offsetting") greenhouse gas emissions is well exemplified by the simple exercise of planting of trees to compensate for the greenhouse gas emissions from personal air travel.

A wide variety of offset methods are in use – while tree planting has initially been a mainstay of carbon offsetting, renewable energy and energy conservation offsets are now becoming increasingly popular. Carbon offsetting as part of a "carbon neutral" lifestyle has gained some appeal and momentum mainly among consumers in western countries who have become aware and concerned about the potentially negative effects of energy-demanding lifestyles and economies on the environment. The Kyoto Protocol has sanctioned official offsets for governments and private companies to earn carbon credits which can be traded on a marketplace. This has contributed to the increasing popularity of voluntary offsets among private individuals and also companies. Offsets may be cheaper or more convenient alternatives to reducing one's own fossil-fuel consumption. However, some critics object to carbon offsets, and many have questioned the benefits of certain types of offsets (such as tree planting), and other projects.

If climate change can ultimately decide whether a human has the right to live, and it indeed can, the whole issue must be approached on the basis of social and moral principles applicable to human rights. There must be global enforcement of measures such as carbon offsets and emissions trading. Alternative fuels should be considered a necessary alternative to fossil fuels. These measures must be taken through a global forum which is inevitably the United Nations. The United Nations is all about "nations". Nations are people, as against countries which are defined geographic areas, and States which are a collection of agencies that form a government. Therefore human rights should incontrovertibly be about nations and their interactions and people helping one another. Rights are generated through human experience, particularly with injustice. Therefore, although such rights are technically entrenched in a Bill of Rights or Constitution, they should not be limited to the law that is written down but be extended whenever required to include new rights if an injustice or wrong is about to be committed.

4.2 Aspirations and Goals

Assembly Resolution A 37-19 of ICAO adopted at its 37th Assembly (Montreal, 28 September–8 October) recognizes the critical importance of providing continuous leadership to international civil aviation in limiting or reducing its emissions that contribute to global climate change. It reemphasizes the vital role which international aviation plays in global economic and social development and the need to ensure that international aviation continues to develop in a sustainable manner and recognizes that the ultimate objective of the United Nations Framework Convention on Climate Change (UNFCCC) is to achieve stabilization of greenhouse gas (GHG) concentrations in the atmosphere at a level that would prevent dangerous anthropogenic interference with the climate system. It also acknowledges that international aviation emissions, currently accounting for less than 2% of total global CO_2 emissions, are projected to grow as a result of the continued development of the sector. Based on these initial premises the Resolution called upon ICAO to continue to take initiatives to promote information on scientific understanding of aviation's impact and action undertaken to address aviation emissions and continue to provide the forum to facilitate discussions on solutions to address aviation emissions.

The Resolution set certain aspirational goals for the air transport industry, calling upon States and relevant organizations to work through ICAO to achieve a global annual average fuel efficiency improvement of 2% until 2020 and an aspirational global fuel efficiency improvement rate of 2% per annum from 2021 to 2050, calculated on the basis of volume of fuel used per revenue tonne kilometre performed. These goals would not attribute specific obligations to individual States, and the different circumstances, respective capabilities and contribution of developing and developed States to the concentration of aviation GHG emissions in the atmosphere will determine how each State may voluntarily contribute to achieving the global aspirational goals; The Resolution also resolved that, without any attribution of specific obligations to individual States, ICAO and its member States with relevant organizations would work together to strive to achieve a collective medium term global aspirational goal of keeping the global net carbon emissions from international aviation from 2020 at the same level, taking into account: (a) the special circumstances and respective capabilities of developing countries; (b) that the different circumstances, respective capabilities and contribution of States to the concentration of aviation GHG emissions in the atmosphere will determine how each State may contribute to achieving the global aspirational goals; (c) that some States may take more ambitious actions prior to 2020, which may offset an increase in emissions from the growth of air transport in developing States; (d) the maturity of aviation markets; (e) the sustainable growth of the international aviation industry; and (f) that emissions may increase due to the expected growth in international air traffic until lower emitting technologies and fuels and other mitigating measures are developed and deployed.

Resolution A37-19 also requested the Council to facilitate the dissemination of economic and technical studies and best practices related to aspirational goals and to provide guidance and other technical assistance for the preparation of States' action plans prior to the end of June 2012, in order for States to conduct their necessary studies and to voluntarily submit their action plans to ICAO, resolving also that a *de minimis* threshold of international aviation activity of 1% of total revenue ton kilometres should apply to the submission of States' action plans on the basis that States below the threshold are not expected to submit action plans towards achieving the global goals; and States below the threshold but that otherwise have agreed to voluntarily contribute to achieving the global goals are expected to submit action plans. A further resolving clause states that there would be a *de minimis* threshold of international aviation activity, consistent with the guiding principles in the Annex, of 1% of total revenue ton kilometres to MBMs as where commercial aircraft operators of States below the threshold should qualify for exemption for application of MBMs that are established on national, regional and global levels; and States and regions implementing market based measures (MBMs) may wish to also consider an exemption for other small aircraft operators.

There were reservations recorded by States to Resolution A 37-19 on the basis that the Resolution calls upon developing States to stultify their progress in air transport by enforcing standards equally on developed and developing States.

4.3 Trading and Market Based Measures

The purchase of carbon credits by industries that emit greenhouse and other gases into the atmosphere and which exceed their emissions allowances is the foundation of emissions trading. However, there is a fundamental conceptual flaw in this process in that trading essentially involves a commodity and carbon credits are not a commodity in the same vein as are other commodities such as gold, precious metals, textiles or agricultural products. This makes the fiscal calculations involved in carbon trading cumbersome and often convoluted. Confusion is worse confounded in the aviation field by the fact that no global emissions trading scheme has been agreed upon by the 190 member States of the International Civil Aviation Organization (ICAO).

The above notwithstanding, ICAO has been at the forefront of aviation and environmental activity as the global forum for aviation and has made vast strides over the past 3–4 years in bringing together a world divided by individual philosophies and interests in the field of aviation and environmental protection. As this article will enumerate, there has been much achieved by ICAO.

ICAO and most of the aviation world is agreed on the fact that a global approach to reduce aviation's greenhouse gas emissions is what is needed now as part of the international climate change negotiations within the United Nations Framework Convention on Climate Change (UNFCC). Aviation is the only global sector to have set itself far-reaching environmental targets: efficiency improvements of 1.5% per year up to 2020, carbon-neutral growth from 2020 onwards, and a 50% reduction of net emissions by 2050 compared to 2005.

The 37th Session of the ICAO Assembly noted that, if the global community were to stabilize greenhouse gas emissions in the atmosphere and maintain it at a level that would prevent dangerous anthropogenic interference with the climate, the increase in global temperature would have to be maintained below 2°C. In order to achieve this target, deep cuts in global emissions would be needed, and all sectors of the economy were being looked to for their contribution – including international aviation, which is well known as representing a significant and growing source of emissions.

One of the economic measures that would be used in this regard is the trading of carbon credits. Therefore, what the aviation community now needs is a clear notion as to how the carbon market works and how credits would be calculated in an emissions trading scheme. This article discusses some aspects in this regard with some focus on the legal and economic aspects of the European Emissions Trading Scheme (ETS) that will come into effect in 2012.

The *Special Report on Aviation and the Global Atmosphere* published by the Inter-Governmental Panel on Climate Change in 2007 states: "The task of detecting climate change is already difficult; the task of detecting the aircraft contribution to the overall change is more difficult because aircraft forcing is a small fraction of anthropogenic forcing as a whole. However, aircraft perturb the atmosphere in a specific way because their emissions occur in the free troposphere and lower stratosphere, and they trigger contrails, so the aircraft contribution to overall climate change may have a particular signature. At a minimum, the aircraft-induced climate change pattern would have to be significantly different from the overall climate change pattern in order to be detected".

Air transport has been, and continues to be, a complex business. It follows therefore, that transactions related to this industry are indeed complex and should be adaptable to modern exigencies. A survey of the International Air Transport Association carried out in 2007 of over 600 companies from 5 countries reflects that 63% confirmed that air transport networks are vital for their investments and business.[1] Thirty percent of those countries said that any constraint placed on the air transport industry would make them invest less.[2]

The High Level Meeting on Aviation and Climate Change, convened by the ICAO from 7 to 9 October 2009 once again brought to bear the vexed issue of ICAO's leadership in the field, only to be re-endorsed by the 56 States which attended the Meeting, that ICAO was indeed the leader and that ICAO should provide guidance on various issues pertaining to the subject. The meeting went on to acknowledge the principles and provisions on common but differentiated responsibilities and respective capabilities, and the fact that developed countries will be taking the lead under the UNFCCC and the Kyoto Protocol. It also acknowledged the principles of non-discrimination and equal and fair opportunities to develop international aviation set forth in the Chicago Convention and re-emphasized the vital role which international aviation plays in global economic and social development and the need to ensure that international aviation continues to develop in a sustainable manner.

The increasing dependency of aviation on fossil fuels, which was exacerbated by the fuel price crisis of 2008, has brought to bear the need to consider the use of alternative fuels in the industry. In this regard, ICAO convened the Conference on Aviation and Alternative Fuels (CAAF) at Rio de Janeiro in Brazil from 16 to 18 November 2009. The Conference was a major event showcasing the state of the art in aviation alternative fuels, and an event at which a Global Framework for

---

[1] *Flying Smart, Thinking Big*, Airbus Industrie's Global Market Forecast, 2009–2028, at 16.
[2] *Ibid.*

4.3 Trading and Market Based Measures

Aviation Alternative Fuels (GFAAF)[3] was considered. The GFAAF was designed as a living document that will be continually updated on the ICAO website that will share information, best practices and future initiatives by ICAO Member States and the air transport industry. The Conference made reference to the High-Level Meeting on International Aviation and Climate Change[4] convened earlier by ICAO which recommended that States and International Organizations actively participate in CAAF with a view to sharing their efforts and strategies to promote such work and to update the 15th Meeting of the Conference of the Parties to the United Nations Framework Convention on Climate Change (UNFCC COP15) to be held in December 2009).[5]

The Conference adopted a Declaration which recognized the urgent need for measures to facilitate access to financial resources, technology exchange, and capacity building specific to aviation alternative fuels and acknowledged that the demand for sustainable fuels extends beyond international aviation, but that aircraft have unique fuel specification requirements It also recognized the need to encourage supply chain stakeholders to ensure that sustainable alternative fuels are made available to aviation and acknowledged that with sufficient incentive and supply, international aviation could deliver a substantial CO_2 reduction benefit from the use of sustainable alternative fuels for aircraft and that, due to its small network of fuel distribution points and its predictable demand international aviation is well suited to becoming a global first adopter of sustainable alternative fuels.

The Declaration recommended that : ICAO and its Member States endorse the use of sustainable alternative fuels for aviation, particularly the use of drop-in fuels in the short to mid-term, as an important means of reducing aviation emissions; ICAO establish a Global Framework for Aviation Alternative Fuels (GFAAF) on

---

[3] See ICAO Doc CAAF/09-WP/23, 18/11/09.

[4] The High Level Meeting on Aviation and Climate Change, convened by the International Civil Aviation Organization from 7 to 9 October 2009 once again brought to bear the vexed issue of ICAO's leadership in the field, only to be re-endorsed by the 56 States which attended the Meeting, that ICAO was indeed the leader and that ICAO should provide guidance on various issues pertaining to the subject. The meeting went on to acknowledge the principles and provisions on common but differentiated responsibilities and respective capabilities, and the fact that developed countries will be taking the lead under the UNFCCC and the Kyoto Protocol. It also acknowledged the principles of non-discrimination and equal and fair opportunities to develop international aviation set forth in the Chicago Convention and re-emphasized the vital role which international aviation plays in global economic and social development and the need to ensure that international aviation continues to develop in a sustainable manner.

[5] At the United Nations Framework Convention on Climate Change 15th Meeting of the Conference of the Parties (COP15) and the Fifth Meeting of the Parties to the Kyoto Protocol (CMP5) which were convened in Copenhagen on 8 December 2009, leaders from the U.S., India, Brazil, South Africa and China came to an agreement to combat global warming. The deal, which was only between five countries, contained no specifics on emissions cuts, but it did commit the countries to look to keep global warming at 2 °C or less and to promise $30 billion in funding to battle climate change by 2012. It also created a framework for international transparency on climate actions for developed and developing nations alike.

aviation and sustainable alternative fuels to communicate what individual and shared efforts expect to achieve with sustainable alternative fuels for aviation in the future for consideration by the 37th Session of the ICAO Assembly. The GFAAF will be continually updated Member States and stakeholders work together through ICAO and other relevant international bodies, to exchange information and best practices, and in particular to reach a common definition of sustainability requirements for alternative fuels; Member States are encouraged to work together expeditiously with the industry to foster the research, development, deployment and usage of sustainable alternative fuels for aviation; Funding efforts that support the study and development of sustainable alternative fuels and other measures to reduce GHG emissions, in addition to the funding for research and technology programmes to further improve the efficiency of air transport, be maintained or improved; Member States are encouraged to establish policies that support the use of sustainable alternative aviation fuels, ensure that such fuels are available to aviation and avoid unwanted or negative side effects, which could compromise the environmental benefits of alternative fuels; The ICAO Council should further elaborate on measures to assist developing States as well as to facilitate access to financial resources, technology transfer and capacity building; There is an urgent need for measures to facilitate access to financial resources, technology exchange, and capacity building specific to sustainable aviation alternative fuels; ICAO takes the necessary steps with the aim of considering a framework for financing infrastructure development projects dedicated to sustainable aviation alternative fuels and incentives to overcome initial market hurdles; ICAO continue to facilitate efforts to develop a lifecycle analysis framework for comparing the relative GHG emissions from sustainable alternative fuels to the lifecycle of conventional fuels for aviation; and ICAO and its Member States should strongly encourage wider discussions on the development of alternative fuel technologies and support the use of sustainable alternative fuels, including biofuels, in aviation in accordance with national circumstances.

The establishment of GFAAF was part of a Declaration adopted at the Rio Conference, which recognized that the introduction of sustainable alternative fuels for aviation will help to address issues of environment, economics, and supply security, and noted *the* very limited availability of qualified alternative fuels for aviation as well as the fact that sustainable alternative fuels for aircraft can be produced from a wide variety of feedstocks for use in global aviation. The Declaration suggested that many regions are candidate production locations and acknowledged that sustainable alternative fuels for aviation may offer reduced lifecycle CO_2 emissions compared to the lifecycle of conventional aviation fuels. It also acknowledged that sustainable alternative fuels for aviation may also offer benefits to surface and local air quality. While the Declaration noted that the technology existed to produce substitute, sustainable fuels for aviation that took into consideration world´s food security, energy and sustainable development needs, it was recognized that the production of sustainable alternative fuels for aviation could promote new economic opportunities.

4.3 Trading and Market Based Measures

At the United Nations Framework Convention on Climate Change 15th Meeting of the Conference of the Parties (COP15) and the Fifth Meeting of the Parties to the Kyoto Protocol (CMP5) which were convened in Copenhagen on 8 December 2009, leaders from the U.S., India, Brazil, South Africa and China came to an agreement to combat global warming. The deal, which was only between five countries, contained no specifics on emissions cuts, but it did commit the countries to look to keep global warming at 2°C or less and to promise $30 billion in funding to battle climate change by 2012. It also created a framework for international transparency on climate actions for developed and developing nations alike.

ICAO now has to tread a thin line of its own diplomacy, in following the principles of sovereignty and equality of States as laid down in the Chicago Convention and balance its leadership role on establishing defining modalities (with the involvement of its member States) for the implementation of the principle of common but differentiated responsibilities.[6] This will be a tough act, but certainly not something that ICAO cannot handle, in the manner in which it has skillfully addressed issues of aviation and the environment in the past.

The adverse effects of aircraft engine emission would inevitably grow with the increasing growth of air transport, which has been, and continues to be, a complex business. Of course, one cannot stultify the growth of this industry as it is vital to the world economy. A survey of the International Air Transport Association carried out in 2007 of over 600 companies from 5 countries reflects that 63% confirmed that air transport networks are vital for their investments and business.[7] Thirty percent of

---

[6] Two basic guidelines used in environmental agreements are The Precautionary Principle, ("PP") and the Common but Differentiated Responsibility Principle ("DR Principle"). The former is a pre emptive measure that addresses possibilities of harm and makes parties to an agreement take measures to anticipate, prevent or minimize the causes of such harm even if there is no scientific certainty regarding the harm. The *Common but Differentiated Responsibilities* principle is enshrined in Principle 7 of the Rio Declaration on Environment and Development which states:

> States shall cooperate in a spirit of global partnership to conserve, protect and restore the health and integrity of the Earth's ecosystem. In view of the different contributions to global environmental degradation, States have common but differentiated responsibilities. The developed countries acknowledge the responsibility they bear in the international pursuit of sustainable development in view of the pressures their societies place on the global environment and of the technologies and financial resources they command.

The core thrust of the principle of *Common but Differentiated Responsibilities* is global partnership and international cooperation. Therefore both developed and developing countries have a common goal of protecting the global environment. Principle 7 also denotes ecological interdependence of all States. A cardinal principle of public international law is sovereignty and the sovereign equality of States. An inevitable corollary to this principle is that legal obligations are based on reciprocity which binds each State equally. However, since States vary in size, capability and economic strength, it is incontrovertible that it becomes difficult to treat equality in substantive terms, thus paving the way for differentiation of responsibility among States.

[7] *Flying Smart, Thinking Big*, Airbus Industrie's Global Market Forecast, 2009–2028, *supra* note 5 at 16.

those countries said that any constraint placed on the air transport industry would make them invest less.

The continuing growth of air transport incontrovertibly means an increase in aircraft engine emissions despite energy efficiency measures employed by the manufacturers of aircraft and components and the airlines themselves.[8] It also means a growing awareness on the part of the aviation industry of the need to introduce and implement measures calculated to reduce emissions and otherwise compensate for the damage that they might cause. An example of such control is the expansion of the European Union's Emissions Trading Scheme (ETS) to aviation, approved by the European Union on 8 July 2008.[9] The European ETS referred to earlier in this article poses an interesting scenario. The key aspects of the proposal are that aircraft operators would be the entities responsible for complying with the obligations imposed by the Scheme. The Scheme would exclude flights by State aircraft, flights under visual flight rules, circular flights ('circuits'), flights for testing navigation equipment or for training purposes, rescue flights and flights by aircraft with a maximum take-off weight of less than 5,700 kg. Under the proposal, each aircraft operator, including operators from third countries, would be administered by one Member State only in order to avoid duplication and an excessive administrative burden on aircraft operators. The Scheme would only cover CO_2 emissions. The Commission will carry out a thorough impact assessment and will put forward a further proposal to address nitrogen oxide (NOx) emissions by the end of 2008. The Scheme requires aircraft operators, like other participants in the Community Scheme, to monitor their emissions of carbon dioxide and report them to the competent authority of their administering Member State by 31 March each year.

For its part, ICAO, at its 36th Assembly, established by Resolution[10] the Group on International Aviation and Climate Change (GIACC). GIACC has its genesis in Appendix K of Assembly Resolution A36-22 which provides *inter alia* that the Assembly requests the Council of ICAO to form a new Group on International Aviation and Climate Change composed of senior government officials representative of all ICAO regions, with the equitable participation of developing and developed countries. Technical support for GIACC is provided by the Committee

---

[8] See generally Abeyratne (2008).

[9] On 8 July 2008, the European Parliament voted to expand the European Union Emissions Trading Scheme to cover aviation emissions from January 2012. What the EU ETS broadly proposes is that operators be allocated allowances each giving them a right to emit 1 tonne of carbon dioxide per year. The total number of allowances allocates a limit on the overall emissions from the activities covered by the Scheme. By 30 April each year operators must surrender allowances to cover their actual emissions. Operators can trade allowances so that emissions reductions can be made where they are most cost-effective.

[10] Resolution A36-22 (Consolidated statement of continuing ICAO poicies and practices related to environmental protection), *Assembly Resolutions in Force* (as of 28 September 2007), Doc 9902, ICAO: Montreal at 1–54.

4.3 Trading and Market Based Measures

on Aviation Environmental Protection (CAEP),[11] for the purpose of developing and recommending to the Council an aggressive programme of action on international aviation and climate change, based on consensus, and reflecting the shared vision and strong will of all ICAO member States. Since its establishment, GIACC has had two meetings, in February 2008 and July 2008 respectively.

The Fourth Assessment report produced by the Inter Governmental Panel on Climate Change (IPCC)[12] released in 2007 bears strong evidence of continuing global warming:

> Warming of the climate is unequivocal, as is now evident from observations of increases in global average air and ocean temperatures, widespread melting of snow and ice, and rising average sea level.[13]

The Report further states that observational evidence from all continents and many natural systems are being affected by regional climate change, particularly through temperature increases.[14] The results of this increase in temperature through

---

[11] ICAO's current environmental activities are largely undertaken through the Committee on Aviation Environmental Protection (CAEP), which was established by the Council in 1983, superseding the Committee on Aircraft Noise (CAN) and the Committee on Aircraft Engine Emissions (CAEE). CAEP is composed of members and observers. CAEP assists the Council in formulating new policies and adopting new Standards on aircraft noise and aircraft engine emissions. CAEP's Terms of Reference and Work Programme are established by the Council. The current structure of the Committee includes five working groups and one support group. Two of the working groups deal with the technical and operational aspects of noise reduction and mitigation. The other three working groups deal with technical and operational aspects of aircraft emissions, and with the study of market-based measures to limit or reduce emissions such as emissions trading, emissions-related charges and voluntary. The support group provides information on the economic costs and environmental benefits of the noise and emissions options considered by CAEP.

[12] The IPCC is a scientific intergovernmental body set up by the World Meteorological Organization (WMO) and by the United Nations Environment Programme (UNEP). The IPCC was established to provide the decision-makers and others interested in climate change with an objective source of information about climate change. The IPCC does not conduct any research nor does it monitor climate related data or parameters. Its role is to assess on a comprehensive, objective, open and transparent basis the latest scientific, technical and socio-economic literature produced worldwide relevant to the understanding of the risk of human-induced climate change, its observed and projected impacts and options for adaptation and mitigation. IPCC reports should be neutral with respect to policy, although they need to deal objectively with policy relevant scientific, technical and socio economic factors. They should be of high scientific and technical standards, and aim to reflect a range of views, expertise and wide geographical coverage.

[13] *Climate Change 2007: Synthesis Report, an Assessment of the Inter-Governmental Panel on Climate Change* topic 1 at 30.

[14] *Id.* 31. The IPCC also states that global atmospheric concentrations of CO_2, CH_4 and N_2O have increased markedly as a result of human activities since 1750 and now far exceed pre-industrial values determined from ice cores spanning many thousands of years. The atmospheric concentrations of CO_2 and CH_4 in 2005 exceed by far the natural range over the last 650,000 years. Global increases in CO_2 concentrations are due primarily to fossil fuel use, with land-use change providing another significant but smaller contribution. The Report further states that it is very likely that the observed increase in CH_4 concentration is predominantly due to

global warming include but are not limited to enlargement and increased numbers of glacial lakes; increasing ground instability in permafrost regions and rock avalanches in mountainous regions; and changes in some Arctic and Antarctic ecosystems.

One of the more ominous statements of the Report is that continued emissions from greenhouse gases[15] at or above current rates would cause further warming and induce many changes in the global climate system during the twenty-first century that would very likely be larger than those observed during the twentieth century.

Warmer surface temperatures and warmer oceans give rise to increased evaporating water, resulting in increases in moisture in the atmosphere. Experts have attributed these increases in moisture and humidity in the recent past to the frequency and intensity of hurricanes and cyclones which occurred in the recent past, particular in reference to the strength and duration of the storms.

It must be noted that the IPCC special report on *"Aviation and the Global Atmosphere"*[16] developed in response to a request from ICAO, addressed sector specific impact of aviation on climate change based on evaluation of consequences of greenhouse gases from aircraft engines and the potential effects from aviation on stratospheric ozone depletion and global climate change. The Report came to the conclusion that the contribution made by aviation to the total radiative forcing (which, in other words is a measure of change in climate) from all human activities was 3.5% and that this percentage which did not take into consideration the effects of possible changes in cirrus clouds was expected to grow.[17]

Although the effects of climate change are not so prominent in the human sphere as in the natural world, there is palpable evidence that problems related to the supply of water will increasingly appear as a result of shrinking glaziers, drought, snow pack, evaporation and the infiltration of salt water in areas below sea level. The IPCC Assessment Report suggests mitigation measures such as demand side management brought to bear through behavioural changes that promote conservation of energy; introduction of technologies that improve energy efficiency; and carbon capture and storage.[18] The potential costs involved in affecting such mitigation is set at 1% of the global GDP.

At its 37th Session, held in Montreal from 28 September to 8 October 2010, The ICAO Assembly considered a proposal[19] from the Secretary General of ICAO

---

agriculture and fossil fuel use. The increase in N_2O concentration is primarily due to agriculture. *Id.* Topic 2 at 37.

[15] Carbon dioxide, methane, nitrous oxide, hydro fluorocarbons, per fluorocarbons and sulphur hexafluoride.

[16] *Aviation and the Global Atmosphere*, Intergovernmental Panel on Climate Change (WMO, UNEP), A Special Report of IPCC Working Groups I and II, Cambridge University Press: Cambridge, 1999.

[17] *Id.* at 21.

[18] *Climate Change 2007...*, *supra*, note 13 at 37.

[19] A37-WP/262 EX/53.

which suggested that the Assembly adopt a resolution that required ICAO to exercise continuous leadership on environmental issues relating to international civil aviation, including greenhouse gas emissions; to continue to study policy options to limit or reduce the environmental impact of aircraft engine emissions and to develop concrete proposals and provide advice as soon as possible to the Conference of the Parties of the UNFCCC, encompassing technical solutions and market-based measures, and taking into account potential implications of such measures for developing as well as developed countries; and to continue to cooperate with organizations involved in policy-making in this field, notably with the UNFCCC. This was an unusual step – for a proposal to come from the Chief Executive of the ICAO Secretariat rather than from the Council of ICAO which reports to the Assembly – and reflected unequivocally that the Council had unprecedentedly failed to reach consensus on a comprehensive approach to aviation and climate change. The underlying reason for this impasse was that developing States could not agree to ambitious emissions reductions suggested by developed States.

The proposed resolution, which was subsequently adopted by the Assembly, also suggested *inter alia* that States and relevant organizations work through ICAO to achieve a global annual average fuel efficiency improvement of 2% until 2020 and an aspirational global fuel efficiency improvement rate of 2% per annum from 2021 to 2050, calculated on the basis of volume of fuel used per revenue tonne kilometre performed. Also suggested was the fact that ICAO and its member States with relevant organizations work together to strive to achieve a collective medium term global aspirational goal of keeping the global net carbon emissions from international aviation from 2020 at the same level, taking into account the special circumstances and respective capabilities of developing countries, the maturity of aviation markets and the sustainable growth of the international aviation industry. In this regard the proposed resolution suggested that the Council of ICAO consider a *de-minimis* exception for States which do not have substantial international aviation activity levels, in the submission of action plans and regular reports on aviation CO_2 emissions to ICAO. It also invited the Assembly to recognize that in the short term, voluntary carbon offsetting schemes constitute a practical way to offset CO_2 emissions, and invited States to encourage their operators wishing to take early actions to use carbon offsetting, particularly through the use of credits generated from internationally recognized schemes such as the Clean Development Mechanism (CDM).[20]

In addition to the 2% annual improvement in fuel efficiency discussed above, the 37th Session of the Assembly also considered a proposal that the feasibility of more

---

[20] The Clean Development Mechanism (CDM) allows a developed country with an emission-reduction or emission-limitation commitment under the Kyoto Protocol to implement an emission-reduction project in developing countries. Such projects can earn saleable certified emission reduction (CER) credits, each equivalent to 1 tonne of CO_2, which can be counted towards meeting Kyoto targets. See http://www.icao.int/icao/fr/env2010/ClimateChange/Finance_f.htm.

ambitious medium and long term goals, including carbon neutral growth and emissions reductions be further explored. There was also a proposal by three States that a more ambitious goal be set – of carbon neutral growth by 2020 compared to 2005 levels. In response, a developing State took the position that ICAO should be guided by the principle of common but differentiated responsibilities (CBDR) under the UNFCCC; the next task for ICAO is to assist States to achieve the goal of 2% annual fuel efficiency improvement; the goal of carbon neutral growth is not realistic and not fair for developing States; and no States should be allowed to take unilateral actions on market-based measures, which drew some support from other developing States.

The main argument of the developing States at the Assembly was that since the larger quantity of GHG emissions was caused by developed States and that developing States should not be called upon to pay for ambitious emissions reduction levels at the same level as developed States. Furthermore developing States claimed that stabilizing the climate should be based on the principles of equity and common but differentiated responsibilities and those obligations under the framework of UNFCC. They concluded that any measure taken should not unduly curb the development of aviation in developing States.

The challenge faced by the Assembly during its discussions was to achieve consensus on establishing guiding principles when designing new and implementing existing market based measures for international aviation, and to engage in constructive bilateral and/or multilateral consultations and negotiations with other States to reach an agreement on issues such as carbon neutral growth and market based measures; as well as on a *de minimis* threshold of international aviation activity, consistent with 1% of total revenue ton kilometres to market based measures as follow:

a) Commercial aircraft operators of States below the threshold should qualify for exemption for application of MBMs that are established on national, regional and global levels; and
b) States and regions implementing MBMs may wish to also consider an exemption for other small aircraft operators.

Notwithstanding the divergence of views between States on the abovementioned issues, and reservations of some developing States, the Assembly was successful in adopting a Resolution which paves the way forward to more understanding and progress in the years to come.

4.4 Emissions Trading

There has been sustained interest in engine emissions ever since commercial turbojet traffic increased substantially in the 1970s. Ground level Ozone, acid rain and climate change can be related to pollutants from aircraft engines such as nitrogen oxides, hydrocarbons and fine particulate matter. Furthermore, air travel,

4.4 Emissions Trading

unlike other modes of travel; involve long distances at relatively short time spans that can generate local, regional as well as global pollution.

The essential philosophy of emissions-trading[21] in environmental protection is based on a certain flexibility allowed to market forces to reach the lowest cost involved in an operation whilst at the same time achieving an environmental target which has been already set. The word "trading" correctly denotes an exchange, and when applied to the aviation context means a certain trade-off between airlines whose fleets pollute more than others and low polluting airlines. The trade-off could take the form of a "purchase" by the high polluting airline of the reduction level of a low polluting airline. Emissions-trading would encourage airlines to seek innovation in technology and to reduce their emission levels.

The subject of emissions-trading is explicitly addressed in Article 6 of the Kyoto Protocol[22] which states that for the purpose of meeting its commitments under Article 3, any Party included in Annex 1 may transfer to or acquire from, any other such Party emission reduction units resulting from projects aimed at reducing anthropogenic emissions by sources or enhancing anthropogenic removals by sinks of greenhouse gases in any sector of the economy provided the parties

---

[21] The subject of emissions-trading falls within the purview of the Intergovernmental Panel on Climate Change (IPCC), which was established in 1988 by the World Meteorological Organization and the United Nations' Environment Programme (UNEP) to assess the scientific basis and impact of climate change. The IPCC's first scientific report was published in 1990 and recommended the negotiation of a framework convention to combat global warming. The United Nations Framework Convention on Climate Change (UNFCCC) was adopted on 9 May 1992 and the treaty entered into force on 21 March 1994. The UNFCCC or FCCC is an international environmental treaty produced at the United Nations Conference on Environment and Development (UNCED), informally known as the Earth Summit, held in Rio de Janeiro in 1992. The treaty aimed at reducing emissions of greenhouse gas in order to combat global warming. The treaty as originally framed set no mandatory limits on greenhouse gas emissions for individual nations and contained no enforcement provisions; it is therefore considered legally non-binding. Rather, the treaty included provisions for updates (called "protocols") that would set mandatory emission limits. The principal update is the Kyoto Protocol, which has become much better known than the UNFCCC itself. The stated objective of UNFCCC is "to achieve stabilization of greenhouse gas concentrations in the atmosphere at a low enough level to prevent dangerous anthropogenic interference with the climate system."

[22] The Third Conference of the Parties to the United Nations Framework Convention on Climate Change (Climate Change Convention) was held from 1 to 11 December 1997 at Kyoto, Japan. Significantly the States parties to the Convention adopted a protocol (Kyoto Protocol) on 11 December 1997 under which industrialized countries have agreed to reduce their collective emissions of six greenhouse gases by at least 5% by 2008–2012. The Kyoto Protocol to the United Nations Framework Convention on Climate Change is an amendment to the international treaty on climate, assigning mandatory emission limitations for the reduction of greenhouse gas emissions to the signatory nations. Article 1 (a) (v) of the Protocol calls each State Party to achieve progressive or phasing out of market imperfections, fiscal incentives, tax and duty exemptions and subsidies in all greenhouse gas emitting sectors that run counter to the objective of the Convention and application of market instruments. See Kyoto Protocol to the United Nations Framework Convention on Climate change, UN Doc. FCCC/CP/1997/L.7/Add.1.

concerned approve of such trading; and, *inter alia*, such trading actually results in a reduction in emission by sources.

Emissions-trading of levels of pollution between airlines differ fundamentally with the existing expectation of each airline maintaining a standard level of emission by its aircraft. When airlines would trade emission levels, the rates at which their aircraft pollute the atmosphere will be taken as a whole and applicable to a whole fleet, so that an airline which is over and above its permitted pollution level could join with another airline which is below the standard level of pollution required of it, thus making the average pollution between the two more acceptable than if taken individually. This mechanism encourages a low polluting airline to achieve even lower standards, in order to trade its levels with high polluting airlines.[23]

At its Seventh Meeting in February 2007 CAEP considered in detail the principles and modalities of market based measures to limit or reduce emissions as required by the Assembly. At this meeting, CAEP followed guidance given by the ICAO Council with regard to emissions trading as a market based option: that emissions trading is one of the three voluntary flexible mechanisms of the Kyoto Protocol; and that the United Nations Framework Convention on Climate Change (UNFCC)[24] and the Kyoto Protocol confer no guidance in relation to emissions trading Schemes that are not provided for in either of these agreements. Therefore the Council mandated CAEP to continue its work to complete guidance material as requested by the Assembly, while acknowledging that emissions trading is an evolving subject in various fora and that in light of further developments this guidance can be reviewed and updated at the discretion of the Council.[25]

It must also be noted that when the Council handed down its decision to CAEP, it also urged ICAO member States to refrain from unilateral action to implement an emissions trading system for international aviation before the Council reports to the 36th Session of the Assembly in September 2007 on its work.[26]

Also at its Seventh meeting, CAEP reviewed the guidance for States developed by its Emissions Trading Task Force and decided to submit them to the Council. Furthermore, CAEP decided to recommend that the Council adopt the guidance on the use of emissions trading for aviation and that the guidance be published prior to the next Session of the Assembly. However, CAEP cautioned the Council that there remained different views with regard to the possible ways of implementing emissions trading Schemes with respect to geographic scope.

In its Report to the Council, CAEP advised that the guidance material for integrating international aviation into States' emissions trading systems focused on those aspects that required consideration with respect to aviation-specific issues:

---

[23] For more information on emissions trading see Hardeman (2007). See also, Abeyratne (1999a,b); Gander and Helme (1999), 12–14, 28–29.

[24] *Supra*, note 21.

[25] CAEP/7-IP/21, Appendix B at B-1.

[26] *Id.* B-2.

4.4 Emissions Trading

furthermore, the guidance identified options and offered potential solutions where possible. It also addressed the aviation-specific options for the various elements of trading systems, such as accountable entities, emissions sources and species (gases) to be covered, trading units, base year and targets, allowance distribution, monitoring and reporting, and geographic scope.

The Council was also advised that one issue that had been particularly difficult to address during the elaboration of the guidance was that of the geographic scope. The central point of disagreement was whether Contracting States could integrate international aviation emissions from aircraft operators of other Contracting States into their emissions trading schemes without the consent of those States. It was recalled that, at CAEP's request, the Council had considered this issue and in November 2006, the Council had recommended[27] that CAEP adopt the same principle used in other key elements of this guidance, i.e. by including the different options of geographic scope, describing their advantages and disadvantages, and starting to address the integration of foreign aircraft operators on a mutually agreed basis, while continuing to analyze further options.

ICAO has embarked upon numerous studies and developed guidance and policy with a view to reducing aircraft engine emissions based on three approaches: reduction of emissions at source through technological innovation; through operational measures (such as efficient air traffic management)[28] and through market based measures.[29] It is also encouraging that ICAO has introduced a carbon calculator, which is a methodology to calculate the carbon dioxide emissions from air travel for use in offset programmes.[30]

The ICAO Carbon Emissions Calculator allows passengers to estimate the emissions attributed to their air travel. It is simple to use and requires only a limited amount of information from the user. The methodology applies the best publicly available industry data to account for various factors such as aircraft types, route specific data, passenger load factors and cargo carried.

ICAO has a distinct charges policy with regard to emissions related charges, the main characteristics of which are that charges should be clearly related to costs and that there should be no fiscal targets behind the charges. ICAO's policy also clearly states that charging authorities should ensure transparency by making available all

---

[27] C-DEC 179/11.

[28] On 12 September 2008, Air New Zealand Flight 8, the first "gate-to-gate optimized flight," landed at San Francisco International Friday. The 777-200ER flight, dubbed ASPIRE 1 (Asia and South Pacific Initiative to Reduce Emissions), started in Auckland and consumed 4,600 L less fuel than normal using a host of strategies to minimize fuel usage. That translated into 12 tons fewer CO_2 emissions. ASPIRE is a joint initiative among US FAA, Airways NZ and Airservices Australia. The 777-200ER was cleared for a "tailored arrival" – a continuous descent at idle thrust and touched 5 min ahead of schedule. San Francisco International Airport playing a leading role in tailored arrivals, a joint project among Boeing, NASA, FAA and the airport. See http://atwonline.com/news/story.html?storyID=14023.

[29] See Hupe (2008), p. 4.

[30] http://www2.icao.int/public/cfmapps/carbonoffset/carbon_calculator.cfm.

financial data required to determine the basis for emissions related charges. Another important ICAO policy is that all funds collected through emissions related charges must be used to mitigate the environment. Furthermore, the establishment of such charges must respect the non-discrimination principle among categories of users enunciated in Article 15[31] of the Convention on International Civil Aviation.[32] The policy goes on to say that the imposition of charges should minimize competition distortions by not penalizing air transport compared with other modes of transport and that when charges are established the interests of all parties concerned should be considered, notably the impact of such charges on the developing world.[33]

4.5 The Carbon Market

The World Bank reports that the carbon market[34] grew in value to an estimated US $30 billion in 2006, which is a threefold increase from 2005.[35] In its Report the World Bank gives two examples concerning Canada and the United States and states that the climate change announcement by the Government of Canada on April 26, 2007 calls for improvements in carbon intensity leading to an emission target of 20% below 2006 levels by 2020.

---

[31] The basic philosophy of Article 15 of the Convention on International Civil Aviation is that every airport in a Contracting State which is open to public use by its national aircraft shall likewise be open under uniform conditions to the aircraft of all the other Contracting States. It also requires that uniform conditions shall apply to the use, by aircraft of every Contracting State, of all air navigation facilities, including radio and meteorological services, which may be provided for public use for the safety and expedition of air navigation. Article 15 subsumes three fundamental postulates:

a) uniform conditions should apply in the use of facilities provided by airports and air navigation services;
b) aircraft operators should be charged on a non-discriminatory basis; and
c) no charges should be levied for the mere transit over, entry into or exit from the departure of a Contracting State.

[32] Convention on International Civil Aviation, signed at Chicago on 7 December 1944. ICAO doc 7300/9 Ninth Edition: 2006.

[33] *ICAO Environmental Report 2007*, ICAO: Montreal at 102.

[34] Emissions trading, as set out in Article 17 of the Kyoto Protocol to the UNFCC, allows countries that have emission units to spare – emissions permitted them but not "used" – to sell this excess capacity to countries that are over their targets. Thus, a new commodity was created in the form of emission reductions or removals. Since carbon dioxide is the principal greenhouse gas, people speak simply of trading in carbon. Carbon is now tracked and traded like any other commodity. This is known as the "carbon market". The Kyoto Protocol is an international agreement linked to the United Nations Framework Convention on Climate Change. The major feature of the Kyoto Protocol is that it sets binding targets for 37 industrialized countries and the European community for reducing greenhouse gas (GHG) emissions. These amount to an average of 5% against 1990 levels over the 5-year period 2008–2012. For details on the Kyoto Protocol, see *supra*, note 30.

[35] *State and Trends of the Carbon Market*, The World Bank, Washington D.C, May 2007 at 3.

4.5 The Carbon Market

The approach incorporates emissions trading and also includes the idea of early action and banking and allows Certified Emission Reductions (CERs) for up to 10% of the projected shortfall. If these assumptions are true, then some demand from Canada could enter the CER market relatively soon.

Also cited are developments in California, the eastern United States and Australia, which show a marked proclivity toward continuous market growth beyond 2012. There is continued debate, especially in California, whether emissions trading, including offsets from overseas will be allowed. Precise rules to be developed will clarify to what extent these emerging carbon markets will seek to maximize value from high quality offsets no matter where they are sourced from. At least two pending pieces of draft federal legislation before the U.S. Senate include provisions that would welcome overseas credits.

The World Bank also reports that the carbon market and associated emerging markets for clean technology and commodities have attracted a significant response from the capital markets and from experienced investors, including those in the United States. Analysts estimated that US$11.8 billion (€9 billion) had been invested in 58 carbon funds as of March 2007 compared to US$4.6 billion (€3.7 billion) in 40 funds as of May 2006.[36] Fifty percent of all capital driven to the carbon value chain is managed from the UK.[37] Most of the newly raised money, of private origin, came to the sell-side (project development and carbon asset creation) which currently represents 58% of the capitalization.

The genesis of carbon trading, at the international plane, was at the United Nations Conference on Environment and Development that was convened in 1992 in Rio de Janeiro. This "Earth Summit" as it was popularly called, attracted a staggering 25,000 delegates, including a large segment of the world's political leaders. The conclusions of the summit were summarized in a document formally titled *The United Nations Framework Convention on Climate Change* (UNFCC).[38] The Convention was to enter into force on the ninetieth day after the date of deposit of the fiftieth instrument of ratification, acceptance, approval or accession. This milestone was passed on March 21, 1994, after Portugal became the fiftieth country to register ratification on December 21, 1993. As of December 10, 1999, as at 22 August 2007, the Convention had been ratified by 192 countries, including the United States.

The Convention calls on developed countries, the so-called Annex I Parties, to return their emissions to 1990 levels by 2000. When countries adopted the Convention, however, they knew that these commitments would not be sufficient to meet its objective. Therefore, at the first Conference of the Parties (COP) to the Convention (March 1995, Berlin), negotiations were launched to agree on stronger commitments for developed countries by the third COP. Intensive and difficult negotiations took place over 31 months and these were finally concluded at COP 3 in Kyoto on 11

---

[36] See R. Bulleid, The Capital Begins to Flow, *Environmental Finance*, April 2006 at 34.

[37] New Carbon Finance, "UK in Pole Position as Carbon Funds Surge – but More Funds required". Press Release 4 April 2007, at www.newcarbonfinance.com.

[38] *Supra* note 21.

December 1997, when the Kyoto Protocol to the UNFCCC was eventually adopted. Article 17 of the Protocol provides that The Conference of the Parties shall define the relevant principles, modalities, rules and guidelines, in particular for verification, reporting and accountability for emissions trading.

The main thrust of the Kyoto Protocol and its ultimate aim is to achieve an overall reduction in the emissions of greenhouse gases by imposing certain quantified limitation and reduction obligations to the member States of the Organization for Economic Cooperation and Development (OECD) and East European countries. Therefore the Protocol has attempted to coalesce various interests under its scope. The main objective of States, both developed and developing, in buying in to the Protocol was to ensure that the Protocol would not adversely affect strong and growing national economies. Some developing States objected to the Protocol and its objectives on the ground that they should not be subject to emissions control under the Protocol since industrialized countries had a proven responsibility for causing the bulk of greenhouse gas emissions in the atmosphere. In order to respond to this claim, the Protocol has set up burden-sharing obligations.

On the implicit recognition that developed States contributed most to global warming, they (the developed States) agreed to reduce their overall emissions of greenhouse gases by at least 5% below 1990 levels in the first commitment period of 2008–2012. It must be noted that developed States stand as the only ones who have accepted quantified emission limitation or reduction commitments under the Protocol, which imposes differentiated emission reduction targets, such as 8% reduction for the European Community and 7% reduction to the United States. This responsibility does not devolve upon developing States in the first commitment period.

Carbon trading[39] essentially involves purchase contracts whereby one party pays another party in return for greenhouse gas (GHG) emissions reductions or for the right to release a given amount of GHG emissions, that the buyer can use to meet its compliance or corporate objectives concerning climate change mitigation.[40]

---

[39] Carbon trading, which is a species of the genus emissions trading is essentially market based and is usually based on a model that first that requires an environmental authority to decide on an acceptable level of overall emissions. The level, or target, once identified, gives rise to permits which are issued consistent with that target, each of which confers the right to release a certain amount of pollution over some period of time. The firms which are issued with these permits, be they airlines or other service and goods providers, may apply these permits to their own emissions, sell excess permits to other pollution sources, or purchase permits from other firms if their emissions exceed their permit holdings.

An emissions permit market could be established provided there is extensive coverage of a permits system and there are no barriers to trading in such a market.

[40] In addition to carbon trading there is also the concept of a carbon tax which is essentially an economic instrument. It is a tax levied on the total quantity of greenhouse gases emitted. The European Union once considered a carbon tax as a viable economic instrument that could be effective in curbing emissions. However, a carbon tax provided less flexibility for member states since E.U. members have varying amounts of pollution and wealth. Since a tax would impose the same levy on all countries, rich and poor, and some members would have been paying a great deal more in taxes than others, the EU's conclusion on a carbon tax was that, using a trading scheme

4.5 The Carbon Market

Trading is conducted by making payment using cash, equity, debt, convertible debt or warrant, or in-kind contributions such as providing technologies to abate GHG emissions. Carbon transactions may either take the form of allowance based transactions where -the buyer purchases emission allowances created and allocated (or auctioned) by regulators under cap-and-trade regimes, or under Project-based transactions, in which the buyer purchases emission credits from a project that can verifiably demonstrate GHG emission reductions compared with what would have happened otherwise.

Another aspect to carbon trading is carbon offsetting, which is the compensating for carbon emissions resulting from human activity. Simple human activity such as putting on a cooker or electric iron involves the production of carbon emissions that contribute to climate change. Offsets are traded in the carbon market. Therefore, simply put, a carbon offset is an emission reduction credit from another organization's project that results in less carbon dioxide or other greenhouse gases in the atmosphere than would otherwise occur. The ICAO carbon calculator, which has already been mentioned, is a tool that assists in calculating the weight of carbon dioxide produced as a result of a passenger's air travel that enables the passenger concerned to take effective steps in compensating for the damage caused by his carbon footprint. David Suzuki, the renowned and prize winning environmental scientist gives the example of wind energy companies that often sell carbon offsets.[41] The wind energy company benefits because the carbon offsets it sells make such projects more economically viable. The buyers of the offsets benefit because they can claim that their purchase resulted in new non-polluting energy, which they can use to mitigate their own greenhouse gas emissions. The buyers may also save money as it may be less expensive for them to purchase offsets than to eliminate their own emissions.

There are numerous types of activities that could give rise to carbon offsets. Installations of solar, small hydro, geothermal, and biomass energy or renewable energy such as the energy produced by wind farms as already mentioned can all create carbon offsets by displacing fossil fuels. Other types of offsets available for sale on the market include those resulting from energy efficiency projects, methane capture from landfills or livestock, destruction of potent greenhouse gases such as halocarbons, and carbon sequestration projects (through reforestation, or agriculture) that absorb carbon dioxide from the atmosphere.

Since the offset market is mostly unregulated, buyers need to select their offsets cautiously and ensure that the offsets they purchase meet recognized standards. Carbon offsetting helps reduce greenhouse gas emissions; and other atmospheric

---

instead would improve flexibility, particularly if, along the lines of current practice in the EU, under a cap-and-trade system the E.U. could set one overall cap on emissions, while allocating allowances to each country based on its individual emissions. Such a trading scheme meant less pressure on industry if the allowances were initially issued for free, whereas a tax (or auctioned allowances) charges industry for all residual emissions See generally, Owen and Hanley (2004).

[41] http://www.davidsuzuki.org/Climate_Change/What_You_Can_Do/carbon_offsets.asp.

pollutants; facilitate in the restoration of degraded lands; improve watersheds and water quality, protect endangered species; creates jobs; and helps save money on electricity and gasoline.

4.6 Consequences of Carbon Trading

There will be palpable trade consequences as a result of the Kyoto Protocol which started the concept of carbon trading. There is little doubt that greenhouse gas emission reduction will affect various sectors in the world economy, notably energy, production and transport. The commitment and action of States coming within Annex 1 to the Protocol that is calculated to meet GHG reduction targets assigned to them will affect the cost of production of traded products which will in turn affect their competitiveness in the World market. Overall trade and investment will be adversely affected as a result of the reduction in Annex I countries of the production of GHG intensive products. The result will be a lowering of the demand for industrial goods and services elsewhere leading to a decrease in the growth of overall trade and investment. Furthermore, the demand in Annex I countries for industrial products from non-Annex I countries that are not facing the Kyoto Protocol's emission reduction commitments will be increased as the non Annex 1 States would be able to produce more cost effectively.

The desired consequences of carbon trading will depend on the rate of compliance and the effectiveness of the regulatory process that implements and monitors the process. In this context a regulatory process which reaches out beyond its territorial boundaries with a trading scheme (such as the EU ETS) may run the risk of being accused of extra territorial application of that scheme. As a legal concept, the issue of emissions trading brings to bear the issue as to whether a single State or a group of States could unilaterally impose a trading scheme on air carriers requiring the acquisition by such carriers of carbon credits pertaining to emissions over territories other than the territories of the imposing State or States. The question that naturally arises is whether such a measure would be an extra territorial application of laws.

Extra territorial jurisdiction is exercised when a State (or in this case a community of States) seeks to apply its laws outside its territory in such a manner as may cause conflicts with other States.[42] It can be justified by the invocation of the effects doctrine or the "effects theory" which goes beyond the principles of sovereignty. This theory relates to a situation where a State assumes jurisdiction beyond its territorial limits on the ground that the behaviour of a party is adversely affecting

---

[42] There is a general common law presumption against the extra territorial application of legislation. See the House of Lords decision in *Holmes* v. *Bangladesh Biman Corporation* [1989] AC 1112 at 1126; 87 ILR 365 at 369. Also, *Air India* v. *Wiggins* [1980] 1WLR 815 t 819; 77 ILR 276 at 27.

4.6 Consequences of Carbon Trading

the interests of that State by producing "effects" within its territory. It does not matter whether all the conduct and practices take place in another State or whether part of the conduct is within the State adopting the legislation. In the latter instance, the conduct of the party would come under the "objective territorial principle" where part of the offence takes place within the jurisdiction. In the case of aircraft engine emissions, the applicable principles would come under both headings as trans-boundary pollution of the environment by an aircraft which flies into Europe may involve the emissions of gases in one State that could cross boundaries and affect Europe.

The effects doctrine has been robustly applied in the United States particularly in the field of antitrust legislation.[43] Judicial recognition of the principle lay in the premise that any State may impose liabilities, even upon persons not within its allegiance, for conduct outside its borders that has consequences within its borders which the State reprehends.[44] This blanket principle was later toned down within the United States to acknowledge growing international protests against the wide ranging and arbitrary manner in which the principle could be applied. The modification involved the need to prove intentional conduct and the fact that the effect should be substantial for the doctrine to be applied.[45] In addition, courts began to insist on a jurisdictional rule of reason that involved consideration of interests of other nations and the nature of relationship between the US and the actors concerned. It is also noteworthy that the *Third Restatement of Foreign Relations Law* provides that a State may exercise jurisdiction based on the effects in the State when the effect or intended effect is substantial and the exercise of jurisdiction is reasonable.[46] Reasonableness is based on the extent the enacting State limited its jurisdiction so as to obviate conflict with the jurisdiction of the State affected to the extent possible.

The 1984 case of *Laker Airways* v. *Sabena*[47] held that once law was declared applicable it could not be subject to qualification or ignored by virtue of comity.[48] However, changes could be effected through diplomatic negotiations. The United

---

[43] See The *US Sherman Antitrust Act 1896* 15 USC paras 1ff.

[44] *US.* v. *Aluminium Company of America*, 148 F.2d 416 (1945).

[45] *Timberlane Lumber Company* v. *Bank of America* 549 F.2d 597 (1976); 66 ILR, 270. Also, *Mannington Mills* v. *Congoleum Corporation*, 595 F.2d 1287 (1979); 66 ILR, 487.

[46] The Third Restatement constitutes a comprehensive revision of the earlier (1965) Restatement, covering many more subjects, and reflecting important developments in the intervening decades. This Restatement consists of international law as it applies to the United States, and domestic law that has substantial impact on the foreign relations of the United States or has other important international consequences.

[47] 731 F.2d 909 (1984).

[48] Comity, at law, refers to legal reciprocity where one jurisdiction will extend certain courtesies to other nations, particularly by recognizing the validity and effect of their executive, legislative and judicial acts. The term refers to the idea that courts should not act in a way that demeans the jurisdiction, laws, or judicial decisions of another country. It is especially important in the application of principles of public international law. Part of the presumption of comity is that other nations will reciprocate the courtesy shown to them.

States Supreme Court ruled in 1993 that US legislation (in this case the *Sherman Act*) could apply to foreign conduct that was meant to produce some substantial effect in the United States.[49] Extra territorial application of laws can be effectively rendered destitute of effect by blocking legislation[50] which a State can enact to preclude the application of a foreign law to citizens of that State.

In several instances, the United States has controlled or influenced activities occurring outside its borders which are calculated to harm the environment. For example: Congress passed a law prohibiting persons and vessels subject to the jurisdiction of the United States from "taking" (killing or injuring) marine mammals on the high seas; The EPA issued subpoenas to American companies demanding information on the use and release of chemicals from companies operating in Mexico, with a view to curbing pollution from the New River in Mexico from flowing into the United States; and Congress passed a law banning the import of ivory from countries that did not have an elephant protection programme, so that the numbers of elephants in Africa and Asia would not decrease due to poaching.

The United States has also used trade and investment measures to influence the conduct of other States. For example, during the 1990s, Congress drew a link between the human rights record of China with most-favoured nations treatment of the World Trade Organization. There have also been instances where goods from States are banned from importation to the United States unless that State complies with certain standards set in U.S. law. Conversely US exports are banned from import into those countries.

In every instance of extra territorial jurisdiction, there are two issues to be considered: the first is whether the State or group of States has the authority to exercise extra territorial jurisdiction; and the second is, whether the exercise of that authority reasonable (taking into consideration the law concerned and the potential foreign policy conflicts).

The fundamental issue that affects aviation in the context of emissions trading is that the air transport industry is quite different from most other trade sectors, particularly in respect of how greenhouse gas emissions of aircraft are quantified and evaluated under the UNFCC. Part of the problem is that aircraft emissions released internationally are often of a trans-boundary nature. Therefore, emissions from aircraft involved in international carriage are not included in the national emissions totals that the parties to the Convention are required to report. Such

---

[49] *Hartford Fiore Insurance Company* v.*California*, 113 S. Ct. 2891 (1993) per Souter J.

[50] The most common instance of blocking legislation concerns the prevention of private information being demanded and obtained from nationals of a State by another State. Several countries have enacted so-called "blocking legislation". Blocking legislation mandates the confidentiality of information and documents and attempts to block foreign efforts to obtain evidence from residents of the enacting jurisdiction. It is often enacted by countries seeking to foster banking and financial industries, such as Switzerland, the Bahamas, Panama and Vanuatu. It generally prohibits residents of those countries and corporations doing business there disclosing confidential business information about others doing business there.

emissions are accordingly not subject to the quantified emissions limitations accepted by the State Parties to Annex 1 of the Kyoto Protocol. This essentially means that the collective flavour permeating the obligations of these parties does not apply to international air transport, which in turn encourages individual and unilateral action on the part of States or groups of States. The consequences of this phenomenon have already been discussed.

The only solution therefore seems to lie with a global mechanism for trading, and that is provided for by the Kyoto Protocol which explicitly provides that States should collectively pursue the limitation or reduction of aviation through ICAO. Principle 12 of the 1992 Rio Declaration on Environment and Development states:

> ...Unilateral actions to deal with environmental challenges outside the jurisdiction of the importing country should be avoided. Environmental measures addressing transboundary or global environmental problems should, as far as possible, be based on international consensus.

Member States of ICAO, at the Organization's 36th Session of the Assembly (Montreal, 18–28 September 2007) adopted Resolution A36-22,[51] which, in Appendix A declares that ICAO is the lead United Nations Agency in matters involving international aviation,[52] and urges ICAO member States, in Appendix L not to implement an emissions trading system on other member States' aircraft operators except on the basis of mutual agreement between those States.[53] It is encouraging that the formal EU position with regard to ICAO is that ICAO is the natural forum to develop a global solution.[54] Mr. Daniel Calleja, Air Transport Director at the European Commission, in an interview with the *ICAO Journal* stated that the EU was committed to processes that have been launched by ICAO's GIACC. Mr. Calleja also stated at the same interview that the EU is ready to change its position if a more global solution can be agreed upon.[55] This could be the ultimate approach for a globally acceptable trading mechanism.

4.7 The Emissions Trading Scheme of the European Union

The trading mechanism works through offsetting. Generally speaking, an offset is a compensating equivalent. Arguably, buying and selling carbon credits is not trading in the real sense as would be in trading in a commodity such as gold or textiles.

---

[51] Consolidated statement of continuing ICAO policies and practices related to environmental protection, *Assembly Resolutions in Force* (As of 28 September 2007), ICAO Doc 9902, 1–54 to 1–74.
[52] *Id*, Appendix A, at 1–55.
[53] *Id*. Appendix L, at 1–73.
[54] Striving Toward Meaningful Solutions, an Interview with Mr. Daniel Calleja, *ICAO Journal*, Issue 04, 2008 14–15 at 15.
[55] *Ibid*.

A credit is not a commodity. The IATA Guidelines and Toolkit on Aviation Carbon Offset Programmes calls offsetting an activity which can mean to balance, cancel out or neutralise.[56] In the context of addressing climate change concerns, offsetting is an action by companies or individuals to compensate for greenhouse gas emissions, in this case arising from their use of commercial aviation. The offset can be equivalent in part or in whole to the associated emissions, by financing a reduction in emissions elsewhere.

Countries, companies or individuals can purchase offsets, in either CO_2 or an equivalent offset by another greenhouse gas, with a view to reducing their net carbon emissions. There are many different ways to achieve CO_2 reductions that can be used as offsets, many of which bring other social, environmental or economic benefits relevant to sustainable development. There are significant differences between offset types. Offsets can either be bought from within the international compliance system under the Kyoto Protocol, or in the voluntary market. In the context of these guidelines, offsets are considered to be a voluntary action by airline passengers.

Simplistically stated, when an activity like air travel produces CO_2 emissions, these emissions can be offset by preventing or reducing a similar amount of emissions somewhere else. This can be achieved either by the airline itself or by its customers. Such offsets can be sourced from various types of project activities – for instance, through forestation or renewable energy projects and can be purchased through specialised offset providers or carbon brokers. The buyer then receives a certificate or record from the seller providing details about the project and the amount of CO_2 reduced.

According to IATA Guidelines there are six fundamental principles of offsetting: Firstly, offsetting should be complimentary in that offsets and trading should be seen as part of wider efforts to reduce emissions alongside technological and operational improvements in fuel efficiency. Offset programmes will only be credible if they are coupled with serious efforts to minimise the company's CO_2 emissions first. Secondly, offsetting should be an activity that is additional to normal business activity. This means that a key requirement for an offset is that the CO_2 reduction or removal used as an offset be 'additional' to business-as-usual activity. Demonstrating additionality is complex, but a number of approaches have been used successfully to ensure the environmental integrity of offsets. Thirdly, Offsetting should be verifiable. This devolves the responsibility on airlines as well as States concerned to maintain records of aircraft CO_2 emissions from operations covered by the offset programme and have them externally verified by an independent third party. The fourth principle pertains to registration whereby CO_2 reductions from offset projects should be recorded and tracked through a central registry, with the amounts purchased progressively subtracted from the total determined for that particular project. The fifth postulate is traceability where the receipt

---

[56] Version 1 – May 2008.

issued to the customer should clearly indicate that the credit has been/will be retired as a result of the purchase and cannot be resold. A receipt may also indicate the type of project that was used to generate the offset, or the quality standard that the offset meets. The final requirement is a guarantee. If an offset is sold where the purchased reduction in CO_2 will be achieved at some future date, then a guarantee that an alternative and equivalent offset will be made if the project fails should be provided. IATA suggests that preferably only offsets already achieved are included.

There are various ways of reducing emissions. An effective way is through operational measures. Improvements in operational performance can deliver substantial emissions reductions and can be implemented in the shorter term. Solid progress in this area was observed in the last few years with the introduction of regional initiatives such as AIRE (Atlantic Interoperability Initiative to Reduce Emissions) and ASPIRE (Asia and South Pacific Initiative to Reduce Emissions) as well as the more close-in procedures such as CDOs (Continuous Descent Operations). In light of the environmental benefits of operational improvements, there is a need to develop methods to calculate and monitor such benefits in a harmonized way, from the operational and environmental perspectives. Several measures have been outlined in this area whereby Regional Planning and Implementation Groups (PIRGS), ANC panels, and CAEP will increase collaboration.

Air operators can cooperate with governments and air navigation service providers to use shorter routes. They can effect continuous descent approaches and other efficacious operational methods. They can maximize on altitudes and engage speeds which are least injurious to the climate. They can operate on appropriate load factors and use the most suited aircraft on particular routes, planning their fleets wisely. Estimation of the carbon footprint is another way of evaluating the quantum of offsets required. The ICAO Carbon Calculator of 2008[57] allows passengers to estimate the emissions attributed to their air travel through a simple interface that requires the user to enter only their origin and destination airports, and their class of travel. The methodology used by the calculator applies the best publicly available industry data to account for various factors such as aircraft types, route specific data, passenger load factors and cargo carried. ICAO, working through CAEP, will continue to improve the fidelity of the Calculator through a transition to more detailed sources of modelled aircraft performance data and ultimately to the integration of measured fuel consumption information.

The ICAO Carbon Emissions Calculator recommends several measures. The first method is user input. Based on user input, the airline's booking system defines the itinerary and it specifies origin, destination and any stopover airports. This will normally include code share and other sectors paid for through that airline. The second is the calculation of the distance of the trip concerned. The Great Circle Distance (GCD)[58] between two airports is calculated using longitude and latitude

---

[57] The ICAO Carbon Calculator can be accessed through ICAO's website at www.icao.int.
[58] The GCD is the shortest path between two points on the surface of the sphere.

coordinates. A correction factor can be used to take account of delays and wind and weather conditions en-route. The third method relates to the aircraft type where it becomes necessary to define the type(s) of aircraft used to fly the specified itinerary. If actual data is not used, it is suggested to use information from flight schedules. The next is to calculate total fuel burn. In order to determine the total fuel burn for the flight(s) the use of actual trip fuel data would give the most reliable results. Another method would be to establish the passenger-related fuel use for the flight the total fuel burn is divided between the number of passengers and the tonnage of mail and freight using load factor data. Unless actual flight data is used, average passenger and freight load factors can be used to establish the ratio to make this division. Seat capacity and load factor also play a role. Here, the passenger-related fuel use for the flight is divided by the actual number of passengers on the flight. If actual numbers are not used, some assumptions will need to be made for the seat capacity and passenger load factor on the flight, using either airline or industry averages. These factors can bring one to a formula that would assist in calculating carbon dioxide emitted per passenger on the formula:

Total Fuel Burn *pax to freight ratio* 3.157/seat capacity*passenger load factor (3.157 being the factor used to convert fuel to carbon dioxide)

The above notwithstanding, it should be noted that carbon dioxide is not the only greenhouse gas and that 30% of greenhouse gases generated by human activity are those other than carbon dioxide. Two leading authorities have stated:

> A scientifically sound solution for the inclusion of non-CO2 effects in an emissions trading progamme (or other approach) would eventually call for something other than a simple multiplication factor. Such a simple multiplication factor would weaken incentives to reduce the total climate impact beyond a reduction of the fuel consumption, which is to say there would be no benefit in reducing non-CO2 effects.[59]

4.8 EU ETS and the Trading Mechanism

On 1 January 2005, the European Emission Trading System (EU ETS) started operation. It represents the spearhead and "one of the most important instruments" of EU climate policy due to its ability to achieve absolute emission reductions in an economically efficient manner. The first phase of the EU ETS (2005–2007) successfully established free trade of emission allowances across the EU, set up the necessary infrastructure for monitoring, reporting, verification including registries and has so far successfully concluded two compliance cycles. It developed into the world's largest single carbon market accounting for 67% in terms of volume and 81% in terms of value of the global carbon market and also worked as the driver of the global credit market and in that triggered investments in emission reduction projects indirectly linking 147 countries to the EU ETS through JI/CDM projects.

---

[59] Sausen and Schumann (2007).

4.8 EU ETS and the Trading Mechanism

Directive 2008/101 which extended the EU ETS to aviation came into effect on 2 February 2009. The total quantity of allowances to be allocated to aircraft operators for the period 1 January to 31 December 2012, will be equivalent to 97% of the mean average of annual emissions in the calendar years 2004, 2005 and 2006, as determined by the European Commission. Fifteen percent of these allowances are to be auctioned, with the remaining allocated as "free" allowances. The total quantity for the 8 year period beginning on 1 January 2013 will be equivalent to 95% of historic aviation emissions. Of these allowances 15% shall be auctioned, 3% set aside in a special reserve (for new entrants or expansions) with the remaining allocated as 'free' allowances.[60] Each Member State is required to determine the total number of free allowances to be distributed to aircraft operators. The amount of allowances apportioned to each operator is to be calculated from the product of the 'benchmark' (allowance per tonne-kilometre) and an operator's verified 'activity' (expressed in tonne-kilometre (TKM)). The benchmark is to be determined by the Commission using the following formula:

Benchmark = Total number of allowances to be allocated free of charge/Sum of all tonne-km data.

It has been estimated that the EU ETS will cause the airlines of the European Regions Airline Association (ERA) an additional cost of €245 million for the period 2012–2013.[61] The European Emissions Trading Scheme (ETS) trading mechanism is quite specific and explicit guidelines have been set by the European Union. These guidelines, developed by the Dutch Emissions Authority in conjunction with the United Kingdom Environment Agency, have as their genesis Directive 2008/101/EC amending Directive 2003/87/EC issued on 13th January 2009 and as published to include the aviation sector into the European Greenhouse Gas Emission Allowance Trading Scheme (EU ETS scheme). As part of the obligations under the European Emissions Trading Scheme, aircraft operators will have to monitor and report data regularly from 2010 onwards. There are EU-wide provisions, adopted by the Commission in April 2009, on what data must be monitored and reported, and how this should be done. These are laid down in the Monitoring and Reporting Guidelines (MRG). One of the requirements from the MRG is for aircraft operators to submit a monitoring plan by 31st August 2009. The Commission will publish templates for monitoring plans and for reporting. These templates may be adapted by EU Member States provided they contain at least the same data input as the templates published by the European Commission.

Under the EU ETS, aircraft operators flying to and from airports in 30 European states from 2012 will surrender allowances in respect of their CO_2 emissions on an annual basis. The large majority (85%) of allowances will be allocated to individual

---

[60] See *Environmental Newsletter*, Clyde And Company, March 2010, at 7.
[61] Lorna Reader, EP Leaves Aviation ETS Unchanged, *Regional International*, February 2009 at 11.

aircraft operators free of charge, based on their respective aviation output (rather than emissions) in 2010, thus rewarding operators that have already invested in cleaner aircraft. The remaining 15% of allowances will be allocated by auction. The scheme also includes a *de minimis* provision under which commercial operators with a low level of aviation activity in Europe are excluded from its scope. This is likely to mean that many aircraft operators from developing countries will be unaffected by the scheme and, indeed, over 90 ICAO states have no commercial aircraft operators included in the scope of the EU ETS.

The EU legislation foresees that, where a third country takes measures of its own to reduce the climate change impact of flights departing from its airports, the EU will consider options available in order to provide for optimal interaction between the EU scheme and that country's measures. In such a case, one option could be that flights arriving from the third country would be excluded from the scope of the EU scheme. The EU therefore encourages other countries to adopt measures of their own and is ready to engage in bilateral discussions with any country that has done so. The legislation also makes it clear that if there is agreement at ICAO on global measures, the EU will consider adapting its ETS accordingly.

No later than 30th April of each year, aircraft operators are required to surrender emission allowances that are equivalent to the amount of emissions that they emitted (and reported to the competent authority) during the previous calendar year. Participants in the ETS scheme who emit more than the emission allowances they have received, can buy allowances from other participating aircraft operators or from operators of stationary installations that also fall within the scope of the EU ETS. Purchasing allowances will ensure that sufficient allowances are surrendered by the 30th of April each year. If participants emit less than the emission allowances allocated to them, the surplus of emission allowances can be sold.

Aircraft operators coming into or going out of European Community aerodromes are required to submit an annual emissions report to the competent authority. This report has to be verified by an independent and accredited verifier prior to submission. In the pre-trading period (2010–2011) you have to submit your first verified emissions report by 31 March 2011 and 2012. You are, however, not required to surrender emissions allowances equivalent to the operator's reported emissions for the years 2010 and 2011. The first trading period starts from 1st January 2012 onwards. Starting from April 2013 aircraft operators will be required to surrender each year emission allowances that cover the verified reported data for the previous year.

Furthermore, before the start of each trading period aircraft operators are required to submit a revised monitoring plan for annual emissions. The first time such a review takes place is before 1 January 2013. In performing the review they would have to assess whether their monitoring methodology can be changed in order to improve the quality of the reported data without leading to unreasonably high costs. The revised monitoring plan for annual emissions needs to be approved by the competent authority of your administering Member State.

4.9 Calculating Emissions

For the first (2012) and the second (2013–2020) trading period each aircraft operator may apply for a free allocation of allowances. Allocation is calculated under the simple formula

$$Allocation\ (tonnes\ CO_2) = Combustion\ emissions\ (tonnes\ CO_2) + Process\ Emissions^{62}\ (tonnes\ CO_2)^{63}$$

For the purpose of applying for an allocation of allowances in accordance with Article 3e(1) or Article 3f(2), the amount of aviation activity shall be calculated in tonne-kilometres using the following formula:

$$tonne - kilometres = distance \times payload$$

where: 'distance' means the great circle distance between the aerodrome of departure and the aerodrome of arrival plus an additional fixed factor of 95 km; and 'payload' means the total mass of freight, mail and passengers carried.

For the purposes of calculating the payload: the number of passengers will be the number of persons on-board excluding crew members, and an aircraft operator may choose to apply either the actual or standard mass for passengers and checked baggage contained in its mass and balance documentation for the relevant flights or a default value of 100 kg for each passenger and his checked baggage.

Emissions will be monitored by calculation. Emissions shall be calculated using the formula:

$$Fuel\ consumption \times emission\ factor$$

Fuel consumption will include fuel consumed by the auxiliary power unit. Actual fuel consumption for each flight shall be used wherever possible and shall be calculated using the formula:

Amount of fuel contained in aircraft tanks once fuel uplift for the flight is complete – amount of fuel contained in aircraft tanks once fuel uplift for subsequent flight is complete + fuel uplift for that subsequent flight.

If actual fuel consumption data are not available, a standardised tiered method shall be used to estimate fuel consumption data based on best available information.

---

[62] The EU does not yet have a formal definition for "process emissions," either. In general terms, though, process emissions refer to those greenhouse gas emissions (or CO_2 emissions in your case) that are the result of a chemical reaction during the processing of a material, as opposed to being the result of combusting a substance.

Combustion emissions are more straightforward, since they are all greenhouse gas (or CO_2) emissions that are the result of combustion (e.g. burning jet fuel).

[63] *EU Calculations Trading Scheme – Calculating the Free Allocation for New Entrants*, A Report Produced for the Department of Trade, November 2004 at 3.

Default IPCC emission factors, taken from the 2006 IPCC Inventory Guidelines or subsequent updates of these Guidelines, shall be used unless activity-specific emission factors identified by independent accredited laboratories using accepted analytical methods are more accurate. The emission factor for biomass shall be zero. A separate calculation will be made for each flight and for each fuel.[64]

The free allowances available to the entire sector will be distributed amongst operators via a benchmarking procedure, where the individual allocation for each operator is proportional to the (reported and verified) tonne kilometres that were flown over the course of the year 2010. Each year from 2012 until 2020 the competent authority will issue the number of allowances allocated to each individual aircraft operator. All aircraft operators falling within the scope of the EU ETS are assigned an "administering Member State". This Member State's national law will apply, and obligations arising from the scheme (e.g. to report annual emissions or surrender allowances) will be towards this state. The administering Member State does not in any way affect the scope of the scheme – the essence of the EU ETS obligations will be the same across the EU.

The EU ETS "cap and trade" scheme is calculated to accord with the principles of the Kyoto Protocol. The scheme, which is currently in the 2008–2012 phase called the "Kyoto Phase" is applicable to approximately 10,000 industries that include, but are not limited to oil drilling and refining, manufacture of ceramic glass and cement products, fossil fuel based energy production pulp, paper and textile manufacture, issues each industry with an allocation or quantity of allowances, each of which allows the concerned industry to emit one million metric tonnes of carbon dioxide. The quantification is based on what are called "historical aviation emissions" that are defined as "the mean average of the annual emissions in the calendar years 2004, 2005 and 2006 from aircraft departing or arriving from a member State of the EU".

The indicator for identifying flights under the ETS is the ICAO designator for airlines. The operator has to monitor all flights under this designator. Where an ICAO designator is not available you are identified as an aircraft operator by the registration markings of aircraft. In that case the operator must monitor all flights under the registration markings. The use of the ICAO designator does not necessarily imply that an aircraft operator is commercially or operationally responsible for a particular flight. This depends in most cases on the type of commercial arrangements between carriers in the aviation sector. Whether code sharing, dry leasing or wet leasing, long or short term leasing is applied by an aircraft operator has no bearing on identifying the aircraft operator. The unique ICAO designator entered in box 7 of the flight plan determines the aircraft operator for a particular flight and thus that operator has to monitor this particular flight. This means that leased-in aircraft and ad hoc or sub charter flights could still be assigned to an operator if his ICAO designator is entered in box 7 of the flight plan. If that operator

---

[64] The 2006 IPCC emission factor for Jet Fuel is 3.157 kg CO_2 per kg of Jet Fuel (http://www.ipcc-nggip.iges.or.jp/public/2006gl/pdf/2_Volume2/V2_3_Ch3_Mobile_Combustion.pdf.

4.9 Calculating Emissions

provides capacity for third parties (e.g. perform ACMI operations, use code sharing or lease in or lease out aircraft), he will have to identify which ICAO designator is used in the flight plans and take only those flights using his own ICAO designator into account. A daughter company does not have to submit a monitoring plan if all flights of the daughter company are performed under the unique ICAO designator of the parent company or sister company, entered in box 7 of the flight plan. The parent or sister company will in that case be the aircraft operator for flights performed by the daughter company and all flights will have to be covered in the monitoring plan and reports of the parent or sister company. An aircraft operator having two Air Operator Certificates but only having one unique ICAO designator should submit one monitoring plan.

Flights that are exempted from the ETS scheme are: flights between aerodromes that are not situated in an EU Member State are not included in the EU ETS scheme and do not have to be monitored or reported; flights of a reigning monarch, heads of state, heads of government and government ministers, of a country other than a Member State; military flights performed by military aircraft; flights related to search and rescue, firefighting flights, humanitarian flights and medical service flights; flights performed exclusively under visual flight rules; circular flights (departing and arriving at the same airport without an intermediate stop); training flights; flights performed exclusively for the purpose of scientific research; flights performed by aircraft with a certified maximum take-off mass of less than 5,700 kg; flights performed in the framework of public service obligations flights performed by a commercial air transport operator who falls below the rule *de minimis non curat lex* (the law does not take into consideration trivialities). This exemption applies to commercial air transport operators only and not to non-commercial air transport operators.

Unlike in the instance of domestic aviation emissions, greenhouse gas emissions from fuel consumption in international aviation are not assigned under the Kyoto Protocol and are consequently not the subject of so-called Assigned Amount Units (AAUs) – particular with reference to the initial commitment period from 2008 to 2012. In addition, the non-CO_2 climate effects, which are not related to fuel burn, from both domestic and international aviation are not covered under the Kyoto Protocol and therefore not covered by AAUs. The quantity of AAUs is based on the commitments laid down in Annex B of the Protocol and specifies a country's permitted greenhouse gas emissions during the first commitment period. These are measured in tonnes of CO_2 equivalent (tCO_2e). Therefore the inclusion of international aviation in the EU ETS might well create problems in accounting that might require the introduction of specific design features between the quantity of emissions covered by the Kyoto Protocol which is in fact emitted and the quantity of Kyoto units which are retired for compliance purposes to cover these emissions. These accounting problems arise because the emissions of international aviation are not underpinned by the AAUs used for compliance control under the Kyoto Protocol. These problems might be overcome by the extension of the scope of the Kyoto Protocol and taking away the exemption of aviation from quantitative obligations; borrowing of AAUs from sectors not covered by the EU ETS. The way this could be accomplished is by using AAUs from sectors not covered by the EU ETS temporarily to underpin any allowances issued for international aviation emissions under the geographical scope with AAUs. Correspondingly, aviation entities are allocated allowances that are fully fungible, i.e. the aviation sector can buy and sell allowances from and to other sectors under the EU ETS without any trade restrictions. Since all allowances will be surrendered at the end of the commitment period, the attached AAUs are only "loaned" to the aviation sector.

Another corrective measure would be to have the aviation sector purchase all the allowances required for compliance from other sectors, with no additional allowances being granted to aviation. Emissions trading in aviation are based on allowances from the EU ETS and Kyoto units only. There could also be an option available to the aviation sector to buy allowances for emissions growth above a baseline. This option would be similar to the previous one, but limits the obligation to surrender allowances to those for emissions growth relative to a base year or base period (baseline). Operators could be also allocated allowances where they can buy additional allowances from non-aviation sectors, but cannot not sell surplus allowances to these entities. Another option would be to give operators allowances from non-aviation sectors with the condition that they can only sell to other sectors as many allowances as they, as a sector as a whole, have already bought from non-aviation sectors during the trading period. The toll that could be used in giving allowances as described is auctioning.

To establish monitoring and reporting protocols, emission inventory activities could rely either on self-reporting by participants or on third parties such as EUROCONTROL. The most accurate monitoring option for CO_2 is for aircraft operators to measure the actual fuel used on each trip flown within the chosen geographical scope of the emission trading system. CO_2 emissions can then be calculated from the carbon content of that fuel. Under current international regulations, the amount of fuel used on each flight must already be registered by airlines. The environmental effectiveness of the emissions trading system would certainly benefit if actual trip fuel were used, as would its economic efficiency, for operational measures to reduce emissions would be duly rewarded. The European airline industry and their association have expressed their preference for a monitoring and reporting method based on actual trip fuel, reported by aircraft operators. They regard this as feasible and fairly straightforward to implement.[65] Calculating the carbon dioxide emission of flights is usually carried out from a known quantity such as units of electricity consumed or fuel burned. Dr. Christian Jardine states: "Combustion of fuel is a stoichiometric chemical reaction, so the mass of CO_2 emissions can be directly related to fuel burn. Thus for example, for every kWh of

---

[65] See *Giving Wings to Aviation*, Inclusion of Aviation Under The European Emissions Trading System (ETS): Design and Impacts, Report of the European Commission, DG Environment No. ENV.C.2/ETU/2004/0907/4r, Delft: July 2005 at 8–10.

4.9 Calculating Emissions   315

energy supplied by gas or fuel oil, the CO_2 emissions are 0.206 or 0.281 kg CO_2, respectively.[66]

With regard to the economic impact of including aviation in the EU ETS, a study conducted in 2004 found that:

> The introduction of emissions trading for the aviation sector, most immediately in respect of its CO2 emissions, while keeping the structure open for including non-CO2 impacts in the future, does not appear to pose many challenges that have not already arisen in the context of the existing EU Emissions Trading Scheme. This suggests that emission trading is a policy option that can be considered alongside other policy instruments to tackle the climate impact of aviation.[67]

Once the methodology for calculating emissions and fixing charges are settled, the aviation community has to face the most critical challenge, which is considering the legalities of emissions trading and carbon offsets, particularly in view of geographic considerations. In view of the significant economic and legal considerations involved the following chapters will be dedicated to a discussion on policy issues.

The air transport industry is full of paradoxes. On the one hand, air transport contributes 10% of the world GDP[68] and employs approximately 80 million people worldwide. Yet, over the decade 1999–2009 the industry lost $56 billion. Given that over that period there were 20 billion passengers carried by air, the industry lost $2.8 per passenger on average. The paradox is that despite, this long history of loss, *Airbus Industrie* forecasts that between 2009 and 2028 there will be a demand for 24,951 passenger and freighter aircraft worth USD 3.1 trillion, and that, by 2028 there will be 32,000 aircraft in service compared with 15,750 in 2009.[69] Another paradox is the open skies concept. On the one hand, progressive thinking favours liberalization of air transport through open skies, which means that air transport must be treated like any other business and should not have market access barriers. On the other hand, airport capacity is finite and a massive injection of capacity will be a severe drain on the process of slot-allocation.

It is also well known that the air transport industry differs in its commercial practices due to its own particular nature and the manner in which it is treated by States and the commercial world. Giovanni Bisignani, the down-to-earth, hard-hitting Director General and CEO of IATA, in his address to the Royal Aeronautical Society, Montreal Branch on 1 December 2009 observed that:

> The aviation industry is in crisis. Since 2001, airlines have lost US$53 billion. That includes losses of US$16.8 billion in 2008, the biggest in our history and US$11 billion this year.

---

[66] Jardine (2009). He also cites DEFRA, *Act on CO2 Calculator: Data, Methodology and Assumptions Paper* V1.2 August 2008.
[67] *Id.* 17.
[68] The gross domestic product (GDP) or gross domestic income (GDI) is a basic measure of a country's overall economic output.
[69] *Flying Smart, Thinking Big*, Airbus Industrie's Global Market Forecast, 2009–2028, at 8–9.

We are forecasting a further US$3.8 billion loss next year. Look into the detail behind these numbers and you see enormous shifts on key parameters".[70]

Mr. Bisignani correctly attributed this continuing struggle of the airline industry to the mandatory bilateral system of the negotiation of air traffic rights between State authorities, stating that no other industry has such restrictions. He observed that shipping companies operated without restrictions and almost all other products enjoyed global markets where an entrepreneur may have to pay an import tax or a duty but he did not need his government to conclude a bilateral agreement. He also stated that, however, those airlines could not sell their product where markets exist and could not merge across borders where it made business sense. The result was a sick industry with too many players.

On an earlier occasion,[71] Mr. Bisignani made similar comments, emphasising that numerous businesses have been buttressed by the lowering of trade barriers, but airlines suffered because they still faced enormous hurdles. Mr. Bisignani pointed out that although the airline industry was among the first to operate globally, it was still waiting for the benefits of globalization. He asked a pointed question: "Do we purchase cars or medications based on the nationality of a company's shareholders?" asking the further question that, if an Egyptian can spend a night at a Singaporean hotel in Hamburg, why couldn't an Australian fly a Brazilian airline from Mexico City to Miami? Mr. Bisignani added that airlines supported 29 million jobs and $2.9 trillion worth of economic activity worldwide and that few industries so vital to the health of the global economy remain so restricted by archaic ownership rules.[72]

Given the above constraints on the air transport industry from a business perspective, there is no room for doubt that the industry, through regulation and the initiatives of aircraft and component manufacturers as well as their own initiatives in greening airlines, has done much to address the issues facing aviation and climate change. Much needs to be done though, in achieving targets set. Perhaps the challenges that lie ahead and what needs to be done can best be summed up in Mr. Bisignani's words:

> Three challenges must be met. The first is to marry the unified approach of the Chicago Convention that guides ICAO with the principle of common but differentiated responsibility (CBDR) that is a cornerstone of the UNFCCC process. The second challenge is to preserve the sectoral approach for international aviation that was established by Kyoto. We don't want exemptions from our responsibility, but the cross-border nature of our international business needs a global approach to avoid competitive distortions. The third challenge is to develop economic measures that are effective in reducing aviation's emissions. That means: Replacing the growing patchwork of green taxes, charges and emissions trading proposals with a global system; allocating the funds to environmental projects; and being fair by treating aviation in proportion to its 2% contribution to global CO2 emissions.[73]

---

[70] http://www.iata.org/pressroom/speeches/2009-12-01-01.htm.
[71] Bisignani (2006).
[72] *Ibid.*
[73] http://www.iata.org/pressroom/speeches/Pages/2009-03-31-01.aspx.

4.9 Calculating Emissions

Indeed Mr. Bisignani is right. The cross border nature of the air transport business needs a global approach. Perhaps the first step would be to try that approach through ICAO. The foundation was attempted at the 36th Session of the ICAO Assembly held in 2007, on the fundamental Principle 12 of the 1992 Rio Declaration on Environment and Development calls for the avoidance of Unilateral actions to deal with environmental challenges outside the jurisdiction of the importing country and states that environmental measures addressing transboundary or global environmental problems should, as far as possible, be based on international consensus. This attempt failed. One can only hope that attempts at future global consensus on environmental regulation will not suffer the same fate.

The bottom line is that our discipline in protecting our environment lies in the way we build relationships with each other and act in cohesion. Someone once said that a person without relationships is like a person without a shadow, and that we all need a light above our heads to cast that shadow. Our shadow is the metaphorical equivalent of what we bestow on others whether in life or in death. Years from now, our children will wonder who we were, with a view to finding themselves, perhaps on the mistaken notion that they must take after us in some way. They might even one day begin to discover their true nature and self and re-create their world and their relationships. However, we do not necessarily become our parents, nor do we as a rule take after our ancestors or their prejudices and notions. In his book "Dreams from my Father"[74] Barack Obama says of his basketball playing and a certain respect and attitude his peers were trying to instill in him that did not only apply to the sport: "that respect came from what you did and not who your daddy was. That you didn't let anyone sneak up behind you to see emotions – such as hurt or fear – you didn't want them to see". This speaks of an individual's strength of character and values.

Years from now, generations to come will analyze the way in which we handled wars and made peace; how we handled our own sustainable development. That will be our shadow. The respect we gain as a generation will, as Obama says, not come from who our fathers and mothers were but how we did things. If we are to leave a good impression behind us, we would have to act as we actually live. Obama says in his other bestselling book *"The Audacity of Hope"*[75] that our politics will have to be "constructed from the best of our traditions and will have to account for the darker aspects of our past...and we'll need to remind ourselves, despite all our differences, just how much we share: common hopes, common dreams, a bond that will not break".

The final question is, how do we apply our traditions and lessons learnt from history to our common goals, particularly when it comes to reducing the negative effects on the climate from an ever increasing air transport system? The first step is to recognize the problem, which we have done. As George Monbiot says, aviation

---

[74] Obama (1995).

[75] Obama (2006).

has been the fastest growing source of greenhouse gas emissions and unless something is done to stop the increasing growth of emissions from aviation, aircraft engine emissions will overwhelm all the cuts we manage to make elsewhere.[76] The second step would be to seek international, or better still, global accord on an acceptable carbon trading scheme or alternative unified system to reach emissions targets.

References

Abeyratne RIR (1999a) Emissions trading as a market-based option in air transport – contractual issues. Environ Policy Law 29(5):226–235 (November 1999)

Abeyratne RIR (1999b) The fuel tax and emissions trading as market-based options in air transport. Ann Air Space Law (Annales de Droit Arien et Spatial) XXIV:1–31

Abeyratne RIR (2008) A review of energy efficiency within the air transport industry. Zeitschrift luft-und Weltraumrecht 57(3):321–334

Bisignani G (2006) Think again airlines. At http://www.foreignpolicy.com/articles/2006/01/04/think_again_airlines?print=yes&hidecomments=yes&page=full, January 2006

Gander S, Helme N (1999) Emissions trading is an effective, proven policy tool for solving air pollution problems. ICAO J 54(7):12–14, 28–29 (September 1999)

Hardeman A (2007) A common approach to aviation emissions trading. Air Space Law XXXII (1):3–18 (February 2007)

Hupe J (2008) En route to Copenhagen: international aviation action on climate change, aviation and the environment. ICAO J 63(4):4–6

Jardine CN (2009) Calculating the carbon dioxide emissions of flights. Environmental Change Institute, Oxford University Centre for the Environment, Oxford, February 2009

Monbiot G (2006) HEAT: how to stop the planet from burning. Doubleday, Canada, p 174

Obama B (1995) Dreams from my father. Random House, New York

Obama B (2006) The audacity of hope. Random House, New York

Owen AD, Hanley N (eds) (2004) The economics of climate change. Routledge, New York, p 115

Sausen R, Schumann U (2007) Aviation carbon offset programmes. ICAO J 6(5):13

---

[76] Monbiot (2006).

Chapter 5
Sustainable Development of Air Transport

5.1 The Nature of Air Transport

Broadly speaking, the phrase "sustainable development" means development that caters to the needs of the present generation without compromising the ability of the future generation to achieve their needs. Usually, this phrase is used in an environmental sense. However, it could also mean – as in the context of this chapter-development that ensures the sustainability of an industry. The initial and fundamental premise of the air transport industry is that the air transport sector has never been continuously profitable and that the industry's profitability has been cyclical. Competition in air transport has been a key factor along with the perceived anomalies that are not seen in other transport modes, such as sovereignty which is a key issue in the award of air traffic rights and the requirement that ownership and control of an airline should be in the hands of nationals of a country I which an airline is registered on a majority basis. All of these factors lead to air transport being subject to the Game Theory, which is a branch of applied mathematics that is used in conformity with economic principles. The Game Theory suggests that strategic interactions among economic agents (air carriers) result in outcomes with respect to their preferences. The success of their choices depends on choices made by others. Apart from air transport economics, the Game theory applies to other areas such as security and environmental protection.

In the words of Giovanni Bisignani, Director General and CEO of IATA who, at a meeting he addressed in February 2011 said:

> Aviation is a unique industry. In 2011 we expect to generate almost $600 billion in revenues with the burden of $205 billion in debt. The industry's activity is critical. Aviation supports 32 million jobs and facilitates the global village by supporting $3.5 trillion in economic activity. Our commitments to tackle climate change are the most ambitious of any global industry. And we are the safest mode of transport. But our margins are pathetic – just 0.1% over the last 40 years. This is not sustainable. We need to look ahead to anticipate

change as we prepare to handle the 16 billion passengers and 400 million tonnes of freight that we will handle in 2050.[1]

Air transport, unlike any other mode of transport, is severely constrained by international treaty. While rail, maritime and road transport do not require the express permission of the State into which they operate services internationally, no scheduled air service can be operated over or into the territory of a contracting State, except with the special permission or other authorization of that State, and in accordance with the terms of such permission or authorization. This anomaly has spawned hundreds of bilateral air services agreements and enabled States to adopt a protectionist attitude in guarding the "market share" of its own national carrier, thus stultifying competition among carriers and depriving the consumer of the most efficient and cost effective air transport product that an otherwise liberalized market would have produced.

To circumvent this obstacle, air carriers have used a multiplicity of tolls, by forming alliances and optimizing operations on routes they are otherwise not entitled to by combining each others' air traffic rights. However, this has proved no permanent panacea, prompting aviation economists and lawyers to suggest that the best way out of the intractable situation would be to include market access in air transport in the GATS Annex of the World Trade Organization. One of the reasons which impelled the author to write this article is that, notwithstanding the numerous assertions made by aviation experts and economists at international *fora*, time and again, that market access in air transport should be brought under the purview of the General Agreement on Trade in Services (GATS) Air Transport Annex of the World Trade Organization, these experts do not seem to have addressed the issue as to how this feat could be achieved within the context of a complex economic and legal environment applicable to air traffic rights in air transport. Another often neglected issue is that the more powerful nations who are prolific in the air transport offered by carriers registered in those States, have resisted any hint at liberalizing market access in air transport through the GATS Annex.

5.2 Market Access

Air transport has been, and continues to be, a complex business. It follows therefore, that transactions related to this industry are indeed complex and should be adaptable to modern exigencies. A survey of IATA carried out in 2007 of over 600 companies from five countries reflected that 63% confirmed that air transport networks are vital for their investments and business.[2] Thirty percent of those countries said that any constraint placed on the air transport industry would make

---

[1] http://www.airtransportnews.aero/article.pl?mcateg=&id=28229.

[2] *Flying Smart, Thinking Big*, Airbus Industrie's Global Market Forecast, 2009–2028, at 16.

them invest less.[3] *Airbus Industrie* forecasts that between 2009 and 2028 there will be a demand for 24,951 passenger and freighter aircraft worth USD 3.1 trillion, and that, by 2028 there will be 32,000 aircraft in service compared with 15,750 in 2009.[4]

Air transport plays an integral part in the tourism industry where 40% of international tourists travel by air. Air transport contributes 10% of the world GDP[5] and employs approximately 80 million people worldwide. Yet, over the decade 1999–2009 the industry lost $56 billion. Given that over that period there were 20 billion passengers carried by air, the industry lost $2.8 per passenger on average. This fact underscores the indispensability of air transport to the global economy on the one hand and the resilience of the industry to resuscitate itself after slow periods of growth followed by periods of losses.

Yet, commercial aviation has been nourished on a strange doctrine, which is based on the premise that aviation should primarily be used to strengthen friendship and understanding among the peoples of the world.[6] This is a doctrine more suited to diplomacy, and should not exclusively be attributed to a commercial activity such as air transport. To make matters worse, the more appropriate doctrine for commercial aviation – competition – has been effectively rejected by the *Convention on Civil Aviation (Chicago Convention)* [7] in its Article 6 which provides that:

> No scheduled international air service may be operated over or into the territory of a contracting State, except with the special permission or other authorization of that State, and in accordance with the terms of such permission or authorization.

This is a most anomalous and strange philosophy and it does not explicitly apply to any other mode of transport.

---

[3] *Ibid.*

[4] Airbus Forecast, *supra*, at 8–9.

[5] The gross domestic product (GDP) or gross domestic income (GDI) is a basic measure of a country's overall economic output.

[6] The Convention on International Civil Aviation, signed at Chicago on 7 December 1944, (supra, note 1) which is the founding document of commercial aviation, in its Preamble, recognizes that the future development international civil aviation can greatly help to create and preserve friendship and international understanding among the nations and peoples of the world, yet its abuse can become a threat to the general security; and it is desirable to avoid friction and to promote that cooperation between nations and peoples upon which the entire peace of the world depends. See ICAO Doc 7300/9 Ninth Edition: 2006, *Preamble*. The most significant modernist construction of the role of civil aviation in securing world peace and security comes from language used in the letters of invitation issued by the United States to the participant States to the Chicago Conference, that, consequent to the war, the restorative processes of prompt communication may greatly facilitate the return to the processes of peace. However, the conscious awareness of the parties to the Convention, that in securing this peace, prudent economic and business principles must not be compromised, should not be forgotten. See Proceedings *of the International Civil Aviation Conference*, Chicago, Illinois, November 1–7 December 1944, US Department of State Volume 1 at 7.

[7] *Id.*

The Chicago Conference of 1944[8] which brought 52 nations together to establish principles of air navigation and foster principles of air transport was aimed at adopting a treaty built on the ruins of war and discord. All that the delegates to the Conference were seemingly interested in were to protect their air power and sovereignty. As a result, a distinct lack of vision of the commercial future of the air transport industry seems to have prevailed, which precluded the adoption of liberal economic principles that would have otherwise allowed for the provision of air transport services to meet the exponential demand in the coming years. Arguably, the most glaring lapse of the Chicago Convention is the tone it sets in its Preamble that "international civil aviation may be developed in a safe and orderly manner and that international air transport services may be established on the basis of equality of opportunity and operated soundly and economically".[9] This prescribes a cautious approach for airlines, depriving them of the opportunity for competition. Worse, it focuses on the interests of the airlines and completely leaves out the important issue of the needs of the user – the passenger – and the consignor of air freight, both of whom depend on the best possible manner in which to use air transport.

One must hasten to add that this myopic attitude only prevailed in the field of air transport and did not affect air navigation. To this end, the Conference circumscribed the functions of ICAO[10] – which came into being through the Chicago Convention – by stating that ICAO's aims and objectives would be to "develop the principles and techniques of air navigation and foster the principles of air transport",[11] thus empowering the Organization to develop and regulate principles of air navigation but take a back seat in the development and regulation of global economic principles of air transport which would enable the user of the service to have the most number of choices of products in his best interest.[12]

---

[8] The Chicago Conference was convened by President Roosevelt of the United States from 1 November to 7 December 1944 and led to the adoption by 52 States (attending the Conference) of the Convention on International Civil Aviation (hereafter, the Chicago Convention). See Proceedings *of the International Civil Aviation Conference*, Chicago, Illinois, November 1–7 December 1944, US Department of State: Volumes I and II.

[9] *Supra*, note 3.

[10] Supra, note 1. In December 2004, following a decision by the 35th Session of the ICAO Assembly, the Council of ICAO approved six Strategic Objectives for 2005–2010: They are: safety; security; environmental protection; efficiency; continuity; and rule of law. At the 37th Assembly of ICAO (2010) these strategic objectives were revised and only four remain: safety; security; environmental protection and sustainability of air transport.

[11] Chicago Convention, *supra*, note 3, Article 44.

[12] Over the past 50 years particularly in the past two decades, commercial air carriers gradually broke the shackles of rigid regulation to form strategic alliances among themselves. These alliances have been formed in the realization that the performance of an airline can be affected by two factors: the average performance of all competitors in the airline industry; and whether the airline concerned is a superior or inferior performer in the industry. Michael Porter encapsulates these two factors in the single premise that any business achieves superior profitability in its industry by attaining either higher prices or lower costs than rivals. Curiously, in the airline industry, it is the latter – lower costs – which has been the cornerstone of strategic alliances.

This unfortunate and self-stultifying approach has subjected commercial aviation to the peccadilloes of protectionist States who consider the concept of sovereignty[13] as preeminent in the provision of air transport to the travelling public and consignors of air freight. Furthermore, this trend has encouraged the protective instincts of States to ensure that their national carriers obtain optimum market share "belonging" to them, based on a now antiquated belief that all passengers, cargo and mail destined to a particular State or leaving that State, is the birth right of the national carrier of that State.

No other known major commercial activity is subject to arbitrary principles of nationality and national ownership, control of businesses and capacity controls as is air transport.

As to why ICAO has been refused by its member States the capacity to lead in the air transport field, and adopt compelling economic principles, and why they continue to do so is a puzzle, particularly when one looks at the robust role given to the Organization in other areas such as safety, security and the environment. For example, the 37th Session of the ICAO Assembly, held from 28 September to 8 October 2010, reaffirmed ICAO's leadership role in the field of environmental protection, and adopted a comprehensive resolution to reduce the impact of aviation emissions on climate change. The agreement provides a roadmap for action through 2050 for the 190 Member States of the Organization. Solidifying its global influence, the Organization signed numerous international agreements, including cooperation agreements with regional civil aviation organizations and bodies from all regions of the world.

Against the backdrop of ICAO's capacity and ability to conduct mandatory safety and security audits of States (which ICAO does) the Assembly endorsed a proactive safety strategy based on the sharing of critical safety information among governments and industry stakeholders. It also endorsed ICAO's plan to establish a multi-disciplinary approach to address the critical issue of runway safety. This will bring together representatives from airlines, airports, air navigation service providers and regulatory authorities. Following a successful diplomatic Conference in Beijing in August 2010, the Assembly built on this achievement by recognizing the need to strengthen aviation security worldwide. In a Declaration, unanimously adopted by participants, international commitment was reaffirmed to enhance aviation security collaboratively and proactively through screening technologies to detect prohibited articles, strengthening international standards, improving security information-sharing and providing capacity-building assistance to States in need. The Assembly also put its full support behind a comprehensive, new ICAO aviation security strategy.

The Resolution adopted by the Assembly on climate change makes ICAO the first United Nations Agency to lead a sector in the establishment of a globally

---

[13] The principle of State sovereignty in airspace is embodied in Article 1 of the Chicago Convention which recognizes that every State has sovereignty over the air space above its territory, the latter being defined in Article 2 as land situated within and water adjacent to the State concerned.

harmonized agreement for addressing its CO_2 emissions. Why is there no such leadership in the economics of air transport?

5.2.1 The Anomaly

The lack of leadership in the economic regulation of air transport and the various "open skies" regimes that have sprung up along with mergers and acquisitions of airlines resulting in airline alliances and other commercial arrangements, is symptomatic of a crisis in law that has to be addressed. *Ex facie*, this would call for a three dimensional approach where a triage of institutions, i.e. the two parties to an agreement on market access would be refereed and regulated by a third party that would guarantee fairness of trade ensure that the conduct of trade in air transport accords with global practices of fair trade from a consumer perspective. This would initially mean that States veer from the parochial dogma of absolute state sovereignty and embrace the concept of sovereignty that is consistent with globalization and accepted trade practices. In its Report to the General Assembly, the International Law Commission recommended a draft provision which required:

> Every State has the duty to conduct its relations with other States in accordance with international law and with the principle that the sovereignty of each State is subject to the supremacy of international law.[14]

This principle, which forms a cornerstone of international conduct by States, provides the basis for strengthening international comity and regulating the conduct of States both internally – within their territories – and externally, towards other States. States are effectively precluded by this principle of pursuing their own interests untrammelled and with disregard to principles established by international law. State Sovereignty thus connotes a responsibility rather than an absolute right to do as a State deems fit within its own territory. The conduct of trade in air transport should be determined by an international political process or a "central market place". Supiot uses the metaphor of the medieval *Marktplatz* of Brussels:

> Its architectural magnificence is imbued with institutional meaning. This ancient market place is bounded by the headquarters of the institutions which endured the smooth operations of the market. The Town Hall housed the municipal authority that saw to the fairness of trade through inspection of weights and measures, while the buildings of various trades (e.g. butchers, bakers, brewers) housed the guilds that upheld the status and quality of labour, without which there could have been nothing valuable to trade. These various buildings also marked out the boundaries of the commercial sphere. If one left the market place say, to go to the courthouse or to the Royal Palace, a different set of rules applied. Indeed if the law of the market had extended to judges or to political leaders, their decisions

---

[14] *Report of the International Law Commission to the General Assembly on the Work of the 1st Session, A/CN.4/13*, June 9 1949, at 21.

would have been up for sale, the city would have been corrupt, and honest traders would have been unable to carry on their business there freely.[15]

Supiot makes a good point which is particularly applicable to the air transport analogy. The lack of central laws and regulations in market access that would otherwise ensure fairness of competition and worn out perception of sovereignty has made States control market access through a certain parochial protectionism, opening the door for various "deals" in the market.

The problem is aggravated by the fact that in modern parlance the various freedoms associated with free trade such as the freedom of establishment; the freedom to provide services; and the freedom to move capital and goods all encourage investors to go "forum shopping" so that they could opt to establish themselves in the jurisdiction that is most conducive to their interests. This allows investors to bypass the jurisdiction that they would be subject to if they were to establish their businesses in their States of nationality and seek less constraining jurisdictions of their choice.[16]

5.2.2 The Solution?

The issue for international air transport is "what is the central market place?" The immediate answer which comes to mind, if air transport were to be treated as a trade, is the World Trade Organization (WTO). One of the most contentious issues in the world of commercial air transport today is the question as to whether the industry should embrace the trade in services regime of the World Trade Organization in preference to the currently restrictive system of the bilateral air transport agreement which entitles States to refuse permission to air carriers who apply to operate commercial air services to and from their territories. The current system is based on national treatment, where States could apply different conditions to carriers operating air services into their territories based on capacity and demand for travel at prices that are deemed acceptable to the State concerned and the travelling public. Corollaries to the current system of bilateral air transport agreement are traditionally seeped in norms requiring capacity to be primarily provided for traffic to and from the two States parties to the agreement, on the basis of "fair and equal opportunity". Under this system, airlines are usually expected to be substantially owned and effectively controlled by a minimum percentage of the Contracting party. This affords a State the opportunity to protect the interest of its own carriers.

Air transport affects world trade in two ways: as a service by itself, directly transporting persons and freight; and as a service feeding other areas of trade mainly

---

[15] Supiot (2010).
[16] See *International Transport Workers' Federation and Finnish Seamens' Union* v. *Viking Line ABP and OU Viking Line Easti*, ECR 2007, 1–10779 (Case C-438/05, 6 December 2007). Also, *Centros Ltd.* v. *Erhvervs-og Selskabsstyrelsen*, ECR, 1-1459case no. C-212/97, 9 March 1999. For an air transport analogy see Abeyratne (2004).

involving tourism and hospitality. However, it cannot be doubted that air transport affects overall economic activities of business, particularly involving cross-border trade. It is also incontrovertible that the world needs an efficient and effective air transport industry if the dual functions of the air transport service were to be sustained over time to cater to the rapidly growing demand for carriage by air of persons and goods. In order to achieve this objective, the air transport industry must be liberalized to the extent that it remains unfettered by commercial constraints. Yet, unlike most other modes of transport, air transport remains rather rigidly controlled by the need for agreements of States to permit carriers of other States into their territories as well as established percentages of national ownership of air carriers which stifle foreign investment in national jurisdictions. While at least one commentator has categorically stated that the trend towards a very liberal open skies international regime is unstoppable,[17] which implicitly gives the industry the assurance that the problem would solve itself in the years to come, others have vigorously advocated that, as a panacea to the problem of rigid regulation, market access in air transport should be in the domain of a liberalized international regime. While the former view cannot be disputed, the latter approach brings to bear the compelling need to address the issue squarely, both in terms of whether the desirable approach would be to bring the industry from the current bilateral structure of air services negotiations into a more generalized regime and if so, what the modalities of such an exercise might entail. As to the former, it is largely a matter of political will. The latter would need some discussion on the legalities involved.

One of the reasons which impelled the author to write this article is the numerous assertions made by aviation experts and economists at international *fora*, time and again, that market access in air transport should be brought under the purview of the General Agreement on Trade in Services (GATS) Air Transport Annex[18] of the World Trade Organization.[19] Of particular inspiration was the fact that these

---

[17] Doganis (2001) at 11.

[18] GATS seeks to liberalize trade regarding the provision of services. It deals with, and provides for two types of obligations – general and specific. The former relates to obligations of States and the latter to commitments made by States. Also, The GATS Agreement identifies and makes provision for specific services through its various annexes. The Air Transport Annex of GATS currently includes aircraft repair and maintenance; the selling and marketing of air services; and computer reservations systems. Notwithstanding the profile of the Annex, a certain ambivalence pervades the interpretation of inclusionary elements in the Annex and their application. This is evident in the fact that only a few WTO members have committed to the three activities.

[19] The World Trade Organization (WTO-OMC) came into being on 1 January 1995 and along with it entered into force the General Agreement on Trade in Services (GATS). The GATS Annex on Air Transport Services applies trade rules and principles such as most-favoured nation (MFN) treatment and national treatment to three specific so-called "soft" rights, namely, aircraft repair and maintenance, selling and marketing of air transport, and computer reservation system services. It excludes from the application of the GATS "services directly related to the exercise of traffic rights". Pursuant to an earlier ministerial decision, the WTO-OMC launched in 2000 the first review of the operation of this Annex with a view to considering possible extension of its coverage in this sector. During the review, which continued into 2002, there was some support to extend the

5.2 Market Access 327

experts do not seem to have addressed the issue as to how this feat could be achieved within the context of a complex economic and legal environment applicable to air traffic rights in air transport. Another often neglected issue is that the more powerful nations who are prolific in the air transport offered by carriers registered in those States, have resisted any hint at liberalizing market access in air transport through the GATS Annex.

The current bilateral air transport agreement system between two States, resulting from Article 6 of the Chicago Convention[20] which lays down the peremptory rule that no scheduled international air service may be operated over or into the territory of a Contracting State, except with the special permission or other authorization of the State concerned, has been criticized as "horse trading".[21] One commentator goes to the extent of saying that there is no defensible reason for not including "the business of air transport" (by which he obviously means market access in air transport) in the GATS Annex, and goes on to give various theoretical justifications in support of his statement, while also suggesting that ICAO[22] should leave all matters of market access in air transport in the hands of the WTO.[23] While it it is not disputed that the current system of negotiations for air traffic rights is primarily grounded on national and regional commercial interests (as the case may be), and that the GATS Annex offers a sound multilateral framework of principles and rules for trade in services which are calculated to expand such trade under conditions of transparency and progressive liberalization, it is not advisable to approach the issue in such dismissive terms – that it would be possible for ICAO to renege on its responsibilities and stand by while the WTO takes over the issue of market access lock stock and barrel, given ICAO's role and responsibilities under an international treaty. The spirit of the Chicago Convention of 1944, an international treaty ratified or otherwise accepted by 190 sovereign states, is subsumed in its *Preamble*, which reflects that governments have agreed on the principles of the treaty, all of which bring to bear the need to develop civil aviation in a safe and orderly manner and the fact that international air transport services should be established on the basis of equality of opportunity and operated soundly and economically. For this purpose, the Chicago Convention has established ICAO and prescribed, as aims and objectives of the Organization, inter alia, that ICAO should insure the safe and orderly growth of international civil aviation throughout the world[24] and meet the needs of the peoples of the world for safe, regular, efficient

---

Annex to include some additional "soft" rights (for example, ground handling) as well as some aspects of "hard rights" (for example, air cargo, non-scheduled and multi-modal transport). However, there is no global consensus at this stage on whether or how this would be achieved.

[20] *Supra*, note 3.
[21] Stadlmeier (1997).
[22] *Supra*, note 1.
[23] See Janda (1995), where the author claims that the failure of ICAO to seize the initiative over multilateral liberalization of air transport services has left the field open to WTO.
[24] Chicago Convention, *supra*, note 3 Article 44 a).

and economical air transport.²⁵ The latter requirement incontrovertibly recognizes the complexity of the air transport system, which cannot be isolated solely from a commercial angle, and emphasizes the need for one Organization, i.e. ICAO, to address the various critical areas of the air transport product. The only dampener on these principles is that, as stated earlier, ICAO is authorized by the Chicago Convention only to *foster* the principles of air transport which does not confer on the Organization the capacity to prescribe economic rules of conduct for air carriers.

Another issue that should be addressed concerns the legality of the applicability of one treaty over another. As discussed earlier, the Chicago Convention, in Article 6, clearly lays down a restriction which would be contrary to the principles of the GATS Agreement regarding the Most Favoured Nations Treatment (MFN) Clause where each member is required to accord immediately and unconditionally to services and service suppliers of any other member treatment no less favourable than that it accords to like services and service suppliers of any other country.²⁶ Would this mean that, for example, the United States and Canada, which, after many decades of "open skies lite", have moved towards a fully liberalized air services agreement,²⁷ would automatically award every country most favoured nations treatment if market access were brought within the GATS Annex? Also relevant is the fundamental question at public international law pertaining to treaties, as to which instrument – the Chicago Convention or the WTO Treaty, might take precedence on the issue of succession of treaties. The need to address this issue is brought to bear by Article 82 of the Chicago Convention which states inter alia that Contracting States should accept the Convention as abrogating all obligations and understandings between them which are inconsistent with its terms *and undertake not to enter into any such obligations and understandings* (author's emphasis). Would this provision mean that Article 6 of the Convention would have to be amended? Could some States go into the GATS umbrella while others opt to stay under Article 6 as it now stands? If this were to happen, would there be a primacy issue between the Chicago Convention and the WTO treaty? Can two contradictory treaties exist hand in hand, equally applicable? And if so, how does this help achieve the overall global objective of universal liberalization of air transport? These questions need to be answered if a meaningful discourse on the future of market access in air transport were to be undertaken.

It is not the purpose of this article to argue the merits and demerits of transferring market access in air transport to the GATS Annex. However, this article will address some issues that might have a practical impact on the subject, along with the more pertinent question of what measures are needed to include market access in the GATS Air Transport Annex, should the world decide to go that way.

---

²⁵ *Id.* Article 44 d).

²⁶ General Agreement on Trade in Services, Uruguay Round Agreement, Article II 1.

²⁷ Field (2005).

5.2.3 *Economic Considerations*

With regard to the question as to whether there is a need to bring the commercial elements of market access in air transport within the GATS Agreement, the answer would initially lie with the States themselves, who, as arbiters of their own destiny and that of their air transport industries, would have to come to terms with relinquishing the element of control now available to them and replacing it with a more liberal free market structure. The free market option would have a direct bearing on national competitive policies of States and overall policies of regional bodies such as the European Union. In this regard, it must be acknowledged that the world is indeed moving toward some form of liberalization, although the question that would remain is whether the entire global air transport industry will be liberalized. This essentially involves a consideration not only of national policy but also of the nature of the air transport product.

The most noteworthy fact in comparing the Chicago regime with that of the WTO, is that the GATS Annex principles are in many ways similar to the principle enunciated by the Chicago Convention. For instance, under each regime the signatories are obligated to conduct business in a non-discriminatory manner, giving each other's carriers a fair and equal opportunity to conduct business. Both have established principles, although the GATS Annex has the flexibility of being open to periodic review that might enhance or progress the liberalization process. However, the two systems part company when it comes to criteria regarding the treatment of contracting parties, where the Chicago principle restricts itself to national treatment and the GATS Annex adheres to the general WTO principle of the "most favoured nation treatment" clause. In practical applicability, the WTO philosophy makes States under the GATS Annex give commitments to all signatories alike, unless exemptions are filed against particular commitments. Under Article 6 of the Chicago Convention, commitments given beyond the meaning and purpose of the Article are encompassed in the bilateral air services agreement, usually given on the basis of reciprocal advantage.

There are two facts of a critical nature that have to be made clear before proceeding any further. Firstly, the delivery of the air transport product, unlike in the case of other products such as textiles or agricultural produce, essentially involves human security and safety. Secondly, ICAO is not a regulator of air transport economics, as often wrongly assumed, and therefore does not, and cannot take the role of multilateral regulator. As a specialized agency of the United Nations, ICAO has aims and objectives, particularly in the economic field, to foster the planning and development of international air transport so as to, inter alia, meet the needs of the peoples of the world for safe, regular efficient and economical air transport.[28] This by no means entitles ICAO, unlike the WTO which has laid down explicit principles of trade in the GATS Agreement, to prescribe rules and

---

[28] Chicago Convention, *supra*, note 3, Article 44 d).

regulations in air transport economics. Both these facts were made abundantly clear at the Fifth Worldwide ICAO Air Transport Conference (ATConf/5: 2003), entitled *"Challenges and Opportunities of Liberalization"*, which was held in Montreal from 24 to 28 March 2003, where participating States adopted a *Declaration of Global Principles for the Liberalization of International Air Transport* which, inter alia, recognized the critical importance of safety and security in international air transport; and reaffirmed the basic principles of sovereignty, fair and equal opportunity, non-discrimination, interdependence, harmonization and cooperation set out in the Chicago Convention which have served international air transport well and continue to provide the basis for future development of international civil aviation. The preeminent consideration in the *Declaration* was safety and security, which was considered of paramount importance in the operation and development of international air transport. The *Declaration* therefore called upon States to accept, in close cooperation with each other, their primary responsibility for ensuring regulatory oversight of safety and security, irrespective of any change in economic regulatory arrangements. However, a caveat is included in the *Declaration* that security measures should to the extent possible not disrupt or impede the flow of passengers, freight, mail or aircraft.

One of the major elements and concerns addressed in the *Declaration* is liberalization of air transport. In this context, there is a statement in the Declaration that ongoing regulatory evolution should create an environment in which international air transport may develop and flourish in a stable, efficient and economical manner without compromising safety and security and while respecting social and labour standards. Prominence is given to the International Air Services Transit Agreement (IASTA), where the *Declaration* urges that States that have not signed the agreement must give urgent consideration to so doing.

One of the critical and thought provoking provisions in the Declaration is found in clause 4.4, which provides that each State will determine its own path and own pace of change in international air transport regulation, in a flexible way and using bilateral, sub-regional, regional, plurilateral or global avenues according to circumstances. Given that the overall approach of the Fifth Worldwide Air Transport Conference was "how to liberalize" (as against "whether to liberalize" which was the preoccupation of the earlier Air Transport Conference of 1994), this provision seems to say that the issue of how to liberalize is very much left to the States themselves, to be done at their own pace.

A repetition of the intent of this clause is found in the provision immediately proceeding clause 4.4, where States are requested that they should, to the extent feasible, liberalize international air transport market access, ensure air carrier access to international capital and air carrier freedom to conduct commercial activities – again a truism and fact of economic reality that had been happening in the aviation community before the Conference. Regarding cargo services liberalization, the same type of "endorsement" is seen in the statement that States should give consideration to liberalizing the regulatory treatment of international air cargo services on an accelerated basis, provided that clear responsibility and control of regulatory safety and security oversight is maintained.

5.2 Market Access

The *Declaration* calls upon ICAO and its Contracting States, together with the air transport industry and other stakeholders in civil aviation, to work to ensure that international air transport continues to develop in a way that ensures high and improving levels of safety and security; promotes the effective and sustainable participation in and benefit from international air transport by all States, respecting national sovereignty and equality of opportunity; and takes into consideration the differing levels of economic development amongst States through maintenance of the principle of "community of interest" and the fostering of preferential measures for developing countries. The Declaration also calls for the providing of adequate supporting infrastructure at reasonable cost, facilitation of the provision of resources – particularly for developing countries, and allowing for growth on a basis that is economically sustainable, supported by adaptation of the regulatory and operating environment, in order to strive to limit its environmental impact. Also considered important is the meeting of reasonable expectations of customers and public service needs, particularly for low traffic or otherwise uneconomical routes, promoting efficiency and minimizes market distortions. Another important dimension is the call for a system which safeguards fair competition adequately and effectively; promotes cooperation and harmonization at the sub-regional, regional and global levels; and has due regard for the interests of all stakeholders, including air carriers and other operators, users, airports, communities, labour, and tourism and travel services providers; with the ultimate purpose of giving international air transport as much economic freedom as possible while respecting its specific characteristics and in particular the need to ensure high standards of safety, security and environmental protection.

The *Declaration* also exhorts the general principle that further economic liberalization must be implemented in a way so as to ensure that there is a clear point of responsibility for each of safety and security in a clearly identified State or other regulatory authority designated by that State for any given aircraft operation.

5.2.4 Some Basic Misconceptions

One reason for the lack of clarity that permeates the issue of market access in air transport services, particularly with regard to its relation to WTO, is the proliferation of misconceptions that tend to obfuscate public policy debates on the possible role of GATS in liberalizing air transport within the parameters of the Air Transport Annex. This ambivalence may partially be due to a lack of familiarity within the aviation community of the complex web of rules governing trade practices within the purview of WTO regulations and also due to the pervading protectionism that has been exhibited by nations irrespective of their resource bases when it came to market access.

The first misconception that has to be addressed is that which pertains to the general belief that inclusion of market access related issues of air transport in the GATS would necessarily lead to liberalization. If air transport services are covered by the Annex, it would indeed remain a voluntary domestic decision of the State concerned and coverage under the GATS Annex would not imply deregulation or any attendant obligation to revise and modify existing regulatory regimes. A commitment to provide air transport services under the GATS Annex would essentially retain for the State concerned its pristine right to enforce regulations in force that bring to bear obligations governing safety, environmental protection and security. Furthermore, committing to a GATS governed market access system would not impel a State to alter or in any manner derogate from entrenched principles regarding foreign ownership of airlines.

Another misconception associated with the GATS system is that it is a rigid, inflexible mechanism that would stultify individual regulatory reform within a State or inhibit a State from initiating its own legislation in trade related issues such as safety, security and environmental protection. The GATS offers the air transport industry, through their States' mechanisms, the flexibility to make choices based on material interests.[29] The GATS would offer member States the right to select opportunities and times to make sector specific market access decisions and national treatment commitments. The GATS admits of progressive liberalization, in accordance with differing levels of development of services. Given the exponential growth of the air transport industry, this system would effectively facilitate consistency between national initiatives toward progress and predatory practices associated with excessive competition.

It must be noted that the most fundamental purpose of GATS is to provide for the liberalization of trade pertaining to the provision of services. GATS seeks to establish a multilateral framework of principles and rules for trade in services with a view to expansion of such trade under conditions of *transparency*,[30] national treatment[31] and *progressive liberalization*.[32] The fundamental principle of GATT is its Most Favoured Nation (MFN) Treatment clause whereby each party to the agreement accords immediately and unconditionally to services and service providers of any other party, treatment no less favourable than that it accords to like services and service providers of any other country. These provisions reflect the basic philosophy of GATS and play a vital role in affecting the decision of the international community on whether or not air transport services should be brought under its purview. Other features of GATS which have attracted discussion in relation to air services are provisions relating to increasing participation of

---

[29] See generally, Market Access: Unfinished Business, Post Uruguay Round Inventory and Issues, World Trade Organization, Special Studies 6:2001 at pp. 99–105.

[30] Article III of GATS requires each party to publish promptly all relevant laws, regulations, administrative guidelines and all other decisions, rulings or measures of general application, by the time of their entry into force.

[31] GATT's national treatment philosophy provides foreign services and services suppliers with treatment no less favourable than that accorded to a country's own services and service suppliers.

[32] Since GATS is an Annex to the GATT agreement it should be noted that the provisions of GATS are governed by those of GATT and that both documents incorporate the same basic principles.

developing countries within GATS[33] and dispute settlement.[34] For the present, the overall purpose of including air transport services in GATS seems to be to apply the broad principles of market access and the MFN philosophy to the selling or marketing of air traffic services. The purview of GATS in controlling air transport services would therefore be considered only in situations where air traffic rights are exercised multilaterally or plurilaterally. GATS would not apply in instances where States elect to use Article 6 of the Chicago Convention which governs all bilateral air services agreements and requires that the permission of a grantor State is necessary for a commercial air transport enterprise to operate air services in to or out of a State. In any event, the Annex on Air Transport Services to GATS does not reflect confidence in itself by providing in Article 6 of the Annex that the operation of the Annex shall be reviewed periodically or at least every 5 years.

5.2.5 Legal Considerations

Arguably, the most complex issues in the shift of market access to GATS lie in the legal field where, not only do conflicting and inconsistent legal issues have to be identified, but they also have to be resolved in the context of applicable provisions. The first issue relates to the principle of sovereignty, which is explicitly provided for in Article 1 of the Chicago Convention. By this provision, Contracting States recognize that every State has complete and exclusive sovereignty over the airspace above its territory.[35] Market access concerns economic sovereignty which is recognized in Article 6 of the Convention, giving each State the right to refuse entry to any carrier operating air services into its territory without permission or authorization. Economic sovereignty is applicable to all areas of the Convention, reflecting the totality of the economic powers of a State as well as its equal status in the international arena. State sovereignty also means an absolute juridical independence from the authority of other States and participants in international economic activity – a principle embodied in Article 2(1) of the United Nations Charter, which recognizes the sovereign equality of all members of the United Nations.

Notwithstanding the above, it must be borne in mind that, although in theory a State has complete, supreme and independent control of its economic policies, in

---

[33] Article II of GATS. Article XVI extends the MFN principle to market access.

[34] Article XXIII on dispute settlement is considered to be well balanced and equitable and provides that if any Party should consider that another Party fails to carry out its obligations or commitments under the agreement, it may make written representations or proposals to the other Party or Parties concerned and the latter should give sympathetic consideration to the representations or proposals so made. If no satisfactory settlement could be arrived at, the GATT agreement provides for a formal dispute settlement procedure in Articles XXII and XXIII.

[35] Article 2 of the Chicago Convention provides that the territory, referred to in Article 1 is deemed to be the land areas and territorial waters adjacent thereto under the sovereignty, suzerainty, mandate and protection of that State.

practice, this right does not extend to giving a State arbitrary powers to be exercised to the detriment of the international community. In practice, any overwhelmingly detrimental conduct by a State would give rise to the risk of retaliation by the international community. The rule of international law overrides any domestic or internal consideration of a State where the well being of other States might be threatened. When applied to commercial aviation, this concept would translate to a State's inability to adopt its own economic policies which are contrary to the principles enunciated in the Chicago Convention, necessitating a certain compliance by all of ICAO 190 member States of prudent aviation economic practices.

It is incumbent upon States that have ratified the Chicago Convention to respect both Articles 1 and 6 of the Chicago Convention. The *Vienna Convention on the Law of Treaties*[36] while recognizing treaties as a source of law, accepts free consent, good faith and the *pacta sunt servanda* as universally recognized elements of a treaty.[37] Article 11 of the Vienna Convention provides that the consent of a State to be bound by a treaty may be expressed by signature, exchange of instruments constituting a treaty, ratification, acceptance, approval or accession, or by any other means agreed upon "ratification", "acceptance", "approval", and "accession" generally mean the same thing, i.e. that in each case the international act so named indicates that the State performing such act is establishing on the international plane its consent to be bound by a treaty. A State demonstrates its adherence to a treaty by means of the *pacta sunt servanda*, which is reflected in Article 26 of the Vienna Convention in that every treaty in force is binding upon the parties and must be performed by them in good faith. The validity of a treaty or of the consent of a State to be bound by a treaty may be impeached only through the application of the Vienna Convention[38] which generally requires that a treaty could be derogated upon only in circumstances the treaty in question so specifies[39]; a later treaty abrogates the treaty in question[40]; there is a breach of the treaty[41]; a *novus actus interveniens* or supervening act which makes the performance of the treaty impossible[42]; and the invocation by a State of the *Clausula Rebus Sic Stantibus*[43] wherein a fundamental change of circumstances (when such circumstances constituted an essential basis of the consent of the parties to be bound by the treaty) which has occurred with regard to those existing at the time of the conclusion of the treaty, and which was not foreseen by the parties, radically changes or transforms the extent of obligations of a State. A State may not invoke the fact that its consent

---

[36] *Vienna Convention on the Law of Treaties*, United Nations General Assembly DocumentA/CONF.39/27, 23 May 1969.

[37] *Vienna Convention, Id,* Preamble.

[38] *Id.* Article 42. 1.

[39] *Id.* Article 57.

[40] *Id.* Article 59.

[41] *Id.* Article 60.

[42] *Id.* Article 61.

[43] *Id.* Article 62.

5.2 Market Access

to be bound by a treaty has been expressed in violation of a provision of its internal law regarding competence to conclude treaties and seek to invalidate its consent unless such violation was manifest and concerned a rule of its internal law of fundamental importance.[44]

States or international organizations which are parties to such treaties have to apply the treaties they have signed and therefore have to interpret them. Although the conclusion of a treaty is generally governed by international customary law to accord with accepted rules and practices of national constitutional law of the signatory States, the application of treaties are governed by principles of international law. If, however, the application or performance of a requirement in an international treaty poses problems to a State, the constitutional law of that State would be applied by courts of that State to settle the problem. Although Article 27 of the Vienna Convention requires States not to invoke provisions of their internal laws as justification for failure to comply with the provisions of a treaty, States are free to choose the means of implementation they see fit according to their traditions and political organization.[45] The overriding rule is that treaties are juristic acts and have to be performed.

In view of the above , It would seem that, with respect to the conflicting philosophies in the Chicago Convention and the WTO instruments, a State cannot validly remain within the Chicago Convention's provisions of bilateralism and enter the realm of multilateralism of the GATS Agreement at the same time, unless measures are taken to rectify the situation. Of course one could argue that the MFN principle allows members to file specific exemptions. However, these exemptions can only be reviewed once every 5 years and would expire in 10 years. Also, The GATS Annex specifies that the Ministerial Conference of the WTO may grant a waiver of obligations regarding any exemptions sought after the entry into force of the GATS Agreement subject to the approval of three quarters of the WTO members. The practicality of this happening in the air transport field is yet to be tested. With regard to primacy between the WTO Agreements and the Chicago Convention, and as stated earlier, the latter, in Article 82, prohibits States from entering into any obligations or understandings which are in conflict with its principles, thus effectively precluding States from embracing the WTO principles while still adhering to the provisions of the Chicago Convention.

One way of resolving this issue might be to amend Article 6 of the Chicago Convention to be consistent with WTO philosophy. This could be done through Article 94, which provides that any proposed amendment to the Convention must be approved by a two thirds vote of attendees at an ICAO Assembly, such amendment to come into force only by ratification by a minimum of a two thirds vote of the ICAO membership of States.[46] For such an amendment to take effect, at

---

[44] *Id.* Article 46.

[45] Reuter (1989).

[46] At the time of writing ICAO had 190 member States.

least 124 States, as the situation stands, should ratify it. Again, it remains to be seen how practically possible it would be for ICAO to muster the ratification of 124 States, particularly in the instance of amending Article 6 regarding which there has so far been no overwhelming outcries by States calling for its amendment.

Inasmuch as the integrity and efficacy of the WTO system could by no means be questioned, there are also many positive features of WTO that would certainly render good guidance to the air transport industry. WTO has shown great enthusiasm and sincerity to act in the best interests of its own objectives of promoting trade and assisting international civil aviation to the best of its ability. Since the responsibility does not devolve upon WTO to assist air services, experts in the aviation field should now decide whether aviation, which has essentially been a symbol of State personality and independence, should be used exclusively as a trading tool. Any attempt at changing the *status quo* of international air transport services should therefore be a carefully thought out one, and a new regime – if one were to be agreed upon – should essentially serve all nations of the world equitably.

As was said earlier, the only way that might could bring market access in air transport within the multilateral framework of the WTO is through the political will of States. If most of ICAO's and WTO's member States (who happen to be largely the same) are united in their requirement to regulate air transport multilaterally, there is nothing to stop them. However, as States have demonstrated, bilateralism, although restrictive, enables States to retain control of market access. Although admittedly some States are giving effect to the liberalization of air transport by entering into open skies agreements with each other (nationally and regionally), it must be noted that reciprocal open skies policies are only cosmetically liberal, as they are almost always carefully crafted with every consideration being given to protecting one's interests while at the same time taking care not to jeopardize such interests through open, untrammelled competition.

There is no doubt that a particular State cannot adhere to or ratify two treaties containing conflicting principles. The two treaties concerned should have some consistency. One way to approach this issue might be, again with the political will of States, to revise Article 6 of the Chicago Convention, from its negative position to a positive one, where the provision could permit airlines of States to freely operate air services into the territories of each other, subject to the requirement that States whose airlines are seeking to operate services should convince the State which agrees for such operation that such services would benefit all concerned, including the consumer, while at the same time giving the latter the right to refuse if there is no convincing for such operations. This would not only preserve the bilateral element as a last resort, but would also encourage competition and, above all, bring some universality to the concept of liberalization which is much vaunted but rarely put in practice.

If States are reluctant to open the market by amending Article 6 of the Chicago Convention to be aligned with the WTO principles, they may still remain under both regimes by simply remaining under the Chicago based system of bilateral air services, but complying with the principles of GATS in so far as they are not inconsistent with the principle embodied in Article 6. States would have to file their

5.2 Market Access

obligations and record existing commitments under both conventions, acknowledging that they abide by the Chicago principle but consider GATS obligations as additional to the fundamental principle of Chicago. However, the overriding principle would be that any State in the GATS Annex regime of market access will have to open out its bilateral obligations already entered into, to all GATS signatories, conditional upon exemptions filed and national commitments made under the GATS.

Although liberalization of air transport cannot be dismissed as a viable prospect for the future, particularly in trading terms, the players concerned must necessarily view air transport in its entirety, as a service composed of critical factors that are inherent in safe and efficient air transport. Perhaps the answer lies in obviating the current approach of selecting one system to the exclusion of the other. It may well be that, since air transport is by no means a normal economic activity, a coordinated approach between the two Organizations would be the only way forward.

The inherent characteristics of air transport, of being a public utility on the one hand and of being confronted with the danger of over regulation on the other, admits of the need for a delicate balance between untrammelled competition and suffocative regulation. While the first approach may give rise to the usual free market inhibitors such as airport, airway and runway congestion, the other approach may ground to a halt the services that may provide air transport commensurate with the demand.

Air transport services, apart from being a public utility, also carry a rich tradition of "behavioural" guidelines. For instance Article 1 of the Chicago Convention provides that Contracting States recognize that every State has complete and exclusive sovereignty over the airspace above its territory. This would effectively preclude an air carrier from another State from operating air services to a particular State without permission. Another compelling factor in favour of regulation of air services is that air transport represents not only a mode of transportation but also reflects political, military, labour, security and environmental significance.[47]

The operative question – as to whether the global community proceeds to consider the operation of air transport services a trading activity along the lines of other conventional trading activities or whether it should continue to consider such services as a public utility which has broader connotations to States politically and therefore should remain regulated – remains one of fundamental importance to the future of international civil aviation. In other words, do air transport services go into the realm of the World Trade Organization or remain the exclusive prerogative of States? In either instance the role of the International Civil Aviation Organization remains a critical one – one which is of paramount importance to the future of commercial aviation.

Finally, it must be underscored that the exclusive principle here is neither "let the market control the activity of air transport" nor "Sovereign States can dictate which

---

[47] For a more detailed explanation of this argument see Wassenbergh (2001).

air transport services operate into and out of their territory". Rather it should be that there must be "effective control with structured and peremptory rules which harmoniously blend human needs with economic and financial interests".

5.3 Open Skies

If one side of the coin presents regulation of market access under the GATS Annex, the other side reflects the concept of open skies which conceptually reflects total liberalization of market access. A discussion of what is involved in such an arrangement bilaterally or multilaterally among States is best seen in the open skies agreement between the European Union and the United States.

There has been some polarization in the discussions that have taken place regarding the advantages and disadvantages of the open skies negotiations between the United States and the European Union (EU), particularly on the US side. This has largely been due to the perennial dichotomy between interests of the legislature, which ensures national control of US airlines, and the aviation industry which is regularly in the red and is looking for openings to get US airlines back on their feet. Both manufacturers (Boeing) and airlines (United) have claimed that the infusion of foreign capital in US airlines could help bail domestic carriers out of financial instability. However, US legislators have argued that an open skies agreement between US and EU, where it has been proposed by the EU that current US restrictions on ownership and control of US airlines, pegged at 25% for foreigners be relaxed, could have serious consequences and could allow foreign interests to restructure the US airline industry.

At the initial stages of the discussions, it was reported that the then Transport Secretary Norman Y. Mineta had mentioned in a statement issued in November 2005, that an open skies agreement between US and EU would provide the US with a historic opportunity to increase travel, reduce fares, expand commerce and bring the two continents closer together and airlines of both US and EU opportunities for healthier competition in a growing travel market and greater connections between cities and towns of all sizes on both sides of the Atlantic. Additionally Jeffrey N. Shane, US Under Secretary for Policy at the Department of Transportation, appearing before the Aviation Sub Committee of the House Transportation and Infrastructure Committee on February 8, 2006, underscored the fact that, to continue to be effective, US carriers required significant capital investments in facilities, technology and a variety of commercial arrangements. In this endeavour, the US airlines should have acces to global capital markets as allowed by law. Mr. Shane assured the Sub Committee that, while an open skies agreement between US and EU would enhance the ability of US airlines to compete and their potential to create employment opportunities, it would by no means allow for the amendment of the current legal structure of having 75% of the voting stock in airlines in the hands of US citizens. Furthermore, Mr. Shane added that the president and two-thirds of the board of directors and other managing officials would be US citizens and an airline company would remain under the control of US citizens.

Mr Shane went on to point out that the department has a statutory mandate to foster a safe, healthy and competitive airline industry that will remain capable of sustaining US economic growth by meeting the peoples' needs in transportation. He further stated that the Department of Transportation's *Proposed Notice of Rulemaking* would also require reciprocity in that, for a non-US investor to enjoy the benefits of the flexibility that would be available, US investors would have equal opportunity and right in the home country of the foreign investor. The main point made by Mr. Shane was that there was well established policy in major industries such as financial services, automobile manufacturing, information technology, steel and pharmaceuticals allowing for capital to flow freely across borders enabling competitors to establish a global market presence, exploit effectively economies of scope and scale and respond to customer demand. The Department's proposal for US airlines was along similar lines, placing maximum reliance on competitive market forces in air transport.

The apprehension of the US legislators in terms of the possible lack of control of US airlines by US citizens under an open skies agreement with the EU is not the only issue at stake for both parties to the negotiations. There are other issues which this article will address along with the ownership and control issue.

The *Economic Briefing* of the International Air Transport Association (IATA) issued in February 2006 notes that over 2000 aircraft were ordered in 2005 from the two largest aircraft manufacturers – Airbus and Boeing[48] These large orders have been placed notwithstanding the airline industry incurring an estimated net loss of $6 billion in 2006. The main concern with such large orders is that the injection of large capacity might adversely affect price competition in the years to come. The only consolation, however, is that the current trend of aircraft orders tends to suggest a managed delivery schedule which will compare favourably with the two previous peak cycles of 1991 and 1999. Another encouraging fact is that a large number of aircraft will be delivered to burgeoning markets in China and India which have already proved the need for more capacity in their markets in the next 5 years.

Still worrying though is the fact that orders from North America and Europe are quite substantial, raising questions as to whether a liberalized regime between the two great regions across the Atlantic might result in capacity dumping and pricing inconsistencies, particularly from 2007 when large orders are scheduled to be delivered. Of particular concern is that, in the event demand growth in services takes a downward path, large scale new deliveries could force airlines to enter into cut-throat price competition just as airlines are beginning to make a profit in 2007.[49]

---

[48] New Aircraft Orders – A Positive Sign but with Some Risks, *IATA Economics Briefing*, IATA: Geneva, February 2006.

[49] The estimates of IATA show net profits for the airlines of $ 6.5 billion in 2007 after incurring $ 46.4 billion losses over the previous 6 years.

IATA acknowledges that the airline industry has not shown much wisdom in the timing of new orders and delivery dates and that the industry, being capital intensive, could therefore be adversely affected by the timing of orders and deliveries which dictate profit and loss in the industry.

It is in this backdrop that the United States and the European Union (EU)[50] have launched a new round of negotiations toward open skies[51] calculated to bring about unfettered competition among their airlines. The fundamental aim of the US/EU negotiations is to do away with the existing tapestry of bilateral air services agreements between individual European Union member States and the United States and set up one system regulating transatlantic aviation. One of the issues on the table is cabotage rights,[52] particularly for European carriers who cannot carry revenue passengers from point to point in the United States. On offer by the United States has been the right for European carriers to fly from anywhere in the EU to any point in the US. However, the US has sought in return beyond fifth freedom rights (i.e. the right for US carriers to carry revenue passengers from a European point to points beyond Europe) and vice versa offered rights beyond the US for European carriers.

Another key issue is ownership and control of carriers, where the US limits foreign voting rights in its airlines to 25% of the stock while the EU has placed a ceiling at 49% foreign voting rights. The United States' compromise to this impasse has been to offer global investors more flexibility in marketing, routing and fleet structures, while retaining the 25% cap on foreign investment in US airlines. The

---

[50] The European Union or the EU is an inter-governmental and supranational union of 27 European countries, known as member States. The European Union was established under that name in 1992 by the *Treaty on European Union* (the Maastricht Treaty). However, many aspects of the Union existed before that date through a series of predecessor relationships, dating back to 1951. The European Union's activities cover all areas of public policy, from health and economic policy to foreign affairs and defense. However, the extent of its powers differs greatly between areas. Depending on the area in question, the EU may therefore resemble: a federation (for example, on monetary affairs, agricultural, trade and environmental policy, economic and social policy); a confederation (for example, on home affairs) an international organization (for example, in foreign affairs) A key activity of the EU is the establishment and administration of a common single market, consisting of a customs union, a single currency (adopted by 12 of the 25 member states), a common agricultural policy, a common trade policy, and a common fisheries policy.

[51] An open skies agreement is defined as a type of agreement which, while not uniformly defined by its various advocates, would create a regulatory regime that relies chiefly on sustained market competition for the achievement of its air services goals and is largely or entirely devoid of a priori governmental management of access rights, capacity and pricing, while having safeguards appropriate to maintaining the minimum regulation necessary to achieve the goals of the agreement. Open skies agreements are believed to provide for more competition, lower prices and higher passenger volumes in markets between signatory nations. See ICAO Doc 9626, *supra*. note 323, at 2.2-2.

[52] A cabotage right or cabotage privilege is a right or privilege granted to a foreign State or foreign carrier to carry revenue traffic from one airport of a State to another in the same contiguous territory of that State. See *Manual on the Regulation of International Air Transport*, ICAO Doc 9626, *supra*, note 2. at 4.1–10.

5.3 Open Skies 341

US has categorically stated that US investment rules cannot and indeed will not be a topic for negotiation.[53] The increased leverage given to foreign investors is meant to facilitate the influx of foreign capital by airlines in the red, such as Delta, United and Northwest Airlines who are facing bankruptcy proceedings.

Despite the setbacks of its major carriers, the United States has taken a courageous step towards liberalization in the belief that deregulation has to continue and that liberalization of air transport between the US and Europe would result in increased market share for US carriers[54] as well as further strengthening already robust competition between North American carriers and their European counterparts. Jeffrey Shane, Under Secretary for Policy at the US Department of Transport, has categorically stated that aviation liberalization is not for the faint of heart,[55] and that a possible breakthrough towards open skies between the US and EU would bring an entirely new level of liberalization to trans-Atlantic air services. The United States believes that open skies and open market access for US and EU carriers will not only bring 750 million people together but it would also be an example and a template for the regions of the world to follow.[56]

The United States claims that one of the issues that both parties are in agreement is that there should be service by every European and every American carrier between all points in Europe and all points in the United States.[57] Although the EU has not confirmed this statement.[58] Broadly, the US is seeking the right for

---

[53] Jeffrey N. Shane, *Aviation Deregulation: A Work in Progress*, speech delivered to the International Aviation Club, Washington DC, November 8, 2005, published in *Moving the American Economy*, US DoT: 2005 at p.7.

[54] The effects of open skies agreements signed by the US with Latin American countries has been beneficial to US carriers. A report released on this subject in 2000 revealed that an open skies arrangement with these countries is particularly advantageous for US carriers. The study showed that from 1997 to 1998 the capacity of US carriers in countries with open skies increased by 24.2%, while local carriers boosted their capacity by only 12.3%. Also, in the same period Origin-Destination (OD) traffic between the US and open skies countries in the region grew by 22.2% while the same traffic rose only 3.5% in countries without open skies. The study mentions that market shares of US carriers in Latin American OD markets began to rise dramatically in the 1990s regardless of whether open skies existed. The study also points out that between 1990 and 1998 the US flag market share between the US and South America jumped from 43.1% to 57.4%, while it climbed eight points to 60.4% in Central America and Mexico. Results of the study also indicates that in 1998, South American airlines lost "more than 1.2 million passengers to their US competitors," while Central American and Mexican carriers lost 1.3 million. See *The impact of open skies between the United States and Latin America, at 78*. (Available from Avman 6355, NW 36th St. Suite 601 Miami, Fl 33166).

[55] At a lecture delivered to the Royal Aeronautical Society, Montreal Branch, on 8 December 2005, on the subject *Air Transport Liberalization: Ideal or Ordeal*, Mr Shane asserted that liberalization begets more liberalization and that liberalization is the classic good deal that will not go unpunished. See www.dot.gov/affairs/briefing.htm at p. 11.

[56] Jeffrey Shane, *supra*, note 53 at p. 2.

[57] EU, US Make Substantial Progress on Open Skies, *Aviation Daily*, Monday, October 24, 2005 at p.2.

[58] *Ibid.*

every US carrier to fly from a European Union member State to another European Union member State and beyond to third countries, which essentially means that all traffic restrictions currently in place at some major European airports, including London Heathrow, should be lifted.

From the European perspective, the hope was that an aviation agreement should be reached in early 2006 once the United States clarified its position on control of airlines.[59] On 23 November 2005, both the US and the EU issued a joint statement which referred to their meeting which took place from 14 to 18 November 2005, and stated that progress was made toward the signing of an aviation pact between the two parties which would authorize every EU and US airline to fly between every city in the EU and every city in the US without restrictions on the number of flights, aircraft types to be used or routes selected, which would also involve unrestricted rights to fly beyond the EU and US to points in third countries.[60] Such an agreement would authorize every EU and every US airline to set fares freely in accordance with market demand and to enter freely into cooperation agreements with other airlines, including code sharing and leasing agreements. A precondition to such flexibility is that there should be a fundamental commitment to the highest standards of aviation safety and security.

5.3.1 The Issues Involved

Individual European States, which up until 1987 were separately chartering their destinies and their carriers' fortunes in their operations of international air services, showed an initial inclination to work towards collective interests by partially liberalizing European pricing policy in 1987. In 1993 the European countries of the European Economic Community agreed to full liberalization of pricing and liberalization of market access to apply on an intra-European basis. The culmination of the unification of European air transport came in 1997 when the European Union agreed to accord cabotage rights to carriers of the EU member States within the Union.

The European Economic Community, which was an economic union of States, had it genesis in the *Treaty of Rome* of 25 March 1957,[61] and became the European

---

[59] EU seeks US Aviation Agreement in 2006, *Air Letter*, Thursday, 22 November 2005, No. 15,873, at p.2.

[60] Text of a Joint Statement by the US and EU, http://www.a2a.aero/news/story.html?story/.

[61] Air transport in the European Community is fundamentally regulated by two treaties, i.e. the Treaty which establishes the European Coal and Steel Community (ECSC Treaty) and the Treaty which establishes the European Economic Community (EEC Treaty, now called the EC Treaty). The former, which was signed in Paris in 1951, addresses issues related to the carriage of coal and steel through the media of rail, road and inland waterways and as such is not directly relevant to aviation. The latter, on the other hand, admits of issues relating to all modes of transport in the carriage of persons and goods and is of some relevance to aviation.

5.3 Open Skies

Union by virtue of the *Treaty of Maastricht* of February 1992, which was amended by the *Treaty of Amsterdam* of October 1997. The EU is a monetary union and not a political union as yet. It is founded on the basic premise that there is no discrimination based on nationality. The application of this premise to air transport can be translated to the fact that only nationally owned carriers of the EU member States could be the subject of bilateral air services agreements negotiated by the Union with third countries. As such, individual States members of the EU cannot separately negotiate bilateral air services agreements with third countries on issues of nationality, as member States have the obligation to honour European Community law when they negotiate air services with third countries. However, as per a decision of the European Court of Justice, handed down in November 2002, sovereign member States of the EU could not be deprived of their power and right to conclude agreements with third countries, due to the fact that the EU common free market only applied to intra-community air transport.[62]

From a legal perspective, air carriers of the member States of the EU cannot have a European nationality since the EU does not have the sovereign status of a State. This notwithstanding, the Council of Ministers of the EU has given the European Commission[63] the mandate to negotiate with the US and other third countries, particularly to have such countries accept the fact that carriers of member States of

---

The EEC Treaty, which was signed in Rome on 25 March 1957, has at its core a Common Transport Policy (CTP) concept which is calculated to achieve the fundamental purposes of the European Community. One of the most salient features of the EEC Treaty is that the tasks of the Community are set out succinctly in Article 2 of the Treaty which provides inter alia for the adoption of a CTP as provided for in Article 3(1) of the Treaty. This provision is linked to Article 74 which in turn provides that the objectives of the Treaty in relation to issues of transportation would be pursued by State Parties within the parameters of the CTP, which is established by the Council of Europe through secondary legislation.

[62] Henri Wassenbergh, Open Skies and a Global Common Air Traffic Market, *Journal LuchtRecht*, Special Edition, *Liber Amicorum* in Honour of Prof. Dr. I.H.Ph. Diederiks-Verschoor, Nr. 9/10, December 2005, 51 at p.53.

[63] Two of the most important EU institutions are the Council of the European Union and the European Commission (the other two being the European Parliament and the European Court of Justice. The Council of the European Union contains ministers of the governments of each of the European Union Member States.. It is sometimes referred to in official European Union documents simply as the Council or the Council of Ministers (which will become its official name if the Treat establishing a Constitution of Europe is adopted). The Council has a President and a Secretary-General. The President of the Council is a Minister of the state currently holding the Presidency of the Council of The European Union, while the Secretary-General is the head of the Council Secretariat, chosen by the member states by unanimity providing general advice, qualified legal advice, translation services and impartial negotiation assistance. The Council of the European Union should be distinguished from the European Council, which meets four times a year in what is informally known as the European Summit EU Summit), and is a closely related but separate body, made up with the heads of state and government of the member states, whose mission is to provide guidance and high level policy to the Council. It is also to be distinguished from the Council of Europe which is a completely separate international organization and not a European institution.

the EU are "Community air carriers" established in the territory of the EU irrespective of their national ownership.

From the perspective of the European carriers, they would like the right to operate between the EU and the US from any point within the EU(which would translate as a seventh freedom[64] right in operating from a country other than other than the carrier's national territory) and extend that service to points within the US (which is the eighth freedom[65] right or consecutive cabotage). Also, the EU carriers are seeking the right to own and control US carriers and therefore be able to operate air services between points in the US, which is identified in the context of air law as ninth freedom[66] or "stand alone cabotage". In order to obtain these rights, the European carriers are seeking the abolition of ownership of US carriers by US nationals so that they (the European carriers) can attract capital from international money markets and enter into merges and acquisitions of foreign carriers. If this were to be at all allowed by the US (which is seemingly an impossibility according to current US policy) the European carriers would still have to operate on the basis that they remain "Community carriers" by their European ownership as they have to be owned in the majority by EU member States or their nationals.

In the US context, what is sought by the US carriers are free access to London Heathrow and seventh freedom carriage within the EU for express carriers. In broad terms, the US interests are focused on turning the North Atlantic aviation market into an open skies area giving rise to a common international air traffic market untrammelled by any conditions on market access, capacity and pricing. This would of course exclude the internal US market and any incursion into current US policy on majority ownership of US carriers by US nationals.

In reality, the US has already acquired for its carriers the rights to operate between European States through current bilateral air services agreements negotiated with individual European States and as such, any demand by the US for fifth freedom rights within the EU cannot be considered cabotage. As Wassenbergh correctly observes, in the absence of a single, unified, sovereign EU airspace, the EU cannot consider operations between sovereign States within the

---

The European Commission (formally the Commission of the European Communities) is the executive body of the European Union. Alongside the European Parliament and the Council of the European Union, it is one of the three main institutions governing the Union.

[64] The seventh freedom of the air is the right or privilege in respect of scheduled international air services, granted by one State to another State, of transporting traffic between the territory of the granting State and any third State with no requirement to include on such operation any point in the territory of the recipient State. i.e. the service need not connect to or be an extension of service to/from the home State of the carrier.

[65] The eighth freedom of the air is the right or privilege, in respect of scheduled international air services, of transporting cabotage traffic between two points in the territory of the granting State on a service which originates or terminates in the home territory of the foreign carrier or (in connection with the seventh freedom of the air) outside the territory of the granting State).

[66] The ninth freedom of the air is the right or privilege of transporting cabotage traffic of the granting State on a service performed entirely within the territory of the granting State.

5.3 Open Skies   345

EU as cabotage.[67] Nonetheless, the open skies judgments of the European Court of Justice[68] of 5 November 2002 were to the effect that the eight EU members, by concluding individual bilateral agreements with the US, had breached EC law in that the individual nationality clauses in all agreements infringed the right of establishment under Article 43 of the EC Treaty as they discriminated on grounds of nationality.[69] The ECJ also held that the agreements infringed the exclusive external competence of the European Commission. The essence of the judgement was that in areas where EC legislation affects third countries, only the EU could enter into international commitments. The new framework of EU air services negotiations are enshrined in Regulation 847/2004 which allows the EC to exercise a "horizontal mandate" to negotiate comprehensive agreements with third countries. This would mean that the third country acknowledges the existence of a single European market and the concomitant fact that EU airlines can operate international flights from any member State where they are established.

The European Council of Ministers' Conclusions of June 2005 introduce three lines of action: namely that the EC could continue to bring existing bilateral agreements between EU member States and third countries into line with Community law through horizontal agreements; there will be established a common aviation area with neighbouring countries by 2010; and global negotiations with key partners would be opened. In this context the EU carriers claim that their negotiations with the US go beyond an open skies regime, leading to total market opening and regulatory convergence. The latter – regulatory convergence – is calculated to establish a level playing field in increasing regulatory cooperation in the fields of competition policy, state aid, aviation security, environmental protection and safety.

The EU is claiming that, with the advantage of a combined negotiating power of 25 member States and a coherent framework for industry within the EU, partners of

---

[67] Wassenbergh, *supra*, note 62 and 945, at p. 55.

[68] The European Court of Justice (ECJ) is formally known as the *Court of Justice of the European Communities*', i.e. the court of the European Union (EU). It is based in Luxembourg, unlike most of the rest of the European Union institutions, which are based in Brussels and Strasbourg. The ECJ is the Supreme Court of the European Union. It adjudicates on matters of interpretation of European law, most commonly: claims by the European Commission that a member state has not implemented a European Union Directive or other legal requirement; claims by member states that the European Commission has exceeded its authority; and references from national courts in the EU member states asking the ECJ questions about the meaning or validity of a particular piece of EC Law. The Union has many languages and competing political interests, and so local courts often have difficulty deciding what a particular piece of legislation means in any given context. The ECJ steps in, giving its ruling which is binding on the national court, to which, the case will be returned to be disposed of. The ECJ is only permitted to aid in interpretation of the law, and not decide the facts of the case itself. Individuals cannot bring cases to the ECJ directly. An individual who is sufficiently concerned by an act of one of the institutions of the European Union can challenge that act in a lower court, called the court of First Instance. An appeal on points of law lies against the decisions of the Court of First Instance to the ECJ.

[69] See generally Abeyratne (2003).

the Union, such as the United States, could gain unrestricted access to the EU market have legal security (through horizontal agreements), achieved through a single negotiation.[70] At the time of writing, the EU was waiting for a final rule from the US as to whether the US rule regarding ownership and control of US airlines would be aligned towards opening US carriers to overseas capital.[71] One commentator has predicted that, should a major US carrier be threatened with Chap. 7 liquidation and a European carrier were to offer investment in that carrier, the US may just be inclined to revisit their existing rules.[72]

The above notwithstanding, and despite the slow pace of negotiation between the US and the EU, both parties have been vigorously forging liberal deals with third countries. At the time of writing, the EC had negotiated and concluded 22 horizontal agreements with third countries, while 59 countries had accepted community clauses. The EU internal market had been recognized in 385 bilateral air services agreements. The EC had also established contact with major partners such as China, Australia, India and the Russian Federation.[73]

5.3.2 Effect of Open Skies Competition

Whatever the outcome of the US/EU negotiations are in terms of cabotage for EU carriers and the US rule on ownership and control of its carriers, it seems certain that the two parties could agree on free market access between points in the US and points in the EU along with beyond fifth freedom rights respectively. As a corollary, free market access along with no limitations on pricing and capacity would certainly open up competition between US and EU carriers.

Competition in the air transport industry is a complex process, as there is no consensus among airline economists as to the exact nature of the industry. The demand for air services, particularly in the context of the airline passenger, is a contrived demand emerging from other demands based on activities such as business and leisure. This calls for a certain segmentation in travel where, in business travel, the passenger does not usually pay for the travel himself, whereas in leisure travel it comes out of his own pocket. Therefore, the leisure market calls for a different kind of competition, primarily based on the fare, whereas in business

---

[70] See *Communication From the Commission, Developing the Agenda for the Community's External Aviation Policy*, Commission of the European Communities, Brussels, 11.03.2005, COM (2005) 79 Final.

[71] US;EU agree on Text of Deal, Await Final Ownership Rule, *Aviation daily*, Monday, November 21, 2005, at p.2.

[72] Baker (2005).

[73] *Ibid.*

travel, although the fare is important, other considerations, such as facilities on board, may also play a considerable role in competition.[74]

Those supporting the retention of regulation argue that the very nature of air transport, being either naturally monopolistic or interdependently oligopolistic, calls for regulation in order that fares are not arbitrarily raised and remain competitive. Another theory in support of regulation is that some form of control should be exercised over "mushroom" airlines that may sprout up to exploit a liberalized market, thus disturbing the existing balance of an integrated network. Of course, each route is a separate market in itself and would require separate consideration. Although principles of economies of scale may apply generally to airline competition, where a fact such as larger aircraft being more efficient than smaller aircraft would apply on a general basis, individual assumptions for different markets have caused the two major aircraft manufacturers, Boeing and Airbus Industrie, to concentrate on manufacturing aircraft with strengths in speed and capacity respectively.

The European Union has expressed some concern as to the possibility of having to face potential dangers stemming from predatory pricing practices, particularly with regard to incumbent airlines dropping their prices in the short term to deter new entrants. However, irrespective of the opening of market access and liberalization of pricing and capacity between the US and EU, there is no room for doubt that neither the US nor the EU would allow an absolute "free for all" as there are strict legal regimes against anti competitive conduct and cut throat practices in pricing in both jurisdictions. The regulation of competition within the European Community is governed by the EC Treaty.[75] Two provisions in particular, Articles 81 and 82, contain principles which outlaw anti-competitive conduct. While the former prohibits the prevention, restriction or distortion of competition, the latter makes itself applicable against abuse by one or more undertakings of a dominant position within the market. The former essentially contains provisions for agreements, decisions or practices with anti competitive effects, and the latter concerns itself with abuses of a dominant marketing position. The aim of these two provisions in particular is to preclude distortion of competition within the Common Market by supplementing the basic principles enshrined in Articles 81 and 82 with substance. The goals of the Treaty in general and Articles 85 and 86, which, in particular promote the free movement of services, goods, persons and capital whilst effectively obviating barriers to trade within the community, is to enforce some regulation. Both these provisions relate generally to all sectors of transport unless explicitly excluded by the Treaty provisions.[76]

---

[74] Deregulation and Airline Competition, Organization for Economic Cooperation and Development, OECD: Paris, 1988 at pp. 20–21.

[75] The EC Treaty, also called the Treaty of Rome, was concluded in 1957 to forge "an even closer union among the people of Europe". See Goh (1997). See *supra*, note 16.

[76] Case 167/73 *Commission* v. *French Republic* [1974] ECR359 at 370.

Article 81 prohibits as incompatible such agreements as directly or indirectly fix purchase or selling prices or any other trading conditions; limit or control production, markets, technical development or investment; share markets or sources of supply; apply dissimilar conditions to equivalent transactions with other trading parties, thereby placing them at a competitive disadvantage; and make the conclusion of contracts subject to acceptance by other parties of supplementary obligations which, by their nature or according to commercial usage, have no connection with the subject of those contracts. These conditions are imposed on agreements between undertakings, which are defined as independent entities performing some economic or commercial activity.

Article 82 provides that any abuse by one or more undertakings of a dominant position within the Common Market or in a substantial part of it shall be prohibited as incompatible with the Common Market insofar as it may affect trade between member States. The Article prohibits: direct or indirect imposition of unfair purchase or selling prices or unfair trading conditions; limitation of production, markets or technical development to the prejudice of consumers; application of dissimilar conditions to equivalent transactions with other trading parties, thereby placing them at a competitive disadvantage; and conclusion of contracts subject to acceptance by the other parties of supplementary obligations which, by their nature or according to commercial usage, have no connection with the subject of such contracts.

In implementing these two provisions, air carriers have to exercise caution in not assuming that purely in view of a bloc exemption on air transport in the Treaty that may pertain to a particular issue, a related practice would be exempt from the prohibitions contained in Articles 81 and 82. In the air transport section of the Treaty, it is abundantly clear that block exemptions may apply only if abuse of dominant position is not evident in a given transaction.[77] Articles 81 and 82 are independent and complementary provisions and any exemption under Article 81 will not necessarily render the provisions of Article 82 nugatory.[78] "Dominant position" was defined in the 1979 decision of *Hoffman-La Roche* v. *Commission*[79] as a position of economic strength enjoyed by an undertaking which enables it to prevent effective competition being maintained on the relevant market by affording it the power to behave to an appreciable extent independently of its competitors, its customers and ultimately of its consumers. Such a position may necessarily preclude some competition except in monopoly or quasi monopoly situations. There is every indication, from existing jurisprudence and EC practice, that an assessment on an abuse of dominant position would not be predicated upon one factor alone or single characteristic but would rather be anchored on numerous factors such as market structure, barriers to entry and conduct of the business enterprise concerned.

---

[77] Adkins (1994).

[78] Case T-51/89, *Tetra Pak Rausing SA.* v. Commission [1990] II E.C.R. 309, [1991] 4C.M.L.R. 334 para 31.

[79] Case 85/76 *Hoffman-La Roche & Co. A.G.* v. *Commission* [1979] E.C.R. 461.

In the United States, the term "antitrust laws" encompasses federal and States legislation (statutes) which regulate competition with a view to wiping out unfair trade practices and preserving competition among sellers and buyers. Needless to say, antitrust laws apply equally to international air services, and are calculated to preclude both conduct and structural changes in business enterprises. A typical example of conduct coming under antitrust laws in the United States is a merger between competitors which would unduly limit competition. These laws are also meant to prevent producers or purchasers of goods from exercising a monopoly in imposing prices which significantly deviate from expected free market competition norms.

Antitrust legislation in the United States goes back to 1890 with the enactment of the *Sherman Act* which makes it criminally illegal for any contract, combination or conspiracy to be formed in restraint of trade. This all encompassing provision prohibits price fixing, anti discounting agreements, divisions of markets by pooling agreements and capacity agreements and exchanges of information that can be considered as competitively sensitive. The Act also prohibits monopolies and conspiracy to monopolize in Sect. 5.2.

In 1914, the United States Congress legislated the *Clayton Act*, primarily to supplement the *Sherman Act*. The *Clayton Act* outlaws certain types of "exclusive dealing" and "tied sales" and prescribes standards for determining the legality of mergers and acquisitions. Both the Acts admit of compensation to persons injured in their trade or business up to three times the amount of their loss plus attorney fees. Courts have also permitted consumer class actions an antitrust activity, allowing for significant recovery of damages.

There is strong precedent against cut-throat pricing in the United States, couched in the judgment of the 1993 *Brooke* case[80] which brings to bear US regulation and judicial policy on predatory practices in an oligopoly setting. The case involved a competitor in the cigarette industry who sold his product below cost, resulting in an action being brought by another cigarette manufacturer under Article 2 of the *Sherman Act*. The Supreme Court held that for an action to succeed the plaintiff must show that the defendant's low prices are below an appropriate level of the plaintiff's costs. The fundamental principle establishing the illegality of predation is that the predator must ultimately be recouped by the act of predation. The Court cited an earlier decision[81] and held that "[R]ecoupment is the ultimate objective of an unlawful predatory pricing scheme. It is a means by which a predator profits from predation. Without it...consumer welfare is enhanced."[82] In both the *Brooke* and *Matsushita* cases, the court found no recoupment and therefore no justification to conclude that the defendants had indulged in predation.[83]

---

[80] *Brooke Group Ltd. v. Brown and Williamson Tobacco Corp.*, 509 U.S. 209 (1993).

[81] *Matsushita Electric Industrial Co. Ltd., et al. v. Zenith Radio Corp. et al.*, 475 U.S. 574 (1986).

[82] *Brooke, supra*, note 80 at 224.

[83] The Areeda-Turner test defines criteria that determine predatory pricing. According to this test, a short run profit maximizing price as well as a price above full costs are non predatory. The test also

In a regulatory context, the International Civil Aviation Organization, in its role as the sole international regulatory body in the field of air transport, has issued clear policy and guidance material on the avoidance or reduction of conflicts over the application of competition laws to international air transport. ICAO has issued these guidelines to address the conflicts that may arise between States which adopt policies, practices and laws relating to the promotion of competition and restraint of unfair competition within their territories. ICAO urges States to ensure that their competition laws, policies and practices, and any application thereof to international air transport are compatible with their obligations under relevant international agreements.[84] Within this guideline, there is a strong recommendation for close consultation between States and all interested parties in order that uniformity in practice be achieved across borders to the maximum extent possible. Accordingly, when a State is adopting laws pertaining to competition, it is expected to give full consideration to views expressed by any other State or States whose interests in international air transport may be affected. States are urged to have full regard to principles of international comity, moderation and restraint. The Guidelines also provide direction on dispute resolution and problem solving.

5.3.3 Commercial Considerations

It is likely that an open skies agreement between the US and EU will lead to increased competition resulting in a wider range of services carried out with greater efficiency and at competitive fares.[85] The main effects of an open skies agreement between the US and EU would be that London's Heathrow Airport will be open to carriers other than the two US carriers – United Airlines and American Airlines – that are allowed to fly there now by treaty. It would also enable broader marketing agreements between European and US carriers and help cargo airlines like FedEx and UPS build larger networks. European carriers would have broader access to US destinations as well. In addition, when the travel and tourism industry combined are contributing tremendously to the world economy, an opening of two of the worlds largest markets to open competition would lead to significant secondary effects in employment opportunities. The air transport industry in Europe and in North America make the greatest contribution to the world GDP in comparison to other

---

goes to consider that a price at or above reasonably anticipated short run marginal costs is non-predatory. Also, a price at or above reasonably anticipated average variable cost should be considered to be within legal limits. see Roos and Sneek (1997).

[84] *Policy and Guidance Material on the Economic Regulation of International Air Transport*, Doc 9587, Second Edition, 1999, Appendix 2 at A2-2. See also *Manual on the Regulation of International Air Transport*, Doc 9626 First Edition 1996, Appendix 5, Guidance Material for Users of Air Transport, at A5-1.

[85] A study conducted in 1986 by the Brookings Institution concluded that consumers in the US saved $ 6 billion annually in fares and the airline industry experienced an average annual profit of $ 2.5 billion since domestic de-regulation in 1978. See Morrison and Winston (1986).

5.3 Open Skies

regions. IATA records that in 2004 North America accounted for 37 per cent of global employment in aviation (4.6 million in direct employment and 0.8 million employed as a result of the catalytic effect of such direct employment) and 50% of the contribution of air transport to the global GDP. In comparison, Europe accounted for 27% of global air transport employment and GDP.[86] the liberalization of the US/EU air transport market will undoubtedly open the doors for both American and European carriers to have increased commercial arrangements such as code sharing[87] that would maximize the utilization of market potential both across the Atlantic and beyond.[88]

The American version of "open skies" is conducive to open competition as it comprises, inter alia, the basic elements of open entry to all routes; unrestricted capacity and frequency on all routes; unrestricted route, traffic rights, double disapproval pricing in third and fourth freedom markets in intra EC markets; liberal charter arrangements; liberal cargo regimes; open code sharing opportunities; and explicit commitment on non-discriminatory operation of access to computer reservation systems.[89] Although in the 1990s Europe thought this concept of open skies and its elements would "endanger the whole process of deregulation of Europe's civil aviation market",[90] the overall approach of the EU to current negotiations with the US has been to accept the philosophy of the US on open skies. However, the scales would tip in favour of the US carriers under this philosophy, as they would be able to consolidate fifth freedom rights through a network of routes whereas the European carriers would not have a comparable variety of points to exercise fifth freedom rights beyond the US.

Another inhibitor for European carriers could be the anti-trust exigencies that could arise under US law which may put European carriers at a substantial disadvantage in the US market[91] For this and the more significant distinction between US ownership and control restrictions (which are placed at 25% for foreign nationals) and leverage given to foreign nationals (49%) in the EU, it would be structurally, economically and legally more advantageous to US carriers to have an open skies agreement with Europe while their European counterparts may not be as

---

[86] *The Economic and Social Benefits of Air Transport*, Air Transport Action Group: 2005, at 7.

[87] One of the commonest uses of code sharing is to signify that two airlines conclude an arrangement according to which two or more connecting flights are offered under a common designator code and flight number or those of both airlines, although individual segments are operated with aircraft of one airline. See Abeyratne (1995).

[88] It will be recalled that, when the US and the Netherlands entered into what was the first open skies agreement between the US and a European country, KLM, who already had access to all US points, consented to the agreement on the condition that both parties accept the KLM/Northwest alliance with anti trust immunity and mutual code sharing. By code sharing, KLM could provide "own" on line services to US points without operating services itself. See Wassenbergh (2000).

[89] . See DoT Order No. 92-8-13, 35 of 1992. Also, I.L.M. 1479 (1996).

[90] Neil Kinnock, Speech to the Association of European Airlines, Luxembourg, April 28, 1995.

[91] The DoT and Department of Justice of the US review anti trust cases on a case by case basis. See DoT order to show cause, 96-5-26, OST-95-618.

well placed under such an agreement.. Therefore an open skies agreement with EU might just help the American carriers in improving their revenues which have shown a $ 9.1 billion loss in 2004, followed by a $10 billion loss in 2005 and a projected $6.5 billion loss in 2006. In contrast, the European carriers have shown profits in the triennium, with $0.6 billion in 2004, $1.3 billion in 2005 and a projected $0.6 billion in 2006.[92]

One of the commercial considerations with regard to achieving enhanced competition through an US/EU open skies agreement would be the extent to which the EU will have the authority to negotiate all aspects of the DoT standard open skies agreement, which include slots. Another would be whether individual EU member States could still, after the EU signs an open skies agreement with the US, negotiate a bilateral air services agreement with the US From the EU perspective, where the Council of Ministers has given a mandate for the Commission to negotiate an agreement with the US, there is no express prohibition so long as there is recognition of the EU as one single area and the Community clause is signed. The consequence of the European Court of Justice decision[93] particularly from a competition angle, was that the core element of the bilateral air services agreement, which is market access involving the award of air traffic rights was untouched by the Court except in instances where an EU member would, in its agreement with the US, explicitly preclude another EU member from operating air services from that member's territory. In other words, Belgium would not be permitted to agree that Air France would not or could not operate services between Brussels and New York. This prohibition is entrenched in the Treaty of Rome which forms the substance of legislative legitimacy of the EU and incorporates the right of equal national treatment for all EU member States. Therefore, if one EU member State were to preclude the right of another member State's airline from having the right to operate air services to the United States from the territory of the first EU State it would tantamount to discrimination by the first State against the second State.

Also, the Court decided that certain specific provisions and areas covered in the questioned bilateral agreements between individual EU members and the US were contrary to EU law since they encroached upon internal EU regulations pertaining to non EU nationals. These laws concern:

a) Provisions pertaining to the allocation of airport slots;
b) Provisions governing pricing, or fares and rates of intra-European air services;
c) Agreements on computer reservation systems insofar as they appear as provisions of the open skies agreements in question; and
d) Provisions which reserved the right to grant permission under the open skies agreements only to airlines substantially owned and effectively controlled by nationals of the EU member States that is party to a particular agreement.

---

[92] IATA Economics, Presentation made to ICAO on 24 January 2006.
[93] *Supra.*

5.3 Open Skies 353

Yet another issue to be considered is the position of the United States and the EU on the issue of public subsidies. This extends both to carriers as well as manufacturers. On the one hand, one recalls the Anglo-French Concorde which sustained its services through subsidies by the British and French governments.[94] On the other, both parties are in dispute over subsidies purportedly given to Boeing and Airbus by the US and EU respectively.[95] Further impediments to competition under a US/EU open skies regime could be the pervading influence of national interest, where member States of the EU could continue to ensure an "inside track" to their carriers, which may include insistence that nationals of a country fly on their national carrier[96]; placement in a computer reservation system where interested parties could give prominence in the system to their carriers; and the use of excessive user fees to discourage foreign carriers.

As to whether there should be absolute, untrammelled competition within the Americas and between the Americas and Europe is a critical issue for the coming years. Of course, one recent suggestion has been to crystallise a "convergence of regulatory principles" between Europe and the United States in competition by establishing a Transatlantic Common Aviation Area (TCAA). This concept, suggested by the Association of European Airlines (AEA) in a policy statement,[97] puts forward detailed and realistic proposals on how to bring about an ideal regulatory convergence between the European region and the United States, addressing three areas:

1. Matters in respect of which harmonization is necessary;
2. Those in respect of which convergence could take the form of mutual recognition; and
3. Those which could in principle be left at the discretion of each party.

The TCAA concept advocates the freedom of the parties to provide services; addresses issues pertaining to airline ownership and the right of establishment; provides recommendations with regard to competition policy; and offers guidelines on the leasing of aircraft.

Since the TCAA aims at replacing traditional governmental regulatory control of such aspects of competition as market entry and pricing, the issues emerging from competition policy become by far the most complex and difficult to deal with, within the parameters of the TCAA. Although the fundamental postulates of competition in Europe (as followed through by European Union regulations) and the United States are broadly similar in intent, and both depend to a certain extent

---

[94] There are other instances, such as Alitalia, where the Italian government saved the carrier by covering its debt into equity and the subsidies given by the Scandinavian countries to SAS.

[95] See Abeyratne (2005).

[96] Doganis cites the example where, under the "fly America Policy, officials or others travelling on behalf of the US Government were and are required to fly on US airlines or US carrier code shared flights irrespective of open skies agreements applying to the sectors flown" Doganis (2001 at p.45).

[97] 52 .Aeropolitical News, IATA,REf.744-18.11.99.

on the application of extra-territoriality in their regulations, there are obvious differences such as those embodied in the different approaches to transatlantic airline alliances. Also, the United States stringently relies on a principle of "public interest" in its air transportation policy, while European competition rules are not as explicit in their policies. The basic essence of a TCAA would therefore establish the principle that matters of route sharing, capacity, pricing and frequency of services should be driven by market forces rather than be determined by governmental intervention. This way a certain commonality might be established between air transport of the two regions.

The United States has, over the past few decades, steadfastly advocated the need for open skies agreements with its partners in aviation. At the bilateral level, 38 "open skies" bilateral air services agreements have so far been concluded by 17 States or areas in the Asia Pacific region, with the United States being one of the partners in 12 cases. With Africa, the United States has concluded open skies agreements with 16 African States[98]: While no U.S. carriers have been directly serving to Africa, they have expanded code share services with European carriers. Also, several African carriers inaugurated services to the United States.

For its part, The EU has also made progress in that the European Commission has concluded horizontal agreements so far with Australia, New Zealand and Singapore, all of which were initialled in 2005. The European Commission has also asked the Council of the EU to grant more comprehensive negotiating mandates for the creation of open aviation areas (OAAs) with Australia, China, and New Zealand.

With regard to Africa, the EU has achieved a significant level of progress. One of the European Commission's negotiating mandates conferred by the Council of the EU is to negotiate, on behalf of all member States, a Euro-Mediterranean aviation agreement with Morocco. This agreement was initialled in December 2005 and will eventually replace all the bilateral air services agreements between Morocco and the EU member States. The European Commission has also been conferred a horizontal mandate to replace certain specific provisions in the existing bilateral agreements declared contrary to Community law. In response to the European Commission's negotiating mandates, African Ministers agreed in May 2005 that it was necessary to adopt a common external policy and recommended to carry out a two phase plan of action for this purpose.

As can be noted, irrespective of the difficulties arising from the transition from a traditional and entrenched bilateral method of negotiation, both the US and EU have forged ahead towards their goal of open skies with an impressive list of precedent. The collective position of these two giants are rife with complex realities of competition and cannot be compared with other nations who might place open skies on a bilateral negotiation table and consider it a done deal if the other party

---

[98] Tanzania (1999), Namibia (2000), Burkina Faso (2000), Ghana (2000), The Gambia (2000), Nigeria (2000), Morocco (2000), Rwanda (2000), Benin (2000), Senegal (2000), Uganda (2002), Cape Verde (2002), Madagascar (2004), Gabon (2004), Ethiopia (2005) and Mali (2005).

accepts. Nor can the US/EU open their territories to unlimited and untrammelled open skies. There has to be a sense of direction where the two parties are headed when capacity, pricing and frequency are open. This direction should address the outcome of open skies and the various exigencies that might follow, such as complexities in slot allocation, national interest, possible carrier alliances, secondary business stemming from open skies.

All inhibitors to open skies notwithstanding, the overall benefits of liberalization must outweigh the consequences of protectionism. As one commentator has said, when all is said and done, "every argument against open skies is an argument in favour of protecting some airline or other against competition...on the flip side of capacity dumping and predatory pricing you find a smashing deal for the markets in and out of the country, more business and tourist travellers, more goods moving by air, hotels flourishing, the overall economy better off and everybody's happy".[99] If the US and the EU were to adopt this same philosophy they could only end the negotiations with a "win-win" deal.

5.4 Slot Allocation

If open skies is a possible way to go in circumventing the obstacles of the existing bilateral system of negotiation air traffic rights, another issue crops up, and that is, how would the airports with their finite capacity and infrastructure service the large numbers of flights that would ensue from open skies?

In addition to the restrictions placed on the free movement of aircraft between States on a commercial basis, airlines face a further hurdle when it comes to obtaining airport slots[100] once they obtain the right to carry traffic in and out of a State. Slots are allocated to aircraft operations in an order of priority as follows: regular scheduled services; ad-hoc services; and other operations. Increasing congestion at major airports brings to bear the increasing difficulties faced by air carriers to obtain slots to land at and depart from airports. Airlines are finding new and innovative ways to bypass the restrictions imposed by Article 6 of the Chicago Convention through various commercial tools. However, there are still the remnants of an unhealthy competition between dominant carriers who retain their

---

[99] Flanagan (1996).

[100] An airport slot is a designated day and time usually within a 15 or 30 min period for an aircraft to arrive at or depart from an airport. See *Regulatory Implications of the Allocation of Flight Departure and Arrival Slots at International Airports*, ICAO Doc 283-AT/119, 2001 at Chapter 1, p.1. European Council Regulation (EEC) No 95/93 of 18 January 1993 on common rules for the allocation of slots at Community airports defines a slot as "the scheduled time of arrival or departure available or allocated to an aircraft movement on a specific date at an airport co-ordinated under the terms of the Regulation".

larger market shares against carriers who do not have as large a competitive profile as their opponents.

The issue of lack of slots at capacity-constrained airports is not new. Over the past decade, growth in air traffic has continued to outstrip available capacity at many airports. According to IATA, the total number of fully-coordinated airports subject to slot allocation under the IATA Schedule Coordination System had increased from 136 in 2000 to 142 in 2008 (Europe has an increase of 16%). Difficulty in obtaining slots can affect the ability of an air carrier to exercise its market access rights granted under relevant bilateral air services agreements.

All formal mechanisms for dealing with the lack of airport capacity are based on the concept of an airport slot, which is the time that an aircraft is expected to arrive at or depart from a capacity-constrained airport. For commercial operations which use airport gates, this time is calculated based on when the aircraft arrives at or leaves the gate. To take into account variations in flight times, unavoidable delays, etc., airport slots may actually be allotted in terms of a time period, such as 16:45 to 17:OO. Airline schedules, of course, are stated in more precise terms and, for example, five flight arrivals in that time period could each appear in the respective airlines' schedules at a specific time, for example 16:45, 16:48, 16:52, 16:53 and 16:58.

However, different airlines may schedule their flight departures at the same time (for commercial or operational reasons), for example on the hour, which at busy airports can exacerbate peaking and often result in aircraft having to wait in line for a take-off clearance. An airport slot should not be confused with an air traffic control (ATC) slot, the take-off or landing time of an aircraft which is assigned by the relevant ATC authority to make optimum use of available capacity at points en route or at the destination airport by sequencing the air traffic to regulate its flow efficiently. Thus, commercial operations may not land or take off in the same order as reflected in their respective schedules, but at times which would enable air traffic control to regulate efficiently the flow of aircraft into or out of the airport and the en-route system. This may involve, for example, interspersing commercial flights with general aviation flights and varying the order of take-off or landing to take account of greater separation requirements for larger aircraft, late arriving aircraft, etc. With the assignment of an airport slot, airlines can build their schedules, taking into account time to taxi to and from gates and customary en-route time, on the assumption that an ATC slot will be made available as close as possible to the time necessary for the flight to operate on schedule. This underlies the importance of close coordination between the coordinator assigning the airport slots and the air traffic control authorities responsible for flowing aircraft into take-off, landing and the en-route system.

An airport slot is essential in order to mount commercial services at an airport which has a slot allocation regime, but it is also part of a multifaceted package of services and facilities provided by different entities, such as gates, air traffic control, ground handling, passenger and cargo processing – all of which require close coordination and cooperation between and among national authorities, airports and airlines In a few slot allocation regimes, procedures differ depending on the type of entity using them. Thus, there can be commuter slots, air carrier slots, new entrant (to a city-pair market) slots, and slots for general aviation, military, domestic or international flights.

International access to airports is governed, inter alia, by Article 15 of the Chicago Convention,[101] the first sentence of which provides,

> Every airport in a contracting State which is open to public use by its national aircraft shall likewise, subject to the provisions of Article 68, be open under uniform conditions to the aircraft of all the other contracting States.

Article 15 establishes a national treatment standard for all contracting States in the context of the use of airports and other air navigation facilities for international air transport. Insofar as the operation of scheduled services is concerned, Article 15 subjects the use of airports to Article 68, which permits each contracting State, subject to the provisions of the Convention, to "designate the route to be followed within its territory by any international air service and the airports which any such service may use". However, Article 68 should not be read individually so as to derogate the national treatment standard established in Article 15.

Article 15 does not itself accord a right to operate international scheduled or non-scheduled air services. The operation of international scheduled services is subject to Article 6 of the Convention, which provides that "no scheduled international air service may be operated over or into the territory of a contracting State, except with the special permission or other authorization of that State, and in accordance with the terms of such permission or authorization." With respect to international non-scheduled operations, although Article 5 of the Convention allows aircraft engaged in non-scheduled commercial flights to overfly or make non-traffic stops in the territory of a Member State without the necessity of obtaining prior permission, authorization is generally required for such commercial flights from any State in which passengers and/or cargo are loaded or unloaded.

The authorization required for international scheduled air services in Article 6 is customarily accorded on the basis of traffic rights exchanged bilaterally or regionally, which either name a specific city at which such rights may be exercised or make a broader, non-specific grant in terms of any city in a State's territory. Although some bilateral and regional agreements include provisions for non-scheduled flights, the general practice has been for the States concerned to approve such flights on the basis of national regulations and policies. Regardless of the underlying source of the authorization to operate international commercial air services, once granted, national treatment and the uniform condition criteria of Article 15 apply.

A combination of this clear standard of treatment and the practice of having common traffic points for national and foreign airlines may explain why bilateral agreements, except in rare instances, do not deal with slot allocation or access to specific airports. However, air service negotiators have to be mindful of the difficulties of obtaining access to capacity-constrained airports and take into

---

[101] *Supra*, note 3.

account that the rights they are seeking for their airlines at those airports may not be able to be exercised for some time.

The uniform treatment principle is important also with respect to conditions on the use of airports, particularly for environmental purposes. Thus curfews, or aircraft noise criteria, as well as any exceptions thereto, must be applied uniformly to both national and non-national aircraft engaged in similar international services. Inter-governmental disputes involving airport access under the uniform treatment rule have been rare; it has been more common for airport access disputes to focus on specific cases where airlines which have the underlying route rights to serve a city have not been able to secure access or increase service to that city's airport because of a lack of available slots. In such cases, States have usually relied on the bilateral provision which requires that designated air carriers have a fair and equal opportunity to operate or compete with respect to the services covered by the agreement.

There has been a continuing debate as to the "ownership" of airport slots, primarily in terms of claims by airlines which have historically used them for long periods of time. However, some formal regulatory regimes either explicitly or implicitly exclude this concept, for example, stating that airlines do not acquire property rights to the slots assigned to them and that the slots must be returned to the aeronautical authority under certain circumstances. The implicit approach ties the continued use of the slot to its use at a specified level (e.g. 80%) and allows the exchange of slots on a one-for-one basis. In one instance in the United Kingdom, a court ruled in March 1999 that financial considerations in connection with an exchange of slots under the European Union (EU) common slot rules did not invalidate the exchange. However, the court did not rule on whether the exchange as such involved real property.[102]

Nevertheless, the obvious value in terms of market access of slots at airports with severe constraints on capacity has led to the treatment of these slots as a de facto financial asset of the airlines holding them. Thus, the purchase by one airline of another will take into account an estimated value of the airport slots involved. However, regulatory authorities have retained the right to approve or disapprove the transfer of airport slots in this manner, primarily through the approval or disapproval of the purchase or merger involved. The only formal pricing of airport slots has occurred in the United States where the purchase, sale and lease of certain domestic slots at the four airports currently subject to the Federal Aviation Administration's High Density Rule has been permitted since 1986. This has led

---

[102] In this case, it was reported that Britain's number two long-haul airline, Virgin Atlantic Airways, and Australia's flag carrier Qantas each paid an alleged 20 million GBP to the small British regional airline Flybe. for six slots at Europe's biggest airport, London Heathrow. Heathrow is known as one of most – if not the most – slot constrained airports in the world, with airlines queuing for years to gain access to the airport. See The Sunday Times, Heathrow Slots Take Off, January #, 2004; *O'Connell*, The Sunday Times, Soaring Costs Of Touching Down, February 22, 2004.

5.4 Slot Allocation

some airlines serving these airports which purchase such slots to reflect their value as assets in their financial accounts.

The slot allocation issue has been extensively addressed by ICAO. In 1999, a detailed study on it was conducted by the Secretariat and submitted for review by the Conference on the Economics of Airports and Air Navigation Services (ANSConf2000) held in 2000. The study, which was published in 2001,[103] analyzed the trends for airports where the demand exceeds capacity supply; the regulatory framework involved; and the means by which governments, airports and airlines have sought to alleviate this problem. It also assessed existing and potential mechanisms for dealing with a chronic shortage of airport capacity and suggested possible improvements of, and alternatives to, the existing systems.

The fifth ICAO Worldwide Air Transport Conference (ATConf/5) held in 2003 addressed the issue of airport constraints in the context of market access. Recognizing that the ability of air carriers to exercise market access rights granted under relevant air services agreements is closely linked to the availability of slots at the airports concerned, the Conference concluded that "in liberalizing market access, due consideration should be given to airport capacity constraints and long-term infrastructure needs. Problems involving air carriers which are unable to exercise their entitled traffic rights at a capacity-constrained airport may, if necessary, be addressed in the context of discussions on the relevant air services agreements. In this regard, sympathetic consideration should be given to the request for preferential treatment from those States whose airports are not slot-constrained but whose air carriers are unsuccessful in obtaining slots at slot-constrained airports, consistent with relevant national legislation and international obligations." The Conference further concluded that "any slot allocation system should be fair, non-discriminatory and transparent, and should take into account the interests of all stakeholders. It should also be globally compatible, aimed at maximizing effective use of airport capacity, simple, practicable and economically sustainable."

As recommended by the ICAO guidance, the most practicable measure States can take in addressing a slot problem is through relevant consultation or dispute settlement mechanism under the bilateral air services agreements, in accordance with the principles of fair and equal opportunity and reciprocity and in a spirit of cooperation and mutual understanding. However, the reality is that some States do not follow or implement ICAO's guidance. This may be largely due to the fact that ICAO's policies and guidance in the economic field, unlike the Standards and Recommended Practices (SARPs) in the technical field which are binding for States, are of a recommendatory nature, for optional use by States. The lack of application of ICAO's policies and guidance is not conducive to the general interest of the Organization and its member States. The question as to how to improve this situation involves a broader policy issue, which will need to be considered by the Council and States at an appropriate time and forum.

---

[103] Circular 283, *Supra*.

Problems of slot availability are mostly caused by congestion at airports and they should be addressed in the most efficient manner. Eliminating exemptions on charges for smaller airlines or airlines that operate fewer frequencies to an airport, and curbing airport monopolies are effective measures to address the congestion issue.[104] One of the ways in which airlines obviate the problem of unavailability of slots is through alliances with each other. Airline alliances are usually created with the intention of putting into place an integrated network of products, services and standards between two or more carriers who have the objective of operating air services more efficiently by eliminating burdensome duplication of costs and achieving better services for the travelling public.[105] These alliances have also been identified as "a distinct form of entry mode that has been used as a low-cost means of gaining access to new markets and local infrastructure".[106] The underlying philosophy of the airline alliances, is not so much an emphasis on the more effective use of resources such as labour, capital and national resources (which are inevitably important factors) but rather an overall reliance on the strategy of location, where the sharing of locations represented by the various airlines have enabled them to produce their goods and services in a consistent manner, thus achieving the status equivalent to a cartel, while still retaining their individual identities. One of the advantages of forming an alliance is that airlines could share their slots within such an alliance to maximize operations and revenue.

5.4.1 Slot Allocation in Europe

The European Economic Community, which was an economic union of States, had its genesis in the *Treaty of Rome* of 25 March 1957[107,108] and became the European Union by virtue of the *Treaty of Maastricht* of February 1992, which was amended

---

[104] Levine (2009).

[105] Kimpel (1997).

[106] Doz et al. (1990). For a detailed categorization of alliances see Rhoades and Lush (1997).

[107] Air transport in the European Community is fundamentally regulated by two treaties, i.e. the Treaty which establishes the European Coal and Steel Community (ECSC Treaty) and the Treaty which establishes the European Economic Community (EEC Treaty, now called the EC Treaty). The former, which was signed in Paris in 1951, addresses issues related to the carriage of coal and steel through the media of rail, road and inland waterways and as such is not directly relevant to aviation. The latter, on the other hand, admits of issues relating to all modes of transport in the carriage of persons and goods and is of some relevance to aviation.

[108] The EEC Treaty, which was signed in Rome on 25 March 1957, has at its core a Common Transport Policy (CTP) concept which is calculated to achieve the fundamental purposes of the European Community. One of the most salient features of the EEC Treaty is that the tasks of the Community are set out succinctly in Article 2 of the Treaty which provides inter alia for the adoption of a CTP as provided for in Article 3(1) of the Treaty. This provision is linked to Article 74 which in turn provides that the objectives of the Treaty in relation to issues of transportation would be pursued by State Parties within the parameters of the CTP, which is established by the Council of Europe through secondary legislation.

5.4 Slot Allocation

by the *Treaty of Amsterdam* of October 1997. The EU is a monetary union and not a political union as yet. It is founded on the basic premise that there is no discrimination based on nationality. The application of this premise to air transport can be translated to the fact that only nationally owned carriers of the EU member States could be the subject of bilateral air services agreements negotiated by the Union with third countries. As such, individual States members of the EU cannot separately negotiate bilateral air services agreements with third countries on issues of nationality, as member States have the obligation to honour European Community law when they negotiate air services with third countries. However, as per a decision of the European Court of Justice, handed down in November 2002, sovereign member States of the EU could not be deprived of their power and right to conclude agreements with third countries, due to the fact that the EU common free market only applied to intra-community air transport.[109]

From a legal perspective, air carriers of the member States of the EU cannot have a European nationality since the EU does not have the sovereign status of a State. This notwithstanding, the Council of Ministers of the EU has given the European Commission the mandate to negotiate with the US and other third countries, particularly to have such countries accept the fact that carriers of member States of the EU are "Community air carriers" established in the territory of the EU irrespective of their national ownership.

A major global issue with regard to slot allocation in Europe is the difficulties airlines of other regions may be facing with regard to night curfews imposed within Europe. Night curfews have created operational problems and financial burden for their air operators, airports and communities around airports. For example, due to night curfews imposed at some airports in Europe, many African airports have to be kept open for operations during the night, since north-bound aircraft must depart late in the night in order to arrive in Europe after the curfew is over (usually after 06:00 A.M.). It has been claimed that night curfews could also contribute negatively to airport congestion and increase delays, which would increase aircraft emissions leading to poorer local air quality at airports.

African States have argued[110] that removal of night curfews at some international airports in Europe would increase airport capacity, thus benefiting the airlines wishing to operate to them, while significantly reducing the night congestion at many African airports. With the widespread use of quieter new generation aircraft,

---

[109] Henri Wassenbergh, Open Skies and a Global Common Air Traffic Market, *Journal LuchtRecht*, Special Edition, *Liber Amicorum* in Honour of Prof. Dr. I.H.Ph. Diederiks-Verschoor, Nr. 9/10, December 2005, 51 at p.53.

[110] At the ICAO Conference on the Economics of Airports and Air Navigation Services (CEANS) held from 15 to 20 September 2008 in Montreal, 53 African States argued that countries have an obligation to provide slots to meet the capacity needs of air carriers operating under air service agreements. Their argument was that denying slots to airlines operating under air services agreements is tantamount to non utilization of the agreements which may be interpreted as a technical way of making the implementation of the Agreement to be unfair and one sided, contrary to the principle of reciprocity. See *CEANS –WP/61*.

aircraft noise, which was the original reason for imposing the curfew, was considered no longer a serious issue. ICAO was therefore called upon to review the continued relevance of night curfews at some international airports. The problem of night curfews is a complex environmental-related issue. Decision on its imposition or removal involves broader considerations often beyond civil aviation (e.g. local social, economic and environmental concerns and policies) and is influenced by the specific situation or conditions surrounding the particular airport, which vary from one another. Thus, it is difficult to find an across-the-board solution to this problem. States suffering from night curfews may address the issue with States concerned through consultation under their air services agreements or other available mechanisms. At the global level, ICAO has addressed the issue within the context of environmental protection and adopted related policies[111] which should be followed by States. It is also desirable for ICAO to continue to keep this issue under review in connection with its work on environmental protection.

The Council of the European Communities, by *Council Regulation (EEC) No 95/93 of 18 January 1993 on common rules for the allocation of slots at Community airports* and acting on Article 84(2)[112] of the Treaty of Rome,[113] identifies "slot-coordinated" airports[114] which are so designated if they suffer from slot-constraints. Member States have, under the Regulation, the discretion to designate any other airport as "slot-coordinated". Consequently, not all "slot-coordinated" airports in the EU actually suffer from slot-constraints. The Regulations require the member states to create the office of a national "slot-coordinator" that allocates slots and monitors the use of allocated slots.

Although Regulation 95/93 does not explicitly mention slot trading, implicitly, as only the slot coordinator can allocate slots, it may seem (according to the mechanism prescribed by the Regulation) that slot trading cannot take place under the regime prescribed therein and a slot market cannot exist. This is because newly-created slots are allocated through the national slot-coordinator and are not traded on a market, and therefore do not leave any room for legal argument as to the

---

[111] See ICAO Assembly Resolution A 36-22 (Consolidated statement of continuing ICAO policies and practices related to environmental protection) adopted at the 36th Session of the ICAO Assembly, Doc 9902 Assembly Resolutions in Force as at 28 September 2007, at 1–54.

[112] Article 84(2) stipulates that the Council may, acting by a qualified majority, decide whether, to what extent and by what procedure appropriate provisions may be laid down for sea and air transport.

[113] Treaty Establishing the European Community as Amended by Subsequent Treaties, Rome, 25 March 1957, *Supra* note 990.

[114] Article 5 of the Regulation prescribes that at a coordinated airport, the Member State responsible shall ensure that a coordination committee is set up. The same coordination committee may be designated for more than one airport. Membership of this committee shall be open at least to the air carriers using the airport(s) in question regularly and their representative organisations, the managing body of the airport concerned, the relevant air traffic control authorities and the representatives of general aviation using the airport regularly.

5.4 Slot Allocation

trading possibilities of slots in the EU. A view has been forwarded[115] that slot trading in the EU therefore is an academic issue that economists can address as to whether the allocation of slots by market forces rather than by an administrative procedure would lead to a more effective and competitive market.

In terms of trading, the most significant provision in the Regulation is Article 8(4) which provides that "slots may be freely exchanged between air carriers or transferred by an air carrier from one route, or type of service to another, by mutual agreement or as a result of a total or partial takeover or unilaterally." The wording in this provision is clear, that only an exchange of slots is permitted. One view is that:

> Since Regulation 95/93 came into force on February 21, 1993, EU officials have repeatedly stressed that despite of the terms "freely exchanged" and "freely transferred", Art. 8(4) of Regulation 95/93 prohibits any form of slot trading...Despite these comments, the wording of Art. 8(4) has given rise to much discussion as to whether it indeed prohibits any form of slot trading from a legal point of view. Art. 8 certainly is no masterpiece of law-making as it merely states that slots may be exchanged, but does not set up rules under which conditions and particularly fails to address the question whether the exchange of slots may be accompanied by financial considerations.[116]

This is indeed a valid point since a transfer of slots could take place in exchange for valuable consideration and arguably still sustain its legitimacy under Article 8(4). Seemingly, with a view to clarifying the issue, and to respond to the increasing problem of congestion at European airports, the European Commission, on 30 April 2008 adopted a document that allows airlines to trade slots both with regard to take off and landing.[117] This measure, which allows secondary trading[118] and sale of airport slots, obviates bottlenecks created by the inflexible "grandfather slots"[119] of Regulation 95/93, was further clarified by the EC. Jacques Barrot, Vice President of the European Commission said:

> At crowded airports, we need to make sure that slots are used as efficiently as possible and that airlines have a fair chance to develop their operations. Slots at airport must be distributed in a fair and non-discriminatory way. Today, we are recognizing for the first time that secondary trading is an acceptable way of allowing slots to be swapped among airlines. We will keep a close eye on the situation across Europe and ensure that secondary trading works to the advantage of consumers, but this system has already shown its value in

---

[115] Matthias Kilian, The development of the regulatory regime of slot allocation in the EU, at http://mitglied.multimania.de/matthias_kilian/vortragberlin.pdf.

[116] *Ibid.*

[117] http://www.euractiv.com/en/transport/eu-allow-sale-airport-slots/article-172103.

[118] Secondary trading is the horizontal transfer of slots between persons allocated, or entitled to hold slots, by agreement between them (the parties), accompanied, if the parties wish, by payment of monetary (or other valuable consideration) from one to the other in respect of such transfer. See *Study on the Impact of the Introduction of Secondary Trading at Community Airports*, Volume 1 Report, November 2006, Mott MacDonald, at 4-1.

[119] Grandfather slots refers to the situation where an airline's historical dominance at an airport has become institutionalized at an airport. This practice has been criticized on the ground that airlines do not own slots and it is the airport operator who grants times of arrival and departure of aircraft.

London, where it has allowed a range of airlines to take advantage of the opportunities provided by the EU-US aviation agreement and to create new levels of competition.[120]

One of the commercial considerations with regard to achieving enhanced competition through an US/EU open skies agreement would be the extent to which the EU will have the authority to negotiate all aspects of the DoT standard open skies agreement, which include slots. Another would be whether individual EU member States could still, after the EU signs an open skies agreement with the US, negotiate a bilateral air services agreement with the US From the EU perspective, where the Council of Ministers has given a mandate for the Commission to negotiate an agreement with the US, there is no express prohibition so long as there is recognition of the EU as one single area and the Community clause is signed. In this context, a decision handed down by the European Court of Justice[121] particularly from a competition angle, was that the core element of the bilateral air services agreement, which is market access involving the award of air traffic rights was untouched by the Court except in instances where an EU member would, in its agreement with the US, explicitly preclude another EU member from operating air services from that member's territory. In other words, Belgium would not be permitted to agree that Air France would not or could not operate services between Brussels and New York. This prohibition is entrenched in the Treaty of Rome which forms the substance of legislative legitimacy of the EU and incorporates the right of equal national treatment for all EU member States. Therefore, if one EU member State were to preclude the right of another member State's airline from having the right to operate air services to the United States from the territory of the first EU State it would tantamount to discrimination by the first State against the second State.

The Court also decided that certain specific provisions and areas covered in the questioned bilateral agreements between individual EU members and the US were

---

[120] Airport Slot Allocation: The Commission Clarifies Existing Rules, *IP/08/672*, Brussels, 30 April 2008.

[121] The European Court of Justice (ECJ) is formally known as the Court of Justice of the European Communities', i.e. the court of the European Union (EU). It is based in Luxembourg, unlike most of the rest of the European Union institutions, which are based in Brussels and Strasbourg. The ECJ is the Supreme Court of the European Union. It adjudicates on matters of interpretation of European law, most commonly: claims by the European Commission that a member state has not implemented a European Union Directive or other legal requirement; claims by member states that the European Commission has exceeded its authority; and references from national courts in the EU member states asking the ECJ questions about the meaning or validity of a particular piece of EC law. The Union has many languages and competing political interests, and so local courts often have difficulty deciding what a particular piece of legislation means in any given context. The ECJ steps in, giving its ruling which is binding on the national court, to which, the case will be returned to be disposed of. The ECJ is only permitted to aid in interpretation of the law, and not decide the facts of the case itself. Individuals cannot bring cases to the ECJ directly. An individual who is sufficiently concerned by an act of one of the institutions of the European Union can challenge that act in a lower court, called the Court of First Instance. An appeal on points of law lies against the decisions of the Court of First Instance to the ECJ.

5.4 Slot Allocation

contrary to EU law since they encroached upon internal EU regulations pertaining to non EU nationals. These laws concern:

a) Provisions pertaining to the allocation of airport slots;
b) Provisions governing pricing, or fares and rates of intra-European air services;
c) Agreements on computer reservation systems insofar as they appear as provisions of the open skies agreements in question; and
d) Provisions which reserved the right to grant permission under the open skies agreements only to airlines substantially owned and effectively controlled by nationals of the EU member States that is party to a particular agreement.

On 10 March 2009, the Commission proposed to revise Regulation 95/93 on the following basis inter alia:

a) The global economic and financial crisis, which is of major severity and scope, was seriously affecting the activities of air carriers. It has led to a significant reduction in air traffic over the winter 2008/09 scheduling season. The summer 2009 scheduling season will also be affected by the economic crisis.
b) The current economic crisis and subsequent readjustments to services (suspension, reduction of frequencies) constitute exceptional circumstances that are having a negative impact on airline companies. Consequently, coordinators should interpret the provisions of Regulation (EEC) No 95/93 on the allocation of slots in such a way that airline companies do not risk losing the slots they fail to use because of the economic crisis ('use-it-or-lose-it' rule).
c) Unless a decision is adopted to maintain the 'grandfather status' of slots, airlines might keep existing capacity in the face of significantly reduced demand, which would aggravate the current economic difficulties.
d) Therefore a new Article should be inserted into the Regulation, laying down that coordinators must accept, against the background of the economic crisis, that air carriers are entitled to the same slots in the summer 2010 scheduling season as allocated to them for the summer 2009 scheduling season (29 March 2009 to 24 October 2009).
e) In order to avoid wasting slot capacity in seasons affected by the crisis, coordinators may reallocate slots not used during the summer 2009 season for the rest of the season as 'ad hoc' slots, which may be used by other air carriers without this conferring 'grandfather rights'.
f) The Commission would continue to analyse the impact of the economic crisis on the air transport sector. Should the situation continue to deteriorate during the winter 2009/10 season (25 October 2009 to 27 March 2010), the Commission may decide to renew these arrangements for the winter 2010/11 season.

Accordingly the EC adopted a new regulation amending Regulation 95/93 by inserting a new provision binding on all EU member States which stated that coordinators will accept that air carriers are entitled to the same series of slots during the summer 2010 scheduling season as were allocated to them during the summer 2009 scheduling season in accordance with the Regulation. The Commission would continue to analyse the impact of the economic crisis on the air transport sector. Should the situation continue to deteriorate during the winter 2009/10 season, The Commission may decide, to renew all or part of these arrangements for the winter 2010/11 season.

It is submitted that if airlines are permitted to purchase, sell and lease airport slots, a certain market value could be attributed to slots, which will in turn allow airlines to have an incentive to trade them and recover their costs. However, such sale or auctioning of slots would have to be on a transparent and non-discriminatory basis. From the standpoint of competition, if airlines could trade slots, those who wish to launch new services need not wait for slots to become available if they wished to start operations into an airport. This would enable such airlines to respond quickly to changes in demand for air services, and leave markets that become unprofitable and embarking new ones that offer trading potential. Above all, the purchase, leasing or selling of slots by and between airlines would make the cost incurred therein a business expense for the airlines concerned rather than it being merely a charge for the use of airport services and facilities.

A note of caution in this scenario is that one has to be cautious of predatory and anti competitive practices. It is encouraging in this context that in Europe, owing to the presence of a mature legal regime on competitive practices, there is a robust guarantee of protection for carriers who may be concerned of the floodgates of market access being opened against them potential dangers stemming from predatory pricing practices, particularly with regard to incumbent airlines dropping their prices in the short term to deter new entrants. However, irrespective of the opening of market access and liberalization of pricing and capacity between the US and EU, there is no room for doubt that neither the US nor the EU would allow an absolute "free for all" as there are strict legal regimes against anti competitive conduct and cut throat practices in pricing in both jurisdictions. The regulation of competition within the European Community is governed by the EC Treaty.[122] Two provisions in particular, Articles 81 and 82, contain principles which outlaw anti-competitive conduct. While the former prohibits the prevention, restriction or distortion of competition, the latter makes itself applicable against abuse by one or more undertakings of a dominant position within the market. The former essentially contains provisions for agreements, decisions or practices with anti competitive effects, and the latter concerns itself with abuses of a dominant marketing position. The aim of these two provisions in particular is to preclude distortion of competition within the Common Market by supplementing the basic principles enshrined in Articles 81 and 82 with substance. The goals of the Treaty in general and Articles 85 and 86, which, in particular promote the free movement of services, goods, persons and capital whilst effectively obviating barriers to trade within the community, is to enforce some regulation. Both these provisions relate generally to all sectors of transport unless explicitly excluded by the Treaty provisions.[123]

---

[122] As mentioned earlier, the EC Treaty, also called the Treaty of Rome, was concluded in 1957 to forge "an even closer union among the people of Europe". See Goh (1997).

[123] Case 167/73 *Commission* v. *French Republic* [1974] ECR359 at 370.

5.4 Slot Allocation

Article 81 prohibits as incompatible such agreements as directly or indirectly fix purchase or selling prices or any other trading conditions; limit or control production, markets, technical development or investment; share markets or sources of supply; apply dissimilar conditions to equivalent transactions with other trading parties, thereby placing them at a competitive disadvantage; and make the conclusion of contracts subject to acceptance by other parties of supplementary obligations which, by their nature or according to commercial usage, have no connection with the subject of those contracts. These conditions are imposed on agreements between undertakings, which are defined as independent entities performing some economic or commercial activity.

Article 82 provides that any abuse by one or more undertakings of a dominant position within the Common Market or in a substantial part of it shall be prohibited as incompatible with the Common Market insofar as it may affect trade between member States. The Article prohibits: direct or indirect imposition of unfair purchase or selling prices or unfair trading conditions; limitation of production, markets or technical development to the prejudice of consumers; application of dissimilar conditions to equivalent transactions with other trading parties, thereby placing them at a competitive disadvantage; and conclusion of contracts subject to acceptance by the other parties of supplementary obligations which, by their nature or according to commercial usage, have no connection with the subject of such contracts.

In implementing these two provisions, air carriers have to exercise caution in not assuming that purely in view of a bloc exemption on air transport in the Treaty that may pertain to a particular issue, a related practice would be exempt from the prohibitions contained in Articles 81 and 82. In the air transport section of the Treaty, it is abundantly clear that block exemptions may apply only if abuse of dominant position is not evident in a given transaction.[124] Articles 81 and 82 are independent and complementary provisions and any exemption under Article 81 will not necessarily render the provisions of Article 82 nugatory.[125] "Dominant position" was defined in the 1979 decision of *Hoffman-La Roche* v. *Commission*[126] as a position of economic strength enjoyed by an undertaking which enables it to prevent effective competition being maintained on the relevant market by affording it the power to behave to an appreciable extent independently of its competitors, its customers and ultimately of its consumers. Such a position may necessarily preclude some competition except in monopoly or quasi monopoly situations. There is every indication, from existing jurisprudence and EC practice, that an assessment on an abuse of dominant position would not be predicated upon one factor alone or single characteristic but would rather be anchored on numerous

---

[124] See Adkins (1994).
[125] Case T-51/89, *Tetra Pak Rausing SA.* v. Commission [1990] II E.C.R. 309, [1991] 4C.M.L.R. 334 para 31.
[126] Case 85/76 *Hoffman-La Roche & Co. A.G.* v. *Commission* [1979] E.C.R. 461.

factors such as market structure, barriers to entry and conduct of the business enterprise concerned.

5.5 Corporate Foresight

Air carriers, airports and air navigation service providers have been called upon in many instances to show cause as to why a weakness in their industry which later resulted in an accident or economic loss was not detected in time and obviated.

From 17 December 2010 for a week or so, heavy snow and ice pounded Europe, grounding air travel across the continent and leaving thousands of passengers stranded as airports struggled to clear a backlog of flights cancelled or delayed by snowfalls. London's Heathrow, the world's busiest international airport, operated a limited schedule as one of its two runways was open and advised passengers not to travel to the airport unless their flight is confirmed. According to airport operator British Airports Authority (BAA) airlines worked to move aircraft and crew back to their normal positions as severe winter weather continued to cause disruption.[127]

Airport operations are systemic and Heathrow's despair had a knock on effect on other European airports. Frankfurt airport, Germany's biggest, was clear of snow and ice but officials cancelled about 300 of 1,340 flights because of problems elsewhere in Europe. During the period of crisis, French civil aviation authorities requested airlines to reduce their flights at the two main Paris airports by 30%. Thousands of travelers were stranded after about 400 flights in and out of Roissy-Charles de Gaulle were scrapped, with some 30,000 travelers' plans disrupted by the cancellations and delays. Throughout Europe hundreds of holiday flights were cancelled, as freezing rain and widespread snowfalls caused travel chaos. Repeated snowfalls stranded travelers in Ireland and Denmark and shut Dusseldorf airport in Germany for hours.[128] Elsewhere, in the Netherlands and Spain the same problem ensued with numerous flights cancelled and passengers stranded.[129] EUROCONTROL, Europe's air traffic supervisory body which supports its member States to achieve safe, efficient and environmentally-friendly air traffic operations across the whole of the

---

[127] Icy conditions also curtailed Europe's train services, left cars skidding through slushy streets and saw major events postponed, including music shows and sporting events. See http://www.ibtimes.com/articles/93795/20101220/travelers-air-passengers-stranded-as-europe-freezes-due-to-heavy-snowfall.htm.

[128] http://article.wn.com/view/2010/12/24/Snow_ice_trap_passengers_in_Europes_airports_d/.

[129] http://www.euronews.net/2010/12/20/arctic-weather-prompts-airline-criticism/.

5.5 Corporate Foresight

European region, reported that approximately 3,000 flights had been cancelled across Europe in a single day on 21 December.[130]

European Union Transport Commissioner Siim Kallas expressed his increasing concern about problems relating to the infrastructure available to airlines – airports and ground handling – and is reported to have said that he was considering forcing airports to provide airlines with a minimum level of infrastructure support during such severe weather.[131] *Airports Council International (ACI)*[132] *at its meeting with the European Commission in January 2011* reported on their extensive planning for weather disruptions and passenger assistance as an integral part of any airport management system. The ACI Executives stated:

> The truly exceptional winter conditions which affected some parts of Europe in December 2010 – with many historic records broken as regards snowfall and temperatures – tested aviation and other transport modes alike. Throughout the difficulties, Europe's airports ensured that over 85% of air traffic was unaffected. Passenger safety was upheld at all times and no accidents occurred despite extremely challenging operating conditions.[133]

The Board of ACI EUROPE emphasized the compelling necessity for sustained and effective cooperation on the ground between all actors and suggested to the Commissioner that airports be empowered by EU law to set minimum service and quality requirements for ground handlers.[134] They added that this would also complement the efficient delivery of the Single European Sky.

Giovanni Bisignani, CEO of the International Air Transport Association (IATA), at a speech delivered to the Aviation Club in London in early February 2011 called for better preparedness for severe weather in the wake of major airport shutdowns in December. Bisignani said:

> The inconvenience to passengers and the paralysis of the UK economy for many days is unacceptable from any perspective. Shovelling snow is not the airline's responsibility. The financial losses they suffered must be compensated, and we must approach next winter with a better plan.[135]

---

[130] http://www.montrealgazette.com/entertainment/serious+about+snow+removal+tells+airports/4011837/story.html.

[131] http://www.canada.com/news/European+freeze+haunts+Christmas+airlines+shops/4007697/story.html.

[132] Airports Council International is the only global trade representative of the world's airports. Established in 1991, ACI represents airports interests with Governments and international organizations such as ICAO, develops standards, policies and recommended practices for airports, and provides information and training opportunities to raise standards around the world.

[133] European Airports and Kallas discuss Industry Challenges, *Transport Policy*, at http://www.transportpolicy.org/europe%E2%80%99s-airports-and-kallas-discuss-industry-challenges/.

[134] Ibid.

[135] UK policies "will destroy the proud legacy" of British Aviation, *Travel daily News*, 28 February 2011. See http://www.traveldailynews.com/pages/show_page/41578-UK-policies-%E2%80%9Cwill-destroy-the-proud-legacy%E2%80%9D-of-British-aviation.

No doubt, the winter freeze and the attendant cancellation of flights caused acute inconvenience to the travelling public. Millions of people faced a struggle to get home in time for Christmas as travel plans were thrown into chaos when planes were grounded, trains cancelled and roads made impassable by snow and ice accumulation. Airport authorities blamed the bad weather for the disruption of flight schedules. A senior Emirates official criticized BAA for refusing to allow two of its aircraft to land in the UK, and warned that the airline faced a massive task in coping with the backlog of passengers now stranded in London.[136]

In the aftermath, some airlines have taken strong objection to the manner in which airports – particularly BAA – handled the heavy snow. Virgin Atlantic was reported to withhold airport fees until the inquiry[137] into BAA was completed.[138] The British Government, on 26 December 2010 considered introducing new laws to allow regulators to fine airports for travel disruption. Transport Minister Philip Hammond, told the Sunday Times that regulators should have tougher powers to punish airports who fail passengers, after thousands were forced to sleep at Heathrow when heavy snow grounded flights. Mr Hammond was of the view that there should be an economic penalty for service failure and that greater weight needs to be given to performance and passenger satisfaction.[139]

5.5.1 Airport Responsibility

The fundamental questions that arise are, should airports be held responsible for service failure brought about by a natural phenomenon; and should airports have had, what in modern business parlance is called "corporate foresight".[140] Firstly, in terms of responsibility, the question could be raised as to whether; irrespective of the

---

[136] http://www.arabianbusiness.com/emirates-lashes-out-at-uk-airports-body-amid-winter-chaos-368334.html?parentID=368939.

[137] BAA's non-executive Director, Sir David Begg is reported to have launched an inquiry just prior to Christmas to inquire into what went wrong. See *infra*, note 14.

[138] Virgin Atlantic Withholds BAA Fees Over Snow Row, *Air Letter*, Monday 10 January 2011, No. 17,147 at 2.

[139] http://www.flyertalk.com/forum/newsstand/1164326-britain-mulls-new-airports-law-after-heathrow-chaos.html.

[140] The functions and responsibilities of an airport will vary according to its size, type of traffic and areas of responsibility. For example, some airports are responsible for air traffic control as well as for meteorological services, while at most other airports such services are provided by separate government entities. Many airports are involved in security functions in varying degrees and in providing facilities for customs, immigration and health authorities. Ground-handling services for the airlines, including terminal handling or ramp handling, or both, are provided by some airports, while at others they are provided by the airlines or by specialized agents or companies. Certain airports also perform functions that exceed the scope of conventional airport activities, such as consultancy services, public works, construction, and real estate development. See Abeyratne (2009).

5.5 Corporate Foresight

business status of the airport (whether it is privatized, autonomous or corporatized) the State in which the airport is situated should bear ultimate responsibility. This responsibility devolves upon the State *in limine* by virtue of Article 28 of the Chicago Convention which stipulates inter alia that each Contracting States to the Convention undertakes, so far as it may find practicable to provide, in its territory, airports, radio services, meteorological services and other air navigation facilities to facilitate international air navigation, in accordance with the standards and practices recommended or established from time to time, pursuant to the Convention. The provision also requires such a State to adopt and put into operation the appropriate standard systems of communications procedure, codes, markings, signals, lighting and other operational practices and rules which may be recommended or established from time to time, pursuant to this Convention. Obviously, the Convention, through an inarticulate premise requires in Article 28 that States provide *functional* airport services among other services prescribed in the provision.

Principally, States bear the ultimate responsibility for aviation safety in their territories. However, airports too owe a duty of care to four categories of users: those in aircraft that are landing and taking off; those in the airport premises; the ground handling staff working on the airside; and other staff employed within the premises of the airport. There is a popular misconception that risks to crew, passengers and staff cease to exist once the aircraft engines are turned off. This is simply not so. One commentator has remarked that it is a common feature at airports that injury and death is caused to ground handling staff during thunderstorms.[141] Adverse weather also portends a serious threat to activities such as refuelling, de-icing and baggage handling operations. There is an increasing burden cast on meteorological information providers to give accurate weather information to airports and many airports are known to have established policy that requires the shut-down of ground handling operations when lightning strikes within three miles of the airport.[142]

The regulatory process governing air navigation services and aerodromes clearly identifies State responsibility. In terms of standard setting at aerodromes, States are required to undertake certification of aerodromes to standards acceptable to the international aviation community through ICAO. Accordingly, ICAO prescribes comprehensive Standards and Recommended Practices calculated to ensure acceptable levels of airport and aerodrome services. Notable among these are specifications on visual aids[143] and aerodrome maintenance.[144]

---

[141] Puempol (2006).

[142] *Ibid.*

[143] See Chaps. 6 and 7 of Annex 14 (Aerodromes) Volume 1 (Aerodrome Design and Operations), Annex 14 to the Convention on international Civil Aviation, Fourth Edition, July 2004, at 6-1 to 8-3. Also Chap. 5 of Annex 14 Volume II (Heliports), Second Edition – July-1995 at p. 30–48 which has similar provisions for the operation of helicopters.

[144] *Id.* Annex 14, Chap. 10.

In addition to the obligation of the State to provide certain services as enumerated in Article 28 of the Chicago Convention, responsibility of the State would also extend to the provision of accurate air traffic control services at the aerodrome. States have to be mindful of the fact that their overall responsibility under the Chicago Convention in providing air navigation services extends to the air traffic controller, whose service is of a unique nature. The special feature in the provision of air traffic control is brought to bear by the nature of the service provided, be it in the relaying of information on meteorology or on traffic. Globally, air traffic control services offer information relayed by people by means of radio communication involving extremely short time periods and using a standard set of terminology in the English language, even in regions of the world where English is not the first language.[145]

The provision of meteorological information to airports and aircraft about to land or take off is also part of State responsibility. Annex 3 to the Chicago Convention provides in Standard 2.1.1. that the objective of meteorological service for international air navigation shall be to contribute towards the safety, regularity and efficiency of international air navigation. This objective shall be achieved by supplying the following users: operators, flight crew members, air traffic services units, search and rescue services units, airport managements and others concerned with the conduct or development of international air navigation, with the meteorological information necessary for the performance of their respective functions.[146]

State responsibility for the provision of meteorological information is provided for in Standard 2.1.4. where each Contracting State is required to ensure that the designated meteorological authority complies with the requirements of the World Meteorological Organization in respect of qualifications and training of meteorological personnel providing service for international air navigation.[147]

It is also provided in the Annex that close liaison shall be maintained between those concerned with the supply and those concerned with the use of meteorological information on matters which affect the provision of meteorological service for international air navigation.[148] Furthermore, States have responsibility establish one or more aerodrome and/or other meteorological offices which shall be adequate for the provision of the meteorological service required to satisfy the needs of international air navigation.[149]

It is incontrovertible that the responsibility of the State is not extinguished merely because an airport is subject to private ownership or private management control. In international air transport, the mere fact that the State has to provide airport

---

[145] Miyagi (2005).

[146] Standard 2.1.2.

[147] Requirements concerning qualifications and training of meteorological personnel in aeronautical meteorology are given in WMO Publication No. 49, Technical Regulations, Volume I – General Meteorological Standards and Recommended Practices, Chapter B.4 – *Education and Training*.

[148] Standard 2.2.1.

[149] Standard 3.3.1.

5.5 Corporate Foresight

services under Article 28 of the Chicago Convention and indeed designate airports within its territory for landing purposes as per Articles 10 and 68 thereof imposes legal responsibility upon the State to be accountable at public international law for any liability incurred as a result of action on the part of airports within its territory.

Irrespective of the responsibility of a State with regard to airports within its territories, which is founded both at customary international law and at private law for liability incurred by airports, a privately run airport may incur tortuous liability on a private basis, as the occupier of the premises. In the instance of a privately managed airport where the entity charged with managing airport services is located within the airport premises, such an entity would be considered as a legal occupier for purposes of liability.

It is a known fact that not many industries face the same challenges with regard to combining social responsibility and business as the airport industry. One of the challenges faced by airports is to cope with adverse weather conditions and ensure that the airport industry remains a safe place to work in. The success of an airport depends on a delicate balance between safety and punctuality and always remains a critical operational challenge for airports. Aviation and weather are strange and unwilling bedfellows and the former has never underestimated the latter and has treated it with respect.

Other hazards that might seriously affect those working on the airside are windstorms (which have, in certain instances lifted baggage containers off the ground) and ice and snow which pose a serious threat of injury to passengers climbing aircraft steps who could slip on accumulated black ice. Catastrophic events such as tsunamis[150] tropical cyclones, snow storms, floods and dust and sand storms are real threats to airports, requiring vigilance and responsibility of both the State concerned and the airport authorities. Early warning systems and emergency and contingency response plans have to be in place, which have to be implemented with precise communications. Aerodrome emergency planning is addressed in detail in Chap. 9 of Annex 14 to the Chicago Convention which provides that every aerodrome must have established an aerodrome emergency plan commensurate with aircraft operations and other activities conducted at an aerodrome.[151]

One of the most critical functions of airports is de-icing of aircraft, since if aircraft have ice on their wings they must be de-iced and protected with anti-icing

---

[150] It is noteworthy that, in response to a request received from IATA, the Secretariat of ICAO has included "tsunami" in the draft amendment to Annex 3 to the Convention on International Civil Aviation which addresses meteorological services, as a phenomenon which should prompt the issuance of an aerodrome warning. It may be expected that within the global warning system for tsunami being established under the Intergovernmental Oceanographic Commission of UNESCO (IOC), the meteorological authorities will receive tsunami warnings in a timely manner and that they will be in a position to implement the proposed amendment, as far as aerodrome warnings are concerned, in time for its applicability date. See AN-WP/8086, Preliminary Review of a Proposed Amendment to Annex 3 and Consequential Amendment to the PANS-ABC.

[151] Standard 9.1.1. Annex 14 to the Convention on International Civil Aviation (Aerodromes), Fourth Edition, July 2004, at p. 9-1.

fluids. An insignificantly small accumulation of ice on the upper structure of the wing could considerably reduce the lift of that surface and all ice must be removed before takeoff.[152] Airports providing ground handling services have to constantly monitor the de-icing process as the fluids applied to de-ice an aircraft are only effective in holding off re-icing for a limited time, after which ice could accumulate again, requiring de-icing. The period between de-icing and re-icing is called the holdover time and would be critical, particularly in the case of a departing aircraft delayed on takeoff.

There are also attendant problems for aircraft stemming from extreme cold temperatures. These could include fuel and hydraulic leaks; difficulty starting the Auxiliary Power Unit (APU); difficulty starting the engines; landing gear tire "cold set"; and difficulty in opening doors. To counter these problems, State regulation[153] and accurate weather forecasts[154] are now provided in many States through computer aided forecasting systems. These modern weather observation systems help the air navigation service provider to improve the quality of weather data and partly replace the human observer in the weather forecasting process.[155]

With regard to looking after passengers stranded at airports due to the cancellation of flights, the airport, as occupier of the premises (be it a lessee or lessor) could be *prima facie* liable on two counts: liability for inadequately maintained property which may cause an accident; and liability for an unsafe premises. Liability under negligent entrustment, be it that of the lessor or lessee, should usually be concurrent with an act of negligent on the part of the entrustee.[156] The fundamental principle under this head of liability is based on the line of argument that one who entrusts his property to another, knowing that such other is incompetent or having the duty to ascertain whether the entrustee is competent or not, is negligent and liable for injury.[157]

---

[152] Stuart and Isaac (1994). See also, Rusmussen et al. (1995–1996), Wagner (1994–1995). Rejected Take-off In Icy Conditions Results in Runway Overrun. *Flight Deck*, No.18; Winter 1995–1996: p. 3–11.

[153] Frank Carlson (1995). The author discusses United States FAA regulations and certain regulations in the CIS States which apply standards for certification of aircraft to ensure they are equipped against extreme cold weather. See also generally, Woolley (1999), McKenna (1994), Freedman (1994).

[154] Arpino (1994).

[155] Steinhorst and van Dijk (1994); also Manningham (1996).

[156] See *Hood* v. *Dealers Transport Co.*, 459F.Supp. 684. It is also noteworthy that one jurisdiction in the United States has followed the approach that the lack of a licence of competence *per se* does not give rise to liability if the absence of licence was not the proximate cause of the injury. See *Laughlin v Rose*, 200 Va. 127, 104S.E. 2d 782 (1958) and *White v Edwards Chevrolet Co.* 186 Va. 669, 43S.E. 2d 870 (1947).

[157] *Department of Water and Power of City of Los Angeles v Anderson* 95F.2d 577. See also, *Cox v Dubois*, 16F.Supp. 2d 861, *Brantley v Vaughn*, 835F. Supp. 258.

5.5.2 Elements of Corporate Foresight Planning

The necessity for corporate foresight stems from the continuing and rapid development of science and technology which are the drivers of social and economic change. Using these two knowledge based and fact intensive fields, airports would be able to obtain a clear picture of challenges and opportunities confronting them. Airports are a complex, big business and their business environment is highly dynamic. Therefore they need proactive measures to respond to the uncertainties of their business as well as a long term orientation to remain stable amidst imponderables. Airports need think tanks to mesh their technology trends and market trends to meet a growing demand for air travel. Foremost in this process is a far reaching and forward looking communications strategy as well as a good team of scientific and economic forecasters.

The first step to corporate foresight is to know what the future is going to be like by adopting a foresight-awareness culture. If, as ACI[158] Director Angela Gittens said at the 20th World Annual General Meeting of ACI in Bermuda in October 2010, airports should transition from the public utility model to the entrepreneurial business model, the key would be customer service excellence. Research and innovation strategies should necessarily be developed through foresight activities. "Foresight" has been defined as:

> [a] participatory, future intelligence gathering and medium-to-long-term vision-building process that systematically attempts to look into the future of science, the economy and society in order to support present-day decision-making and to mobilise joint forces to realise them.[159]

Corporate foresight is a process of formulation and should not be confused with a set of techniques. Through a sustained consultative process, corporate foresight involves the examination of a series of future scenarios and ideally prescribes solutions. Foresight shares common ground with risk management and evaluation of risk,[160] and addresses the nature of the particular business and the uncertainties of the business environment. For example, in the airport industry, a grave uncertainty is the weather, as in the context of an unexpected winter storm or the eruption of a volcano which spews ash into the atmosphere. Both these events occurred in 2010 in Europe.

In every instance of a natural disaster the focus revolves around those who are affected by such events. Therefore, it is difficult not to discuss the merits and

---

[158] *Supra.*

[159] Becker (2002).

[160] Blaise Pascal, in his book *Ars Cogitandi* states that fear of harm ought to be proportional not merely to the gravity of the harm but also to the probability of the event. It is also a fact of risk management that, under similar conditions, the occurrence (or non-occurrence) of an event in the future will follow the same pattern as was observed in the past. For a discussion on risk assessment and risk management see Ferguson (2008) and Bobbitt (2008).

demerits of a strategy that would bring about the least damage based on a balance of probability. The process of foresight should commence with airport builders and managers who should look at airports as business enterprises that contribute to efficient air travel. The terminal is used to enable passengers and freight to connect with aircraft for their transportation on departure and to connect with ground transportation on arrival. As such airport business planners must take into account elements that are not only exclusively related to the carrying on of a business, but also those principles that are essential for the safer, secure and efficient running of an airport. In this context airport planning becomes a necessary element to the airport business. An airport administration should take into account when planning for the injection of additional aircraft capacity in an airport the responses of the international community in the form of Standards and Recommended Practices as promulgated by ICAO, in order that international civil aviation retains a certain consistency and uniformity in its global activity. For instance, ICAO has in use an Airport Planning Manual[161] in two parts setting out in detail, all aspects of airport planning. ICAO has in this document developed a master planning process which involves plans, programmes and stringent policy that go to make a viable airport. The document serves as a basis for providing for the orderly and timely development of an airport adequate to meet the present and future air transportation needs of an area or State.[162] The manual starts from the fact that early aviation history recognized the need for some public control of land in the vicinity of an airport[163] and bifurcates this need to reflect airport needs i.e. obstacle limitation areas and future airport development etc. and the need to ensure 60 minimal interference with the environment and the public[164]. By this dual approach ICAO has introduced a whole new area of thought into airport development. What was once a concern to merely provide easy facilities for the fluid movement of air traffic has now become in addition an ecological concern. By this process, airport development now falls into three main areas which are:

a) The development of airport capacity and facilities;
b) The balancing of airport development with necessary security measures; and,
c) The balancing of airport development with ecology i.e. city planning, noise pollution avoidance etc.

The ICAO Airport Planning Manual ensures a balance between airport development and ecological considerations.

Cooperation in technical and economic areas would have to be further expanded to include safety and ecological factors in the technical field and all economic research in city planning and infrastructural development in the economic field.

---

[161] *Airport Planning Manual*, Doc. 9184-AN/902, Parts 1 and 2.
[162] *Id*. Part 1, 2.9.1 (a).
[163] Doc. 9184-AN 902 Part 2, 1.3.1.
[164] *Id*. 1.3.2.

5.5 Corporate Foresight

These studies would have to be done in the form of committed and in depth country studies by individual States taking into consideration futuristic studies of a country's outlook and the financial outlay that the country would be prepared to make for an airport expansion programme. The outcome of these studies could then form legislation for the planning of airports in a State. Such legislation would present, for the first time, a cohesive and enforceable set of laws in that State that would meet the airport congestion problem.

Although the concept of airport planning laws can be summed up easily as above, the three broad areas of ecology, safety and infrastructural planning need a sustained approach of study before such are incorporated into laws. For a start, ICAO's *Airport Planning Manual* is geared to provide information and guidance to those responsible for airport planning,[165] where information on a comprehensive list of planning subjects such as sizes and types of projects[166] task identification,[167] preparation of manpower and cost budgets,[168] selection of consultants[169] and standard contract provisions[170] are given. With these guidelines each State can start its planning process.

5.5.2.1 The Link with Air Transport

The airline and airport business are interlinked and inter-connected and, since air transport is a growth industry, so is the airport industry. According to the global market forecast of Airbus Industrie forecasts that from 2009 to 2028, some 25,000 new passenger and freighter aircraft valued at US$3.1 trillion will be delivered.[171] This rapidly evolving demand is driven by emerging economies, evolving airline networks, expansion of low cost carriers and the increasing number of mega-cities as well as traffic growth and the replacement of older less efficient aircraft with more eco-efficient airliners. These are factors driving demand for new aircraft.

The forecast also attributes the demand for larger aircraft to the compelling need to ease aircraft congestion and to accommodate growth on existing routes and to

---

[165] Airport Planning Manual, Doc 9184-AN/902, Part 3.
[166] *Id.* 1.3.1–1.3.5.
[167] *Id.* Chapter 2.2.1.
[168] *Id.* 2.4.
[169] *Id.* 3.1.
[170] *Id.* Appendix.
[171] http://www.airbus.com/en/gmf2009/appli.htm?onglet=&page=. The forecast anticipates that in the next 20 years, passenger traffic RPK's will remain resilient to the cyclical effects of the sector and increase by 4.7% per year or double in the next 15 years. This will require a demand for almost 24,100 new passenger aircraft valued at US$2.9 trillion. With the replacement of some 10,000 older passenger aircraft, the world's passenger aircraft fleet of 100 seats or more will double from some 14,000 in 2009.

achieve more with less. Needless to say, this exponential growth in air traffic[172] will place a burden on airport capacity and consequent demands upon the airport industry.

Therefore, in a manner of speaking, the future of airports is linked with the future of airlines. Naval Taneja, academic and industry strategist is of the view that the airline business model will change in the future and game changing strategies will have to be in place in the air transport industry. Taneja ascribes to unconventional thinking and innovative technology the foremost place and states that these two that will bring about change management. One example he gives is the enhancement by airlines of the customer experience on their websites by offering a virtual assistant to answer questions and direct travellers to the information that they are seeking.[173] Another expert is of the view that airports will invariably follow suit by preparing for similar technological innovations such as the maximizing the use of the internet, and the provision and availability of information on passengers' connecting flights, airport maps, information on their destinations and other phone or PDA applications which will all be built into their eyeglasses.[174]

5.5.2.2 Corporate Foresight

Corporate foresight hinges on the early identification of markets and technologies. However, it takes a long time to restructure a corresponding system along the lines of market forecasts and the development of technology. Building competence to a level that would correspond to effective risk management. One commentator identifies a period of 10 years as preparatory for effective corporative foresight to be built.[175] A salient feature in foresight is leadership, which is linked inextricably to the quality of not being surprised at any unexpected event or circumstance. Essential to this process is the prompt identification of developments in the areas of science, technology and society that are likely to ensure future benefits both from a business and social perspective. Foresight has been categorized as (a) *anticipatory intelligence*, i.e. providing background information and an early warning of recent developments; (b) *direction-setting*, i.e. establishing broad guidelines for the corporate strategy; (c) *determining priorities*, i.e. identifying the most desirable lines

---

[172] *Id*. The forecast states that the greatest demand for passenger aircraft will be from airlines in Asia-Pacific and emerging markets. The region that includes the People's Republic of China and India accounts for 31% of the total, followed by Europe (25%) and North America (23%). In terms of domestic passenger markets, India (10%) and China (7.9%) will have the fastest growth over the next 20 years. The largest by volume of traffic, will remain domestic US.

[173] See Naval Taneja, Technology enabled game changing models, http://www.airlineleader.com/_webapp_1098704/Technology-enabled_game-changing_airline_business_models.

[174] Michael Rogers' Commentary on the 20th ACI World Annual General Assembly at http://www.airtransportnews.aero/analysis.pl?acateg+reports.

[175] Patrick Becker, *Corporate Foresight in Europe*: A First Overview, supra, note 30 at 8.

5.5 Corporate Foresight

of R & D as a direct input into specific (funding) decisions; (d) *strategy formulation*, i.e. participating in the formulation and implementation of strategic decisions; and (e) *innovation catalysing*, i.e. stimulating and supporting innovation processes between the different partners.

From an airport perspective, *anticipatory intelligence* should be extended to the various disaster scenarios that could be: (a) inability to accommodate flights due to frozen runways; (b) straining of infrastructure in accommodating a passenger influx greater than the airport could handle at any given time; (c) a possible security threat that would slow down traffic and clog the system; and (d) adverse weather systems (such as storms) that could halt ground handling systems. In this context, most problems that an airport could face are weather related. In *direction setting* the airport management should have a corporate strategy that would also be an adequate contingency plan. Such a plan should include adequate staff training for contingencies; providing back-up support staff who have pre-organized schedules for exigencies and adequately compensating staff who have to cope with a stressful situation and irate customers. The most important factor in such situations is an effective method of communicating with everyone and providing extra services. A major factor in *determining priorities*, is the assurance at all times, whether good or bad, of an effective and efficient safety and security system at the airport. *Strategy formulation* would include developing a range of alternative routes and timings and alternative means of transport; having adequate and additional accommodation ready to make customers comfortable; having joint plans with airlines for the provision of meals through ground handlers or other caterers at the airport; and ensuring that constant contact be maintained with the Consulates involved, and the local police.

Innovation catalyzing is arguably the most creative element of corporate foresight. Here, the starting point is to look at existing planning documentation. ICAO's *Airport Planning Manual* is geared to provide information and guidance to those responsible for airport planning,[176] where information on a comprehensive list of planning subjects such as sizes and types of projects[177] task identification,[178] preparation of manpower and cost budgets,[179] selection of consultants[180] and standard contract provisions[181] are given. With these guidelines each State can start its planning process. Regrettably, a study conducted on corporate foresight reflects that the transportation sector does not consider innovations as important in corporate foresight.[182] It is all the more reason that, as further action, corporate

---

[176] Airport Planning Manual, Doc 9184-AN/902, Part 3.
[177] *Id.* 1.3.1–1.3.5.
[178] *Id.* Chapter 2.2.1.
[179] *Id.* 2.4.
[180] *Id.* 3.1.
[181] *Id.* Appendix.
[182] *Supra* note 176 at p. 22.

foresight in the airport industry needs an un-fragmented process of corporate development and strategic planning. An initiative could be taken by ACI to gather a team of forecasters, technical experts and airport planners to develop guidelines for disaster management.

The above discussion focuses on the incontrovertible fact that there are two aspects to the issue of preparedness for natural disasters by airports. The first is airport responsibility and the second is the need for corporate foresight. As for airport responsibility, in addition to the overall responsibility of the State for the provision of airport services in Article 28 of the Chicago Convention, Annex 14 to the Chicago Convention, pertaining to aerodromes and their operations also contains requirements which devolve responsibility on States to provide weather information and take necessary steps. For example, there is an umbrella provision requiring States to provide information on the condition of the movement area and the operational status of related facilities to enable units to provide services to arriving and departing aircraft.[183] Recommendation 2.9.9. of the Annex suggests that whenever a runway is affected by snow, slush or ice and it has not been possible to clear the precipitant fully, the runway should be assessed and the friction coefficient measured. The follow up provision appears in Standard 10.2.8 which requires that the surface of a runway shall be maintained in such a condition so as to provide good friction characteristics and low rolling resistance. Snow, slush, ice, standing water, mud, dust, sand, oil, rubber deposits and other contaminants are required to be removed as rapidly and completely as possible to minimize accumulation. There is also a recommendation in the Annex that a taxiway should be kept clear of snow, slush, ice etc. to the extent necessary to enable the aircraft to be taxied to and from an operational runway.[184] The order of priority for clearance of products of adverse weather, as prescribed by the Annex is: runways in use; taxiways serving runways in use; aprons; holding bays; and other areas.[185]

This notwithstanding, airports also have a responsibility to persons adversely affected by cancellation of flights. Irrespective of the responsibility of a State with regard to airports within its territories, which is founded both at customary international law and at private law for liability incurred by airports, a privately run airport may incur tortuous liability on a private basis, as the occupier of the premises. In the instance of a privately managed airport where the entity charged with managing airport services is located within the airport premises, such an entity would be considered as a legal occupier for purposes of liability.[186]

However, airports remain entitled to levy charges for services rendered irrespective of any inconvenience caused to airlines as a result of a snowstorm or

---

[183] Standard 2.9.1, Annex 14 to the Convention on International Civil Aviation, Aerodromes, Volume 1 Fourth Edition: July 2004 at 2–4.

[184] Recommendation 10.2.9.

[185] Recommendation 10.2.11.

[186] For a discussion on this subject, See Abeyratne, *Airport Business Law*, *Supra*, note 15, at pp. 106–108

other natural phenomenon which compels an airport to take measures with regard to flights coming in and going out of that airport. Airport charges are based on cost recovery for services provided and States are responsible for policy pertaining to such charges. The fundamental postulate of ICAO's policies on airport charges lies in Article 15 of the Chicago Convention which states that every airport in a contracting State which is open to public use by its national aircraft shall likewise, subject to the provisions of Article 68,[187] be open under uniform conditions to the aircraft of all the other contracting States. The provision goes on to say that like uniform conditions shall apply to the use, by aircraft of every contracting State, of all air navigation facilities, including radio and meteorological services, which may be provided for public use for the safety and expedition of air navigation. Any charges that may be imposed or permitted to be imposed by a contracting State for the use of such airports and air navigation facilities by the aircraft of any other contracting State shall not be higher: as to aircraft not engaged in scheduled international air services, than those that would be paid by its national aircraft of the same class engaged in similar operations, and; as to aircraft engaged in scheduled international air services, than those that would be paid by its national aircraft engaged in similar international air services.

Article 15 sets out the following three basic principles:

- Uniform conditions shall apply to the use of airports and air navigation services in a Contracting State by aircraft of all other Contracting States;
- The charges imposed by a Contracting State for the use of such airports or air navigation services shall not be higher for aircraft of other Contracting States than those paid by its national aircraft engaged in similar international operations; and
- No charge shall be imposed by any Contracting State solely for the right of transit over or entry into or exit from its territory of any aircraft of a Contracting State or persons or property thereon.

Specific regulatory provisions applicable to charges levied by airports have their genesis in ICAO and are contained in Doc 9082.[188]

As regards corporate foresight, an airport has to start with a culture of corporate foresight. Emergency management is a dynamic process. Planning, though critical, is not the only component. Training, conducting drills, testing equipment and coordinating activities with the community are other important functions. More importantly, airports should work jointly, and in partnership with airlines and air navigation service providers in furthering their corporate foresight.

---

[187] Article 68 provides that Each contracting State may, subject to the provisions of this Convention, designate the route to be followed within its territory by any international air service and the airports which any such service may use.

[188] *ICAO's Policies on Charges for Airports and Air Navigation Services* Doc 9082/8 Eighth Edition-2009.

5.6 The Use of Alternative Fuels as an Economic Measure

Over the past decades, market forces have always ensured that fuel burn (and associated CO2 emissions) has been kept to a minimum for efficiency reasons. As a result of permanent fleet modernization, with new aircraft achieving unmatched efficiency performance, fuel burn has been reduced by about 70% over the last 40 years.

Improvements in aircraft fuel efficiency are inextricably linked to how engine, aircraft and systems manufacturers design their products. The concepts, the design criteria, the design optimization and the technology transition processes are all tightly interconnected, and the interactions usually increase as a product is developed. Generation after generation, aircraft have shown impressive weight reductions, aerodynamics improvement and engine performance increase, thus reducing drastically the amount of fuel burn (and of CO2 emitted) to perform the same or further improved operational mission.

Simultaneously, product innovations are permanently introduced through design, simulation, modelling, testing and validation tools. The optimization process and the challenging tradeoffs involve iterative loops at the technology, design and product levels. As a consequence, it takes approximately a decade to design and develop an aircraft. In order to make the appropriate decisions, when investing in future technologies, aircraft engine and airframe manufacturers need a stable regulatory framework, based on dependable scientific knowledge, and consistent funding to sustain the current and future extensive research programmes.

Some commitments have been made by the manufacturing industry to keep that improvement trend: in Europe, the goals set by the Advisory Council for Aeronautics Research in Europe (ACARE) are targeting an additional 50% improvement in fuel burn and associated CO2 emissions in 2020, compared to 2000 performance. This should be done while reducing the perceived noise levels by 50% and the emissions of NOx by 80% over the same period. Comparable objectives are set in the US, through the different programmes running with the NASA for instance.

Associated research programmes, development clusters to foster synergies through appropriate partnerships, have been set up, thus enabling to better take in consideration the challenges associated to the interdependencies between environmental improvement and other parameters (performance, economics...) and within the environmental criteria themselves between noise, local air quality and climate change-related issues. Some choices for future technology engine and aircraft configuration will be based on the decisions the society will make relative to these challenges.

In addition to the traditional embedded technological improvement pattern that aircraft engine and airframe manufacturers are continuously supporting, some further opportunities to further reduce the emissions from aviation may arise. They are related to the design, development, validation and production of sustainable alternative fuels. This new avenue must be further explored to identify the environmental benefit it can generate, on top of the expected technological improvements.

5.6.1 The Use of Fossil Fuel and Its Effects

Fuel efficiency is a major consideration for aircraft operators as fuel currently represents around 20% of total operating costs for modern aircraft.[189] One of the options being considered for aviation, where emissions are released in high altitude affecting the environment and impacting local air quality, is the use of alternative fuels. The most serious concern lies in the impact of emissions caused by the currently used fossil fuels to climate change, which has prompted the Kyoto Protocol (1997)[190] to the United Nations Framework on Climate Change (UNFCC) to require developed countries in particular to decrease their collective green house gas emissions by 5% by the period 2005–2012,[191] with 1990 as a bench mark. The demand for air travel continues to increase, making the concern of aircraft engine emissions caused by fossil fuel burn even more compelling.[192] Preliminary results from ICAO's Committee on Aviation Environmental Protection (CAEP) [193] indicate

---

[189] See IPCC Fourth Assessment Report, *infra*, note 1078 at p. 352.

[190] The Protocol will enter into force 90 days after "not less than 55 Parties to the [Climate Change] Convention, incorporating Parties included in Annex 1 which accounted in total for at least 55% of the total carbon dioxide emissions for 1990 of the Parties included in Annex 1" have ratified (Art. 24 of the Protocol).

[191] Report of the Committee on Aviation Environmental Protection, Fifth Meeting, Montreal, Jan. 8–17, 2001, ICAO Doc 9777, CAEP/5 at 1A-1.

[192] Currently, the world uses 3,917 Mega tonnes (Mt) of liquid fuel annually 3, including approximately 0.02 Mt of biofuel, very little of which is consumed by international aviation. Most fuel use is for direct combustion, emitting carbon dioxide (CO_2) in direct proportion to fuel burn. Preliminary estimates from ICAO's Committee on Aviation Environmental Protection (CAEP) indicate that global aviation fuel burn is expected to grow from approximately 200 Mt in 2006 to between 450 and 550 Mt in 2036. Not accounting for the impact of alternative fuels, but considering the effects of improved efficiency and aircraft technologies, CO_2 is predicted to grow from 632 Mt in 2006 to the range of 1,422–1,738 Mt in 2036. At the time of writing, these results have not yet been reviewed or accepted by CAEP and should therefore be considered preliminary.

[193] The current environmental activities of the International Civil Aviation Organization (ICAO), see *infra* note 25, are largely undertaken through the Committee on Aviation Environmental Protection (CAEP), which was established by the Council in 1983, superseding the Committee on Aircraft Noise (CAN) and the Committee on Aircraft Engine Emissions (CAEE). CAEP is composed of members and observers. In 1998, the Assembly requested that States from regions that are not represented or under-represented in CAEP participate in the Committee's work. Some progress has been made in this regard and efforts continue to attract new participants. CAEP assists the Council in formulating new policies and adopting new Standards on aircraft noise and aircraft engine emissions. CAEP's Terms of Reference and Work Programme are established by the Council. The current structure of the Committee includes five working groups and one support group. Two of the working groups deal with the technical and operational aspects of noise

that the demand for air travel is expected to continue to grow through at least 2036 and on a per-flight basis; efficiency is expected to continue to improve throughout that period. The anticipated gain in efficiency from technological and operational measures is not expected to completely offset the predicted growth in demand driven emissions, leaving a potential "mitigation gap" to achieving sustainability. Commercializing sustainable alternative fuels for aircraft can be an essential strategy for closing this gap. While there are no significant quantities of such fuels available for commercial aviation today, it is anticipated that these fuels will become an essential component of the future aircraft fuel supply.[194]

The world today uses 3,917 mega tonnes (Mt) of liquid fuel annually, of which 0.02 Mt is biofuel.[195] Very little of this biofuel is used by international aviation. By 2036, international aviation could need a substantial contribution from sustainable alternative fuels for aircraft in order to reduce its overall GHG footprint. One of the many significant benefits in the use of sustainable alternative fuels in aircraft would be the drastic lowering of emissions containing sulphur oxides (SOx) and particulate matter (PM),[196] as a result of which local air quality would be improved.

Removing sulphur from conventional jet fuel to produce a low sulphur jet fuel will significantly reduce PM and SOx emissions from aircraft. Currently, jet fuel has a specification maximum of 3,000 parts per million (ppm) for sulphur; however, jet fuel in the market has a lower sulphur content. Worldwide surveys conducted during 2007 found that annual weighted average jet fuel sulphur content ranged from 321–800 ppm.[197] Hydrodesulphurization, which could be applied to remove the fuel sulphur, is a common process in petroleum refineries, and low sulphur diesel fuel is already widely used internationally. Low sulphur jet fuel has sulphur content less than 15 ppm.

As regards PM, According to a recent analysis of the impacts of aviation emissions on human health,[198] primary PM emissions are responsible for 13% of

---

reduction and mitigation. The other three working groups deal with technical and operational aspects of aircraft emissions, and with the study of market-based measures to limit or reduce emissions such as emissions trading, emissions-related charges and voluntary measures. The support group provides information on the economic costs and environmental benefits of the noise and emissions options considered by CAEP.

[194] ICAO doc. CAAF/09-WP/3, 16/10/09.

[195] *Id*. at 4. In April 2007 in its Fourth Assessment Report, the IPCC stated: "The introduction of biofuels could mitigate some of aviation's carbon emissions, if biofuels can be developed to meet the demanding specifications of the aviation industry, although both the costs of such fuels and the emissions from their production process are uncertain at this time". See the Report at http://www1.ipcc.ch/pdf/assessment-report/ar4/wg3/ar4-wg3-chapter5.pdf at p. 352.

[196] Particulate matter (PM) from fuel combustion is a mixture of microscopic solids, liquid droplets, and particles with solid and liquid components suspended in air. Solid particles, such as soot or black carbon, are referred to as non-volatile particles. Volatile PM is comprised of inorganic acids (and their corresponding salts, such as nitrates and sulphates), and organic chemicals from incomplete fuel combustion.

[197] Taylor (2009).

[198] Brunelle-Yeung (2009).

total PM impacts. Secondary PM is much more significant with sulphur-related PM emissions responsible for 33% and NOx-related PM emissions responsible for 54%. With low sulphur jet fuel, SOx emissions would be significantly reduced, which in turn would result in a significant reduction in secondary PM. Overall, volatile primary PM emissions are reduced due to reduced fuel sulphur content. Hydrodesulphurization causes other fuel modifications that also reduce non-volatile PM emissions.

The current understanding of experts with regard to PM pollution is insufficient to fully evaluate the magnitude of health and environmental effects of exposure. However, indications are that the size of PM is a significant factor. Coarse particles can be inhaled but tend to remain in the nasal passage. Smaller particles are more likely to enter the respiratory system. Health studies have shown a significant association between exposure to fine and ultra fine particles and premature death from heart or lung disease. Fine and ultra fine particles also have been linked to effects such as cardiovascular symptoms, including cardiac arrhythmias and heart attacks, and respiratory symptoms such as asthma attacks and bronchitis. These effects can result in increased hospital admissions, emergency room visits, absences from school or work, and restricted activity days. Individuals that may be particularly sensitive to fine particle exposure include people with heart or lung disease, older adults, and children.[199] Standard setting organizations specify requirements that jet fuel must meet for physical properties, chemical content, contaminant limits, and overall performance requirements. To limit PM and SOx fuel combustion emissions, fuel standards have maximum limits on the sulphur content of fuels.

5.6.2 *The Availability of Sustainable Alternative Fuels and Costs Involved*

5.6.2.1 Availability

There are currently no sustainable alternative fuels for aircraft in commercial production. However, this is expected to change in the near future. Planning is underway for producing new fuels with low life-cycle emissions. When these fuels enter the market, their costs will be high and they may require subsidies or production incentives in order to make them economically viable. As industry gains more experience producing these fuels their costs will decrease, as will their life-cycle greenhouse gas (GHG) emissions. In the long-term, industry may design new aircraft and engines to take advantage of unconventional aircraft fuels with extremely low life-cycle CO_2 emissions.

---

[199] U. S. Environmental Protection Agency, *Review of the National Ambient Air Quality Standards for Particulate Matter; Policy Assessment of Scientific and Technical Information*, http://www.epa.gov/ttn/naaqs/standards/pm/data/pmstaffpaper_20051221.pdf, December 2005.

New sustainable alternative fuels for aircraft may be better suited for regional and local production in countries around the world in light of the variety of potential feedstocks. Once refined into fuel, the feedstock used is irrelevant to the aircraft. Most of the feedstocks studied to date also produce by-products that may be of value locally. Communities may be able to develop new businesses or other sources of income from alternative fuel production. To meet these objectives, significant investment will be needed in regions where States desire to become producers of sustainable alternative fuels for aircraft.

In the short-term, sustainable alternative fuels for aircraft may be available in limited quantities and have a life-cycle CO_2 footprint equal to, or less than, conventional jet fuel. It will be necessary to blend these alternative fuels with conventional jet fuel at up to a maximum of 50% to produce a drop-in fuel. Drop-in jet fuels are completely interchangeable with conventional jet fuel, and so will not require modification of fuel handling and distribution systems, including gauges, meters, fuelling vehicles, and hydrant systems, as well as aircraft engines, once the fuels have been blended.

GHG emission reductions in the short-term will be limited as sustainable alternative fuels for aircraft are initially introduced. For example, assuming the lifecycle CO_2 footprint of sustainable alternative fuels for aircraft provides a 20% reduction compared with conventional jet fuel, and a 50% fuel blend makes up 10% of the total jet fuel market, the GHG emissions reduction would be 1% compared to forecast emissions without the new fuels. However, reductions in particulate matter and sulphur oxides will be more significant. Achieving air quality benefits from the use of these fuels is independent of production life-cycle considerations.

In the medium-term, it is possible that sustainable alternative fuels for aircraft will be available in much larger quantities. The significant research and development activity currently underway is expected to lead to a number of commercial scale production facilities. The Commercial Aviation Alternative Fuels Initiative (CAAFI) is currently seeking to ensure that at least ten alternative jet fuel production facilities are built and in operation within 5 years.[200] Also, these new fuels will have been certified for greater use in blends, possibly up to 100% alternative fuel, thus moving from drop-in blend fuels to drop-in neat fuels according to industry roadmaps. During this time, the fuels may have reached cost parity, especially if the value of their carbon reduction benefits is accounted for.

As the aviation industry increases its use of sustainable alternative fuels for aircraft, ongoing efforts will be applied to reduce the life-cycle impacts of these fuels. These may include: thoroughly exploring and identifying feedstock resources; enhancing resource quality; improving oil recovery and extraction; finding markets for co-products; creating higher value co-products; improving the efficiency of converting raw biofuels to jet fuel; and reducing the length and number of transport links.

---

[200] Altman (2009).

5.6 The Use of Alternative Fuels as an Economic Measure

As for long term availability of these fuels, the aviation industry may explore more radical fuels that require redesigned engines and airframes. Fuels such as liquid hydrogen and liquid methane might be used to significantly reduce GHG emissions.[201] Managing these cryogenic liquids on an aircraft will require heat exchangers to vaporize and heat the fuel prior to use on-board[202] and the fuel supply infrastructure will need to be substantially redesigned or replaced. While new aircraft designs and new fuel production pathways may be required, these new approaches may result in more energy efficient and environmentally benign air travel.

Sustainable alternative fuels for aircraft can be produced from a wide variety of feedstocks. Sources of oil-producing plants such as camelina, moringa, babacu, macauba, jatropha, halophytes, and algae are already being evaluated for fuel production. These plants and others can be grown in countries around the world. It is quite likely that different feedstocks can be optimally grown in different countries, suggesting that many regions are candidate production locations. Once refined into fuel, the feedstock used is irrelevant to the aircraft. Additionally, since these feedstocks have a relatively low energy density, especially compared to crude oil, it is uneconomic to ship them over long distances. As a result, sustainable alternative fuels for aircraft may be better suited to production on a local scale.

The by-products or secondary products from sustainable alternative fuels for aircraft production can become valuable inputs to local economies. These materials, such as animal feed or solid residues that can be used as fuel for cook stoves, may be valued locally even if the bio-oils are shipped out of the region for refining. Communities may be able to develop new businesses or other sources of income from alternative fuel production. While sustainable alternative fuels for aircraft can be produced from a wide variety of feedstocks and processes, only those that attract sufficient investment will achieve a market presence. Identifying resources, producing test quantities of oil, certifying the fuels, conducting performance studies, and investing in plot plant production are all expensive. Significant investment will be required in any region where States desire to become producers of sustainable alternative fuels for aircraft.

5.6.2.2 Costs Associated with Sustainable Alternative Fuels

Today, estimated costs for sustainable alternative fuels for aircraft range from 2–5 times the cost of conventional jet fuel[203] and in some cases, even higher. However, these values are estimates, as the fuels are not commercially available. These

---

[201] Daggett et al. (2007).

[202] Walther et al. (1995).

[203] Caldecott and Tooze (2009) (notes "...best estimates of current minimum production costs are approximately US$100–130 per barrel ..."; Start-up biofuel firm Solix currently producing fuel from algae at $32.81/gallon but expects costs to come down as better design improves production integration.

estimates also do not include a value for carbon credits. Until more significant fuel quantities are available, the cost of sustainable alternative fuels for aircraft will remain highly uncertain. As a result, these fuels will likely require subsidies or incentives, at least initially, to encourage their production given the risks involved in moving from pilot scale to commercialization.

The development of new processes for producing sustainable alternative fuels for aircraft can quickly reduce costs of fuels to compete with conventional jet fuel in the mid-term. Over time, as new fuel production becomes commercialized and the processes are improved, fuel costs are expected to decrease. Larger plants will bring economies of scale; capital costs will be reduced through the application of advanced production technologies; operating and maintenance costs will be reduced through process refinement, improved control systems, and greater experience; transportation links will be shortened; and the total value of products may increase. The International Energy Agency has projected costs for 2nd generation biofuel plants producing a biodiesel to fall an average of 2.5%/year between 2010 and 2030, or be reduced by approximately 40%.5 Since there are no sustainable alternative fuels for aircraft being produced today, a specific cost reduction curve cannot be developed or confirmed for these fuels. There are, however, many projects in various stages of development whose investors believe they can quickly reach cost parity with conventional fuels, although their financial plans and forecasts have not been made public.

With most of the feedstocks presently being evaluated, a substantial portion of the mass will end up as a by-product of fuel production. Finding markets for by-products (or co-products) will be important for overall process economics. Some high-value markets are available today for select bio-oils, such as nutraceuticles, which are used as nutritional supplements and feedstocks for pharmaceutical production. These materials represent a small part of the total oil produced, and large-scale fuel production may have a significant impact on their market value. Other by-products, such as meal, may be used for animal feed, solid fuel, or similar low value products; however, these materials will be an important consideration in overall process economics.

The conclusions that could be arrived at from the foregoing discussion are: that development of new sustainable alternative fuels for aircraft production processes are able to reduce costs of fuels to compete with conventional jet fuel in the mid-term; sustainable alternative fuels for aircraft can be produced from a wide variety of feedstocks for use in global aviation, suggesting that many regions are candidate production locations; sustainable alternative fuels for aircraft are well suited to production on a local scale because unlike crude oil, the energy density of currently-proposed feedstocks is too low to support economical shipment over long distances; the by-products or secondary products from sustainable alternative fuels for aircraft production are valuable inputs to local economies; and that sustainable alternative fuels for aircraft can be produced from a wide variety of feedstocks and processes, yet, only those that attract sufficient investment will achieve a market presence.

5.6 The Use of Alternative Fuels as an Economic Measure

5.6.3 *The Rio Conference of 2009*

ICAO convened the Conference on Aviation and Alternative Fuels (CAAF) at Rio de Janeiro in Brazil from 16 to 18 November 2009. The Conference was a major event showcasing the state of the art in aviation alternative fuels, and an event at which a Global Framework for Aviation Alternative Fuels (GFAAF)[204] was considered. The GFAAF was designed as a living document that will be continually updated on the ICAO website that will share information, best practices and future initiatives by ICAO Member States and the air transport industry. The Conference made reference to the High-Level Meeting on International Aviation and Climate Change[205] convened earlier by ICAO which recommended that States and International Organizations actively participate in CAAF with a view to sharing their efforts and strategies to promote such work and to update the 15th Meeting of the Conference of the Parties to the United Nations Framework Convention on Climate Change (UNFCC COP15) to be held in December 2009).[206]

The establishment of GFAAF was part of a Declaration adopted at the Rio Conference, which recognized that the introduction of sustainable alternative fuels for aviation will help to address issues of environment, economics, and supply security, and noted *the* very limited availability of qualified alternative fuels for aviation as well as the fact that sustainable alternative fuels for aircraft can be produced from a wide variety of feedstocks for use in global aviation. The Declaration suggested that many regions are candidate production locations and

---

[204] See ICAO Doc CAAF/09-WP/23, 18/11/09.

[205] The High Level Meeting on Aviation and Climate Change, convened by the International Civil Aviation Organization from 7 to 9 October 2009 once again brought to bear the vexed issue of ICAO's leadership in the field, only to be re-endorsed by the 56 States which attended the Meeting, that ICAO was indeed the leader and that ICAO should provide guidance on various issues pertaining to the subject. The meeting went on to acknowledge the principles and provisions on common but differentiated responsibilities and respective capabilities, and the fact that developed countries will be taking the lead under the UNFCCC and the Kyoto Protocol. It also acknowledged the principles of non-discrimination and equal and fair opportunities to develop international aviation set forth in the Chicago Convention and re-emphasized the vital role which international aviation plays in global economic and social development and the need to ensure that international aviation continues to develop in a sustainable manner.

[206] At the United Nations Framework Convention on Climate Change 15th Meeting of the Conference of the Parties (COP15) and the Fifth Meeting of the Parties to the Kyoto Protocol (CMP5) which were convened in Copenhagen on 8 December 2009, leaders from the U.S., India, Brazil, South Africa and China came to an agreement to combat global warming. The deal, which was only between five countries, contained no specifics on emissions cuts, but it did commit the countries to look to keep global warming at 2°C or less and to promise $30 billion in funding to battle climate change by 2012. It also created a framework for international transparency on climate actions for developed and developing nations alike.

acknowledged that sustainable alternative fuels for aviation may offer reduced lifecycle CO_2 emissions compared to the lifecycle of conventional aviation fuels. It also acknowledged that sustainable alternative fuels for aviation may also offer benefits to surface and local air quality. While the Declaration noted that the technology existed to produce substitute, sustainable fuels for aviation that took into consideration world´s food security, energy and sustainable development needs, it was recognized that the production of sustainable alternative fuels for aviation could promote new economic opportunities.

The Declaration suggested, inter alia, the following:

a) That ICAO and its Member States endorse the use of sustainable alternative fuels for aviation, particularly the use of drop-in fuels in the short to mid-term, as an important means of reducing aviation emissions;
b) That ICAO establish a Global Framework for Aviation Alternative Fuels (GFAAF) on aviation and sustainable alternative fuels to communicate what individual and shared efforts expect to achieve with sustainable alternative fuels for aviation in the future for consideration by the 37th Session of the ICAO Assembly[207];
c) That GFAAF will be continually updated;
d) That Member States and stakeholders work together through ICAO and other relevant international bodies, to exchange information and best practices, and in particular to reach a common definition of sustainability requirements for alternative fuels;
e) That Member States are encouraged to work together expeditiously with the industry to foster the research, development, deployment and usage of sustainable alternative fuels for aviation;
f) That funding efforts that support the study and development of sustainable alternative fuels and other measures to reduce GHG emissions, in addition to the funding for research and technology programmes to further improve the efficiency of air transport, be maintained or improved;
g) That Member States are encouraged to establish policies that support the use of sustainable alternative aviation fuels, ensure that such fuels are available to aviation and avoid unwanted or negative side effects, which could compromise the environmental benefits of alternative fuels;
h) That the ICAO Council should further elaborate on measures to assist developing States as well as to facilitate access to financial resources, technology transfer and capacity building;
i) That there is an urgent need for measures to facilitate access to financial resources, technology exchange, and capacity building specific to sustainable aviation alternative fuels;

---

[207] The ICAO Assembly is convened once every 3 years by the ICAO Council to discuss the policy of the Organization. The 37th Session of the Assembly was scheduled for 28 September to 8 October 2010.

5.6 The Use of Alternative Fuels as an Economic Measure 391

j) That ICAO takes the necessary steps with the aim of considering a framework for financing infrastructure development projects dedicated to sustainable aviation alternative fuels and incentives to overcome initial market hurdles;
k) That ICAO continue to facilitate efforts to develop a lifecycle analysis framework for comparing the relative GHG emissions from sustainable alternative fuels to the lifecycle of conventional fuels for aviation; and
l) That ICAO and its Member States should strongly encourage wider discussions on the development of alternative fuel technologies and support the use of sustainable alternative fuels, including biofuels, in aviation in accordance with national circumstances.

It is worthy of note that one of the recommendations calls for ICAO to organize a meeting of States, financial institutions, fuel producers, feedstock producers, aircraft manufacturers and operators to consider the critical issues of cost and financing infrastructure projects dedicated to aviation alternative fuels and incentives to overcome initial market hurdles. ICAO pledged to carry out the following key activities with a view to supporting the objectives of GFAAF:

a) Providing forums for education and outreach on sustainable alternative fuels for aviation;
b) Providing forums for facilitating the exchange of information on financing and incentives for sustainable alternative fuels for aviation programmes working with the relevant UN and regional financial entities;
c) Facilitating development of standardized definitions, methodologies and processes to support the development of sustainable alternative fuels for aviation, taking into consideration the work that has been done so far in this area; and
d) Supporting a platform for access to research roadmaps and programmes.

Some notable achievements of the Conference were that it provided a global forum for the 190 Member States of ICAO to discuss this issue and prepare for UNFCC COP15; it adopted the Fuel Readiness Level (FRL) developed by Commercial Aviation Alternative Fuels Initiative (CAAFI), as a best practice; defined drop-in jet fuel blend and drop-in neat jet fuel; recommended the use of life-cycle analysis as the appropriate means for comparing the relative emissions from alternative jet fuels to conventional jet fuel; endorsed the use of the existing industry qualification and certification processes as the appropriate means for approving a new alternative jet fuel; and made every effort to ensure the consideration of aviation alternative fuels within relevant international, regional and State efforts to develop sustainability criteria for all alternative fuels.

ICAO has offered the view (as a post-Conference thought) that sustainable drop-in alternative fuels produced from biomass or renewable oils offer the potential to reduce life-cycle greenhouse gas emissions and therefore reduce aviation's contribution to global climate change. The statement further said that sustainable alternative fuels could be an important tool in the efforts to close the mitigation gap while allowing the sector to respond to growing demand, and that the use of such fuels may also offer reduced emissions of particulate matter, lessening aviation's impact on air quality as the result of the significantly lower fuel sulphur content.[208]

One of the more important topics which were discussed at UNFCC COP15 was the continued relationship between the UNFCCC, the International Civil Aviation Organization (ICAO) and the International Maritime Organization (IMO) with regard to emissions from international aviation and shipping. Both ICAO and IMO were of the strong view that they should continue to be responsible for dealing with reductions in emissions from their respective industries and sought guidance from UNFCC COP 15. Within the discussions, Ethiopia proposed to introduce levies on international aviation and shipping, calculated to help developing countries raise funds in the nature of billions of dollars needed to cope with issues of climate change. This proposal was backed by Australia, Britain and France. The Conference, which fizzled out into a watered down agreement between five countries did not offer any specific guidelines in the context of aviation or maritime affairs.

Irrespective of the failure of UNFCC COP15-which was a non-event-to come up with a multilateral treaty on climate change,[209] it is interesting to note that nearly 2 months before COP15 the High Level Meeting provided a sense of direction for States to consensually agree that ICAO should continue to have a leading role on all aviation matters, including issues relating to international aviation and climate change. There was also general agreement that a comprehensive approach would be necessary to achieve reductions in aviation's emissions. States broadly expressed their support for ICAO to continue the development of a basket of measures, including measures to provide assistance to developing countries; a process to develop a framework for market-based measures in international aviation; and facilitation of the development and use of alternative fuels for aviation.

One might well ask, if ICAO were to maintain this leadership role, what it would accomplish in the area of sustainable alternative fuels. The answer seems to lie in the Recommendations of CAAF. For a start, the ICAO Council has to present to the 37th Session of the Assembly input based on the three events, the High Level Meeting, CAAF and UNFCC COP 15 that would encourage Member States to develop policy actions to accelerate the appropriate development, deployment and use of such fuels and to work together through ICAO and other relevant international bodies, to exchange information and best practices, and in particular to reach a common definition of sustainability requirements and to ;work together expeditiously with

---

[208] http://www.greenaironline.com/news.php?viewStory=673.

[209] Commentators are of the view that the agreement reached is far from perfect – and a long way from what environmentalists were hoping from the Copenhagen summit just a few months ago. However, some believe it is a start. President Obama is reported to have stated that for the first time, all major economies have come together to accept their responsibility to combat climate change and that the Copenhagen agreement is a consensus that will serve as the foundation for global action against climate change for years to come. See http://www.time.com/time/specials/packages/article/0,28804,1929071_1929070_1948974,00.html?cnn=yes.

the industry to foster the development and implementation of sustainable alternative fuels for aviation. Integral to this approach would be the need to ensure that policy recommendations and decisions considered by ICAO and individual States consider environmental, social and economic sustainability aspects, while also taking into account technical requirements including safety aspects. Member States should also be urged to invite industry to actively participate in further work on sustainable alternative fuels for aviation facilitated by ICAO and the Assembly should be advised of initiatives by States and other organizations related to sustainable alternative fuels for aviation. The ICAO Council should, as recommended by CAAF, commit to further elaborate on measures to assist developing States as well as to facilitate access to financial resources, technology transfer and capacity building.

The above notwithstanding, it would be incorrect to assume that the task of minimizing the deleterious effects of the environment caused by aviation should devolve upon ICAO alone. This should be a collective effort on the part of the entire aviation industry. Government, industry and the travelling public will need to invest in order to deliver technological innovation, development and carbon free flight.

References

Abeyratne RIR (1995) Legal and regulatory issues of computer reservations systems and code sharing agreements in air transport. Editiones Frontieres, p 119

Abeyratne RIR (2003) The decision of the European court of justice on open skies – how can we take liberalization to the next level? J Air Law Commerce 68(3):485–518

Abeyratne RIR (2004) The decision in the Ryanair case – the low cost carrier phenomenon. Eur Transport Law XXXIX(5):585–601

Abeyratne RIR (2005) The airbus-boeing subsidies dispute – some preliminary legal issues. Air Space Law XXX(6):379–395

Abeyratne RIR (2009) Airport business law. PublishAmerica, Bloomington, IN, p 13

Adkins B (1994) Air transport and E.C. competition law. European Competition Law Monologues. Sweet and Maxwell, London, p 81

Altman R (2009) Landmark synthetic jet fuel specification action creates opportunities for airports. Int Airport Rev 4:62–64

Arpino R (1994) Automated weather observation systems find increasing acceptance at major airports. ICAO J:15

Baker C (2005) Back to the table. Airline Business:36 at 38

Becker P (2002) Corporate foresight in Europe: a first overview. Institute of Science and Technology Studies, Germany, p 7

Bobbitt P (2008) Terror and consent: the wars for the twenty first century. Knopf, New York, pp 98–179

Brunelle-Yeung E (2009) The impacts of aviation emissions on human health through changes in air quality and UV irradiance. Thesis, Master of Science in Aeronautics and Astronautics, Massachusetts Institute of Technology, Boston, MA

Caldecott B, Tooze S (2009) Green skies thinking: promoting the development and commercialization of sustainable bio-jet fuels, Policy Exchange, research note

Daggett DL, Hendricks RC, Walther R, Corporan E (2007) Alternate fuels for use in commercial aircraft. The Boeing Company, published by the American Institute of Aeronautics and Astronautics, Inc. ISABE-2007-1196

Doganis R (2001) (2001) The airline business in the 21st century. Routledge, London

Doz Y, Prahalad CK, Hamel G (1990) Control, change and flexibility: the dilemma of transnational collaboration. In: Bartlett C, Doz Y, Hedlund G (eds) Managing the global firm. Routledge, London, pp 17–143 at 33

Ferguson N (2008) The ascent of money. The Penguin Press, New York, p 188

Field D (2005) True open skies? Airline Business:8

Flanagan M (1996) Open skies and the survival of the fittest. Aerospace:16 at p 18

Frank Carlson E (1995) Brrrr! Airplanes operating at extremely low ground temperatures. Airliner:9

Freedman JE (1994) Applicability of E-scan technology for future airport aviation weather radars. Air Traffic Control Quarterly 2(1):53–78

Goh J (1997) European air transport law and competition. Wiley, New York, p 15

Janda R (1995) Passing the torch: why ICAO should leave economic regulation of international air transport to the WTO. Ann Air Space Law XX(Part 1):409–430, at 430

Kimpel S (1997) Antitrust considerations in international airline alliances. J Air Law Commerce 63:475–513 at 476

Levine ME (2009) Airport congestion: when theory meets reality. Law and economics research paper series, working paper no. 08-55. Yale J Regulation 26(1):37 at 76–87

Manningham D (1996) The keys to using weather radar. Business Commercial Aviation:60–62

McKenna JT (1994) Airlines seek uniform winter safety rules. Aviation Week Space Technol:46–47

Miyagi M (2005) Serious accidents and human factors. American Institute of Aeronautics and Astronautics, Virginia, p 143

Morrison S, Winston C (1986) The economic effects of domestic deregulation, pp 1–2

Puempol H (2006) Airports and aviation weather services: a new alliance forming? Int Airport Rev (4):75 at 76

Reuter P (1989) Introduction to the law of treaties. Pinter Publishers, London and New York, p 16

Rhoades DL, Lush H (1997) A typology of strategic alliances in the airline industry: propositions for stability and duration. J Air Transport Manage 3(3):109–114

Roos HB, Sneek NW (1997) Some remarks on predatory pricing and monopolistic competition in air transport. Air Space Law XXII(3):154

Rusmussen R, Cole J, Knight K, Moore RK, Kupperman M (1995–1996) How heavy is that snowfall. Flight Deck 18:24–26 at 24

Stadlmeier S (1997) International commercial aviation: from foreign policy to trade in services. In: Benko M, de Graaf W (eds) Forum for air and space law, vol 3. Gif-sur-Yvette Cedex, Editions Frontieres, France, p 318

Steinhorst G, van Dijk WCM (1994) Modern weather observation systems in use at airports can improve the quality of weather data. ICAO J 49(8):12–14, at 13

Stuart RA, Isaac GA (1994) Archived weather data provides new information on ground based icing. ICAO J:5

Supiot A (2010) A legal perspective on the economic crisis of 2008. Int Labour Rev 149(2):151–162, at 152

Taylor WF (2009) Survey of sulphur levels in commercial jet fuel. CRC Aviation Research Committee of the Coordinating Research Council, Alpharetta, GA

Wagner GA (1994–1995) Take-off and landing on icy conditions. Flight Deck 14:21–25

Walther R et al (1995) Aero engines for alternative fuels, in hydrogen and other alternative fuels for air and ground transportation. European Commission, Brussels, published by Wiley

Wassenbergh H (2000) Common market, open skies and politics. Air Space Law XXV(4–5):174 at 176

Wassenbergh H (2001) Towards global economic regulation of international air transportation through interregional bilateralism. Ann Air Space Law XXVI:237 at p. 243

Woolley D (1999) Winter operations evolve in Europe. GSE Today:23–26

Chapter 6
Conclusion

The air transport industry is full of paradoxes. On the one hand, air transport contributes 10% of the world GDP[1] and employs approximately 80 million people worldwide. Yet, over the decade 1999–2009 the industry lost $56 billion. Given that over that period there were 20 billion passengers carried by air, the industry lost $2.8 per passenger on average. The paradox is that despite, this long history of loss, *Airbus Industrie* forecasts that between 2009 and 2028 there will be a demand for 24,951 passenger and freighter aircraft worth USD 3.1 trillion, and that, by 2028 there will be 32,000 aircraft in service compared with 15,750 in 2009.[2] Another paradox is the open skies concept. On the one hand, progressive thinking favours liberalization of air transport through open skies, which means that air transport must be treated like any other business and should not have market access barriers. On the other hand, airport capacity is finite and a massive injection of capacity will be a severe drain on the process of slot-allocation.

Air transport has been, and continues to be, a complex business. It follows therefore, that transactions related to this industry are indeed complex and should be adaptable to modern exigencies. A survey of IATA carried out in 2007 of over 600 companies from five countries reflects that 63% confirmed that air transport networks are vital for their investments and business.[3] 30% of those countries said that any constraint placed on the air transport industry would make them invest less.[4]

Air transport plays an integral part in the tourism industry where 40% of international tourists travel by air. This fact underscores the indispensability of air transport to the global economy on the one hand and the resilience of the

---

[1] The gross domestic product (GDP) or gross domestic income (GDI) is a basic measure of a country's overall economic output.
[2] A *Flying Smart, Thinking Big*, Airbus Industrie's Global Market Forecast, 2009–2028, at 8–9.
[3] *Id.*, at 16.
[4] *Ibid.*

industry to resuscitate itself after slow periods of growth followed by periods of losses.

It is also well known that the air transport industry differs in its commercial practices due to its own particular nature and the manner in which it is treated by States and the commercial world. Giovanni Bisignani, the down-to-earth, hard-hitting Director General and CEO of IATA, in his address to the Royal Aeronautical Society, Montreal Branch on 1 December 2009 observed that:

> The aviation industry is in crisis. Since 2001, airlines have lost US$53 billion. That includes losses of US$16.8 billion in 2008, the biggest in our history and US$11 billion this year. We are forecasting a further US$3.8 billion loss next year. Look into the detail behind these numbers and you see enormous shifts on key parameters.[5]

Mr. Bisignani correctly attributed this continuing struggle of the airline industry to the mandatory bilateral system of the negotiation of air traffic rights between State authorities, stating that no other industry has such restrictions. He observed that shipping companies operated without restrictions and almost all other products enjoyed global markets where an entrepreneur may have to pay an import tax or a duty but he did not need his government to conclude a bilateral agreement. He also stated that, however, those airlines could not sell their product where markets exist and could not merge across borders where it made business sense. The result was a sick industry with too many players.

On an earlier occasion,[6] Mr. Bisignani made similar comments, emphasising that numerous businesses have been buttressed by the lowering of trade barriers, but airlines suffered because they still faced enormous hurdles. Mr. Bisignani pointed out that although the airline industry was among the first to operate globally, it was still waiting for the benefits of globalization. He asked a pointed question: "Do we purchase cars or medications based on the nationality of a company's shareholders?" asking the further question that, if an Egyptian can spend a night at a Singaporean hotel in Hamburg, why couldn't an Australian fly a Brazilian airline from Mexico City to Miami? Mr. Bisignani added that airlines supported 29 million jobs and $2.9 trillion worth of economic activity worldwide and that few industries so vital to the health of the global economy remain so restricted by archaic ownership rules.[7]

The problem referred to by Mr. Bisignani has its genesis in Article 6 of the Chicago Convention which states that no international air service may be operated over and into the territory of a contracting State, except with the special permission or other authorization of that State, and in accordance with the terms of such permission or authorization. The state in which the air transport industry operates today, in having to enjoy market access through bilateral or multilateral agreements for air traffic rights carved out between States, is related to the discussions that took

---

[5] http://www.iata.org/pressroom/speeches/2009-12-01-01.htm.
[6] Bisignani (2006).
[7] *Ibid.*

place at the Chicago Conference of 1944 which led to the adoption of the Chicago Convention.

The restrictive influence of Article 6 has pervaded more than six decades where liberalization of market access in air transport has been stifled through bilateral negotiations between States who have flittered between pre determination of capacity and definition of market share. The result has been the progressive imbalance of the market and the gradual realization that the current bilateral structure is a hindrance to liberalization of air transport. For their part, States are trying to figure out how to liberalize the market.

Ironically, one could place these facts against the backdrop of pronouncements of high level policy makers who are increasingly reaching the conclusion that aviation is making a substantial contribution to the global economy. *Oxford Economics*, an economic forecasting agency, in a recent report recognizes the wide range of benefits that air transport brings to economies and societies globally.

This report,[8] issued in June 2009, also suggests that the world's future prosperity may depend on a growing and thriving aviation industry, which currently supports nearly 8% of the world's economy, and questions the environmental benefits and social impacts of limiting that growth. One wonders, then, as to why an indispensable economic tool such as the air transport industry which contributes so substantially to the global economy,[9] is shackled by restrictions such as limitations placed on air traffic rights and free trading in airport slots which other similar businesses are not subject to. It is hoped that the discussion below would shed some light on the issue.

A complex industry will bring about complex issues which require diligent responses. These responses cannot be isolated, particularly when they relate to issues of safety, security, environmental protection and sustainability of air transport. For instance, Security Council Resolution 1368, adopted immediately after the events of 11 September 2001 called on all States to work together urgently to bring to justice the perpetrators, organizers and sponsors of these terrorist attacks and stressed that those responsible for aiding, supporting or harbouring the perpetrators, organizers and sponsors of these acts will be held accountable. Resolution 1373

---

[8] Oxford Economics (2009).

[9] ICAO has estimated that the direct contribution of civil aviation, in terms of the consolidated output of air carriers, other commercial operators and their affiliates, was U.S. $370 billion. Direct employment on site at airports and by air navigation services providers generated 1.9 million jobs while production by aerospace and other manufacturing industries employed another 1.8 million people. Overall, the aviation industry directly employed no less than 6 million persons in 1998. These direct economic activities have multiplier effects upon industries providing either aviation-specific and other inputs or consumer products. In simple terms, every U.S. $100 of output produced and every 100 jobs created by air transport trigger additional demand of U.S. $325 and in turn 610 jobs in other industries. The total economic contribution of air transport, consisting of the direct economic activities and the multiplier effects, is estimated at U.S. $1,360 billion output and 27.7 million jobs worldwide. See ICAO Circular (2004). Also *The Economic Benefits of Air Transport*, 2000 Edition, ATAG:IATA, at 7. Also, *ICAO Circular 219, Id.*, at 4.

adopted by the Security Council on 28 September 2001 inter alia called on all States to work together urgently to bring to justice the perpetrators, organizers and sponsors of these terrorist attacks and stressed that those responsible for aiding, supporting or harbouring the perpetrators, organizers and sponsors of these acts will be held accountable. There have been similar collective responses for global action on environmental protection.

The example of Haiti is a glowing example of collective State response and ICAO action. On Tuesday 12 January 2010, at 16.53 local time, an earthquake with a magnitude of 7.0 Mw on the Richter scale[10] hit Haiti at a depth of 13 km. The International Red Cross estimated the people affected by the earthquake at around 3 million.[11] It was estimated that around 100,000–200,000 people would have perished as a result of the catastrophe.[12]

The quake damaged many important buildings including the Presidential Palace, The National Assembly and the Cathedral of Port-au-Prince. The headquarters of the United Nations Stabilization Mission in the city collapsed killing many UN workers including the Chief of Mission. The disaster also caused severe damage to communication systems, air, land, and sea transport facilities, hospitals, and electrical networks, drastically hampering rescue and aid efforts.

Many countries responded to appeals for humanitarian aid, pledging funds and dispatching rescue and medical teams, engineers and support personnel. Communication systems, air, land, and sea transport facilities, hospitals, and electrical networks had been damaged by the earthquake, which hampered rescue and aid efforts.[13] From an aeronautical perspective, there was considerable confusion over who was in charge of the relief efforts and of flights in and out of Haiti, and the problem was compounded by air traffic congestion and problems with prioritization of flights which in turn hampered relief work. ICAO engaged in a sustained effort to render assistance in the area of air traffic flow management and the operation of

---

[10] The Richter scale is the best known scale for measuring the magnitude of earthquakes. The magnitude value is proportional to the logarithm of the amplitude of the strongest wave during an earthquake. A recording of 7, for example, indicates a disturbance with ground motion 10 times as large as a recording of 6. The energy released by an earthquake increases by a factor of 30 for every unit increase in the Richter scale. See http://www.matter.org.uk/schools/content/seismology/richterscale.html.

[11] CBS News (2010).

[12] Haiti Quake Death toll may hit 200,000-Minister, *Reuters Alertnet* (Reuters). http://www.alertnet.org/thenews/newsdesk/N15143632.htm. Retrieved 15 January 2010.

[13] Initially, relief flights from around the world were delayed as a two-tier system was put in place to make sure that the most urgently needed supplies and personnel were getting in first, and the rest followed using a pre-arranged arrival and take-off schedule. Once a ground control team from the Air Force's first Special Operations Wing began landing aircraft on January 13, a team from the U. S. Federal Aviation Administration and Haitian authorities began working to restore navigation and communication capabilities. A team of U.S. military and civilian aviation experts working with Haitian officials began sorting flights, allowing some to land immediately based on what was needed most.

relief flights. This is what would happen if a catastrophe were to affect any State which would require assistance from an aeronautical or air transport perspective.

References

Bisignani G (2006) Think again airlines. Available at http://www.foreignpolicy.com/articles/2006/01/04/think_again_airlines?print=yes&hidecomments=yes&page=full, January 2006

Oxford Economics (2009) Aviation, the real world wide web. Available at http://www.oxfordeconomics.com/free/pdfs/ox_econ_aviation_report/main.html, June 2009

ICAO Circular (2004) The economic contribution of civil aviation 219:1

CBS News (2010) 13 January 2010. http://www.cbsnews.com/stories/2010/01/13/world/main6090601.shtml?tag=cbsnewsSectionContent.4. Retrieved 13 January 2010

APPENDIX

Beijing Convention of 2010
CONVENTION
on the Suppression of Unlawful Acts Relating to International Civil Aviation
Done at Beijing on 10 September 2010

**CONVENTION
ON THE SUPPRESSION OF UNLAWFUL ACTS
RELATING TO INTERNATIONAL CIVIL AVIATION**

THE STATES PARTIES TO THIS CONVENTION,

DEEPLY CONCERNED that unlawful acts against civil aviation jeopardize the safety and security of persons and property, seriously affect the operation of air services, airports and air navigation, and undermine the confidence of the peoples of the world in the safe and orderly conduct of civil aviation for all States;

RECOGNIZING that new types of threats against civil aviation require new concerted efforts and policies of cooperation on the part of States; and

BEING CONVINCED that in order to better address these threats, there is an urgent need to strengthen the legal framework for international cooperation in preventing and suppressing unlawful acts against civil aviation;

HAVE AGREED AS FOLLOWS:

Article 1

1. Any person commits an offence if that person unlawfully and intentionally:

2.

(a) performs an act of violence against a person on board an aircraft in flight if that act is likely to endanger the safety of that aircraft; or

(b) destroys an aircraft in service or causes damage to such an aircraft which renders it incapable of flight or which is likely to endanger its safety in flight; or

(c) places or causes to be placed on an aircraft in service, by any means whatsoever, a device or substance which is likely to destroy that aircraft, or to cause damage to it which renders it incapable of flight, or to cause damage to it which is likely to endanger its safety in flight; or

(d) destroys or damages air navigation facilities or interferes with their operation, if any such act is likely to endanger the safety of aircraft in flight; or

(e) communicates information which that person knows to be false, thereby endangering the safety of an aircraft in flight; or

(f) uses an aircraft in service for the purpose of causing death, serious bodily injury, or serious damage to property or the environment; or

(g) releases or discharges from an aircraft in service any BCN weapon or explosive, radioactive, or similar substances in a manner that causes or is likely to cause death, serious bodily injury or serious damage to property or the environment; or

(h) uses against or on board an aircraft in service any BCN weapon or explosive, radioactive, or similar substances in a manner that causes or is likely to cause death, serious bodily injury or serious damage to property or the environment; or

(i) transports, causes to be transported, or facilitates the transport of, on board an aircraft:

(1) any explosive or radioactive material, knowing that it is intended to be used to cause, or in a threat to cause, with or without a condition, as is provided for under national law, death or serious injury or damage for the purpose of intimidating a population, or compelling a government or an international organization to do or to abstain from doing any act; or

(2) any BCN weapon, knowing it to be a BCN weapon as defined in Article 2; or

(3) any source material, special fissionable material, or equipment or material especially designed or prepared for the processing, use or production of special fissionable material, knowing that it is intended to be used in a nuclear explosive activity or in any other nuclear activity not under safeguards pursuant to a safeguards agreement with the International Atomic Energy Agency; or

(4) any equipment, materials or software or related technology that significantly contributes to the design, manufacture or delivery of a BCN weapon without lawful authorization and with the intention that it will be used for such purpose; provided that for activities involving a State Party, including those undertaken by a person or legal entity authorized by a State Party, it shall not be an offence under subparagraphs (3) and (4) if the transport of such items or materials is consistent with or is for a use or activity that is consistent with its rights, responsibilities and obligations under the applicable multilateral non-proliferation treaty to which it is a party including those referred to in Article 7.

2. Any person commits an offence if that person unlawfully and intentionally, using any device, substance or weapon:

(a) performs an act of violence against a person at an airport serving international civil aviation which causes or is likely to cause serious injury or death; or

(b) destroys or seriously damages the facilities of an airport serving international civil aviation or aircraft not in service located thereon or disrupts the services of the airport, if such an act endangers or is likely to endanger safety at that airport.

3. Any person also commits an offence if that person:

(a) makes a threat to commit any of the offences in subparagraphs (a), (b), (c), (d), (f), (g) and (h) of paragraph 1 or in paragraph 2 of this Article; or

(b) unlawfully and intentionally causes any person to receive such a threat, under circumstances which indicate that the threat is credible.

4. Any person also commits an offence if that person:

(a) attempts to commit any of the offences set forth in paragraph 1 or 2 of this Article; or

(b) organizes or directs others to commit an offence set forth in paragraph 1, 2, 3 or 4(a) of this Article; or

(c) participates as an accomplice in an offence set forth in paragraph 1, 2, 3 or 4(a) of this Article; or

(d) unlawfully and intentionally assists another person to evade investigation, prosecution or punishment, knowing that the person has committed an act that constitutes an offence set forth in paragraph 1, 2, 3, 4(a), 4(b) or 4(c) of this Article, or that the person is wanted for criminal prosecution by law enforcement authorities for such an offence or has been sentenced for such an offence.

2. Each State Party shall also establish as offences, when committed intentionally, whether or not any of the offences set forth in paragraph 1, 2 or 3 of this Article is actually committed or attempted, either or both of the following:

(a) agreeing with one or more other persons to commit an offence set forth in paragraph 1, 2 or 3 of this Article and, where required by national law, involving an act undertaken by one of the participants in furtherance of the agreement; or

(b) contributing in any other way to the commission of one or more offences set forth in paragraph 1, 2 or 3 of this Article by a group of persons acting with a common purpose, and such contribution shall either:

(i) be made with the aim of furthering the general criminal activity or purpose of the group, where such activity or purpose involves the commission of an offence set forth in paragraph 1, 2 or 3 of this Article; or

(ii) be made in the knowledge of the intention of the group to commit an offence set forth in paragraph 1, 2 or 3 of this Article.

Article 2

For the purposes of this Convention:

(a) an aircraft is considered to be in flight at any time from the moment when all its external doors are closed following embarkation until the moment when any such door is opened for disembarkation; in the case of a forced landing, the flight shall be deemed to continue until the competent authorities take over the responsibility for the aircraft and for persons and property on board;

(b) an aircraft is considered to be in service from the beginning of the preflight preparation of the aircraft by ground personnel or by the crew for a specific flight until twenty-four hours after any landing; the period of service shall, in any event, extend for the entire period during which the aircraft is in flight as defined in paragraph (a) of this Article;

(c) "Air navigation facilities" include signals, data, information or systems necessary for the navigation of the aircraft;

(d) "Toxic chemical" means any chemical which through its chemical action on life processes can cause death, temporary incapacitation or permanent harm to humans or animals. This includes all such chemicals, regardless of their origin or of their method of production, and regardless of whether they are produced in facilities, in munitions or elsewhere;

(e) "Radioactive material" means nuclear material and other radioactive substances which contain nuclides which undergo spontaneous disintegration (a process accompanied by emission of one or more types of ionizing radiation, such as alpha-, beta-, neutron particles and gamma rays) and which may, owing to their radiological or fissile properties, cause death, serious bodily injury or substantial damage to property or to the environment;

(f) "Nuclear material" means plutonium, except that with isotopic concentration exceeding 80 per cent in plutonium-238; uranium-233; uranium enriched in the isotope 235 or 233; uranium containing the mixture of isotopes as occurring in nature other than in the form of ore or ore residue; or any material containing one or more of the foregoing;

(g) "Uranium enriched in the isotope 235 or 233" means uranium containing the isotope 235 or 233 or both in an amount such that the abundance ratio of the sum of these isotopes to the isotope 238 is greater than the ratio of the isotope 235 to the isotope 238 occurring in nature;

(h) "BCN weapon" means: "biological weapons", which are:

(a) (1)microbial or other biological agents, or toxins whatever their origin or method of production, of types and in quantities that have no justification for prophylactic, protective or other peaceful purposes; or (ii) weapons, equipment or means of delivery designed to use such agents or toxins for hostile purposes or in armed conflict.

(b) "chemical weapons", which are, together or separately:

(i) toxic chemicals and their precursors, except where intended for:

(A) industrial, agricultural, research, medical, pharmaceutical or other peaceful purposes; or

(B) protective purposes, namely those purposes directly related to protection against toxic chemicals and to protection against chemical weapons; or

(C) military purposes not connected with the use of chemical weapons and not dependent on the use of the toxic properties of chemicals as a method of warfare; or

(D) law enforcement including domestic riot control purposes,

as long as the types and quantities are consistent with such purposes; munitions and devices specifically designed to cause death or other harm through the toxic properties of those toxic chemicals specified in subparagraph (b)(i), which would be released as a result of the employment of such munitions and devices; (iii) any equipment specifically designed for use directly in connection with the employment of munitions and devices specified in subparagraph (b)(ii). nuclear weapons and other nuclear explosive devices. "Precursor" means any chemical reactant which takes part at any stage in the production by whatever method of a toxic chemical. This includes any key component of a binary or multicomponent chemical system;

the terms "source material" and "special fissionable material" have the same meaning as given to those terms in the Statute of the International Atomic Energy Agency, done at New York on 26 October 1956.

APPENDIX 405

Article 3
Each State Party undertakes to make the offences set forth in Article 1 punishable by severe penalties.

Article 4
1. Each State Party, in accordance with its national legal principles, may take the necessary measures to enable a legal entity located in its territory or organized under its laws to be held liable when a person responsible for management or control of that legal entity has, in that capacity, committed an offence set forth in Article 1. Such liability may be criminal, civil or administrative.

2. Such liability is incurred without prejudice to the criminal liability of individuals having committed the offences.

3. If a State Party takes the necessary measures to make a legal entity liable in accordance with paragraph 1 of this Article, it shall endeavour to ensure that the applicable criminal, civil or administrative sanctions are effective, proportionate and dissuasive. Such sanctions may include monetary sanctions.

Article 5
1. This Convention shall not apply to aircraft used in military, customs or police services.

2. In the cases contemplated in subparagraphs (a), (b), (c), (e), (f), (g), (h) and (i) of paragraph 1 of Article 1, this Convention shall apply irrespective of whether the aircraft is engaged in an international or domestic flight, only if:

(a) the place of take-off or landing, actual or intended, of the aircraft is situated outside the territory of the State of registry of that aircraft; or

(b) the offence is committed in the territory of a State other than the State of registry of the aircraft.

3. Notwithstanding paragraph 2 of this Article, in the cases contemplated in subparagraphs (a), (b), (c), (e), (f), (g), (h) and (i) of paragraph 1 of Article 1, this Convention shall also apply if the offender or the alleged offender is found in the territory of a State other than the State of registry of the aircraft.

4. With respect to the States Parties mentioned in Article 15 and in the cases set forth in subparagraphs (a), (b), (c), (e), (f), (g), (h) and (i) of paragraph 1 of Article 1, this Convention shall not apply if the places referred to in subparagraph (a) of paragraph 2 of this Article are situated within the territory of the same State where that State is one of those referred to in Article 15, unless the offence is committed or the offender or alleged offender is found in the territory of a State other than that State.

5. In the cases contemplated in subparagraph (d) of paragraph 1 of Article 1, this Convention shall apply only if the air navigation facilities are used in international air navigation.

6. The provisions of paragraphs 2, 3, 4 and 5 of this Article shall also apply in the cases contemplated in paragraph 4 of Article 1.

Article 6
1. Nothing in this Convention shall affect other rights, obligations and responsibilities of States and individuals under international law, in particular the purposes

and principles of the Charter of the United Nations, the Convention on International Civil Aviation and international humanitarian law.

2. The activities of armed forces during an armed conflict, as those terms are understood under international humanitarian law, which are governed by that law are not governed by this Convention, and the activities undertaken by military forces of a State in the exercise of their official duties, inasmuch as they are governed by other rules of international law, are not governed by this Convention.

3. The provisions of paragraph 2 of this Article shall not be interpreted as condoning or making lawful otherwise unlawful acts, or precluding prosecution under other laws.

Article 7

Nothing in this Convention shall affect the rights, obligations and responsibilities under the Treaty on the Non-Proliferation of Nuclear Weapons, signed at London, Moscow and Washington on 1 July 1968, the Convention on the Prohibition of the Development, Production and Stockpiling of Bacteriological (Biological) and Toxin Weapons and on Their Destruction, signed at London, Moscow and Washington on 10 April 1972, or the Convention on the Prohibition of the Development, Production, Stockpiling and Use of Chemical Weapons and on Their Destruction, signed at Paris on 13 January 1993, of States Parties to such treaties.

Article 8

1. Each State Party shall take such measures as may be necessary to establish its jurisdiction over the offences set forth in Article 1 in the following cases:

(a) when the offence is committed in the territory of that State;

(b) when the offence is committed against or on board an aircraft registered in that State;

(c) when the aircraft on board which the offence is committed lands in its territory with the alleged offender still on board;

(d) when the offence is committed against or on board an aircraft leased without crew to a lessee whose principal place of business or, if the lessee has no such place of business, whose permanent residence is in that State;

(e) when the offence is committed by a national of that State.

2. Each State Party may also establish its jurisdiction over any such offence in the following cases:

(a) when the offence is committed against a national of that State;

(b) when the offence is committed by a stateless person whose habitual residence is in the territory of that State.

3. Each State Party shall likewise take such measures as may be necessary to establish its jurisdiction over the offences set forth in Article 1, in the case where the alleged offender is present in its territory and it does not extradite that person pursuant to Article 12 to any of the States Parties that have established their jurisdiction in accordance with the applicable paragraphs of this Article with regard to those offences.

4. This Convention does not exclude any criminal jurisdiction exercised in accordance with national law.

Article 9

1. Upon being satisfied that the circumstances so warrant, any State Party in the territory of which the offender or the alleged offender is present, shall take that person into custody or take other measures to ensure that person's presence. The custody and other measures shall be as provided in the law of that State but may only be continued for such time as is necessary to enable any criminal or extradition proceedings to be instituted.

2. Such State shall immediately make a preliminary enquiry into the facts.

3. Any person in custody pursuant to paragraph 1 of this Article shall be assisted in communicating immediately with the nearest appropriate representative of the State of which that person is a national.

4. When a State Party, pursuant to this Article, has taken a person into custody, it shall immediately notify the States Parties which have established jurisdiction under paragraph 1 of Article 8 and established jurisdiction and notified the Depositary under subparagraph (a) of paragraph 4 of Article 21 and, if it considers it advisable, any other interested States of the fact that such person is in custody and of the circumstances which warrant that person's detention. The State Party which makes the preliminary enquiry contemplated in paragraph 2 of this Article shall promptly report its findings to the said States Parties and shall indicate whether it intends to exercise jurisdiction.

Article 10

The State Party in the territory of which the alleged offender is found shall, if it does not extradite that person, be obliged, without exception whatsoever and whether or not the offence was committed in its territory, to submit the case to its competent authorities for the purpose of prosecution. Those authorities shall take their decision in the same manner as in the case of any ordinary offence of a serious nature under the law of that State.

Article 11

Any person who is taken into custody, or regarding whom any other measures are taken or proceedings are being carried out pursuant to this Convention, shall be guaranteed fair treatment, including enjoyment of all rights and guarantees in conformity with the law of the State in the territory of which that person is present and applicable provisions of international law, including international human rights law.

Article 12

1. The offences set forth in Article 1 shall be deemed to be included as extraditable offences in any extradition treaty existing between States Parties. States Parties undertake to include the offences as extraditable offences in every extradition treaty to be concluded between them.

2. If a State Party which makes extradition conditional on the existence of a treaty receives a request for extradition from another State Party with which it has no extradition treaty, it may at its option consider this Convention as the legal basis for extradition in respect of the offences set forth in Article 1. Extradition shall be subject to the other conditions provided by the law of the requested State.

3. States Parties which do not make extradition conditional on the existence of a treaty shall recognize the offences set forth in Article 1 as extraditable offences between themselves subject to the conditions provided by the law of the requested State.

1. Each of the offences shall be treated, for the purpose of extradition between States Parties, as if it had been committed not only in the place in which it occurred but also in the territories of the States Parties required to establish their jurisdiction in accordance with subparagraphs (b), (c), (d) and (e) of paragraph 1 of Article 8, and who have established jurisdiction in accordance with paragraph 2 of Article 8.

5. The offences set forth in subparagraphs (a) and (b) of paragraph 5 of Article 1 shall, for the purpose of extradition between States Parties, be treated as equivalent.

Article 13
None of the offences set forth in Article 1 shall be regarded, for the purposes of extradition or mutual legal assistance, as a political offence or as an offence connected with a political offence or as an offence inspired by political motives. Accordingly, a request for extradition or for mutual legal assistance based on such an offence may not be refused on the sole ground that it concerns a political offence or an offence connected with a political offence or an offence inspired by political motives.

Article 14
Nothing in this Convention shall be interpreted as imposing an obligation to extradite or to afford mutual legal assistance if the requested State Party has substantial grounds for believing that the request for extradition for offences set forth in Article 1 or for mutual legal assistance with respect to such offences has been made for the purpose of prosecuting or punishing a person on account of that person's race, religion, nationality, ethnic origin, political opinion or gender, or that compliance with the request would cause prejudice to that person's position for any of these reasons.

Article 15
The States Parties which establish joint air transport operating organizations or international operating agencies, which operate aircraft which are subject to joint or international registration shall, by appropriate means, designate for each aircraft the State among them which shall exercise the jurisdiction and have the attributes of the State of registry for the purpose of this Convention and shall give notice thereof to the Secretary General of the International Civil Aviation Organization who shall communicate the notice to all States Parties to this Convention.

Article 16
1. States Parties shall, in accordance with international and national law, endeavour to take all practicable measures for the purpose of preventing the offences set forth in Article 1.

2. When, due to the commission of one of the offences set forth in Article 1, a flight has been delayed or interrupted, any State Party in whose territory the aircraft or passengers or crew are present shall facilitate the continuation of the journey of the passengers and crew as soon as practicable, and shall without delay return the aircraft and its cargo to the persons lawfully entitled to possession.

Article 17
1. States Parties shall afford one another the greatest measure of assistance in connection with criminal proceedings brought in respect of the offences set forth in Article 1. The law of the State requested shall apply in all cases.
2. The provisions of paragraph 1 of this Article shall not affect obligations under any other treaty, bilateral or multilateral, which governs or will govern, in whole or in part, mutual assistance in criminal matters.

Article 18
Any State Party having reason to believe that one of the offences set forth in Article 1 will be committed shall, in accordance with its national law, furnish any relevant information in its possession to those States Parties which it believes would be the States set forth in paragraphs 1 and 2 of Article 8.

Article 19
Each State Party shall in accordance with its national law report to the Council of the International Civil Aviation Organization as promptly as possible any relevant information in its possession concerning:

(a) the circumstances of the offence;

(b) the action taken pursuant to paragraph 2 of Article 16;

(c) the measures taken in relation to the offender or the alleged offender and, in particular, the results of any extradition proceedings or other legal proceedings.

Article 20
1. Any dispute between two or more States Parties concerning the interpretation or application of this Convention which cannot be settled through negotiation, shall, at the request of one of them, be submitted to arbitration. If within six months from the date of the request for arbitration the Parties are unable to agree on the organization of the arbitration, any one of those Parties may refer the dispute to the International Court of Justice by request in conformity with the Statute of the Court.
2. Each State may at the time of signature, ratification, acceptance or approval of this Convention or accession thereto, declare that it does not consider itself bound by the preceding paragraph. The other States Parties shall not be bound by the preceding paragraph with respect to any State Party having made such a reservation.
3. Any State Party having made a reservation in accordance with the preceding paragraph may at any time withdraw this reservation by notification to the Depositary.

Article 21
1. This Convention shall be open for signature in Beijing on 10 September 2010 by States participating in the Diplomatic Conference on Aviation Security held at Beijing from 30 August to 10 September 2010. After 27 September 2010, this Convention shall be open to all States for signature at the Headquarters of the International Civil Aviation Organization in Montréal until it enters into force in accordance with Article 22.
2. This Convention is subject to ratification, acceptance or approval. The instruments of ratification, acceptance or approval shall be deposited with the

Secretary General of the International Civil Aviation Organization, who is hereby designated as the Depositary.

3. Any State which does not ratify, accept or approve this Convention in accordance with paragraph 2 of this Article may accede to it at any time. The instrument of accession shall be deposited with the Depositary.

4. Upon ratifying, accepting, approving or acceding to this Convention, each State Party:

(a) shall notify the Depositary of the jurisdiction it has established under its national law in accordance with paragraph 2 of Article 8, and immediately notify the Depositary of any change; and

(b) may declare that it shall apply the provisions of subparagraph (d) of paragraph 4 of Article 1 in accordance with the principles of its criminal law concerning family exemptions from liability.

Article 22

1. This Convention shall enter into force on the first day of the second month following the date of the deposit of the twenty-second instrument of ratification, acceptance, approval or accession.

2. For each State ratifying, accepting, approving or acceding to this Convention after the deposit of the twenty-second instrument of ratification, acceptance, approval or accession, this Convention shall enter into force on the first day of the second month following the date of the deposit by such State of its instrument of ratification, acceptance, approval or accession.

3. As soon as this Convention enters into force, it shall be registered with the United Nations by the Depositary.

Article 23

1. Any State Party may denounce this Convention by written notification to the Depositary.

2. Denunciation shall take effect one year following the date on which notification is received by the Depositary.

Article 24

As between the States Parties, this Convention shall prevail over the following instruments:

(a) the Convention for the Suppression of Unlawful Acts Against the Safety of Civil Aviation, Signed at Montreal on 23 September 1971; and

(b) the Protocol for the Suppression of Unlawful Acts of Violence at Airports Serving International Civil Aviation, Supplementary to the Convention for the Suppression of Unlawful Acts Against the Safety of Civil Aviation, Done at Montreal on 23 September 1971, Signed at Montreal on 24 February 1988.

Article 25

The Depositary shall promptly inform all States Parties to this Convention and all signatory or acceding States to this Convention of the date of each signature, the date of deposit of each instrument of ratification, approval, acceptance or accession, the date of coming into force of this Convention, and other relevant information. IN WITNESS WHEREOF the undersigned Plenipotentiaries, having been duly authorized, have signed this Convention. DONE at Beijing on the tenth day of

September of the year Two Thousand and Ten in the English, Arabic, Chinese, French, Russian and Spanish languages, all texts being equally authentic, such authenticity to take effect upon verification by the Secretariat of the Conference under the authority of the President of the Conference within ninety days hereof as to the conformity of the texts with one another. This Convention shall remain deposited in the archives of the International Civil Aviation Organization, and certified copies thereof shall be transmitted by the Depositary to all Contracting States to this Convention.

Index

A
ACIP. *See* AFI Comprehensive
 Implementation Programme (ACIP)
ACSA. *See* The Central American Agency
 for Aeronautical Safety (ACSA)
Advance passenger information (API)
 Aviation and Transportation
  Security Act, 221
 carrier to capture passport details, 220–221
 compelling information tool, 221
 data, 4
 electronic data interchange system, 221
 PNR data elements, 222
Aerial photography, 119, 121
Aerotoxic syndrome
 causes, 79–80
 certificate of airworthiness, 81–82
 Montreal and Warsaw Conventions, 84
 no bleed architecture, 81
 pilots and, 80
 significance, 78–79
 Southwest Airlines Co., 82–83
 Turner v. Eastwest Airlines Limited, 83–84
AFI Comprehensive Implementation
 Programme (ACIP), 144–145
Africa-Indian (AFI) Ocean Region, 144–145
Airline Pilots Association (ALPA), 88
Airlines blacklisting. *See* Safety oversight
 audits
Air navigation services provision
 Annex 2 and Annex 11, 27
 over the high seas, 26
 regional plans, 27–28
 transit passage right, 28–29
Air Operator Certificates (AOCs), 2
Airport and aviation security
 Beijing Convention of 2010, 248–261
 civil unrest, 246
 extraordinary rendition, 261–276
 MANPADS, 243–245
 missile attacks, 245
 perimeter security, 245–246
 political and economic interests, 246–247
 Resolution A15-7, 247
 Resolution A17-1, 247–248
 Resolution A20-2, 248
 SAMs, 245–246
 success rate, Western States, 246
Airport responsibility
 aircraft stemming, 376
 air traffic control services, 374
 Article 28 of the Chicago
  Convention, 373
 aviation safety, 373
 catastrophic events, 375
 de-icing of aircraft, 375–376
 duty of care, 373
 international air transport, 374–375
 liability for inadequately maintained
  property and unsafe premises, 376
 meteorological information, 374
 regulatory process, 373
 service failure by natural
  phenomenon, 372
 windstorms, 375
Airport slot allocation in Europe
 Articles 81 and 82 of anti-competitive
  conduct, 368–369
 Art. 8(4) of Regulation 95/93, 364–365
 dominant position, 369–370
 EU regulations to non EU nationals,
  366–367
 European Economic Community treaty,
  362–363

Airport slot allocation in Europe (*cont.*)
  night curfews, 363–364
  open skies agreement, 366
  revised Regulation 95/93, 367–368
  slot-coordinator, 364
  slot trading, 368
Airspace, safety issues
  ATS and military authorities, 74–75
  CANSO, 67
  civil aircraft and military purposes, 70–71
  co-ordination of activities, 65–66
  FIR responsibility, 70
  global air traffic management operational concept, 73–74
  IATA management, 66–67
  ICAO guidelines, 76
  military and civil aviation, 64, 68–69
  missile testing, 76–77
  NOTAM, 75
  nuclear missile firings, 65
  recent developments, 72–73
  SES and SESAR, 67–68
Air traffic control (ATC), 120, 358, 374
Air traffic management (ATM), 7, 22
Air traffic services (ATS), 8
  Annex 11 and Annex 2, 31–32
  economic liberalization, 34
  modern treaty law, 25
  situations, 32–33
  State responsibility, 33–34
Air transport
  air traffic rights, 322
  competition, 321
  Game Theory, 321
  international treaty, 322
Air Transportation Safety and System Stabilization Act (ATSAA), 197, 249
ALPA. *See* Airline Pilots Association (ALPA)
Alternative fuels
  availability, 387–389
  costs associated with, 389–390
  fossil fuel and its effects, 385–387
  product innovations, 384
  reduced fuel burn/fuel efficiency, 384
  The Rio Conference of 2009 (*see* The Rio Conference of 2009)
American Airlines, 284
API. *See* Advance passenger information (API)
APU. *See* Auxiliary Power Unit (APU)
Arctic Climate Impact Assessment (ACIA), 281
Area Coordination Centre (ACC), 102, 109
Article 6 of the Chicago Convention, 329–330, 398–399

Asia and South Pacific Initiative to Reduce Emissions (ASPIRE), 309
Assembly Resolution A 37–19 of ICAO, 285–287
Assigned Amount Units (AAUs), 315–316
Association of Civil Aviation Authorities of the Caribbean (ACAAC), 147–148
A Strategic Action Plan for Future Aviation Safety, 36
ATC. *See* Air traffic control (ATC)
Atlantic Interoperability Initiative to Reduce Emissions (AIRE), 309
ATM. *See* Air traffic management (ATM)
ATS. *See* Air traffic services (ATS)
ATSAA. *See* Air Transportation Safety and System Stabilization Act (ATSAA)
Auxiliary Power Unit (APU), 79, 313, 376
Aviation and Transportation Security Act, 221
Aviation medicine
  accidents, 94–95
  aerotoxic syndrome (*see* Aerotoxic syndrome)
  ALPA, 88–89
  Chicago Convention, 92
  EU Commission, 91–92
  FAA, 89–91
  incrementalism, 97–98
  medical certification, 88
  negligence, 86–88
  pilot responsibility, 93–94
  plaintiff rights, 85–86
  pre-existing medical conditions, 95–97
Aviation Security Manual, 225
Aviation Security Panel of ICAO
  Annex 17, 212, 213
  Recommended Practice, 213
  security culture
    Article 2, Convention for the Suppression of the Financing of Terrorism, 217
    audit results, 216
    bilateral exchanges of information, 215
    downfall, air traffic, 214
    international terrorism, 217–218
    Resolution A 36–20 adoption, 215–216
    Secretariat Audit Results Review Board (ARRB), ICAO, 216–217
    threats, 214
    UASP, 214–216
  security regime, 212–213
  sharing information
    API, 220–222

Index 415

Assembly Resolution A 35-1, 218
"no fly" lists, 222–225
PNR, 219–220

B
BAG Regional Accident Investigation Agency (BAGAIA), 146–147
BAG Regional Safety Oversight Organization (BAGASOO), 146–147
Banjul Accord Group (BAG), 146
Beijing Convention of 2010
 Article 1, 403–405
 Article 2, 405–406
 Article 3, 4 and 5, 407
 Article 6, 407–408
 Article 7 and 8, 408
 Article 9, 10 and 11, 409
 Article 12, 409–410
 Article 13, 14, 15 and 16, 410
 Article 17, 18, 19 and 20, 411
 Article 21, 411–412
 Article 22, 23 and 24, 412
 Article 25, 412–413
 ATSAA, 249
 Aviation Security Panel of ICAO, 250–251
 aviation security regime, 251
 Convention on the Suppression of Unlawful Acts Relating to International Civil Aviation, 252
 cyber attacks threat, 251
 international air law instruments, 248–249
 offences under the Convention
  act of violence against person, 258–259
  Article 31.1 of the Vienna Convention on the Law of Treaties, 259–260
  bioterrorism attack, 257–258
  communicate false information, 256
  cyber terrorism, 256
  destroys/damages air navigation facilities, 255
  device/substance, 254–255
  extradition, offenders, 260
  ICAO initiatives, 260–261
  purpose of causing death, 256–257
  RODS, 258
  service/causes damage, 254
  unlawfully and intentionally performance, 253–254
 Preamble, 249
 States Parties, 403
 37th Session of the ICAO Assembly, 251–252

types, threat
 cyber-terrorism, 252
 dangerous pathogens, 253

C
CAA. *See* Civil Aviation Authority (CAA)
CAPS. *See* Computer Assisted Passenger Screening System (CAPS)
CAPSCA project. *See* Cooperative Arrangement for the Prevention of Spread of Communicable disease through Air travel (CAPSCA) project
Carbon market/trading
 Annex I Parties, 301
 carbon offsetting, 303–304
 Certified Emission Reductions, 301
 Earth Summit, 301–302
 extra territorial jurisdiction, 304–306
 Kyoto Protocol, 301–302
 overall trade and investment, 304
 purchase contracts, 302–303
 quantified emissions limitation, 306–307
 regulatory process, 304
 Resolution A36–22, 307
 World Bank report, 300–301
Carbon neutral growth, 295–296
Carbon offsetting, 303–304, 307–309
Caribbean Aviation Safety and Security Oversight System (CASSOS), 148–150
Caribbean Community (CARICOM) Secretariat, 147–148, 150
CARICOM Council for Trade and Economic Development (COTED), 148
Causal indicators, 40
CBDR. *See* Common but differentiated responsibilities (CBDR)
Certified Emission Reductions (CERs), 301
Civil Air Navigation Services Organization (CANSO), 67
Civil Aviation Authority (CAA), 80
Civil unrest and aviation
 airport and aviation security (*see* Airport and aviation security)
 keeping airports open
  Article 27 of Vienna Convention, 243
  Chicago Convention, 242
  Strategic Objectives, 241
 mass civil unrest, 239–240
 peace prevailing level, 240
 tourism in Tunisia and Egypt, 240–241
Clean development mechanism (CDM), 14
Climate change

Climate change (*cont.*)
  Arctic, 281
  carbon dioxide level, 282–283
  carbon offsetting, 285
  glazier, 281–282
  human right to life, 283–284
  1997 Kyoto Protocol, 282
  optimal use of fuel and alternative fuel sources, 284
  solar energy, 283
Commercial considerations, open skies
  anti-trust exigencies, 353–354
  bilateral air services agreement, 354, 356–357
  convergence of regulatory principles, 355
  DoT standard open skies agreement, 354
  global GDP, 352–353
  laws, 354
  open competition, 353
  protectionism, 357
  public subsidies, 355
  TCAA concept, 355–356
Commercial practices of air transport industry, 398–399
Committee on Aviation Environmental Protection (CAEP), 4, 292–293
Common but differentiated responsibilities (CBDR), 14
Communications, Navigation and Surveillance (CNS) system, 7
Competition, open skies
  antitrust laws, 351
  Articles 81 and 82, 349–350
  business and leisure travel, 348–349
  Clayton Act, 351
  economies of scale principle, 349
  Hoffman-La Roche v. Commission, 350
  policies, practices and laws, 352
  predatory pricing practices, 349, 351
  Sherman Act, 351
Compliance examination (COMPEX), 269
Complicity theory
  Blackstone's view, 180–181
  de Vattel's view, 180
  notion, *patienta* or *receptus*, 180
  principal–agent dilemma, 180
  State responsibility, 180
Comprehensive Systems Approach (CSA), 2
Computer Assisted Passenger Screening System (CAPS), 268
Condonation theory
  based on, 181
  categories, support, 183
  causality principle, 183
  Convention for the Suppression of the Financing of Terrorism, 182
  defined, war crime, 184
  International Law Commission, 183–184
  Jane case, 181
  separate delict theory, 183
Continuous Descent Operations (CDOs), 309
Continuous Monitoring Approach (CMA), 2
Convention for the Unification of Certain Rules Relating to the Precautionary Attachment of Aircraft (Rome, 1933), 273
Convention on International Civil Aviation, 2
Convention on International Civil Aviation of 1944 (Chicago Convention), 199–200
Convention on the Suppression of Unlawful Acts Relating to International Civil Aviation, 5, 15, 171–172. *See also* Beijing Convention of 2010
Cooperative Arrangement for the Prevention of Spread of Communicable disease through Air travel (CAPSCA) project, 9
Corporate foresight
  airport responsibility (*see* Airport responsibility)
  infrastructure, 371
  planning (*see* Planning, corporate foresight)
  weather disruptions and passenger assistance, 371–372
  winter weather, 370–371
CSA. *See* Comprehensive Systems Approach (CSA)
"Cyber Corps" strategy, 169
Cybercrime Convention
  Article 2, 173–174
  Article 7, 174
  biometric information, 174
  goals, 173
  PKD, 174
Cyber terrorism
  advantage, 165–166
  attacks on cyberspace, 167
  "Cyber Corps" strategy, 169
  defined, 166–167
  digital environment, 166
  "fear factor," 171
  fear of harm, 168
  ICAO's Recommended Practice, adoption and implementation, 165
  international and national reliance, 165

Index 417

international efforts
   Convention on the Suppression of Unlawful Acts Relating to International Civil Aviation, 171–172
   Cybercrime Convention, 173–174
   offences against civil aviation, 171
   United Nations Resolution on Millennium Goals, 172–173
national efforts
   Amendment 12 to Annex 17–Security to the Convention on International Civil Aviation, 177
   Canadian Criminal Code, 175–176
   1986 Electronic Communications Privacy Act, US, 176–177
   interception of data, 175
   2000 Regulation of Investigatory Powers Act, UK, 176
   Telecommunications (Interception and Access) Act adoption, Australia, 175
National Strategy to Secure Cyberspace, 169
11 September 2001 attack, 168
Threat Assessment, 170
threat on Federal Aviation Administration (FAA) computer systems, 170

D

Declaration of Global Principles for the Liberalization of International Air Transport
   clause 4.4, 332
   economic development amongst States, 333
   safety and security, 332–333
Declaration on Aviation Security, 12
de minimis threshold, 15, 286–287, 296
Democratic Peoples' Republic of Korea (DPRK), 65, 76–77
Department of Transportation Office of Inspector General (DOT-OIG), 89

E

EASA. *See* European Aviation Safety Agency (EASA)
ECAC. *See* European Civil Aviation Conference (ECAC)
ECCAIRS. *See* European Center for Civil Aviation Incident Reporting System (ECCAIRS)

Economic liberalization and safety
   ATConf/5, 49
   codesharing arrangement, 44, 57
   commercial aviation affecting, 52–54
   complex situations, 59–64
   European Community, 51
   factors, 51–52
   foreign registered aircraft, 42–43, 54–55
   international carriers, 49–50
   KPIs, 46–47 (*see also* Performance management and accountability, SMS)
   off-shore operations, 43–44, 55–56
   outsourcing activity, 44–45, 56–57
   regionalism and, 45, 57–58
   SARPs and guidance material, 58
   and security regulation, 52
   types of, 50
Electronic data interchange system, 221
Emission calculation
   AAU, 315–316
   airport capacity/slot allocation, 317
   allocation of allowances, 313–314
   allowances from non-aviation sectors, 316
   aviation and climate change, 318–319
   commercial practices of air transport industry, 317–318
   emission factors, 314
   EU ETS cap and trade scheme, 314–315
   fuel consumption, 313
   ICAO designator, 314–315
   monitoring and reporting protocols, 316–317
   open skies concept, 317
   sustainable development., 319
Emission trading. *See also* European Emissions Trading Scheme
   CAEP guidance material, 298–299
   ICAO Carbon Emissions Calculator, 299
   ICAO emissions related charge policy, 299–300
   pollution levels, 298
   subject of, 297–298
Environmental issues
   aspirations and goals, 285–287
   calculating emission (*see* Emission calculation)
   climate change (*see* Climate change)
   emission trading (*see* Emission trading)
   market based measures (*see* Trading and market based measures)
Environmental protection, ICAO
   climate change, 13
   greenhouse gas emissions, 13–14

Environmental protection (*cont.*)
  market based measures, 15
EUROCONTROL. *See* European Organization for the Safety of Air Navigation
European and North Atlantic Volcanic Ash Task Force (EUR/NAT VATF), 98
European Aviation Safety Agency (EASA), 24, 45, 92. *See also* Regional safety
European Center for Civil Aviation Incident Reporting System (ECCAIRS), 149
European Civil Aviation Conference (ECAC), 45, 142, 151
European Emissions Trading Scheme, 288
  aircraft operator and emission allowance, 311–312
  carbon offsetting, 307–309
  Directive 2008/101/EC, 311
  free trade of emission allowances, 310–311
  ICAO Carbon Emissions Calculator, 309–310
  Monitoring and Reporting Guidelines, 311
  operational measures, 309–310
  passenger-related fuel use, 310
European Low Fares Airline Association (ELFAA), 117
European Organization for the Safety of Air Navigation, 22, 67
European Regions Airline Association (ERA), 311
European Union (EU) Commission, 91–92
Extraordinary rendition
  air transport use, peaceful purposes
    Chicago Conference, 267
    "general security," 266–267
    legitimate profiling process, 268
    Preamble, 267
    racial profiling, 268–269
    11 September 2001 events, 266
  Article 13, ICCPR, 262
  definition, 261–262
  false arrest and imprisonment (*see* False arrest and imprisonment)
  nature of air transport
    Article 3 (c), Chicago Convention, 269
    civil/military investigation, 274
    loss, insurance coverage, 274
    military purposes, 270–272
    over-flight permission, 274–275
    post-Chicago air law instruments, 273
    public aircraft, state aircraft and civil aircraft, 271–273
    Resolution A10–19, 271
    Resolution A14–25, 269–270
    Resolution A21–21, 270
    Resolution A36–13, 270
    status, airline pilots and crew, 275
  Third Geneva Convention on Prisoners of War of 1949, 263

F
False arrest and imprisonment
  Christy v. Leachinsky case, 264–265
  Fourth Amendment to the U.S. Constitution, 263
  Murray v. Ministry of Defence case, 265
  physical detention, 263
  Terry v. Ohio case, 264
FATF. *See* Financial Action Task Force
Federal Aviation Administration (FAA), 25, 89, 157
Federal Aviation Administration (FAA) computer systems, 170
Fifth Meeting of the Parties to the Kyoto Protocol (CMP5), 291
Fifth Worldwide Air Transport Conference (ATConf/5), 49, 332, 361
Financial Action Task Force (FATF), 236–237
Financing of terrorism, preventive measures
  Article 2, International Convention for the Suppression of the Financing of Terrorism, 225–226
  definition, terrorism, 226–227
  international terrorism
    acts, 229–236
    factors, 228
  money laundering (*see* Money laundering)
  providing/collecting funds, 226
Flight information region (FIR), 23, 70
Flight NW 253 bombing attempt
  communiqué, ICAO, 210–211
  human errors, 210
  Resolution A33–1, 211
  unlawful transport, biological, chemical, nuclear weapons, 211–212
Foresight, 377. *See also* Corporate foresight
Fossil fuel
  impact of emissions, 385
  mitigation gap, 385–386
  particulate matter, 386–387
  sulphur jet fuel, 386
Full body scanners in aviation security
  ATSAA, 197
  AVSEC Panel, 212–225
  Convention on International Civil Aviation, 197
  costs, 198
  displacement theory, 198

Index 419

Flight NW 253 bombing attempt, 209–212
Fourth Amendments, 199
right of passenger's privacy (*see* Rights of passenger's privacy)
State security
    Article 2, Convention for the Suppression of the Financing of Terrorism, 207
    categories, support, 208
    complicity theory, 206
    condonation theory, 206–207
    ICAO, 208–209
    principal–agent dilemma, 205–206
    threats, 197–198

G

Game theory, 321
General Agreement on Trade in Services (GATS), 322
Global Air Navigation Plan (GANP), 6
Global Air Traffic Management Forum on Civil and Military Cooperation, 72
Global Aviation Safety Plan (GASP), 6
Global Aviation Safety Roadmap (GASR), 6
Global economy, 398–399
Global Framework for Aviation Alternative Fuels (GFAAF), 288–289, 391–392
Global warming, 283–284
Great Circle Distance (GCD), 309
Grotius' theory, 180
Group on International Aviation and Climate Change (GIACC), 292

H

Haiti earthquake, ICAO action, 400–401
Historical aviation emissions, 314

I

IAOPA. *See* International Council of Aircraft Owner and Pilot Association (IAOPA)
IASA Programme. *See* International Aviation Safety Assessments (IASA) Programme
IATA. *See* International Air Transport Association (IATA)
IATA on work related to Training and Qualifications Initiatives (ITQI), 2
IATA Operational Safety Audit (IOSA) Programme, 156
IAVW. *See* International Airways Volcano Watch (IAVW)
ICAO. *See* International Civil Aviation Organization (ICAO)
ICCPR. *See* International Covenant on Civil and Political Rights (ICCPR)
IFFAS. *See* International Financial Facility for Aviation Safety (IFFAS)
ILC. *See* International Law Commission (ILC)
Implementation Support And Development-Safety (ISD-Safety) Programme, 10
Implementing the Global Aviation Safety Roadmap, 37
Industry Safety Strategy Group (ISSG), 36
Instrument flight rules (IFR), 29
Integrated Safety Trend Analysis and Reporting System (iSTARS), 2
International Air Services Transit Agreement (IASTA), 332
International Air Transport Association (IATA), 22
International Airways Volcano Watch (IAVW)
    components, 101
    establishment, 99–100
    ICAO's role, 104–106
    observatories, 102
    three-area approach, 104
    uses, 100
    VAAC and, 101–102
    zero tolerance, 103–104
International Aviation Safety Assessments (IASA) Programme, 157
International Civil Aviation Organization (ICAO), 1. *See also* Strategic issues and regulation, ICAO
    ACIP (*see* Regional safety)
    airspace use (*see* Airspace, safety issues)
    Annexes and USOAP, 39
    business plan, 38–39
    IFFAS, 41–42
    indicators, 40
    key factors, 42
    leadership values, 38
    philosophical adjustment, 41
    responsibilities, 39–40
    UAS (*see* Unmanned aerial systems (UAS))
    volcanic ash effects mitigation (*see* Meteorological issues)
International Council of Aircraft Owner and Pilot Association (IAOPA), 125
International Covenant on Civil and Political Rights (ICCPR), 262
International Financial Facility for Aviation Safety (IFFAS), 41–42

International Law Commission (ILC), 138–139
International terrorism
    acts of defence, 230–231
    aid and abet national terrorism, 232–233
    deterrence problems, 233–234
    factors, 228
    nonviolent acts, 231
    practical solution, 235–236
    random acts of violence, 231–232
IOSA Programme. *See* IATA Operational Safety Audit (IOSA) Programme
ISSG. *See* Industry Safety Strategy Group (ISSG)

J
Joint Aviation Authorities (JAA), 155

K
Key performance indicators (KPIs), 46
1997 Kyoto Protocol, 282

L
Laker Airways v. Sabena, 305–306
Latin American Civil Aviation Commission (LACAC), 45
Law of the Sea *UNCLOS*, 28
Legal considerations, market access
    Article 94, 337–338
    bilateral air services, 338–339
    economic policies, 335–336
    political will of States, 338
    principle of sovereignty, 335
    untrammelled competition and suffocative regulation, 339
    Vienna Convention on the Law of Treaties, 336–337

M
Machine Readable Passport (MRP), 187
Machine readable travel document (MRTD)
    biometric identification, person, 187–188
    ensure uniformity, 188
    technical reliability, 188
MANPADS. *See* Man Portable Air Defense Systems (MANPADS)
Man Portable Air Defense Systems (MANPADS)
    as effective weapons, 244
    threat of, 243

Market access
    air transport network, 322–323
    Article 6 of the Chicago Convention, 329–330
    aviation safety and security strategy, 325
    central laws and regulations, 327
    central market place, 326, 327
    Chicago Conference of 1944, 324
    climate change, 325–326
    commercial aviation, 323
    conduct of trade, 326–327
    economic considerations, 331–333
    free trade, 327
    GATS Annex, 328–330
    GDP, 323
    international comity and conduct of States, 326
    International Law Commission, 326
    lack of leadership, 326
    legal considerations (*see* Legal considerations, market access)
    misconception, 333–335
    Most Favoured Nations Treatment Clause, 330
    State sovereignty concept, 325, 326
    world trade, 327–328
    World Trade Organization, 327
Market based measures (MBM). *See* Trading and market based measures
Meteorological information provision
    authority, 30
    objectives, 29
    State responsibility, 30–31
Meteorological issues
    aircrafts affected, 108
    airlines liability, 117–118
    airspace closure, 110–111
    breach of statutory duty, 112–114
    Chicago Convention, 109–110
    EUR/NAT VATF, 98
    IAVW, 99–100 (*see also* International Airways Volcano Watch (IAVW))
    Icelandic volcano eruption, 106–107
    objectives and determination, 112
    Regulation 261/2004, 115–117
Meteorological Service for International Air Navigation, 128
Meteorological watch office (MWO), 102, 109
Money laundering
    cutting off funding, 238
    definition, 236
    electronic surveillance, passports, 238
    fundamental measures *vs.*, 236–237
    on global scale, 236

Index 421

mandatory confiscation, 237
media, role, 238–239
Most Favoured Nations Treatment (MFN) Clause, 330, 334–335
MRP. *See* Machine Readable Passport (MRP)
MRTD. *See* Machine readable travel document (MRTD)

N

National Strategy to Secure Cyberspace, 169
NavCanada, 25
Next Generation Air Transport System (NextGen), 68
Next Generation of Aviation Professionals (NGAP), 3
"No fly" lists
  Aviation Security Manual, 225
  freedom of speech and right to access information, 224
  maintenance, 222
  right to privacy, 223
  storage and use, personal data, 223
  subject of privacy, 223
  technology role, 223–224

O

Online Aircraft Safety Information System (OASIS), 2
Open skies, 317, 340–357
  cabotage rights, 342
  combined negotiating power, 347–348
  commercial considerations (*see* Commercial considerations, open skies)
  community air carriers, 345–346
  competition (*see* Competition, open skies)
  concept, 397
  European pricing policy, 344
  European treaty, 344–345
  foreign capital, 342–343
  lines of action, European Council of Ministers, 347
  market share for US carriers, 343
  ninth freedom of air/stand alone cabotage, 346
  price competition, 341–342
  Regulation 847/2004, 346–347
  seventh freedom carriage, 346
  unrestricted rights to fly, 344
  US economic growth, 340–341
Operation Safe Pilot, 89
Outcome indicators, 40
Output indicators, 40

P

Pacific Aviation Safety Organization (PASO), 45
Passenger Name Record (PNR), 4
  data elements, 222
  PNR Data Guidelines, 219
  Recommended Practice, 219–220
Passport, security of
  diplomatic relations
    extradition issues, 196–197
    falsification of national passport, 195
    Mahmoud al-Mabhouh killing, 195–196
    principles for State responsibility, 192
    quantum of proof, 191
    Rainbow Warrior case, 193
    Resolution A 56/83, 192–193
    Russian intelligence officers expulsion, 194
    sending and receiving States, 195
    US decision, People's Burea's personnel withdrawl, 193–194
    Vienna Convention on Diplomatic Relations, 194–195
  materials and techniques, 190
  personalization process, 189–190
  software packages, 190
  threats
    Annex 9, Convention on International Civil Aviation, 191
    photo-substitution and deletion/alteration of text, 190
Performance Based Navigation (PBN), 3, 8
Performance management and accountability, SMS
  goals and standards, 47–48
  levels and components, 46
PGMs. *See* Precision-guided munitions (PGMs)
PKD. *See* Public Key Directory (PKD)
Planning and Implementation Regional Groups (PIRGs), 8
Planning, corporate foresight
  aircraft capacity, 378
  airport charges, 382–383
  airport planning laws, 379
  anticipatory intelligence, 381
  Article 15 of the Chicago Convention, 383
  direction setting and determining priorities, 381
  fight cancellation, 382
  foresight-awareness culture, 377
  foresight category, 380–381

Planning (*cont.*)
  guidelines for diaster management, 381–382
  innovation catalyzing, 381
  link with air transport, 379–380
  markets and technologies identification, 380
  natural disaster, 377–378
  Recommendation 2.9.9. of the Annex, 382
  strategy formulation, 381
  technical and economic field, 378–379
PNR. *See* Passenger Name Record (PNR)
Point of Contact (PoC), 3
Precision-guided munitions (PGMs), 245
Procedures for Air Navigation Services, 31
Protocol Supplementary to the Convention for the Suppression of Unlawful Seizure of Aircraft, 5, 15
Public Key Directory (PKD), 174

R
Racial profiling, 268–269
Radiotelephony communications, 10
Real-Time Outbreak Disease Surveillance (RODS), 258
Regional Aviation Safety Oversight System for the Caribbean (RASOS), 45, 147–149
Regional safety
  ACIP, 144–145
  ACSA, 147
  Africa, 145–146
  BAG, 146–147
  Caribbean, 147–151
  Europe, 151–154
  ICAO's role, 141–142
  necessity, 139–140
  Regional Safety Oversight Manual, 142–144
  roles played, 141
  RSOOs, 140–141
  South Asia, 151
Regional Safety Oversight Organizations (RSOOs), 7, 140–142
Regional Supplementary Procedures, 27
Regulatory convergence, 347
Remotely-piloted aircraft (RPA), 126
Required Navigation Performance (RNP), 8
Resolution 1373, 399–400
Rights of passenger's privacy
  body scanners
    data protection legislation, 204–205
    as trans-border storage, 204
  concept, 202–203
  Convention on International Civil Aviation of 1944 (Chicago Convention), 199–200
  data subject, 200–201
  display and storage, personal data use, 200
  protection of human rights, 205
  self-determination, 203
  technological activity, 201–202
  Universal Declaration of Human Rights, 202
The Rio Conference of 2009
  achievements, 393
  CAAF Recommendations, 394–395
  Declaration, 391–393
  GFAAF establishment, 391
  key activities of ICAO, 393
  post-Conference thought, 393
  UNFCC COP15, 393–394
Rules of the Air, 26, 29
Ryanair Airlines, 284

S
Safety Assessment of Foreign Aircraft (SAFA) programme, 155
Safety, ICAO
  communicable diseases prevention, 9
  GASP objectives, 6
  global framework, 5–6
  language proficiency, 10
  PBN global goals, 8
  regional, 9–10
  runway, 7
  SMS implementation, 5
  up-to-date consolidated statement, 7–8
Safety issues
  airspace use (*see* Airspace, safety issues)
  aviation medicine (*see* Aviation medicine)
  meteorological (*see* Meteorological issues)
  oversight audits (*see* Safety oversight audits)
  regional (*see* Regional safety)
  SMS (*see* Safety management systems (SMS))
  UAS (*see* Unmanned aerial systems (UAS))
Safety management systems (SMS), 5
  air navigation services (*see* Air navigation services provision)
  Annexes and need for, 19
  ATS provision (*see* Air traffic services (ATS))
  economic liberalization (*see* Economic liberalization and safety)
  key elements and, 21

Index 423

meteorological information provision
    (*see* meteorological information
    provision)
requirements and steps involved, 20
Resolution A35-7, 21–22
roadmap (*see* Safety roadmap)
state responsibility (*see* State
    responsibility)
Safety oversight audits
    airlines blacklisting, 155–156
    ICAO and, 154
    policy aspects, 156–160
    regulatory, 160–162
Safety roadmap
    ICAO role, 37–38 (*see also* International
        Civil Aviation Organization (ICAO))
    incontrovertible truths, 35–36
    ISSG and part 1, 36–37
    part 2, 37
SARI. *See* South Asian Regional Initiative
    (SARI)
SARPs. *See* Standards And Recommended
    Practices (SARPs); Standards and
    recommended practices (SARPs)
Security, ICAO
    audit programme, 12
    declaration, 12–13
    unlawful interference acts, 10–12
Security issues
    attacks on travel documents,
        177–197
    civil unrest and aviation, 239–276
    cyber terrorism, 165–177
    full body scanners, 197–225
    suppression, financing of terrorism,
        225–239
Separate delict theory, 183
SESAR. *See* Single European Sky Air Traffic
    Management Research (SESAR)
SIGMETs, 101
Single European Sky Air Traffic Management
    Research (SESAR), 68
Single European Sky (SES) legislation,
    67–68
Slot allocation, 317
    aircraft operations, 357
    air traffic control slot, 358
    capacity-constrained airports, 358
    in Europe (*see* Airport slot allocation
        in Europe)
    ICAO policies and guidance, 361
    international access, 359
    market access, 360–361
    ownership, 360
    slot availability problem, 362
    type of entity, 358–359
    uniform treatment principle, 359–360
Slot-allocation, 397
SMS. *See* Safety management
    systems (SMS)
Social Security Administration Office of
    Inspector General (SSA-OIG), 89
Soft law, 16
South Asian Regional Initiative (SARI), 151
Standards and Recommended Practices
    (SARPs), 6, 11, 20, 26, 209
State responsibility. *See also* Airport
    responsibility
    ATM system, 23
    Chicago Convention, 22–23
    service providers, 25
    unified strategy and Resolution
        A35-7, 24
State Safety Programme (SSP), 5
St. Elmo's fire, 108
Strategic issues and regulation, ICAO
    activities, 1
    Beijing convention and protocol, 5
    environmental protection, 4 (*see also*
        Environmental protection, ICAO)
    law, 15–17
    safety, 2–3 (*see also* Safety, ICAO)
    security, 3–4, ICAO (*see also* Security,
        ICAO)
Sustainable development
    air transport nature, 321–322
    alternative fuels (*see* Alternative fuels)
    corporate foresight (*see* Corporate
        foresight)
    market access (*see* Market access)
    open skies (*see* Open skies)
    slot allocation (*see* Slot allocation)
System-wide Information Management
    System (SWIM), 68

T
Technical Cooperation Bureau, 41
Technical Co-operation Programme, 39
15th Conference of Parties (COP/15), 4
The Central American Agency for
    Aeronautical Safety (ACSA), 147
The Civil Air Navigation Services
    Organisation (CANSO), 5
The Eighth Meeting of the Committee on
    Aviation Environmental Protection
    (CAEP/8), 4

Third Geneva Convention on Prisoners of War of 1949, 263
Total fuel burn, 310
Tourism industry, 397–398
Trading and market based measures. *See also* Emission trading
  aircraft engine emission and ozone depletion, 293–294
  aircraft-induced climate change, 288
  aviation alternative fuels, 288–289
  carbon credits, 287–288
  carbon neutral growth and emission reduction, 295–296
  Emission Trading Scheme, 291–292
  fuel efficiency improvement, 294–295
  GIACC, 292–293
  sustainable alternative fuels, 289–290
Transatlantic Common Aviation Area (TCAA) concept, 355–356
Travel documents, attacks on
  abuse, Australian passport system, 179
  complicity (*see* Complicity theory)
  condonation (*see* Condonation theory)
  diplomatic outcry, 178
  forged foreign passports, 177–178
  fraudulent EU member states' passports, 179
  knowledge
    Article 5 of the ILC document, 186
    biometric identification, 188–189
    Corfu Channel case, 184–185
    draft provision, International Law Commission, 186
    ePassport, 187
    ICAO Facilitation Division, 186–187
    imputability or attribution, 185
    Island of Palmas case, 184
    MRP and MRTD, 187–188
    Pan Am case, 186
    physiological biometrics, 188
  security, passport (*see* Passport, security of)
Tree planting, 285

U

UAS. *See* Unmanned aerial systems (UAS)
UASP. *See* Universal Security Audit Programme (UASP)
UAV. *See* Unmanned Aerial Vehicle (UAV)
Unified strategy, 21

United Nations Framework Convention on Climate Change (UNFCCC), 4
United Nations Framework Convention on Climate Change 15th Meeting of the Conference of the Parties (COP15), 291
United Nations Manual on Cybercrime, 172
United Nations Resolution on Millennium Goals, 172
Universal Declaration of Human Rights, 202
Universal Safety Oversight Audit Programme (USOAP), 2, 22, 39. *See also* Safety oversight audits
Universal Security Audit Programme (UASP), 214–215
Unmanned aerial systems (UAS)
  accident cases, 137–138
  aerodrome operations, 128
  airworthiness regulations, 119–120
  ATC functions, 120–121
  ATM, 127
  ATS, 132
  civil aircraft and, 121
  Conventions, 135–137
  IAOPA, 125
  ICAO circular, 123–125
  insurance coverage, 137
  meteorological information, 128–129
  operations over the high seas, 129–131
  personnel licensing, 121–122
  principles of international law, 138–139
  Resolution A 35–14^{397}, 133–134
  Resolution A 37–15, 134
  Resolution A14–25 and A21–21, 133
  RPA, 126–127
  rules of the air, 122
  security, 129
  state aircraft issue, 132–133
  UAV, 118–119
  unmanned balloons, 122–123
  uses, 119
Unmanned Aerial Vehicle (UAV), 118
USOAP. *See* Universal Safety Oversight Audit Programme (USOAP)

V

Visual flight rules (VFR), 29
Volcanic Ash Advisory Centres (VAAC), 99

W

World Meteorological Organization (WMO), 100

CPSIA information can be obtained at www.ICGtesting.com
Printed in the USA
LVOW090211280612

288006LV00007B/18/P

Übungsbuch Medien- und Internetmanagement